NEUROBIOLOGY OF SOCIAL BEHAVIOR

NEUROBIOLOGY OF SOCIAL BEHAVIOR

TOWARD AN UNDERSTANDING OF THE PROSOCIAL AND ANTISOCIAL BRAIN

MICHAEL NUMAN

ELSEVIER

AMSTERDAM • BOSTON • HEIDELBERG • LONDON
NEW YORK • OXFORD • PARIS • SAN DIEGO
SAN FRANCISCO • SINGAPORE • SYDNEY • TOKYO
Academic Press is an imprint of Elsevier

Academic Press is an imprint of Elsevier
32 Jamestown Road, London NW1 7BY, UK
225 Wyman Street, Waltham, MA 02451, USA
525 B Street, Suite 1800, San Diego, CA 92101-4495, USA

British Library Cataloguing-in-Publication Data
A catalogue record for this book is available from the British Library

Library of Congress Cataloging-in-Publication Data
A catalog record for this book is available from the Library of Congress

ISBN: 978-0-12-416040-8

For information on all Academic Press publications
visit our website at elsevierdirect.com

Typeset by TNQ Books and Journals
www.tnq.co.in

15 16 17 18 10 9 8 7 6 5 4 3 2 1

Dedicated to my wife, Marilyn, and in memory of Jay Rosenblatt

Contents

Preface ix
About the Author xi

1. An Introduction to Neural Systems

1.1 Introduction 1
1.2 A Schematic Overview of the Mammalian Brain 2
1.3 Functional Neuroanatomy 3
1.4 Conclusions 41

2. Basic Genetics and Epigenetics

2.1 Introduction 43
2.2 Basic Genetics 43
2.3 Basic Epigenetics 52
2.4 Epigenetic Effects on Physiological and Behavioral Development 54
2.5 Conclusions 61

3. Aggressive Behavior

3.1 Introduction 63
3.2 Neural Systems of Offensive Aggression in Nonhuman Mammals 65
3.3 Neurobiology of Impulsive or Affective Aggression in Humans 86
3.4 Early Adverse Life Experiences, Genes, and the Development of Aggression 94
3.5 Conclusions 105

4. Sexual Behaviors and Sexual Differentiation

4.1 Introduction 109
4.2 The Hormonal Basis of Sexual Behavior 110
4.3 The Neurobiology of Male Sexual Behavior in Nonprimate Animals 111
4.4 The Neurobiology of Female Sexual Behavior in Nonprimate Animals 123
4.5 The Neurobiology of Sexual Behavior in Primates 134
4.6 Sexual Differentiation and the Development of Sexual Behavior 143
4.7 Conclusions 163

5. Parental Behavior

5.1 Introduction 165
5.2 Neural Systems Regulating Maternal Motivation in Rats 169
5.3 Neural Systems and Maternal Behavior in Sheep: Maternal Responsiveness and the Formation of Selective Attachments 203
5.4 Maternal Behavior in Nonhuman Primates 207
5.5 Paternal and Alloparental Behavior 209
5.6 Developmental Influences on the Maternal Behavior of Mammals 214
5.7 The Neurobiology of Maternal Behavior in Humans 224
5.8 General Conclusions 232

6. Monogamy and the Formation of Enduring Social Attachments between Mating Partners

6.1 Introduction 235
6.2 The Vole Model System 236
6.3 OT and Vasopressin Neural Systems and the Regulation of Social Monogamy in Birds and Primates 257
6.4 OT, Vasopressin, and Pair Bonding: Conclusions 262
6.5 The Relationship of OT Neural Systems to Autism Spectrum Disorder 263

7. Human Sociality

7.1 Introduction 271
7.2 An Evolutionary Perspective 271
7.3 Important Points from Previous Chapters 274
7.4 The Insular Cortex and Empathy: A Potential Starting Point through Which Feeling States Influence Prosocial Behaviors 276
7.5 Oxytocin and Human Prosociality 284
7.6 Human Prosociality: Conclusions 289
7.7 Psychopathy: A Breakdown of Prosocial Neural Circuits that Results in Antisocial Behavior 290
7.8 Epilogue 298

References 301

Index 341

Preface

In writing this book on the *Neurobiology of Social Behavior*, my goal was to present a broad and comprehensive review of the neural basis of mammalian social behaviors, both prosocial and antisocial. This book is meant to be accessible and important for scholars and researchers who study the neurobiology of social behavior from different levels of analysis. Therefore, the book has much to offer for molecular and developmental neurobiologists, behavioral neuroscientists, and social-affective-cognitive neuroscientists, and it is suitable for both graduate students and advanced researchers. Since the neural basis of social behavior pathologies is discussed in most chapters, the book will also be valuable for clinicians. Although most neuroscientists have particular areas of expertise, an underlying premise of this book is that the integration of genetic, developmental, neural circuit, and functional magnetic resonance imaging (fMRI) research in the analysis of social behaviors promotes a more complete understanding of the neural underpinnings of such behaviors in animals and humans.

The book begins with chapters on functional neuroanatomy and molecular genetics and epigenetics. Chapters on aggression, sexual behaviors and sexual differentiation, parental behavior, monogamy, and human sociality follow these introductory chapters. All chapters integrate animal and human studies, but the final chapter deals with highly developed human social characteristics, such as empathy, trust, cooperation, and altruism.

In each chapter on particular social behaviors, a detailed analysis of the neural systems regulating normal aspects of the behavior is presented, and this is then followed by an examination of how dysfunctions within these neural systems can give rise to abnormal social behavior. The role of genes, gene–environment interactions, and/or epigenetics in the development of social behaviors is also discussed in most chapters. In integrating animal and human studies, an overarching principle is that there are core allocortical and subcortical circuits that influence social behavior across mammalian species. Research on the neural basis of social behavior in animals is shown to provide basic and important information about the neurobiology of social behavior in humans.

I am a behavioral neuroscientist, and my own research program examines the neural and neurochemical systems that regulate maternal behavior in rodents. As a result of an invitation by Stephanie Brown, in 2012 I wrote a chapter entitled "Neural circuits regulating maternal behavior: Implications for understanding the neural basis of social cooperation and competition" (see reference [698] in the References section of this book for the full citation). Writing this chapter and interacting with Stephanie and other colleagues at a symposium at the University of Michigan made me think about the broader implications of my research; this experience was one of the driving forces that led me to write this current book on the neurobiology of social behavior. Indeed, one of the themes within the *Neurobiology of Social Behavior* is that the neural circuits underlying mother–infant bonding may have provided the primordial neural scaffold upon which other types of strong prosocial bonds are built.

Another theme that runs through this book is that there are distinct prosocial and antisocial neural circuits, and that social stimuli can be routed over one or the other circuit, leading to either prosocial or antisocial actions and feeling states. Importantly, just in which circuit a social stimulus is processed through can be influenced by several factors that include evolutionary forces and genetics, early life experiences, adult experiences and cognitive factors, and physiological state.

When comparing a neural circuitry analysis derived from animal research with fMRI data derived from human research, although both sets of data may point to a certain brain region as being important for a particular aspect of social behavior, the animal research can offer additional information about the neurochemical microcircuitry within a particular brain region of interest that fMRI and other imaging data are not yet able to provide. Throughout this book, I make the point that by understanding such microcircuitry, one gains a fuller appreciation of how social stimuli can be processed over either prosocial or antisocial circuits, and how such processing can be modulated by a variety of factors.

I have been inspired, encouraged, and supported by many mentors, colleagues, and students throughout my career. During my doctoral graduate studies at the University of Chicago, Howard Moltz, Eckhard Hess, and Pete Grossman initiated my interests in understanding how evolutionary forces, genetics, and developmental processes affect the neural circuits that control social behavior. These interests flourished during my postdoctoral research with Jay Rosenblatt at the Institute of Animal Behavior at Rutgers University. During my independent academic career at Boston College, the National Institutes of Health and the National Science Foundation provided support for my research program, and special thanks are due to Kathie Olsen. The way I think about how the brain works to regulate behavior has been strongly influenced by the research and scholarship of Donald Pfaff and Larry Swanson. So many other scientists have greatly impacted my thinking, and this is amply demonstrated by the detailed analysis and discussion of their research findings in this book. My students have also been a tremendous source of support for me, and they have provided independent insights that strongly influenced the direction of our research program. In particular, my former graduate students, Teige Sheehan and Danielle Stolzenberg, created great research under my mentorship.

My wife and colleague, Marilyn, has been the most important source of intellectual and social support throughout my career. In addition, Marilyn read, edited, and commented on each chapter in this book and she also prepared all of the original figures. This book would not be as good as it is without her important input and efforts.

Finally, the staff at Elsevier produced what I think is a beautiful book, and they supported me each step of the way during the long process of writing and organizing the *Neurobiology of Social Behavior*.

About the Author

Michael Numan received his PhD from the University of Chicago. He is a Fellow of the American Association for the Advancement of Science (AAAS) and the Association for Psychological Science (APS). Most of his research, which focuses on the neural mechanisms regulating maternal behavior, was conducted while he was a Professor of Psychology and Behavioral Neuroscience at Boston College. He is the author of a previous book, with Thomas Insel, *The Neurobiology of Parental Behavior*. He lives with his wife, Marilyn, in the Santa Fe-Albuquerque region of New Mexico.

CHAPTER

1

An Introduction to Neural Systems

1.1 INTRODUCTION

The purpose of this book is to provide a comprehensive understanding of the neurobiology of social behavior in mammals, including humans. Social behavior, broadly defined, comprises those behaviors that are exhibited by conspecifics as they interact and includes both affiliative (prosocial) and antagonistic (antisocial) interactions. The social behaviors and processes that will be examined include aggression and competition, sexual behaviors, parental behaviors, the formation of social attachments, cooperation, and altruism. Each of these social behaviors will be analyzed across the different levels of investigation that have been used to study the behavior, from molecular neurobiology to neural circuits to functional magnetic resonance imaging (fMRI) data. The first two chapters of the book provide the necessary background in functional neuroanatomy (Chapter 1) and molecular biology and genetics (Chapter 2) that will serve as a foundation for a full appreciation of the neurobiology of social behavior that will be presented in the subsequent chapters.

One important aspect of this book is to uncover the neural mechanisms that determine whether social stimuli are assigned either a positive or a negative valence by the perceiver, with positive social stimuli activating neural pathways that cause contact seeking behaviors, acceptance, caregiving, and other prosocial behaviors, while negative social stimuli activate neural pathways that cause avoidance, rejection, competition, or even attack (antisocial behaviors). In other words, it may be valuable to view social behaviors as being influenced by two elementary neural networks: social stimuli can be processed by either an aversion/rejection network or an attraction/acceptance network, with such differential neural processing leading to different social outcomes [698]. Further, social stimuli may be automatically (innately) routed over either attraction or aversion networks, or such valence properties may be acquired through learning mechanisms. These core neural processes may then set the foundation for more complex social interactions. But what are the neural systems that regulate core approach and avoidance processes, processes that are embedded within and influence more complex social interactions? In what follows, an overview will be presented of functional neuroanatomy with an aim to explain such approach and avoidance systems in mammals. Animals approach or avoid a broad range of stimuli, not just social stimuli. Research findings will be presented from both social and nonsocial contexts with the understanding, which will be borne out in subsequent chapters, that there is overlap in the neural systems influencing social and nonsocial motivational processes (see [82]).

Neurobiology of Social Behavior
http://dx.doi.org/10.1016/B978-0-12-416040-8.00001-8

1.2 A SCHEMATIC OVERVIEW OF THE MAMMALIAN BRAIN

Based on the neuroanatomy in Swanson [951], Figure 1.1 depicts the general organization of the mammalian brain. Moving from rostral to caudal, the brain is broken down into two great divisions: the cerebral hemispheres (telencephalon) and the brainstem. Caudal to the brainstem lies the spinal cord. The cerebral hemispheres are divided into the cerebral cortex and the underlying, and therefore subcortical, cerebral nuclei. The cerebral cortex can be divided into isocortex (neocortex) and allocortex, with the former containing six well-defined cellular layers, while the latter contains less than six well-differentiated layers [951]. Isocortex makes up most of the occipital, parietal, temporal, and lateral prefrontal parts of the cortex. Examples of allocortex include the olfactory (piriform) cortex, parts of the amygdala, the hippocampal formation, and parts of the medial and orbital prefrontal cortex. Some important subcortical cerebral nuclei include the caudate nucleus (dorsal striatum), nucleus accumbens (ventral striatum), globus pallidus (dorsal pallidum), ventral pallidum, septal area, and some nuclei within the amygdala. The brainstem, upon which sits the cerebral hemispheres, is composed of the thalamus and hypothalamus (diencephalon), and the lower brainstem, which includes the midbrain, pons,

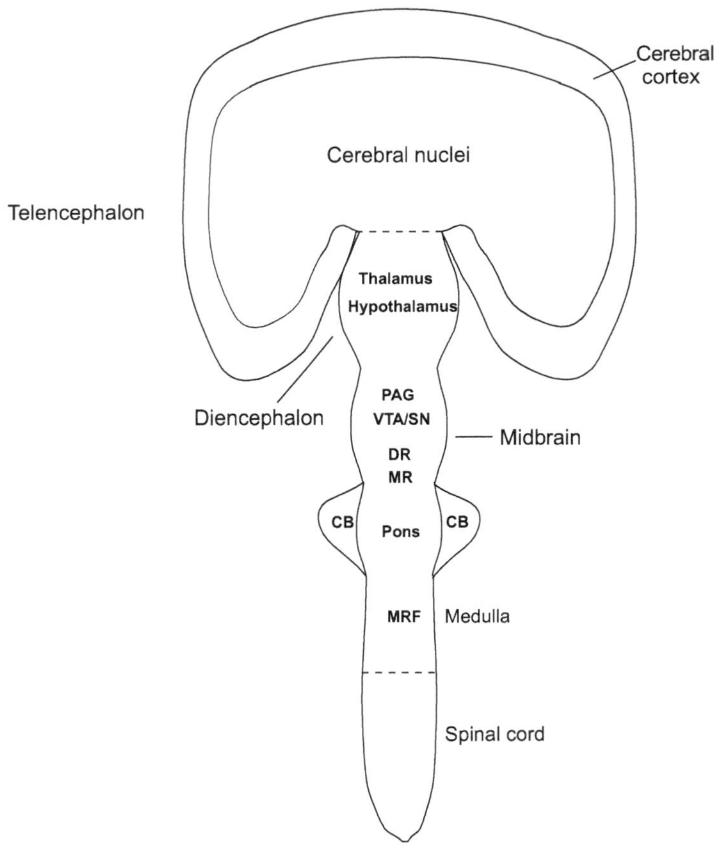

FIGURE 1.1 A schematic overview of the mammalian brain. The telencephalon (cerebral hemispheres) is composed of outer cortical layers and underlying cerebral nuclei. Caudal to the telencephalon is the brainstem, containing the diencephalon, midbrain, pons, cerebellum, and medulla. The spinal cord lies caudal to the medulla. Each section of the brainstem lists certain structures that are emphasized in this book. This listing is not meant to represent the exact anatomical location of the labeled structures but simply indicates their general location. For example, within the diencephalon, the thalamus is dorsal, not anterior, to the hypothalamus. Abbreviations: CB = cerebellum; DR = dorsal raphe nucleus; MR = median raphe nucleus; MRF = medullary reticular formation; PAG = periaqueductal gray; SN = substantia nigra; VTA = ventral tegmental area.

cerebellum, and medulla. Some of the lower brainstem nuclei that will be shown to play important roles in social behavior are the midbrain periaqueductal gray (PAG), dopamine neurons within the ventral tegmental area of the midbrain, and the serotonin neurons of the raphe nuclei located in the midbrain and pons. The term *brainstem motor area* (BSMA) will be used to refer to a group of nuclei in the lower brainstem with indirect and direct connections to cranial and spinal motor neurons. Most relevant with respect to the BSMA, the PAG and the midbrain locomotor region (located lateral to the PAG) both project to the medullary reticular formation, whose axons project to cranial and spinal motor neurons [361,420,1069]. As will become important, the descending projection of the midbrain PAG to the medullary reticular formation is an important route through which PAG output affects the display of reflex-like defensive and aggressive responses, such as behavioral immobility, escape responses, or biting, and reflexive responses related to sexual and parental behaviors.

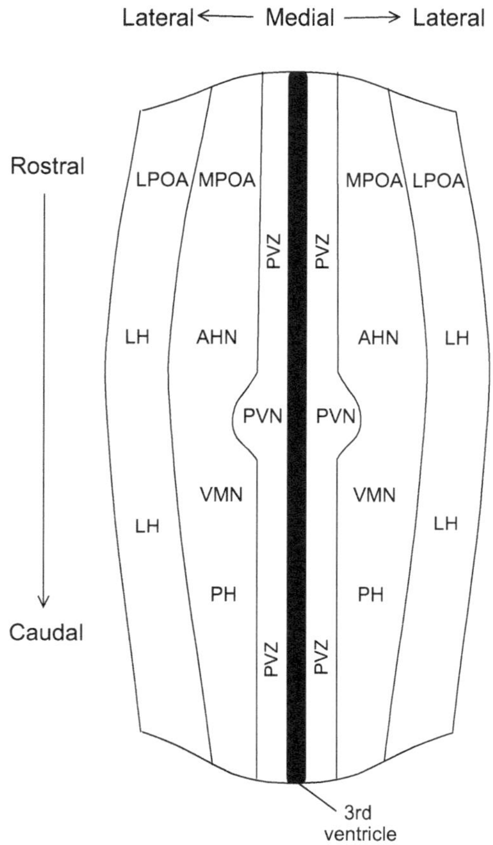

FIGURE 1.2 Horizontal section through the hypothalamus. The third ventricle is shown as a thick black line down the midline. Moving from medial to lateral, the hypothalamus contains a periventricular zone (PVZ), a medial zone, and a lateral zone. Other abbreviations: AHN = anterior hypothalamic nucleus; LH = lateral hypothalamus; LPOA = lateral preoptic area; MPOA = medial preoptic area; PH = posterior hypothalamus; PVN = paraventricular nucleus; VMN = ventromedial nucleus.

1.3 FUNCTIONAL NEUROANATOMY

1.3.1 The Hypothalamus

1.3.1.1 Introduction

The three major functions of the hypothalamus are its regulatory influences over the autonomic nervous system and the pituitary gland, and its involvement in the control of a variety of motivated behaviors, including social behaviors [716,813,950]. Figure 1.2 shows a horizontal section through the hypothalamus, displaying its rostral-to-caudal and medial-to-lateral organization. With respect to its organization from medial to lateral, the hypothalamus contains a periventricular zone (which surrounds the third ventricle) and a medial and lateral zone. The periventricular zone contains those neurons primarily involved in neuroendocrine and autonomic regulation, while it is the nuclei of the medial and lateral zones that play dominant roles in the control of motivated behaviors, which involve influences over the somatic motor mechanisms that control both the reflexive and voluntary aspects of these behaviors.

The social behaviors influenced by the hypothalamus include reproductive behaviors (sexual and parental) and aggressive and defensive behaviors. As will be seen, it is likely that separate and distinct neuronal populations within the hypothalamus regulate different social behaviors.

Given the involvement of the hypothalamus in social and other motivated behaviors, one would expect that it would be a recipient of significant sensory inputs, and indeed this is the case [759,813]. The hypothalamus receives olfactory inputs from the amygdala and other olfactory areas. Afferents from the brainstem carry tactile and pain inputs, and the hypothalamus receives multimodal sensory inputs from the prefrontal cortex and from the hippocampus, the latter arriving either directly or indirectly via the septal area. Finally, an organism's internal state importantly influences its social behavior, and primary among these internal factors are hormones. Neurons in the hypothalamus contain receptors for prolactin, estradiol, testosterone, progesterone, and adrenal corticosteroids [813,950]. A simple view is that hormones and other internal physiological stimuli bias how various sensory inputs are processed by the hypothalamus, which in turn affects the hypothalamic efferent pathways that are activated. With respect to social behavior, such effects would allow an organism to respond in one way or another to particular social stimuli, depending on the current hormonal milieu that is affecting the operation of specific neural circuits.

1.3.1.2 *Motivation: Appetitive, Avoidance/Rejection, and Consummatory Behaviors*

Several definitions of motivation exist [90,716,764,1021]. In its simplest definition, motivation is an internal process that modifies an organism's responsiveness to a constant stimulus. That is, if an organism shows a change in the way it responds to a constant stimulus, some internal alteration must be mediating the behavioral change. As examples, food deprivation increases an organism's responsiveness to food-related cues, the hormonal events associated with pregnancy termination increase a female's responsiveness to infant stimuli, and gonadal steroids influence the occurrence of male and female sexual responses to sexual stimuli. Another definition of motivation refers to those internal processes that arouse and direct behavior toward a particular goal, giving rise to the term *goal-directed behavior*. Two major types of goal-directed responses are approach responses (also called appetitive or reward-seeking responses) toward a desired or pleasant stimulus, and avoidance or rejection responses toward aversive or noxious stimuli [868]. As will be described below, rejection responses can include approaching an aversive/unpleasant stimulus in order to attack it. The mechanisms underpinning such responses would define appetitive motivation and aversive motivation. Therefore, I am broadening the typical view of aversion to mean more than avoidance or withdrawal. An aversive stimulus is one that an organism does not like, and goal-directed responses can either avoid/escape from that stimulus or approach the stimulus to actively reject it.

In describing motivated behavior with respect to a desired or rewarding stimulus (one with a positive valence), such behavior can be separated into an appetitive goal-directed phase and a consummatory or terminal phase. During the appetitive or reward-seeking phase, the organism searches its environment to acquire the particular goal object. For example, a hungry animal will search for food, a sexually motivated organism will search for a mate, and a maternal female will seek out her infants or will search for displaced infants in order to transport them to a secure area. The consummatory phase is composed of those behaviors that occur once the desired goal is obtained, and the behaviors that occur during this phase are elicited by proximal cues from the goal object. Examples of consummatory responses toward a desired stimulus are eating food, copulating with a mate, and nursing infants. Therefore, appetitive goal-directed behaviors reflect an underlying motivation, drive, or desire to engage in a behavioral interaction with a specific goal stimulus, while consummatory behaviors reflect the ability to perform specific behavioral responses once the goal object is attained.

Appetitive reward-seeking behaviors are variable, flexible, and can be influenced by learning and higher cognitive processes—the organism needs to search and possibly manipulate its environment in order to gain access to the desired goal. In contrast, consummatory responses are more reflexive in nature and are elicited by proximal stimuli from the goal object. A typical example is hunger and food intake. In response to the internal changes that result from food deprivation (lower energy supplies), food-seeking behaviors and food intake are activated. For the appetitive component, the organism will search its environment for food, relying on previous experience. The consummatory phase would be made up of the actual oral motor responses involved in ingestion.

In describing motivated behavior with respect to a noxious or aversive stimulus (one with a negative valence), it can similarly be separated into a goal-directed phase and a consummatory or terminal phase. In the case of defensive aggression, where an individual is threatened or attacked by another individual, the animal may attempt to use goal-directed responses to avoid or escape from the situation and to reach a safe location, but if cornered, she/he will engage in relatively stereotyped defensive/aggressive behaviors in response to the aggressive acts of the opponent. As will be elaborated upon in the chapter on aggression, male offensive aggression (for example, the aggression shown by a resident territory owner toward an intruder) can be similarly characterized as a response to an aversive (disliked) social stimulus that gives rise to both goal-directed and consummatory rejection responses. In this case, the territory owner approaches and chases the intruder (goal-directed rejection responses) and ultimately attacks the intruder (consummatory rejection responses) if proximal contact occurs.

Note that the definition of motivation as a change in responsiveness to a constant stimulus applies to both goal-directed and consummatory responses. A satiated animal will not search for food and will not eat food that is placed in its mouth. In contrast, the definition of motivation as a process that arouses and directs behavior toward a particular goal only applies to the proactive voluntary goal-directed phase of motivated behavior.

At this point, I would like to briefly indicate what I mean when I use the term *emotion*. On the one hand, this term will be used to refer to emotional behaviors, which are basically goal-directed and consummatory responses to aversive or noxious stimuli. However, for humans, this term will also be used to refer to those affective feeling states that occur throughout the pleasant-aversive continuum, with these experiential states being associated with goal-directed and consummatory responses to *both* pleasant and aversive stimuli. A discussion of feeling states will be reserved for humans, since we cannot measure such states in nonhuman animals.

In thinking about the neural underpinnings of the goal-directed and consummatory phases of motivated behaviors, because the goal-directed phase can be variable, flexible, and influenced by learning processes, it should involve telencephalic mechanisms. In contrast, the consummatory phase, which is stereotyped, reflexive in nature, and based on reactions to proximal stimuli, might be regulated primarily by lower brainstem mechanisms. Given the importance of the hypothalamus for motivated behaviors, to the degree that it is involved in both goal-directed and consummatory responses, one might predict that hypothalamic interactions with the telencephalon (i.e., cerebral hemispheres) regulate goal-directed responses, while hypothalamic interactions with the brainstem regulate consummatory responses. A schematic diagram of these possibilities is shown in Figure 1.3. The involvement of telencephalic mechanisms would allow for more flexible, adaptive, voluntary responding through the use of higher integrative and cognitive processes. In conclusion, the hypothalamus may monitor and respond to an organism's internal state and relay this information to the telencephalon, which then regulates strategic responses to the external environment based, in part, on this information.

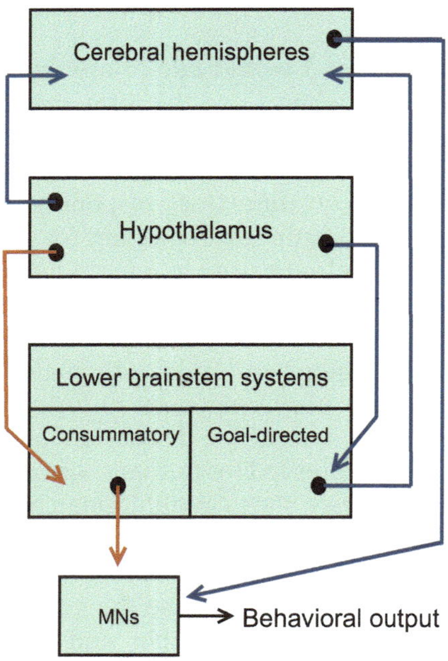

FIGURE 1.3 A general depiction of the neural circuits through which the hypothalamus can regulate goal-directed and consummatory aspects of motivated behaviors. Hypothalamic projections to lower brainstem neurons, such as the periaqueductal gray, which then regulate the output of cranial and spinal motor neurons (MNs), would be a descending route that controls consummatory responses (shown in red). Goal-directed behaviors require telencephalic control, and the hypothalamus can influence telencephalic mechanisms through direct ascending projections (shown in blue on the left side of the diagram) or by descending projections to lower brainstem neurons, such as the dopamine neurons of the ventral tegmental area, which then ascend to exert modulatory influences on the telencephalon (these projections are shown in blue on the right side of the diagram).

Once a particular goal is achieved, proximal stimuli may activate hypothalamic neurons with descending projections to the brainstem that regulate specific consummatory responses.

1.3.1.3 The Hypothalamus and Neuroendocrine Regulation

Because of the important role of hormones in social behavior, a brief introduction to neuroendocrinology will be provided [677]. Figure 1.4 shows a schematic sagittal section through the basomedial hypothalamus and the attached pituitary gland. The periventricular zone of the hypothalamus contains most of the neurons that regulate the pituitary gland, which is divided into the anterior and posterior (neural) pituitary. When action potentials occur in those neurons that regulate the anterior pituitary, their neurochemicals (called neurohormones) are secreted into the primary capillary plexus located at the base of the hypothalamus. These neurochemicals then travel down the hypothalamic-pituitary portal veins to reach the secondary capillary plexus located in the anterior pituitary. From there, these neurohormones can reach cells in the anterior pituitary to affect the synthesis and release of additional hormones. Two examples of hypothalamic regulation of anterior pituitary function, which are relevant to the content of this book, will be presented. Gonadotropin-releasing hormone (GnRH; considered a hormone because even though it is released from axon terminals of a hypothalamic neuron in response to action potentials, such release occurs into the blood to affect nonproximal target cells in the anterior pituitary) affects the synthesis and release of luteinizing hormone (LH) and follicle-stimulating hormone (FSH) from anterior pituitary cells. LH and FSH then circulate in the blood to reach the gonads where they regulate the synthesis and release of testosterone from the testes and estradiol and progesterone from the ovaries. The importance of the brain's regulation of the pituitary is that it allows neural stimuli to affect the endocrine

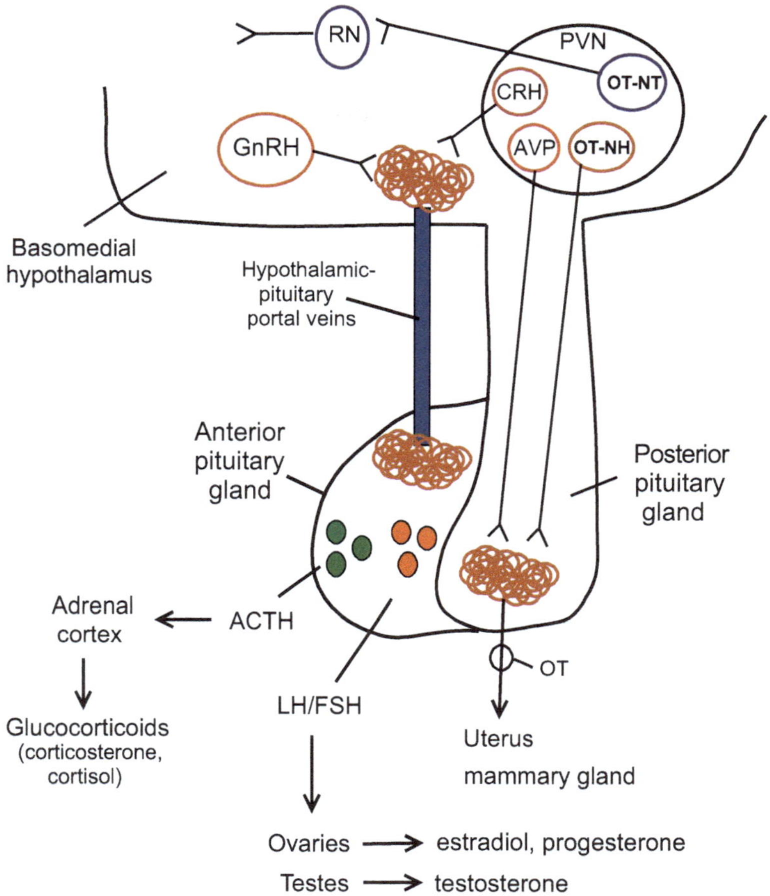

FIGURE 1.4 The hypothalamus and neuroendocrine regulation. With respect to the posterior pituitary (neural lobe), oxytocin (OT) and vasopressin (AVP, arginine vasopressin) neurons in the paraventricular nucleus (PVN) of the hypothalamus send their axons directly to a capillary plexus (red overlapping circles) in the posterior pituitary. Since OT and AVP are released into the blood at this site, they act as neurohormones (NH). OT is shown as acting on the uterus, where it stimulates uterine contractions, and on the mammary glands, where it stimulates milk ejection. Other OT neurons, by synapsing on regular neurons (RN) within the brain, can release OT as a neurotransmitter (NT) or neuromodulator. Although not shown, AVP can also act as a neurotransmitter within the brain. With respect to hypothalamic control of the anterior pituitary, hypothalamic neurons secrete neurohormones into a primary capillary plexus located within the basomedial hypothalamus. These neurohormones then travel down the hypothalamic-pituitary portal veins to reach a secondary capillary plexus in the anterior pituitary gland, where they then influence the release of particular anterior pituitary hormones. In one of the examples of such neurohormone action, gonadotropin-releasing hormone (GnRH) stimulates the release of luteinizing hormone (LH) and follicle-stimulating hormone (FSH) from the anterior pituitary. LH/FSH, in turn, stimulate estradiol and progesterone synthesis and release from the ovaries and testosterone synthesis and release from the testes. In the other example, CRH stimulates the release of adrenocorticotropic hormone (ACTH) from the anterior pituitary. ACTH, in turn, stimulates the synthesis and release of glucocorticoids from the adrenal cortex. Within the hypothalamus, neurons that release neurohormones are shown in red, while regular neurons that synapse on other neurons are shown in blue. The green- and red-filled circles in the anterior pituitary represent cells that secrete either ACTH or LH/FSH, respectively.

system. A good example relates to breeding seasons. Some species are reproductively active only during certain times of the year. Very generally, changes in day length regulate the function of GnRH neurons. For species that breed in the spring and summer, increases in day length ultimately affect neural processes that activate GnRH neurons, which in turn stimulate LH and FSH release. Subsequent increases in testosterone and estradiol in the blood then enter the brain to activate the neural systems involved in sexual motivation and behavior. Another neurohormone that regulates the anterior pituitary is corticotropin-releasing hormone (CRH). CRH is produced by neurons in the paraventricular nucleus (PVN) of the hypothalamus (as well as by other neurons), and when it is released into the hypothalamic-pituitary portal veins, it affects the synthesis and release of adrenocorticotropic hormone (ACTH) from the anterior pituitary. ACTH, in turn, acts on the adrenal cortex to stimulate the synthesis and release of glucocorticoids (corticosterone and cortisol). One function of glucocorticoids is their action on the liver to promote gluconeogenesis, which increases blood glucose levels. However, corticosterone and cortisol can also enter the brain to affect neural function and behavior. Stressful (aversive) stimuli (psychological stressors such as fear and anxiety-inducing stimuli and physical stressors such as noxious stimuli that cause pain) activate CRH neurons and the HPA axis (hypothalamo-pituitary-adrenal axis). The subsequent rise in blood glucose provides the organism with additional energy resources in such emergency situations. Not too surprisingly, glucocorticoid action in the brain also influences aggression-related neural systems.

The posterior or neural lobe of the pituitary is regulated differently from the anterior pituitary. Some neurons in the PVN and the supraoptic nucleus (SON) of the hypothalamus (SON is not shown in Figure 1.4), which manufacture oxytocin or vasopressin, send their axons directly to a capillary plexus located in the neural lobe. When PVN and SON neurons are activated, their action potentials release oxytocin and vasopressin directly into the blood. For the purpose of this brief introduction, the discussion will be limited to oxytocin. When oxytocin is released into the blood, it can act on peripheral targets such as uterine cells or myoepithelial cells in the mammary gland. Oxytocin is a hormone closely tied to the maternal condition. It acts on the uterus to cause uterine cells to contract and therefore aids the birth process. It is also crucial for the milk-ejection reflex—when a baby sucks on its mother's nipple, neural pathways are activated that stimulate PVN and SON neurons to release oxytocin. The increases in blood levels of oxytocin reach the mammary gland, where they act on oxytocin receptors to cause the milk ducts to contract, which squirts milk into the baby's mouth. The milk-ejection reflex is a perfect example of how neurohormones, in this case oxytocin, link the nervous system with the endocrine system.

In Figure 1.4, some neurons are labeled as regular neurons (RN). These are the typical neurons that we are most familiar with—they are activated by neurons and they release their neurotransmitter locally onto other neurons. The figure indicates that some PVN oxytocin neurons are regular neurons in that they synapse on other regular neurons instead of projecting to the capillary plexus in the posterior pituitary. That is, oxytocin can be released into the brain as a neurotransmitter or released into the neural lobe capillary plexus as a hormone. The same is true for vasopressin. Therefore, chemical structure does not define the difference between a hormone and a neurotransmitter, function does. If a neurochemical is released locally at a synapse, it is referred to as a neurotransmitter (or neuromodulator), but if the same chemical is released into the blood to act at a distant target, then it is called a hormone or neurohormone. As will be seen, oxytocin action in the brain plays a major role in regulating maternal behavior and other prosocial behaviors [703]. Oxytocin's action in parturition, milk ejection, and maternal behavior is a great example of an integrated neurochemical system—the peripheral effects of oxytocin as a hormone coincide with and support its role as a brain neurotransmitter that promotes maternal behavior.

1.3.2 Ernst and Fudge (2009): A Neural Model of Goal-Directed Motivational Processes

A review by Ernst and Fudge [272] will provide a starting point for a discussion of the role of the cerebral hemispheres in the regulation of goal-directed behaviors. In their Triadic Model, they propose that the ventral striatum (nucleus accumbens) plays a primary role in regulating goal-directed appetitive behaviors, while the amygdala regulates goal-directed avoidance responses. The prefrontal cortex is viewed as modulating or regulating the output from the amygdala and the ventral striatum so that an adaptive balance between the two systems is achieved. Starting with this simplified model, which proposes distinct functions for each of these areas, they acknowledge several qualifications, admitting that the amygdala and the striatum are heterogeneous structures and that each can regulate both appetitive approach and goal-directed avoidance responses. It is these qualifications that I want to emphasize in order to present a more complex view of the control of motivation. In what follows, a discussion of amygdala function will be presented, followed by a review of the ventral striatum and associated structures, and concluding with an analysis of the role of the prefrontal cortex in motivation.

1.3.3 The Amygdala

1.3.3.1 Basic Anatomy

The major components of the amygdala are similar in primates, cats, and rats. Figure 1.5 presents a cross-section through the rat amygdala showing some of its most important nuclei. With respect to the anatomy, a few points will be emphasized that are very relevant to social behavior [501,621,733,952]. First, the medial amygdala (MeA) receives strong olfactory inputs, and it, in turn, projects strongly to the hypothalamus via the stria terminalis. Different MeA neurons project to different parts of the hypothalamus, and these projection neurons have been found to either contain gamma-aminobutyric acid (GABA) or glutamate, as well as certain neuropeptides [95]. The central nucleus of the amygdala (CeA) receives a variety of sensory inputs, and it has important projections to the lateral hypothalamus (LH) and to the periaqueductal gray (PAG) in the midbrain. CeA is divided into a lateral and medial nucleus (CeAl and CeAm), and its neurons use GABA as a neurotransmitter. CeAm is the major projection nucleus of CeA, with efferents to LH and PAG. The lateral, basolateral, and basomedial nuclei (LA, BLA, BMA) of the amygdala receive olfactory, gustatory, auditory, visual, and somatic sensory inputs from isocortex and allocortex, although simple visual and auditory stimuli may reach the amygdala directly from the dorsal thalamic sensory relay nuclei. The major projection neurons of the lateral and basal amygdala nuclei use glutamate as their neurotransmitter. One projection of BLA is to the CeAm, allowing for a BLA-CeAm-LH or BLA-CeAm-PAG circuit. However, the BLA and BMA also have strong projections to the ventral striatum (VS) and associated areas. To the extent that BLA/BMA neurons influence motivation, one might predict that BLA-CeAm-PAG projections mediate consummatory type responses, while BLA/BMA-VS projections mediate goal-directed responses. That is, projections from the amygdala that go directly to the brainstem may regulate simple, reflex-like responses, while projections to other nuclei within the cerebral hemispheres, such as VS, may regulate more strategic and voluntary goal-directed responses. Finally, the amygdala has significant reciprocal connections with the

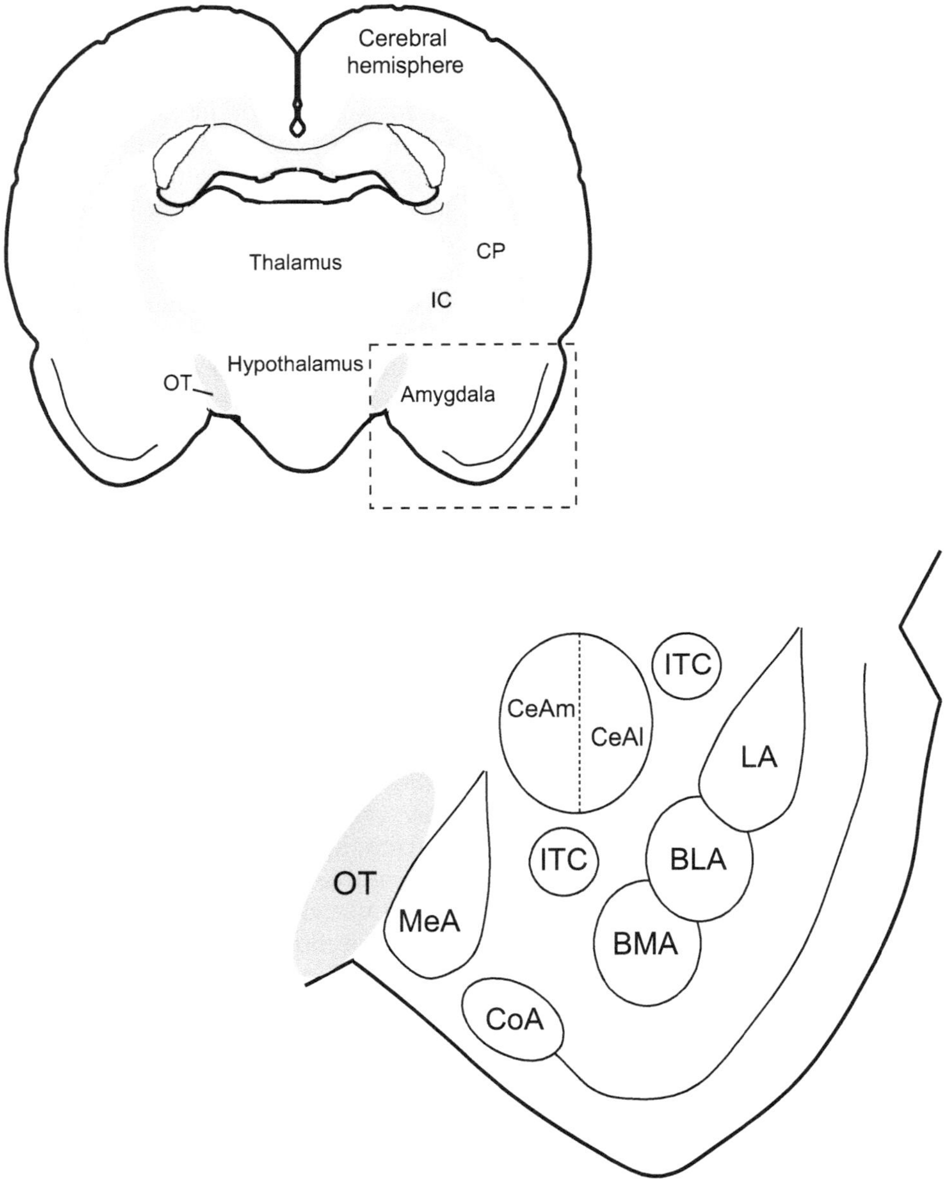

FIGURE 1.5 The major nuclei of the amygdala. The upper diagram shows a frontal section through the rat brain, with the amygdala and nearby regions outlined by a dashed box. An expanded view of the amygdala is shown in the lower diagram. Abbreviations: BLA = basolateral amygdala; BMA = basomedial amygdala; CeAm = medial part of the central nucleus of the amygdala; CeAl = lateral part of the central nucleus of the amygdala; CoA = cortical nucleus of the amygdala; CP = caudate/ putamen; IC = internal capsule; ITC = intercalated nuclei of the amygdala; LA = lateral nucleus of the amygdala; MeA = medial nucleus of the amygdala; OT = optic tract. *The upper frontal section is modified from Swanson LW. Brain maps: Structure of the rat brain, 2nd ed. Amsterdam: Elsevier, 1998/1999.*

prefrontal cortex (PFC), and the nature and importance of these connections will be a major focus of this chapter and other parts of this book.

In addition to these major connections, there are also groups of local inhibitory GABAergic neurons within the amygdala, which when activated, can restrain the efferent outputs just described. The axons of these local inhibitory neurons do not leave the amygdala, and their cell bodies are contained within LA, BLA, BMA, CeAl, and in the intercalated nuclei (ITC) of the amygdala [918].

Note the apparent difficulty that this anatomy presents for the Triadic Model of Ernst and Fudge [272]. If BLA/BMA neurons are involved in avoidance and the VS is involved in appetitive approach, what might be the function of BLA/BMA glutamatergic projections to VS?

1.3.3.2 Functional Anatomy of the Amygdala: Reflexive Fear Responses

The work of LeDoux [530] has emphasized the role of the amygdala in fear-related processes. The primary focus of LeDoux's research has been on consummatory (reflexive) defensive responses rather than goal-directed avoidance responses, and this research has examined the neural circuitry underlying the conditioned fear response (CFR). The CFR involves a Pavlovian learning procedure—a conditioned stimulus (CS), such as a neutral tone, is paired with an aversive or noxious unconditioned stimulus (US), such as foot shock. After several pairings of the CS with the US, the CS acquires the ability to elicit a CFR. The particular CFR that has been the object of many studies in rodents is the freezing response—initially the tone does not cause inhibition of movement, but after several CS–US pairings, the CS begins to elicit the freezing response, where the animal becomes immobile. There is a tremendous amount of recent research of the neurobiology of the CFR, but only certain aspects will be highlighted here [20,190,267,407,545,1076]. It has been shown that plasticity within an LA-BLA-CeA-PAG circuit is necessary for the ability of a tone that has been paired with shock to inhibit motor activity in an experimental animal. The classic research on fear conditioning indicated that LA was the primary site where associations between the CS and the US are formed, while CeAm output projections to PAG were necessary for the behavioral expression of CFRs. Although LA does not project directly to CeAm, it influences CeAm indirectly through projections to BLA. Figure 1.6(A) shows a simplified neural circuit for the formation of the CFR of freezing. As a result of CS–US pairings, synaptic plasticity mechanisms strengthen the synapse from the CS sensory input arriving from either the cortex or the thalamus, onto LA neurons, which then results in the CS activating an LA-BLA-CeAm-PAG circuit that results in somatomotor inhibition (the PAG has descending projections to the medullary reticular formation; [420]). Figure 1.6(A) also indicates the neural circuits that, if active, could restrain the output of CeAm and therefore suppress the CFR [20,190]. Certain BLA glutamatergic neurons project to either CeAl or ITC, and these neurons then send GABAergic projections to CeAm that inhibit its output. Three points are worth noting. First, different populations of BLA neurons can either promote or depress reflexive fear responses mediated by CeAm output. Second, any stimulus that activates those ITC and/or CeAl neurons that synapse on CeAm projection neurons is in position to depress fear responses. Third, uncovering the microcircuitry within a neural region is essential for an understanding of the ways in which its projection neurons, which connect to other brain regions (such as PAG), are regulated.

When the role of oxytocin in social behavior is examined in several future chapters, it will show that one of the effects of oxytocin when it is released into the brain as a neurotransmitter is that it has anxiolytic effects—it can decrease fearfulness. In this context, the impressive work of Stoop and colleagues [431,932,999] is relevant to the current discussion of amygdala microcircuitry. Oxytocin receptors are located in CeAl, and oxytocin activates CeAl neurons that, in turn, inhibit CeAm neurons that project to PAG. In behavioral studies, they found that oxytocin microinjection into the CeA suppresses the CFR in rats.

The studies reviewed above have analyzed the amygdala's involvement in CFRs. A recent study has explored the amygdala microcircuitry underlying an unconditioned or innate anxiety response in mice that were tested on the elevated plus maze [982]. The elevated plus maze consists of a central start box that connects to either closed arms or alleys (the alleyway has walls) or to open arms without walls. Rodents display anxiety-like responses in open spaces; if a mouse spends a lot of time in the open arms of the maze, this is usually interpreted as indicating a low level of anxiety or fear of open novel spaces, while a decreased amount of open arm exploration time would be interpreted as an enhancement of fearfulness. Tye et al. [982], using optogenetic techniques (see Box 1.1), found that when mice received photostimulation of the BLA, they showed decreased open arm exploration, suggesting an increase in fearfulness. In contrast, selective photostimulation of BLA axon terminals in CeAl resulted in an anxiolytic effect, with such mice spending more time in the open arms compared to control mice. Subsequent in vitro brain slice optical stimulation coupled with electrophysiological recordings indicated that two separate populations of neurons in BLA project to CeA—one to CeAl and one to CeAm. Direct photostimulation of the entire BLA had a net excitatory effect on CeAm, which would presumably activate CeAm projections to PAG, leading to movement suppression. In contrast, photostimulation of BLA axon terminals in CeAl would activate GABAergic inhibitory projections from CeAl to CeAm, in this way depressing the unconditioned fear response to the open arms. The authors conclude that the

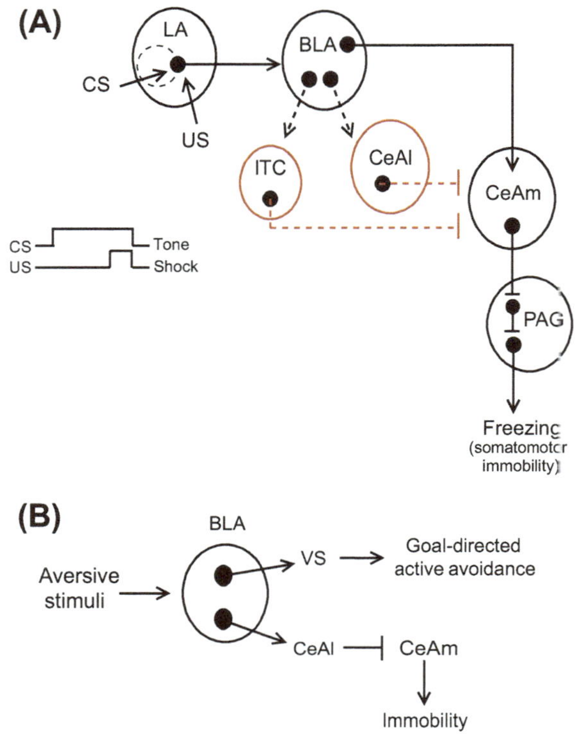

FIGURE 1.6 (A) The classic neural model of the conditioned fear response (CFR). Initially, a neutral tone does not cause an animal to become immobile (freezing response). However, after a tone conditioned stimulus (CS) is paired with shock (the unconditioned stimulus; US) over several trials (a time line depicting CS–US presentations is shown on the left under the lateral nucleus of the amygdala = LA), the CS acquires the ability to elicit freezing. Neural activity derived from the CS and US converge on LA, with the result that the synapse that relays sensory input from the CS to LA is strengthened (surrounded by a dashed line). LA neurons then excite neurons in the basolateral amygdala (BLA), which in turn excite neurons in the medial part of the central nucleus of the amygdala (CeAm). CeAm activates the periaqueductal gray (PAG), which, through its descending projections, causes the freezing response. Since CeAm neurons are GABAergic (inhibitory), CeAm is shown as activating the output of PAG to the lower brainstem by inhibiting local inhibitory neurons within PAG. The neural pathways shown in dashed lines are NOT active during the CFR. However, if they were active, they would inhibit the CFR; additional neurons in BLA project to intercalated nuclei (ITC) and to the lateral part of CeA (CeAl), and these latter nuclei (shown in red), when active, inhibit CeAm output, with the result that the CFR would be depressed. Axons ending in a bar

balance between the direct and indirect inputs of BLA to CeAm regulate the final output of CeAm and the level of fearfulness that is exhibited.

Most of the research on these reflexive-type fear responses has been performed in rodents. However, the CeA also appears to be an important output region regulating simple fear responses in rhesus monkeys. For example, monkeys with bilateral excitotoxic amino acid lesions of CeA reach for a preferred food item in the presence of a snake with a much shorter latency then do intact control monkeys [470]. An important human example is the case of SM, a 44-year-old woman (in 2010) with extensive bilateral amygdala damage resulting from Urbach–Wiethe disease, which is a genetic disorder that causes abnormal calcification of the amygdala, resulting in lesions [287]. The exact age of onset of her amygdala damage is not known with certainty, although it is likely to have been a long-term lesion. SM was examined under a variety of naturalistic test conditions that were expected to arouse fearfulness, such as visits to an exotic pet store that had snakes and spiders, and to a commercial "haunted" house. Compared to controls, SM showed a relative absence of fear responses (she approached and handled snakes and she did not scream when surprised by "monsters" in the haunted house). Importantly, she also reported that she did not *experience* fear in these situations. When neuroscientists study nonhuman animals, we can only examine and measure their behavior, and we have no access to what they may be experiencing. In contrast, with people, we can interview them and ask them about their affective states. I will have more to say about SM later, and I will also discuss the putative neural underpinnings of emotional experience.

represent inhibitory connections, and those ending in an arrow are excitatory. (B) The potential neural circuitry underlying goal-directed active avoidance or escape responses. The CFR is a reflexive consummatory response, and the elicited freezing response does not avoid or escape shock. To avoid or escape from shock, neural pathways would have to inhibit freezing while also promoting goal-directed active responses. This could occur if an aversive or noxious stimulus activated BLA neurons with excitatory projections to CeAl, which would then inhibit CeAm. Further, aversive stimuli could also activate BLA neurons that project to the ventral striatum (VS = nucleus accumbens), which would lead to goal-directed avoidance or escape responses. Axons ending in a bar represent inhibitory connections, and those ending in an arrow are excitatory.

1.3.3.3 Functional Anatomy of the Amygdala: Goal-Directed Avoidance Responses

Up to this point, I have been primarily examining the role of the amygdala in simple reflexive-like conditioned and unconditioned fear responses, and I have emphasized the importance of amygdala output projections to the brainstem in mediating such responses (i.e., CeAm to PAG). What about goal-directed avoidance or escape responses? The amygdala, not surprisingly, is involved in these as well, and the important projections appear to connect the amygdala with other parts of the cerebral hemispheres. BLA projections to the cerebral hemispheres include projections to ventral striatum and prefrontal cortex; perhaps these projections are important for goal-directed avoidance and escape responses. In this context, consider the following scenario. You are in a dark alley, and in the distance you see a suspicious individual carrying a knife. Initially, you might attempt to avoid this person and escape from the situation. However, if the person chases and catches you, you might engage in reflexive defensive behaviors. In this scenario, perhaps the dominant neural activity would switch from BLA projections to VS or PFC to CeAm projections to PAG as the behavior changed from goal-directed avoidance to defensive responses. In an interesting human study, there is evidence that this switch in neural activity does indeed occur under such a situation [654]. With respect to the role of the amygdala in fearfulness, while BLA projections to CeAm may regulate consummatory or simple fear and defensive responses, BLA projections to CeAl may inhibit consummatory defensiveness

BOX 1.1

OPTOGENETIC METHODS AND NEURAL CIRCUITRY ANALYSIS

The development and use of optogenetic methods over the past decade have greatly increased our understanding of how specific phenotypically defined neural circuits regulate behavior. Excellent overviews of this method have been written [981,1056,1057], and only a brief introduction will be provided here. (Aspects of this presentation will be more fully appreciated after reading the sections on the amygdala and basal ganglia in this chapter and Chapter 2 on genetics.)

This method combines the use of optical and genetic techniques so that specific neural circuits can either be stimulated or inhibited by light. In a typical experiment, a genetically modified virus carrying a microbial opsin gene is stereotaxically injected through an implanted guide cannula into a specific neural region. The virus is then incorporated into the genome of the neurons within the injection site. Where possible, the promoter region of the opsin gene is constructed so that the opsin is only expressed within certain types of neurons. However, some promoter sequences that would be needed to limit opsin expression are too large to be packaged in viral vectors. In such cases, more advanced strategies, which involve the use of transgenic mice, are needed to restrict expression to specific neurons.

The two most commonly used opsin genes that have been injected into the brain are channelrhodopsin 2 and halorhodopsin. When channelrhodopsin 2 is expressed within a neuron, it produces a cation ion channel that is sensitive to blue light; when activated by blue light, the ion channel opens, allowing Na^+ ions to enter the neuron to cause an excitatory depolarization. In contrast, halorhodopsin produces a Cl^- ion pump protein that is incorporated in the cell membrane of neurons and is activated by yellow light. When activated, the pump moves Cl^- ions into the neuron, hyperpolarizing and therefore inhibiting neural activity. Importantly, when these opsins are expressed, they are incorporated into the cell membrane throughout a neuron and, therefore, they are not only present in cell bodies and dendrites but are also expressed within axons and axon terminals. Since any one neuron may synapse in several brain regions, optogenetic techniques allow one to selectively excite or inhibit specific axon terminals by directing a light stimulus at the terminal rather than the cell body. A light stimulus is usually introduced into the brain by lowering a thin fiber-optic wire that is coupled to a light source, such as a laser, into the selected brain region through a previously implanted guide cannula.

Three studies will be described that have employed optogenetics to understand the neural circuits that underlie behavior. These studies relate to the content of this chapter, and relevant figures in this chapter will be utilized to explain the rationale and method of each study. In reference to Figure 1.6, research on the CFR proposes that the site of neural plasticity is within LA. That is, when a tone is paired with a shock, the ability of tone synaptic inputs to activate LA projection neurons (pyramidal neurons) is strengthened. Using optogenetic techniques, Johansen et al. [458] (also see [459]) set out to test this hypothesis. A modified viral vector was stereotaxically injected into LA of rats. The virus contained the channelrhodopsin 2 gene with a CaMKII (calcium modulated kinase II) promoter sequence, which was meant to restrict expression of the gene to pyramidal cells and avoid expression within inhibitory GABA interneurons. Subsequently, the attempt was made to condition the rats using blue light stimulation of LA as the US. A tone was paired with light stimulation of the LA

BOX 1.1 *(cont'd)*

over several trials, and it was determined whether a CFR (behavioral immobility) occurred to the tone stimulus alone. The results indicated that this paradigm did lead to the development of a CFR but that the strength of the conditioning (amount of time spent freezing) was weak in comparison to the normal conditioning paradigm where the tone would be paired with shock. The weakness of the conditioning might be explained as follows. Although channelrhodposin 2 expression may have been restricted to the LA pyramidal cells, it would be expressed in all such cells, and therefore would have included both positively and negatively valent neurons and LA neurons that were part of more than one circuit. Therefore, the stimulated LA neurons may not only have been part of an aversive LA-to-BLA-to-CeAm-to-PAG circuit, but could also have been part of aversive or appetitive circuits that ultimately projected to the NA-VP circuit. This lack of specificity probably diluted the observed effects.

In an outstanding study, Knobloch et al. [502] explored the anxiolytic effects of oxytocin within CeAl of the amygdala by using optical stimulation of OT axon terminals in this region. A viral vector that contained the channelrhodopsin 2 gene coupled to the OT gene promoter sequence was injected into the PVN of the hypothalamus. This promoter sequence restricted the expression of the opsin to oxytocin neurons. These opsin-containing OT neurons were shown to project to a variety of brain regions, including the amygdala. In vitro studies showed that blue light stimulation of axons in the CeAl region was associated with OT release and with increased neural excitation in CeAl neurons. This activation subsequently inhibited CeAm neurons (see Figure 1.6). Recall that the output of CeAm to PAG mediates the CFR. In vivo studies, in rats that had previously been conditioned, showed that blue light stimulation of CeAl inhibited the CFR.

The above studies used rats and the injection of a modified viral vector into a specific brain region. More complex studies have combined microinjections of modified viral vectors into the brains of transgenic mice (see Chapter 2). Such approaches are used when viral vector modification alone cannot fully limit the expression of an opsin to a specific neuron type. In reference to Figure 1.8, which describes the operation of the dorsal basal ganglia, research has suggested that activation of the direct pathway stimulates, while activation of the indirect pathway inhibits, movement. Kravitz et al. [509] used optogenetic techniques with transgenic mice to prove this mechanism of operation. This study utilized Cre recombinase transgenic mouse lines. The D1-Cre strain expressed the Cre recombinase protein under the control of the D1 dopamine receptor gene promoter, and therefore would only be produced in neurons that contained the D1 receptor. The D2-Cre transgenic mouse line only expressed the Cre recombinase protein in neurons that contained the D2 dopamine receptor. For each strain, a modified viral vector containing the channelrhodopsin 2 gene was injected into the dorsal striatum (caudate-putamen). The opsin gene was flanked by a pair of DNA sequences referred to as lox sites. The Cre/lox recombination mechanism works as follows: in those neurons that express Cre recombinase, the recombinase protein recognizes the lox sites and operates to splice the lox-flanked gene into the genome of the Cre recombinase-expressing neuron. Therefore, in the D1-Cre strain and the D2-Cre strain, channelrhodopsin 2 was expressed in either the direct pathway D1-containing neurons or the indirect pathway D2-containing neurons, respectively. As predicted, blue light stimulation of the dorsal striatum of the D1-Cre strain stimulated movement, while blue light stimulation of the dorsal striatum inhibited motor activity in the D2-Cre transgenic mice.

via a depression of CeAm output to PAG, which would then allow BLA projections to either VS or PFC to regulate goal-directed and complex avoidance and escape strategies [352,654]. See Figure 1.6(B).

In support of the above neural model, several studies in rodents that have employed neuron-specific excitotoxic amino acid lesions of different amygdala nuclei have shown that lesions of BLA disrupt a previously learned instrumental avoidance response, while lesions of CeA are without effect [183,482]. In the Choi et al. [183] study, rats were trained in a two-way active avoidance paradigm where they learned to shuttle back and forth in an alleyway in order to avoid a shock that was predicted by a tone. Most rats acquired this response, although about 20% showed poor learning because they tended to freeze when the tone came on and therefore did not run down the alley to avoid the shock. These latter rats appeared to only show the CFR (Pavlovian learning), which interfered with the learning of the instrumental goal-directed active avoidance response. For the 80% of rats that learned well, post-learning lesions of BLA, but not CeA, disrupted the avoidance response to the tone. Importantly, for the 20% of rats that were poor learners of the active avoidance response, subsequent lesions of CeA allowed the instrumental response to be learned. Killcross et al. [482] suggest that the processing of aversive stimuli through a BLA-to-VS circuit may allow for goal-directed instrumental avoidance responses. Figure 1.6(B) summarizes the potential neural circuitry through which threatening or aversive stimuli might lead to goal-directed avoidance and escape responses.

1.3.3.4 Functional Anatomy of the Amygdala: Goal-Directed Appetitive Responses

The involvement of the amygdala in goal-directed avoidance behavior fits with the model presented by Ernst and Fudge [272]. But there is also evidence that the amygdala is involved in goal-directed appetitive responses [501]. Research shows that BLA damage produces deficiencies in the ability of stimuli that have been paired with a primary reward to maintain instrumental or operant responding when the primary reward is remote in time. This, to me, is a good definition of goal-directed appetitive behavior, where an organism responds to stimuli that predict rewards in order to ultimately obtain the reward, which then leads to consummatory responses. Everitt, Cador, and Robbins [276] showed that excitotoxic amino acid lesions of BLA disrupted a male rat's instrumental responding (an operant bar press response) to gain access to a sexually receptive female. In a study by Simmons and Neill [904], male rats were trained to obtain a food reward on an FR (fixed ratio) 16 operant conditioning schedule of reinforcement. Once a stable baseline rate was established (learning had occurred), bilateral injections of muscimol were administered into the BLA. Muscimol is a GABA-A receptor agonist that temporarily inactivates neurons by mimicking the inhibitory neural effects of GABA. The animals were then placed in the operant chamber and their responses were measured over 20 min. Muscimol produced a dose-related decrease in lever presses. In contrast, muscimol injections into the BLA did not affect the amount of food ingested during a 30-min free feeding session. These findings conform to the idea that muscimol in BLA depressed appetitive goal-directed reward seeking behavior but did not affect the consummatory response that is regulated by direct contact with proximal food cues. Therefore, the BLA appears to be part of a neural network that controls whether an organism will work, and how much energy will be devoted, to achieve a desired goal: appetitive or reward-seeking behavior (also see [24]). As will be shown shortly, such appetitive amygdala circuits connect the amygdala to other regions in the telencephalon, such as the ventral striatum and related structures.

1.3.3.5 Functional Anatomy of the Amygdala: Distinct Neural Circuits That Respond to Either Positively or Negatively Valent Stimuli

Because of the involvement of the amygdala in both goal-directed avoidance and appetitive behaviors, I take a labeled line view of the amygdala, which proposes that distinct neural circuits within the amygdala carry specific types of information. A variety of sensory stimuli reach the amygdala from the cerebral cortex. Some stimuli are innately noxious or threatening, and others are innately attractive and appealing. Still other stimuli gain aversive or appetitive characteristics as a result of being paired with innate positive or negative stimuli. Aversive, noxious, or unpleasant stimuli are proposed to activate one population of amygdala neurons, the efferent projections of which can give rise to goal-directed avoidance, escape, or rejection responses. I will refer to these amygdala neurons as those with a negative valence, since they code for unpleasant stimuli. In contrast, reward-related or appetitive stimuli activate positively valent amygdala neurons whose efferents can give rise to goal-directed appetitive responses. In this view, the amygdala regulates both goal-directed appetitive and avoidance responses, based on the circuits activated by particular stimuli. Some interesting neurophysiological data support these views.

Belova, Paton, and Salzman [84] recorded the neural activity of single neurons within the amygdala from rhesus monkeys as they were exposed to different visual stimuli that were paired with either liquid rewards or aversive air puffs into the eye. Monkeys were trained to focus on the center of a screen for 1 s, and then a neutral visual image was presented for 300 ms. Then 1.5 s later, a liquid reward for some stimuli or an aversive air puff for other stimuli was presented. After the monkeys learned the initial value or valence (+ or −) of each visual image (when the monkeys expected the liquid reward they engaged in anticipatory licking behavior; when they expected the air puff, they blinked), the researchers reversed the image value assignments. During reversal, the initial positive image was followed by an air puff, and the initially negative image was followed by a reward. The basic findings can be outlined as follows:

1. With respect to the US, liquid reward activated one population of neurons in the amygdala, while air puffs activated a separate population of amygdala neurons. One might call these innate + and − stimuli that activate positively or negatively valent amygdala neurons, respectively.
2. Once the monkey learned the relation between a visual stimulus and a reward, that visual stimulus activated the positively valent neurons. When conditions reversed, so that the previously rewarded visual stimulus was followed by an aversive air puff, the response of the positively valent neuron to that visual stimulus declined. In contrast, a separate population of amygdala neurons responded to visual stimuli paired with air puffs, but the response of these negatively valent amygdala neurons to that visual stimulus declined once reversal training occurred so that the visual image was followed by a reward rather than an air puff. These results indicate that neurons in the amygdala encode stimulus value or valence, and neurons that respond to positive or pleasant stimuli are separate from those that respond to negative or aversive stimuli.
3. Reconstruction of electrode location via magnetic resonance imaging (MRI) analysis indicated that neurons that responded to either positive or negative stimuli were located in BLA as well as other amygdala regions.

Schoenbaum, Chiba, and Gallagher [869] performed a similar single neuron recording study on rats. Water-deprived rats had to sample an odor presented in a port on each trial (odor sampling) in order to decide whether to respond (go response) at a nearby fluid well. A go response resulted in delivery of a rewarding sucrose solution after the presentation of a "positive" odor, or an aversive quinine solution after the presentation of a "negative" odor. Rats would begin each session by responding on every trial, irrespective of whether a positive or negative odor was presented. Learning was evident when a rat began to withhold responses (no-go) after sampling the negative odor in order to avoid quinine delivery. Neural activity was recorded during odor evaluation trials after the rat had learned the associations (responding to the + odor and not to the – odor). Within the BLA, some neurons had higher rates of firing during evaluation of odors that predicted sucrose delivery when compared to their rates when "negative" odors were sampled, whereas other neurons responded more strongly during evaluation of cues that signaled quinine delivery. Importantly, during reversal learning, if a neuron responded to a "positive" odor that was reversed to a "negative" odor, over a series of trials that neuron ceased to respond strongly to the now negative odor. Therefore, a neuron's selectivity was not tied to the sensory features of a particular odor but rather depended on the associated outcome. Importantly, when BLA neurons developed selective responding during odor sampling, this selectivity (difference between rate of firing to positive versus negative odor) was not present during early training trials but developed rapidly, well before accurate choice performance was achieved, which suggests that neural activity was a measure of stimulus valence rather than a measure of a particular motor response (go versus no-go).

In a related study, Shabel and Janak [883] recorded from single neurons in the rat amygdala and found three classes of neuronal responses. As in the above studies, some neurons responded selectively to appetitive stimuli, while others responded selectively to aversive stimuli. A third class of neurons responded to both appetitive and aversive stimuli. This latter class might be considered arousal neurons that signal motivational and emotional salience independent of valence.

Most of the discussion so far has concentrated on the BLA and CeA. However, MeA also contains separate populations of neurons that respond to stimuli with either a positive or negative valence. MeA receives strong olfactory input, and research on rodents shows that predator odors activate one group of MeA neurons that project to hypothalamic regions that regulate defensive and escape responses, while sexual pheromones activate another group of MeA neurons that project to different hypothalamic regions that regulate reproductive behaviors [184]. When I argue for a labeled line point of view, it is basically a localization of function point of view. Within the brain regions that influence motivation and emotion, different circuits have different functions, in a manner similar to the sensory systems, where, for example, the function of lateral geniculate nucleus projections to visual cortex is distinct from medial geniculate nucleus projections to the auditory cortex.

The results of these amygdala studies are important for several reasons. Not only do they indicate that the amygdala contains separate neurons that respond to aversive or appetitive cues, but they also show that experience with an initially neutral stimulus, due to learning processes, can result in that stimulus acquiring either a positive or a negative valence. Let's put this in the framework of understanding social behavior. When you first meet someone, your emotional/motivational response may be neutral. However, after some social interactions, that person's stimuli may gain the ability to activate either positive or negative amygdala neurons, depending on the outcomes of your interactions with that person.

Although I have mainly discussed the role of the amygdala in motivational and emotional contexts that did not involve social behavior, these studies are clearly related to the role the amygdala might play in social behaviors, and in subsequent chapters this view will be strongly affirmed. At this point, I will note that fMRI studies (Box 1.2) in humans show that the amygdala can be activated by both positive and negative social stimuli, such as angry or happy faces [251]. Further, the human patient SM, who has bilateral amygdala damage, shows deficits in social behaviors, although these deficits are not extreme [6,398,771]. SM has a long-standing amygdala lesion, and Phelps and LeDoux [771] point out that she may not show pervasive and severe social deficits because of compensatory mechanisms resulting from a lifetime of social interactions, which may have allowed other parts of the brain to regulate a certain level of social responsiveness. Further, social knowledge obtained through social interactions that occurred prior to the formation of the calcified amygdala lesion may have allowed for certain aspects of appropriate social behavior, particularly if such social information was stored in brain regions outside the amygdala. Interestingly, however, in a study by Heberlein and Adolphs [398], when SM and normal control subjects were shown films of inanimate objects (triangles and squares) interacting, normal subjects spontaneously created social narratives of these interactions (anthropomorphizing), while SM did not. One interpretation is that SM has permanent deficits in automatic or immediate social processing. Finally, the social behavior of individuals with Williams syndrome (WS) is relevant. WS is a human genetic disorder caused by the hemizygous deletion of about 25 genes from chromosome 7 [447]. In comparison to normal controls, WS individuals are characterized by hypersociality—they are gregarious and empathic and show no fear of strangers [461]. Haas, Mills, Yam, Hoeft, Bellugi, and Reiss [372], in an fMRI study, found that WS subjects, in comparison to controls, showed increased amygdala activation to happy faces and decreased amygdala activation to fearful or angry faces (also see [638]). Since amygdala activation to threatening nonsocial stimuli was not altered, these findings appear specific to social stimuli. The genetic deletion appears to have affected the functional activity of positively and negatively valent amygdala neuron connectivity with other parts of the brain [372,447], resulting in a facilitation of amygdala prosocial circuits and a depression of amygdala circuitry related to social aversion, withdrawal, and rejection.

1.3.4 The Dorsal and Ventral Basal Ganglia and the Influence of the Nigrostriatal and Mesolimbic Dopamine Systems

1.3.4.1 Introduction

In the Ernst and Fudge model [272], the ventral striatum (nucleus accumbens) was given a special role in the regulation of goal-directed appetitive responses. However, just as the amygdala was shown to be involved in both appetitive and aversive motivation, the same appears to be true for the ventral striatum. In order to critically evaluate this proposal, an overview of two important neural systems within the subcortical cerebral hemispheres is needed: the dorsal basal ganglia and the ventral basal ganglia, each of which receives significant dopaminergic input from the brainstem [434].

The midbrain gives rise to several major ascending DA systems that terminate in the cerebral hemispheres, and two of these are the nigrostriatal and the mesolimbic DA systems. Although there is some overlap in the organization and function of these systems, the nigrostriatal DA system primarily modulates functional activity within the dorsal basal ganglia, while the mesolimbic DA system primarily modulates activity within the ventral basal ganglia

BOX 1.2

UNDERSTANDING THE FMRI BOLD SIGNAL

Much research dealing with human social, cognitive, and emotional neuroscience utilizes the indirect measurement of neural activity in the brain under specified conditions through functional magnetic resonance imaging (fMRI; [734]). In a typical cognitive or affective neuroscience fMRI scanning procedure, a subject is placed in a scanner and a magnetic field is passed through the head region and brain while the subject performs a specified task, such as observing visual scenes (for example, angry, happy, or neutral faces). The scanner measures a hemodynamic response in the brain referred to as the blood-oxygen-level dependent (BOLD) signal. The BOLD signal is based on the differential magnetic properties of oxygenated and deoxygenated hemoglobin. The following sequence of events is usually considered to result in an increased BOLD signal [42,43]: as neural activity increases in a particular brain region, increases in local cerebral blood flow occur in that region. This increased blood flow results in an increase in the oxygenated-to-deoxygenated hemoglobin ratio, which results in an increase in the measured BOLD response.

Given that the BOLD signal is an indirect measure of neural activity within a brain region, an important question is the nature of the neural activity that is correlated with the BOLD signal. Research on primates and rodents, which has combined optogenetics, electrical recording, and fMRI procedures, has provided some answers [32,536,798]. The BOLD signal is positively correlated with the amount of synaptic input to a neural region, the degree of neural processing within both excitatory and inhibitory local circuit neurons within the region, and the amount of neural activity in the output or projection neurons through which the region of interest connects to other neural regions. Importantly, it has been clearly shown that an increased BOLD signal can occur in a brain region even though there is no increase in the activity of the projection neurons [32,798]. The following figure shows two cases where an increased BOLD signal (above a baseline control value) would be detected in two different brain regions, but in one case (Area 1) the output of a region would be increased and in the other (Area 2) the output would be decreased. In this figure, axons with a solid line are highly active (HI), while those with a dashed line are exhibiting low activity. Therefore, an increased BOLD signal in each of the boxed areas does not necessarily mean that the output of each brain region has increased. Indeed, its output to other brain regions may have actually decreased.

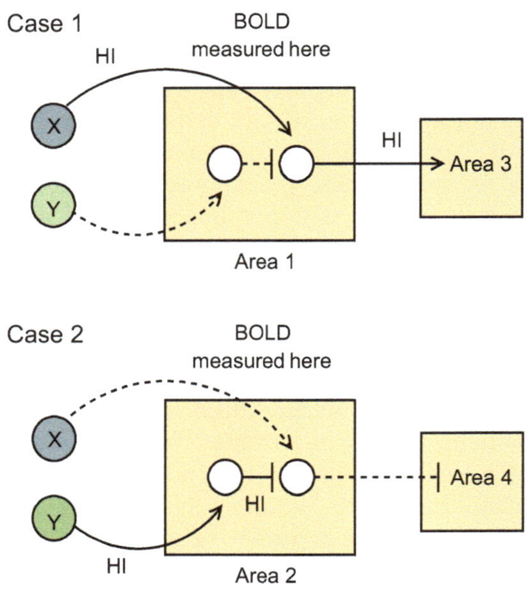

In the context of this understanding, note that statistical procedures have been applied to fMRI data in order to get a measure of the

BOX 1.2 *(cont'd)*

degree of connectivity between two or more brain regions during a particular cognitive task, for example. These statistical procedures, which are correlational in nature, provide measures of functional and effective connectivity [734]. In functional connectivity, the data simply provide evidence that the BOLD signals in two (or more) regions are correlated, but cause–effect or directional information is not determined. With effective connectivity measures, statistics are used to provide evidence of the direction of a neural circuit, for example, that an increased BOLD signal in one area causes an increased BOLD in another area. Based on the figure in this box, note that such positive effective connectivity could result from Area 1 projections directly stimulating Area 3 (Case 1) or from Area 2 projections inhibiting Area 4 (Case 2). In Case 2, the increased BOLD response in Area 2 results in a disinhibition of Area 4, which increases the Area 4 BOLD signal, because Area 2 projection neurons are inhibited. In the figure, axons ending in a bar are inhibitory, and those ending in an arrow are excitatory.

The analysis presented indicates that the interpretation of fMRI data is complicated and that a full appreciation of the meaning of an increased BOLD signal in a brain region requires an understanding of the underlying microcircuitry. Although this is usually not possible from fMRI data alone, the integration of nonhuman animal recording studies with human fMRI data may help guide an appropriate interpretation of the fMRI data.

[434,881]. As shown in the Figure 1.7, the nigrostriatal DA system originates in the substantia nigra pars compacta (SNc) and terminates in the caudate nucleus and putamen, which are collectively referred to as the dorsal striatum. Nigrostriatal DA input to the dorsal striatum facilitates an organism's behavioral reactivity to sensory stimuli that are primarily of isocortical origin. In contrast, the mesolimbic DA system originates from neurons in the ventral tegmental area (VTA) of the midbrain, and one of its major sites of termination is the nucleus accumbens (NA) or ventral striatum. One of the functions of the mesolimbic system is to facilitate an organism's behavioral reactivity to sensory stimuli that are primarily being relayed to the NA from neural regions that have been referred to as components of the limbic system: amygdala, hippocampus, and allocortical parts of the prefrontal cortex [91]. Mogenson [655] referred to this system as the limbic motor system, and DA input to NA was viewed as increasing an organism's responsiveness to stimuli that have motivational and emotional significance. For example, since the BLA projects to NA [1044], one might conclude that DA input to NA facilitates responding to stimuli that NA receives from BLA, and that such behavioral reactivity might include either goal-directed appetitive responses or goal-directed avoidance responses, depending on whether BLA is relaying stimuli with a positive or negative valence.

1.3.4.2 Detailed Analysis of the Operation of the Nigrostriatal DA System and the Dorsal Basal Ganglia

Although my focus will be on the ventral basal ganglia, which includes the ventral striatum, because much more research has been done on the organization and function of the dorsal basal ganglia, an overview of its operation will be presented first and then compared to the ventral basal ganglia. Figure 1.8 shows a schematic of dorsal basal ganglia neural circuits

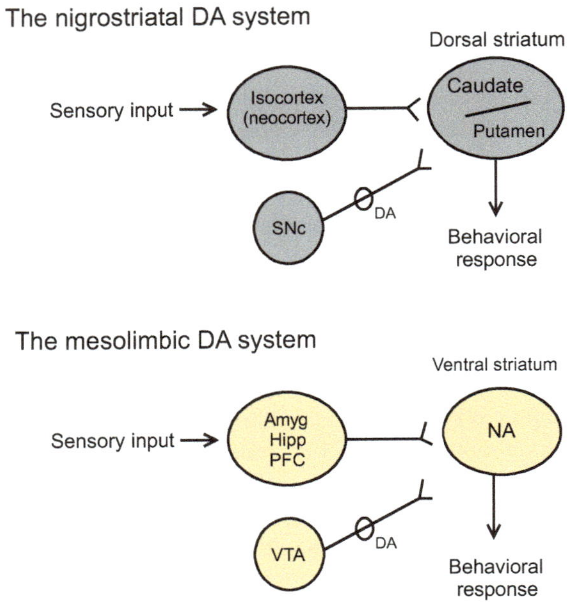

FIGURE 1.7 A simplified differentiation between the functional neuroanatomy of the nigrostriatal dopamine (DA) system and the mesolimbic DA system. For the nigrostriatal system, DA input to the caudate/putamen (dorsal striatum) that originates from the substantia nigra pars compacta (SNc) facilitates an organism's behavioral reactivity to sensory stimuli primarily of isocortical origin. For the mesolimbic system, DA input to the nucleus accumbens (NA = ventral striatum) that originates from the ventral tegmental area (VTA) facilitates an organism's behavioral reactivity to sensory stimuli derived from the amygdala (Amyg), hippocampus (Hipp), and allocortical parts of the prefrontal cortex (PFC). *Modified from Figure 1.3 in Numan and Stolzenberg [713] with permission from Elsevier.*

[310,331,361,434]. Sensory association isocortex projects to the caudate and putamen, or dorsal striatum, and uses glutamate as an excitatory neurotransmitter. The main projection or output neurons of the dorsal striatum are the GABAergic medium spiny neurons (MSNs), and their efferents form two major output pathways: the direct pathway, whose activity facilitates movement, and the indirect pathway, whose output suppresses movement. The MSNs of the direct pathway contain the D1 class of DA receptors, while the MSNs of the indirect pathway contain the D2 class of DA receptors. MSNs have high K^+ permeability, and their resting membrane potential is very negative, near the K^+ equilibrium potential of about −80 mV. Therefore, it is hard for excitatory cortical input to activate these neurons to the threshold for action potentials, and they are typically viewed as being in a downstate or hyperpolarized. When DA input from the SNc acts on D1 receptors on the MSNs of the direct pathway, in conjunction with excitatory cortical input, it acts to *facilitate* the depolarization of the MSNs of the direct pathway, bringing them to an upstate of about −65 mV. This upstate makes the MSNs more easily activated by strong sensory inputs from the isocortex. Once the direct pathway is activated, the GABAergic MSNs *inhibit* the internal segment (medial) of the globus pallidus (GPi). Since the GABAergic output of GPi inhibits the brainstem motor area and the ventrolateral thalamic nucleus (VL), two brain regions whose efferents facilitate movement, activation of the direct pathway stimulates movement through a process of disinhibition (inhibition is removed from BSMA and VL).

With respect to the indirect pathway, very strong cortical input can bring these MSNs into the upstate and activate them. These MSNs project to and inhibit the GABA output neurons of the external segment of the GP (GPe). When active, GPe functions to inhibit the subthalamic nucleus (STN). STN is also excited by glutamatergic afferents from the isocortex. Therefore, activation of the indirect pathway stimulates the STN through disinhibition, which potentiates the effects of isocortical stimulation of STN. Excitatory glutamatergic neurons of the STN *activate* GPi, which results in movement inhibition (because BSMA and VL are inhibited). This would be a mechanism that would allow cortical mechanisms to terminate a movement or depress unwanted movements. Importantly, DA action on D2-like receptors in the dorsal striatum *depresses* the ability of the isocortex to activate the MSNs that contribute to the

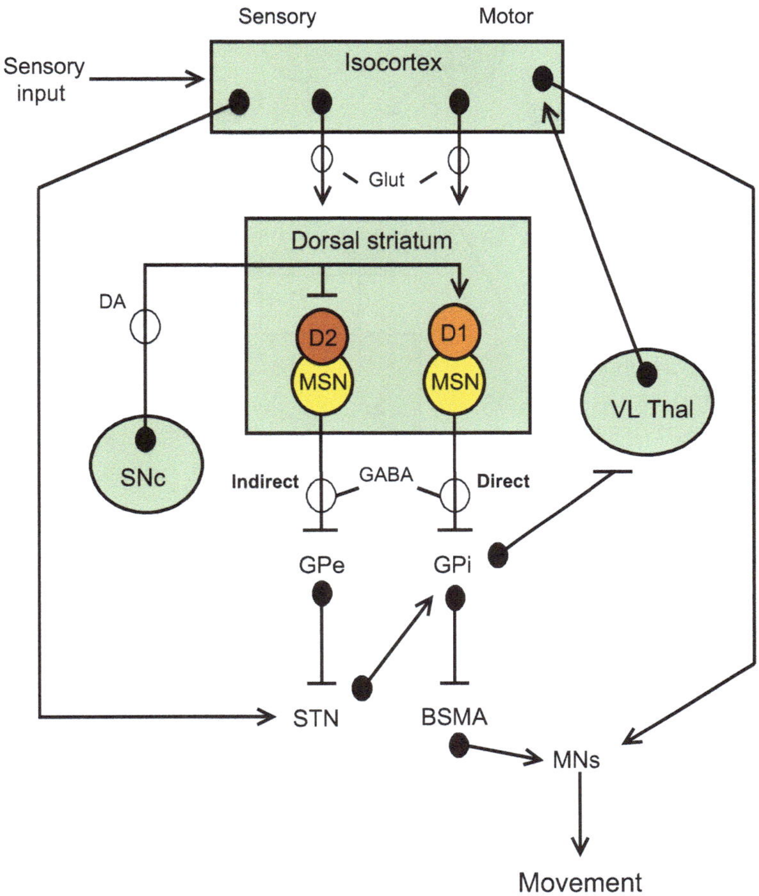

FIGURE 1.8 Diagrammatic representation of the functional organization of the dorsal basal ganglia. Isocortical inputs stimulate the outputs of medium spiny neurons (MSN) within the dorsal striatum (caudate/putamen). Two MSN output pathways to different parts of the globus pallidus (GP) are defined. MSNs that contain D1 dopamine (DA) receptors and project to the internal segment of GP (GPi) give rise to the direct pathway that facilitates behavioral responses. MSNs that contain D2 dopamine (DA) receptors and project to the external segment of GP (GPe) give rise to the indirect pathway that inhibits or restrains behavioral responses. Substantia nigra pars compacta (SNc) DA input to the dorsal striatum facilitates movement because DA action on D2 receptors depresses neural activity in the indirect pathway, while DA action on D1 receptors enhances neural activity across the direct pathway. See text for other details. Axons ending in a bar represent inhibitory connections, and those ending in an arrow are excitatory. Additional abbreviations: BSMA = brainstem motor area; Glut = glutamate; MNs = motor neurons; STN = subthalamic nucleus; VL Thal = ventrolateral nucleus of the thalamus.

indirect pathway. Therefore, activity within the nigrostriatal DA pathway and DA action on both D1 and D2 receptors facilitates movement by suppressing the indirect pathway and facilitating the direct pathway. One can therefore describe the following motor mechanism in the dorsal basal ganglia: when the cortex orders a movement, in conjunction with DA release, the direct pathway is activated and the indirect pathway is inhibited. To terminate the movement, DA levels decline, and the indirect pathway is activated by strong cortical input.

There is recent experimental evidence that has offered excellent support for the operational characteristics of the dorsal basal ganglia as described above [411,509]. For example, in mice, optogenetic stimulation restricted to MSNs of the direct pathway (that contain D1 receptors) has been shown to facilitate movement, while optogenetic stimulation of D2 receptor containing MSNs (which would be inhibited by dopamine (DA)) depresses locomotion (see Box 1.1).

1.3.4.3 Operation of the Mesolimbic Dopamine System and the Ventral Basal Ganglia

The operation of the mesolimbic DA system is less well understood [434,881,914]. The ventral striatum, which contains the nucleus accumbens (NA), receives excitatory glutamatergic inputs from the basolateral and basomedial amygdala, hippocampus, and prefrontal cortex, and DA input from the VTA. NA has been divided into a medial or shell part (NAs) and a lateral or core part (NAc). Individual GABAergic MSNs in NA core and shell, which for the most part contain either D1 or D2 receptors, project to a variety of target regions, including a dominant projection to the ventral pallidum (VP), which forms the NA-VP circuit [575,815,974,984,1075]. It is not yet entirely clear, however, whether distinct direct and indirect pathways emanate from NA, with activity in a putative direct pathway promoting goal-directed behaviors, and activity in a putative indirect pathway inhibiting such behaviors. It is also not clear whether DA acts on D1 receptors to stimulate a direct pathway and on D2 receptors to suppress an indirect pathway.

Figure 1.9 shows a simplified schematic of the ventral basal ganglia. Both D1 receptor containing and D2 receptor containing NA MSNs project to VP [434,575,815,914,1075], and VP targets include the subthalamic nucleus, brainstem motor areas, and mediodorsal thalamus (MDT) [362]. Drawing an analogy from the dorsal basal ganglia, one could argue that VP presents a mixture of both the external and internal segments of the globus pallidus, with VP projections to BSMA and MDT representing aspects of the direct pathway, while VP projections to STN would be a component of the indirect pathway. However, to my knowledge, it has not been shown that NA D2-containing MSNs only project to those parts of VP that project to STN while NA D1-containing MSNs only project to those parts of the VP that project to either BSMA or MDT. It is also not known whether STN stimulates those VP neurons that project to BSMA and MDT. Finally, while the predominant output neurons of the globus pallidus are GABAergic, exerting an inhibition over their targets (see Figure 1.8), the VP output neurons are not only GABAergic but also contain a significant population of excitatory glutamatergic projection neurons [328].

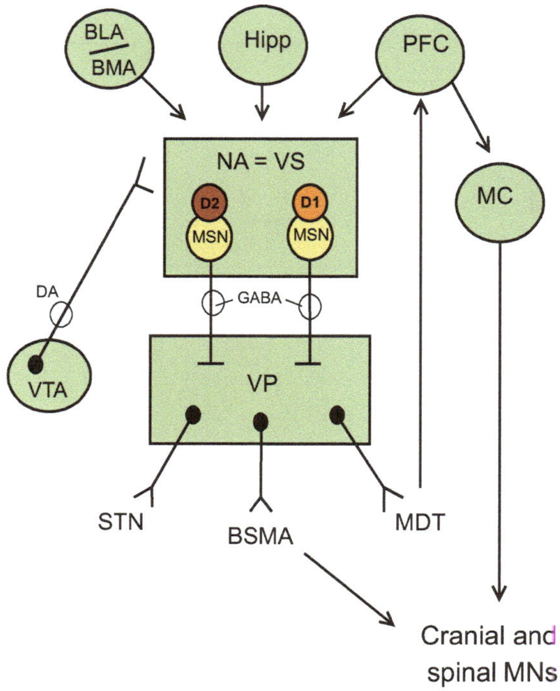

FIGURE 1.9 A simplified representation of the ventral basal ganglia. The functional organization of the ventral basal ganglia is not as well understood as that for the dorsal basal ganglia (Figure 1.8). It is not entirely clear whether distinct direct and indirect pathways result from the projections of nucleus accumbens (NA) D1 and D2 dopamine (DA) receptor-containing medium spiny neurons (MSN) to the ventral pallidum (VP). NA receives excitatory inputs from the basolateral and basomedial amygdala (BLA/BMA), hippocampus (Hipp), and prefrontal cortex (PFC). NA MSNs contain GABA and inhibit VP. VP outputs to the subthalamic nucleus (STN), brainstem motor area (BSMA), and mediodorsal thalamus (MDT) influence an organism's behavioral reactivity. Ventral tegmental area (VTA) DA input to NA modulates activity across the NA-VP circuit. Axons ending in a bar represent inhibitory connections, and those ending in an arrow are excitatory. When an axon ends with typical axon terminals (shown as Y), the excitatory or inhibitory nature of the synapse is left undefined. Other abbreviations: MC = motor cortex; MNs = motor neurons; VS = ventral striatum.

With respect to the operation of the NA-VP circuit, the seminal research of Mogenson

[655] led to the proposal that the activity of GABAergic NA MSNs restrains goal-directed motivated behavior, and that the function of DA action on NA was to depress the output of these MSNs, in this way disinhibiting VP, the activity of which was considered to promote goal-directed behavior. There is a substantial amount of current research that supports this view [696,803,805,876,915,956,1042]. In opposition, however, there is also a literature that supports the view that DA action on NA stimulates GABAergic MSN output, which might then inhibit target regions such as VP, and that such actions activate goal-directed behavior [24,411,562,939]. These two bodies of research could be conceived as being consistent with the existence of both an indirect and direct pathway, respectively, emanating from NA.

In an interesting study by Wirtshafter and Stratford [1040], rats were trained on a progressive ratio 6 (PR6) schedule of reinforcement using food as a reward. On this schedule, the first response on the operant lever is rewarded, but then the number of responses required to earn each subsequent food pellet is increased by six after each reinforcement, so that seven responses are required for the second reward, 13 for the third, and so on. The break point is defined as the number of lever presses the rat makes for a reward before it ceases responding for 3 min. The idea here is that a higher break point indicates greater motivation or greater goal-directed behavior. In animals that received saline injections into NAs, the break point was about 60 operant responses. When amphetamine was injected into NAs, which would increase DA release and increase activation of both D1 and D2 receptors, the average break point increased to about 80 bar presses. Most importantly, the injection of muscimol into NAs also increased the break point to about 80. This last result suggests that a *global inhibition of NAs activity*, without any activation of a supposed direct pathway, can increase motivation or goal-directed approach responses to rewarding stimuli. To the extent that the NAs GABAergic MSNs were inhibited, this should cause increased activity in VP. Therefore, increased VP activity may have increased the break point in the Wirtshafter and Stratford study. In support, Farrar et al. [284] reported that rats with VP inactivation had diminished willingness to work hard on an operant conditioning task to obtain sucrose reward.

In order to resolve these controversies, perhaps different functional populations of neurons exist within the NA-VP circuit [752]. Not only may NAs be distinct in function from NAc [71,434], but subcircuits within each of these broad NA regions may also have distinct input–output relations that have different operational rules. These different populations may regulate separate and distinct larger neural networks; for certain goal-directed behaviors, it might be important for DA action on NA to primarily suppress NA and disinhibit VP, while for other behaviors, the reverse might be the case [127].

1.3.4.4 The Numan Model

I have presented a neural model of ventral basal ganglia function [696,713,714] based on the early ideas of Mogenson [655] and my research on maternal behavior, which will be described in Chapter 5. Aspects of this model are shown in Figure 1.10(A) and may be most accurate with respect to the operation of only certain circuits within NAs. The model shows the basal amygdala (BLA/BMA) providing excitatory sensory inputs to both the NA and VP [757,760]. The output of VP is conceived as being essential for goal-directed responses. Without DA release into NA, the functional effects of BLA/BMA projections to NA and VP cancel each other out. This happens because BLA/BMA activation of NA causes GABA release into VP, which blunts the VP response to its inputs. However, when DA is released into NA, the model proposes that DA acts to suppress the response of NA to input from the amygdala; therefore, less GABA is released into VP, which opens a gate allowing VP to respond to BLA/BMA inputs and promote goal-directed responses. Counter intuitively (based on the operation of the dorsal basal ganglia), my model actually proposes that such DA inhibition of NA responsiveness is caused by DA action of D1 receptors. Therefore, this model does not include a supposed direct pathway, where enhanced NA output would be essential for goal-directed maternal responses.

Since a large body of data on the dorsal basal ganglia suggests that DA action on D1 receptors stimulates MSN output from the dorsal striatum, how could DA action on D1 receptors restrain NA MSN output to VP? Unlike the dorsal striatum, there is a significant population of D1 receptors that are located on the axon terminals of glutamatergic afferents to NA [260]. This anatomical relationship is shown in Figure 1.10(B). Importantly, DA action on presynaptic D1 receptors results in presynaptic inhibition of glutamate release, which would depress NA MSN activation by incoming afferents [177,260].

This analysis indicates that the organization of the ventral basal ganglia is complex and that different types of motivated behaviors may be regulated in different ways by this system. Therefore, for a general statement, it might be best to propose that DA action on the NA-VP circuit promotes goal-directed behaviors.

1.3.4.5 The Role of the Hypothalamus in VTA-DA Activation

What causes DA to be released into NA so it can act on D1 and D2 receptors? Given the role of the hypothalamus in motivation, it makes sense that it is one of the sources of mesolimbic DA activation [716]. The hypothalamus is conceived as monitoring an organism's internal state while also being responsive to the external environment. When appropriate, the hypothalamus would activate DA release into NA through its known projections to the

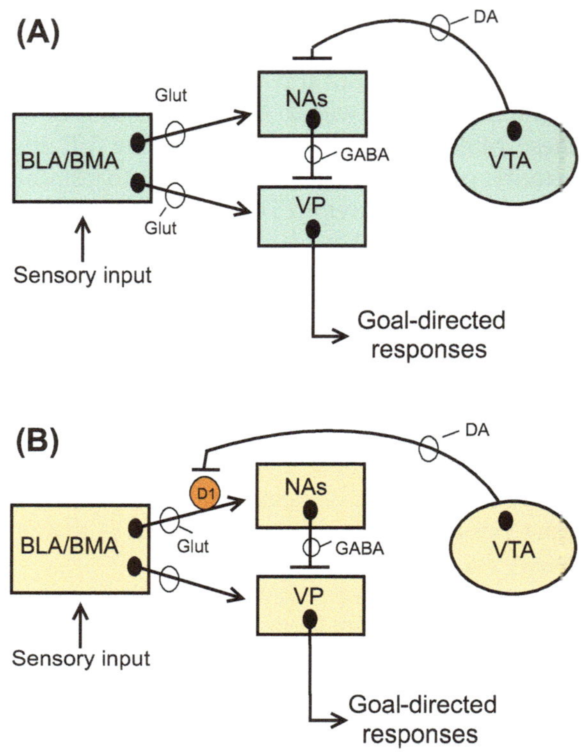

FIGURE 1.10 (A) A model of ventral basal ganglia organization derived from the research of Mogenson [655], Numan [696], and Numan et al. [707]. The basolateral and basomedial amygdala (BLA/BMA) are shown as providing excitatory glutamatergic (glut) input to both the shell region of nucleus accumbens (NAs) and the ventral pallidum (VP). The output of the VP is proposed to mediate goal-directed behavior in response to inputs from the amygdala. Without dopamine (DA) action on NAs, goal-directed responses do not occur because the stimulatory effects of BLA/BMA input to VP are counteracted by the inhibitory effects of NAs GABA input to VP. DA input to NAs from the ventral tegmental area (VTA) is proposed to promote goal-directed responses by suppressing NA activity, which increases the effectiveness of BLA/BMA excitation of the VP. *(Modified from Figure 1.6(B) in Numan and Stolzenberg [713] with permission from Elsevier.)* (B) An elaboration of the model shown in (A), which presents a hypothesis about where DA might act to dampen NA inhibition of VP. D1 DA receptors are shown as located presynaptically on the axon terminals of BLA/BMA neurons that synapse within NAs. DA action on these D1 receptors is proposed to exert a presynaptic inhibitory effect with the result that less glutamate stimulation occurs within NAs. Such an effect would then decrease NA inhibition of VP. (Presynaptic D1 receptors are actually located on

VTA [328] in order to promote adaptive goal-directed responses. Different hypothalamic nuclei, related to different motivational states [950], might be responsive to specific aspects of an organism's internal environment and to specific external stimuli. Under the right conditions, such nuclei might activate the mesolimbic DA system. Let's take two examples (see [716]): goal directed food-seeking responses and goal-directed maternal responses. For food intake or hunger, the lateral hypothalamus (LH; a region that regulates food intake) may respond either directly or indirectly to glucose or fat levels. When energy supplies are low due to a period without food intake, stimuli that have previously been associated with food, or distal food stimuli, may become capable of activating LH, which then stimulates VTA to promote food-seeking behavior. Once food is obtained, LH projections to the brainstem and spinal cord may regulate consummatory responses. In the absence of food deprivation, the LH would not be responsive to food-related stimuli, and food seeking and eating would not occur at a high level. With respect to maternal behavior, pregnancy hormone action on the medial preoptic area (MPOA; a hypothalamic region that regulates maternal responsiveness; see Chapter 5) may render the MPOA responsive to infant-related stimuli; MPOA projections to the mesolimbic DA system would then facilitate maternal infant-seeking behaviors so a mother would be able to contact her infants. MPOA projections to the brainstem and spinal cord may regulate consummatory nursing behavior, once the mother has gained contact with her infants (see Figure 1.3). For each of these cases, specific goal-directed responses are the result of specific hypothalamic neurons, with unique functions, activating DA release into NA, with the result that the NA-VP circuit becomes responsive to stimuli associated with a specific motivational state.

This analysis indicates that the hypothalamus is positioned to play an important role in the regulation of the specificity of goal-directed motivation. The mesolimbic DA system and the NA–VP circuit can be conceived as a nonspecific motivational system in the sense that DA release into NA regulates a variety of goal-directed behaviors [272,716]. However, just which goal-directed response occurs depends on whether DA is released into NA in the presence of particular stimuli. Given that different nuclei in the hypothalamus appear to be involved in specific motivational processes, as described above (also see [950]), hypothalamic output to VTA-DA neurons may direct the types of goal-directed responses that occur. In other words, hypothalamic connectivity with the mesolimbic DA system may mediate an interaction between specific and nonspecific motivational systems, with the hypothalamus influencing the particular stimuli that are processed by the NA-VP at any one point in time.

With respect to the idea that the hypothalamus may activate VTA-DA neurons, a neuroanatomical study by Geisler et al. [328], which combined the injection of a retrograde tracer into the VTA with detection of mRNA for the vesicular glutamate transporter in neuronal cell bodies, found that several hypothalamic nuclei provided glutamatergic (excitatory) inputs to VTA. These nuclei included: MPOA, LPOA, LH, and VMN. Although the AHN did not have a major projection to VTA, the AHN does project to PAG, which in turn has a glutamatergic projection to VTA. Although this research supports the idea that the hypothalamus may stimulate the mesolimbic DA system, this study did not examine the VTA neuron type that received the glutamatergic inputs. In addition to DA neurons, the VTA also contains GABA and glutamate neurons [724,1047].

◀ BLA/BMA axon terminals within NAs, but for clarity of presentation, the BLA/BMA axon terminals are drawn outside NAs.) Axons ending in a bar are inhibitory, and those ending in an arrow are excitatory.

Other research, however, indicates that glutamatergic input to VTA is a major stimulator of VTA-DA neurons [881,1078].

1.3.4.6 The Mesolimbic DA System and Goal-Directed Approach and Avoidance Responses

In line with the examples that have been presented, most investigators view the mesolimbic DA system as a reward pathway [881]—DA release into NA affects the way the NA-VP circuit responds to appetitive or reward-related stimuli, leading to goal-directed reward-seeking behaviors. This view, of course, is the one taken by Ernst and Fudge [272]. Research has clearly shown that DA release into NA promotes food-seeking behaviors and the appetitive aspects of sexual and maternal behaviors [50,716,904,929]. This reward view of mesolimbic DA function is shown in Figure 1.11.

Although the neural model of Ernst and Fudge [272] labels the amygdala as primarily involved in avoidance and the ventral striatum as involved in approach, a conflict arises because the operation of the mesolimbic DA system in the context of appetitive (approach) motivation is based in part on the neural inputs that the NA-VP circuit receives from amygdala neurons with a positive valence. For example, Simmons and Neill [904] found that when muscimol was injected into the BLA on one side of the brain and a DA receptor antagonist was injected into the nucleus accumbens on the opposite side of the brain, a rat's operant responding for a food reward was significantly suppressed. This finding and the logic behind it can also be used to question the exclusive role of the mesolimbic DA system in reward; since BLA also contains neurons that respond to aversive stimuli, couldn't BLA input to NA-VP also play a role in goal-directed avoidance behaviors, and couldn't the effects of such input be potentiated by DA?

Therefore, a broader view of the functional role of the mesolimbic DA system is based on the facts that individual BLA/BMA neurons respond to stimuli with either a positive or a negative valence and that BLA/BMA projects to NA-VP circuit. From this perspective, DA release into NA should be able to affect subcircuits in NA-VP that are engaged by either appetitive or aversive stimuli. When DA is released into NA in the context of appetitive stimuli, reward-seeking behaviors occur, but when DA is released into NA in the context of aversive stimuli, goal-directed avoidance or rejection responses occur. This view is based on the idea that the limbic motor system is not likely to regulate only positive approach responses. Would one assume that the "cognitive motor system", that is, nigrostriatal DA regulation of the dorsal basal ganglia, only affected moving forward or to the right but not backwards or to the left?

A neural model that incorporates ideas about how the mesolimbic DA system might work is shown in Figure 1.12. This model, which focuses on BLA/BMA input to NA-VP, takes a

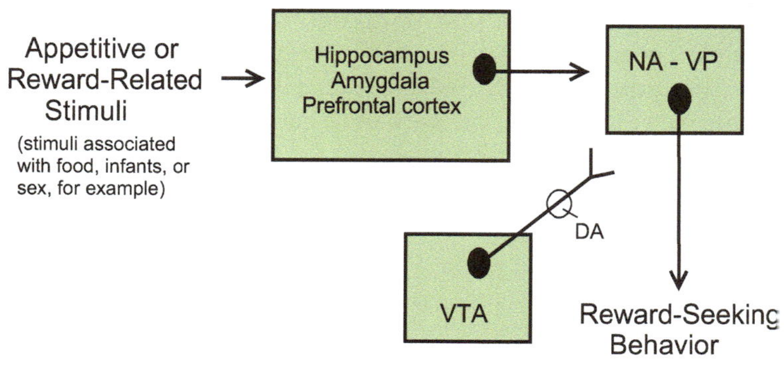

FIGURE 1.11 The standard view of the mesolimbic dopamine (DA) system as a reward circuit that mediates goal-directed reward-seeking behaviors in response to appetitive stimuli. Axons ending in an arrow are excitatory, and those ending with typical axon terminals (shown a Y) are left undefined with respect to excitation or inhibition. Other abbreviations: NA-VP: nucleus accumbens–ventral pallidum circuit; VTA = ventral tegmental area.

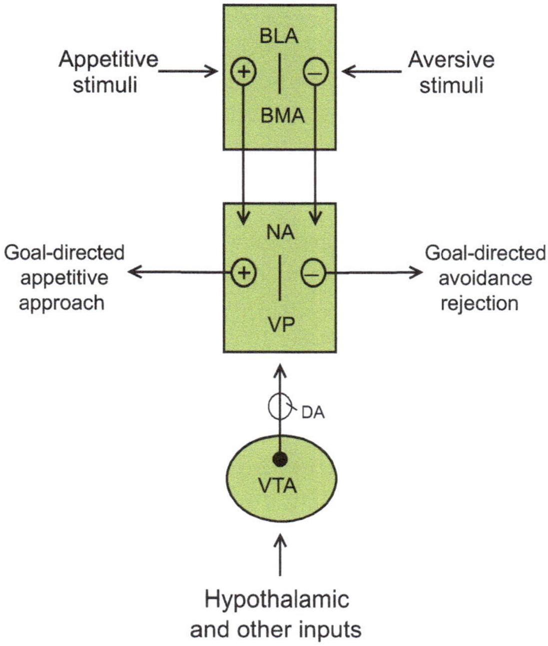

FIGURE 1.12 Distinct neural circuits within the mesolimbic dopamine (DA) system that regulate either reward-seeking responses to appetitive stimuli or active avoidance/rejection responses to aversive stimuli. Basolateral amygdala (BLA) and basomedial amygdala (BMA) contain separate neurons that respond to either appetitive or aversive stimuli. These neurons are labeled with a positive or negative sign, respectively. These positive or negative inputs are then relayed to distinct neurons within the nucleus accumbens-ventral pallidum circuit (NA-VP), the outputs of which give rise to either goal-directed approach or avoidance/rejection responses, respectively. DA input to these NA-VP circuits can potentiate either goal-directed appetitive or avoidance/rejection responses, depending on the particular circuits that are active. The hypothalamus (along with other areas) provides stimulatory inputs to ventral tegmental area (VTA) DA neurons.

labeled-line point of view where positively or negatively valent neurons in the limbic system project to distinct parts of the NA-VP circuit in order to regulate either goal-directed appetitive responses or goal-directed avoidance responses. This view argues that noxious or aversive stimuli, like reward-related stimuli, should activate DA release into NA. Anatomically, it has already been suggested that hypothalamic projections to VTA may be one of the neural inputs that activate DA release into NA. Since the hypothalamus contains neurons involved in defensive, avoidance, and aggressive behaviors, as well as appetitive behaviors, it should be possible for the hypothalamus to be involved in activating DA release into NA in situations involving either approach or avoidance.

Some evidence that supports the model shown in Figure 1.12 is described below.

1. Neurophysiological research indicates that different VTA-DA neurons can be activated by either appetitive or aversive stimuli, respectively [126,130,612]. Although some avoidance responses may be regulated by VTA projections to regions outside the NA-VP circuit [526], I want to focus on the research that supports the view that VTA-DA projections to NA-VP can influence both goal-directed appetitive and aversive responses. Most of the research described below focuses on the role of the mesolimbic DA system in goal-directed avoidance and rejection, since the role of this system in reward seeking has already been reviewed.
2. Anstrom, Miczek, and Budygin [33] reported that in aggressive encounters between a resident rat and an intruder, when the intruder is being defeated and shows defensive and submissive postures, DA is released into NA and action potential frequency concomitantly increases in the VTA. These results indicate that aversive social conditions, not obviously linked to rewarding stimuli, are associated with DA release into NA. Similarly, Badrinarayan et al. [46] have reported that the presentation of a conditioned aversive stimulus (a CS that had been paired with shock) activated DA release into NAs.
3. McCullough, Sokolowski, and Salamone [620] have shown that DA input to NA is necessary for the performance of an instrumental active avoidance response.

In the instrumental task lasting 45 min, a shock was presented to rats for 5 s every 30 s, but the rats could escape or avoid the shock for 30 s by pressing a lever. Performance on the avoidance task (in well-trained rats) led to significant increases in extracellular DA levels, as measured by microdialysis. There was a significant positive correlation (0.78) between DA increases and number of avoidance responses. In a second experiment, groups of rats were trained on the lever press avoidance procedure. After training, rats received intra-accumbens injections of 6-hydroxydopamine (6-HD) or vehicle. The 6-HD is a dopaminergic neurotoxin that destroys DA neurons and therefore would destroy DA axon terminals in NA. Dopamine depletion in NA led to a substantial decrease in lever pressing to avoid or escape shock. However, these rats did not appear to lose sensitivity to shock, as they were seen to flinch or vocalize with shock presentation. In contrast to controls, DA-depleted rats were all observed to freeze or become immobile during shocks, while controls, if they did not avoid, would engage in active instrumental lever pressing escape responses. It is interesting to speculate that as a result of DA depletion, rats showed consummatory reflexive defensive responses to shock but were not able to show goal-directed avoidance responses.
In line with these early results [620], additional recent work supports the view that DA action in NA is essential for goal-directed avoidance responses [215,1034]. Further, fMRI research on humans has reported an increased NA blood-oxygen-level dependent (BOLD) response, presumably due to enhanced DA release into NA [872], during the performance of an active avoidance response where a button press after a warning signal prevented the appearance of an aversive visual image [547].
In the amygdala section, evidence was presented that BLA/BMA is involved in active avoidance responding. Given that BLA/BMA projects to the NA-VP circuit, the above results fit with the model shown in Figure 1.12 with respect to goal-directed avoidance.

4. Research from Berridge's group has presented important results that show that there is a topographic organization of NA function with respect to appetitive and aversive motivation [285,803,805]. They showed that inhibition of neural activity in the rostral NAs, with either muscimol or a glutamate receptor (GluR) antagonist, promoted appetitive motivation, while inhibition of neural activity in the caudal NAs promoted aversive motivation. Importantly, DA action on D1 receptors in the rostral NAs was necessary for the promotion of appetitive behavior by GluR antagonist-induced inactivation of rostral NAs, while DA action on both D1 and D2 receptors in the caudal NAs was necessary for the potentiation of aversive motivation by GluR antagonist-induced inactivation of caudal NAs. One way to interpret these results is that DA action in NAs facilitates the effects of NAs inactivation on motivated behaviors. These results are important because they show that suppression of NA output, which presumably releases VP from inhibition, can promote either approach or avoidance responses depending on the particular neural circuits that are affected, and that DA action on NA is involved in both of these motivational effects.

1.3.5 The Prefrontal Cortex

1.3.5.1 Introduction

Ernst and Fudge [272] proposed that the prefrontal cortex (PFC) serves to modulate or regulate the output of the amygdala and ventral striatum so that adaptive and appropriate goal-directed responses occur. Given the strong neural inputs that the PFC receives from sensory association neocortex [61], one might view the PFC as exerting executive, rational, or cognitive control over basic goal-directed approach and avoidance responses. Many researchers view PFC control mechanisms as downregulating or restraining basic aversive and appetitive responses controlled by the amygdala and the NA-VP circuit [397], in this way dampening overly fearful or aggressive responses to potential threats and risky appetitive responses to rewarding stimuli. Since cortical projection neurons are glutamatergic, any inhibitory effects that the PFC might have on either amygdala or ventral striatum output would have to be mediated by projections to inhibitory interneurons. In addition to this perspective of a PFC restraining influence, evidence will be presented that PFC connections to the amygdala and NA-VP circuit can also enhance aversive and appetitive responses. In other words, the PFC is positioned to downregulate or upregulate amygdala and NA-VP output, depending on the particular PFC neurons that are exerting such modulatory actions.

To offer an illuminating proposal with respect to human social behavior, political and religious ideologies often construct our social world into in-groups and out-groups. To the extent that the socialization processes through which ideologies are learned are incorporated within the isocortex and PFC, then one might assume that PFC output to the amygdala and NA-VP circuit would enhance social avoidance and rejection responses to out-group members, depress such responses to in-group members, and enhance acceptance responses to one's in-group members (see [698] and Chapter 7).

The anatomy of the PFC is complex, and I will only present a brief overview for primates and rats [60,61,210,319,320,726,786,787,799]. In primates, the PFC can be divided into three major regions: lateral PFC, orbital PFC, and medial PFC. The lateral PFC and some parts of the medial and orbital PFC are isocortex, while the remaining parts of the medial and orbital PFC are allocortical in nature and a well-developed granular layer 4 is not present. In rats, all major parts of the PFC are allocortical and agranular (lacking a layer 4). Although rats do not have an isocortex equivalent to the lateral PFC, it has been suggested that an area referred as the precentral cortex (the area rostral to the primary motor cortex, also referred to as the frontal pole area), may represent a rudimentary granular (isocortical) PFC [787]. Because of species differences in the cytoarchitecture of the PFC, a current definition that is applicable across species is that the PFC is composed of those parts of the cortex in the lateral, orbital, and medial frontal lobe that are rostral to the primary motor cortex and that also receive significant neural inputs from the mediodorsal thalamic nucleus (MDT; [726]). As an example, since the anterior part of the cingulate cortex (ACC) receives significant input from MDT, it has been included as part of the medial prefrontal cortex in both rats and primates.

Figure 1.13 shows a schematic of the lateral, orbital, and medial PFC in primates. Based on cytoarchitectonic differences between PFC regions, different areas have been assigned different numbers in order to differentiate the regions. Only some of these area numbers are shown in the figure. The orbital PFC can be divided into medial, central, and lateral parts. Important regions in the medial PFC include area 24 (dorsal ACC), area 25 (ventral or subgenual ACC), and area 32 (the medial PFC region rostral to areas 24 and 25). All of these medial PFC regions are allocortical in nature, and it is these particular medial PFC regions that will be emphasized throughout this book because homologous areas also exist in rodents, which will permit an integration of human and animal studies with respect to medial PFC function. Figure 1.14 shows a frontal section through the rat brain that defines the various regions of the PFC. As indicated, the precentral cortex (PrC) of the rat PFC may be homologous

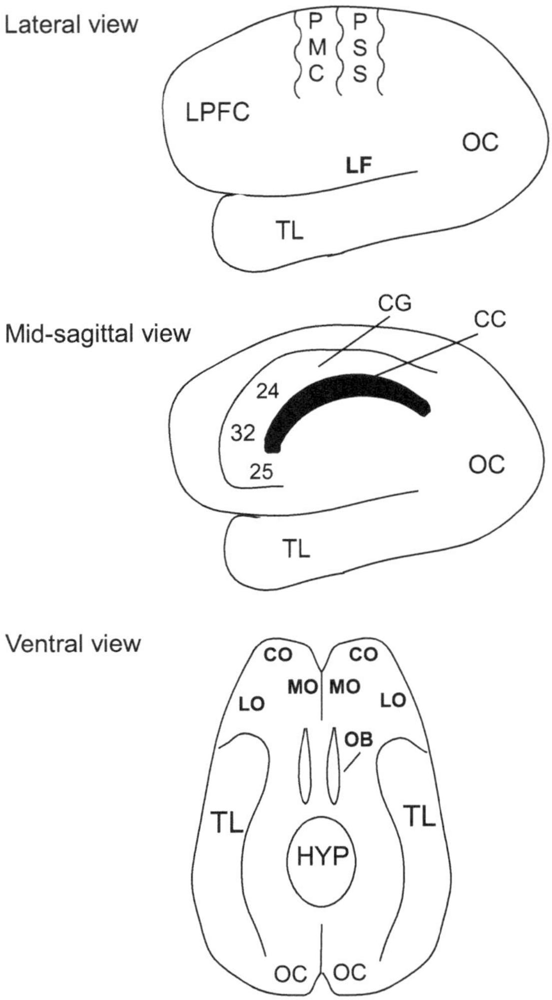

FIGURE 1.13 Lateral, mid-sagittal, and ventral views of the primate brain, emphasizing the locations of different parts of the prefrontal cortex. The lateral prefrontal cortex (LPFC) is shown in the frontal lobe rostral to the primary motor cortex (PMC). The mid-sagittal section shows three important areas within the medial prefrontal cortex, labeled as areas 24, 25, and 32. These areas are located in the anterior parts of the cingulate cortex (CG = cingulate gyrus). The orbital prefrontal cortex is shown in the ventral view, and contains the medial (MO), central orbital (CO), and lateral orbital (LO) regions. Other abbreviations: CC = corpus callosum; HYP = hypothalamus; LF = lateral fissure; OB = olfactory bulb; OC = occipital lobe; PSS = primary somatic sensory cortex; TL = temporal lobe. *Modified from Figure 2.8 in Fuster [319], with permission from Elsevier.*

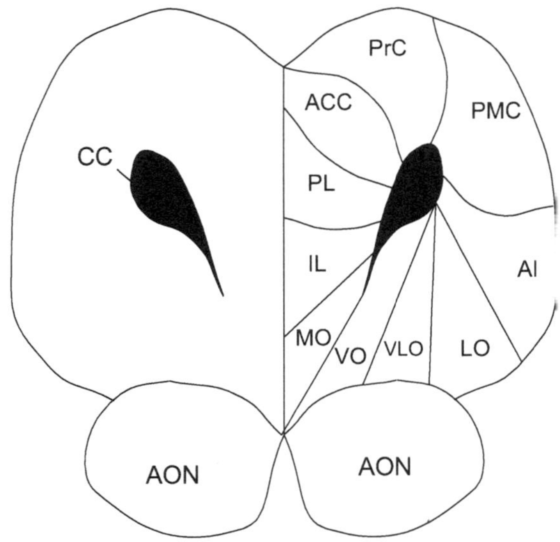

FIGURE 1.14 A frontal section showing the major parts of the rat prefrontal cortex. These parts include the anterior cingulate cortex (ACC), prelimbic cortex (PL), infralimbic cortex (IL), medial orbital (MO), ventral orbital (VO), ventrolateral orbital (VLO), and lateral orbital (LO) cortices, and the anterior insular cortex (AI). The precentral cortex (PrC) may be homologous to the primate lateral prefrontal cortex. Also shown are the anterior olfactory nucleus (AON), the primary motor cortex (PMC), and corpus callosum (CC). *Modified from Figure 1(b) in Dalley et al. [210], with permission from Elsevier.*

to the primate lateral PFC. With respect to the rat medial PFC, the ACC, prelimbic cortex (PL), and infralimbic cortex (IL) are considered to be homologous to primate areas 24, 32, and 25, respectively [320,787]. As in primates, the rat orbital regions can also be divided into medial (medial and ventral orbital), central (ventrolateral orbital), and lateral regions (lateral orbital and anterior agranular insular regions). (I will have more to say about the insular cortex in subsequent chapters. In both rats and primates, the insular cortex has agranular and granular components, with agranular regions being located in the anterior insular cortex. Those parts of the agranular insular cortex in the posterior-lateral orbital cortex represent a rostral extension of the insular cortex onto the orbital surface [787].)

Based on neuroanatomical connectivity, Price [787] and Ongur and Price [726] have divided parts of the PFC into a medial network and an orbital network. To simplify, the medial network consists of the medial PFC and most medial orbital PFC regions, while the orbital network is composed of most parts of the central and lateral orbital PFC. Table 1.1 provides a summary of the PFC areas in the primate and rat and their likely homologies. A block diagram of some of the important neural connections within the prefrontal cortex is shown in Figure 1.15 [61,320,622,726,786,787,799]. The lateral PFC and the orbital network are the main recipients of processed sensory inputs from the sensory association cortex. Sensory inputs can reach the medial network via its connections with the orbital network and with the entorhinal cortex. Finally, both the medial and orbital networks have reciprocal connections with the amygdala, while the medial network provides the major efferent pathways to the ventral striatum (NA) and to the hypothalamus.

TABLE 1.1 Comparisons of Some of the Prefrontal Cortical Regions in Primates and Rodents

Primates	Rodents
MEDIAL NETWORK	
Area 24 (dorsal ACC)	ACC
Area 32	PL
Area 25 (subgenual ACC)	IL
Medial orbital	MO/VO
ORBITAL NETWORK	
Central orbital	VLO
Lateral orbital	LO/AI
OTHER AREAS	
Lateral PFC	PrC

Abbreviations: ACC = anterior cingulate cortex; AI = anterior insular cortex; IL = infralimbic cortex; LO = lateral orbital cortex; MO = medial orbital cortex; PFC = prefrontal cortex; PL = prelimbic cortex; PrC = precentral cortex; VLO = ventrolateral orbital cortex; VO = ventral orbital cortex.

In viewing this organization, one can conceive of how the reciprocal interactions between the lateral PFC and sensory association cortex, along with the connections of the lateral PFC with the primary motor cortex, may be involved in working memory, attentional processes, and movement planning [61], while the connections of the lateral PFC-sensory association complex with the orbital and medial networks would allow cognitive processes to regulate neural activity within the amygdala, striatum, and hypothalamus. This latter circuit mechanism would allow a sort of hierarchical top-down regulation, where neocortical cognitive processes influence allocortical and subcortical neural events.

The projections from the amygdala to the PFC are also worth considering. Barbas et al. [61] have emphasized the connection between the basal nuclei of the amygdala with the posterior orbital PFC. The amygdala not only projects directly to the PFC but also indirectly via connections with MDT. They compare the amygdala-to-MDT-to-orbital PFC connection with those of the major sensory systems, such as vision, where the optic tract projects to the lateral geniculate nucleus, which in turn projects to the primary visual neocortex. This comparison suggested to them that the relay of amygdala input to the posterior orbital PFC via the MDT might be the route through which emotions are subjectively experienced. Amygdala projections to the anterior insular cortex also seem to be important in this regard.

With respect to involvement of such amygdala-orbital PFC connections in emotional experience, a disruption in such pathways may explain why SM is unable to experience fear. Interestingly, a recent study [288] has reported that SM and two other patients with amygdala lesions caused by Urbach–Wiethe disease can experience fear under certain conditions. The inhalation of CO_2, which caused an oxygen deficit, evoked fear and panic attacks in these patients. Therefore, the amygdala-to-orbital PFC connection may be important for the experience of fear that is induced by exteroceptive aversive

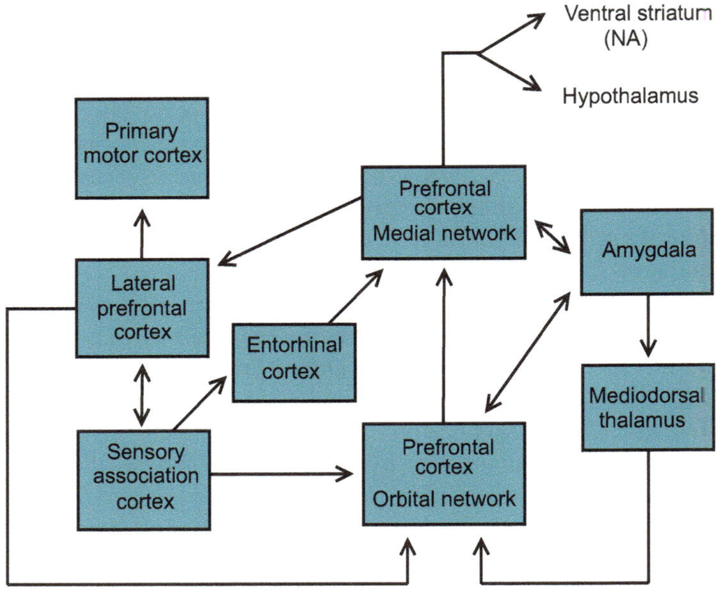

FIGURE 1.15 A block diagram outlining some of the major neural connections of the prefrontal cortex, which is divided into the lateral prefrontal cortex and the orbital and medial networks. Interactions within sensory association cortex-lateral prefrontal cortex-primary motor cortex circuits mediate the more cognitive aspects of behavior regulation. Sensory inputs from the sensory association cortex can also reach the orbital and medial networks, and these two networks are reciprocally connected with the amygdala. The main prefrontal cortex output to the hypothalamus and ventral striatum (nucleus accumbens = NA) is via the medial network. See text for other important details.

stimuli, while interoceptive visceral (autonomic) stimuli that evoke fearfulness appear to reach the orbital PFC (or other brain regions) through circuits that do not require the amygdala.

For a final consideration, the facts that the amygdala projects to the medial and orbital networks, and that the medial network projects to lateral PFC, also suggest routes over which the valence characteristics of stimuli processed through the amygdala (emotional and motivational processes) might influence cognitive processes and decision making within the lateral prefrontal-sensory association cortex complex.

In order to fully appreciate the interactions of the prefrontal cortex with the amygdala and ventral striatum in the control of motivational and emotional processes, a more detailed evaluation of the particular nuclei and microcircuitry involved is necessary. I will present several examples from rodents and primates to help elucidate such interactions, and I will start with simpler processes to show how PFC output to the amygdala can either decrease or increase CFRs that result from amygdala projections to the brainstem. The important role of PFC interactions with the hypothalamus will not be discussed in detail in this chapter but will be fully discussed in subsequent chapters.

1.3.5.2 PFC Regulation of Conditioned Fear Responses and Goal-Directed Responses

In reference to Figure 1.6, recall that the CFR occurs when a neutral CS, such as a tone, is paired with an aversive US (shock). The neural circuitry analysis presented showed that CS–US pairings strengthened the ability of the CS to activate an LA-to-BLA-to-CeAm-to-PAG circuit. Recent research has shown that the output of the medial PFC can modulate this amygdala-based CFR. In fact, different regions of the medial PFC have been shown to have opposing effects—the output of IL area exerts an inhibitory effect on the CFR while the output of the PL area has a potentiating effect [645]. The effects of the medial PFC on the CFR have often been studied in the context of extinction learning [995]. Fear extinction describes the decrease in the CFR that occurs after repeated presentations of the CS without the US. The experiments by Vidal-Gonzalez et al. [995] used the following paradigm. First, rats were conditioned: a tone

CS was presented for 30 s and was coterminated with a 1 s shock. The CFR was measured by the percentage of time the rats were immobile during the 30 s CS interval. Once the criterion for CFR acquisition was achieved (>20% immobility), on the following day, rats were exposed to partial extinction training composed of eight tone presentations without shock. During tone presentation, rats received microstimulation of PL cortex or IL cortex, or were unstimulated. IL stimulation enhanced extinction, and PL stimulation delayed extinction. Based on other neuroanatomical and neurophysiological data, the microcircuitry shown in Figure 1.16 has been

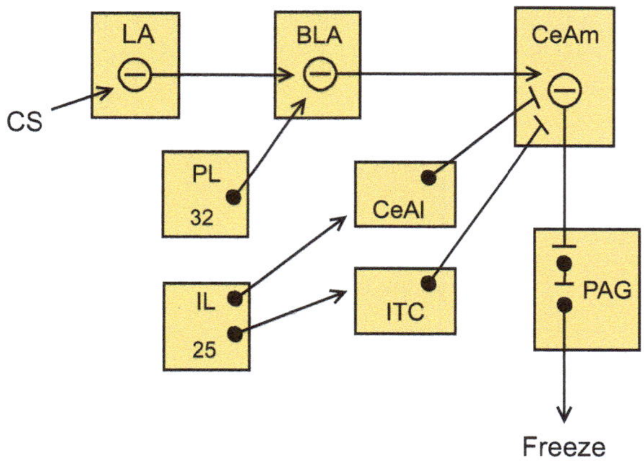

FIGURE 1.16 Medial prefrontal cortex projections to the amygdala can either upregulate or downregulate the conditioned fear response (CFR). Amygdala neurons that respond to aversive, negatively valent stimuli are shown with a negative sign within their cell bodies. The classic CFR circuit (LA-BLA-CeAm) is shown as projecting to the periaqueductal gray (PAG) to cause the conditioned freezing response. The rat prelimbic cortex (PL), which may be homologous to area 32 in primates, is shown as promoting the CFR by stimulating BLA projections to CeAm. The infralimbic cortex (IL), which may be homologous to area 25 in primates, is shown as depressing the CFR via excitatory projections to CeAl and ITC. (IL may also suppress the fear response by projecting to those BLA neurons that stimulate CeAl and ITC: see Figure 1.6(A)) Axons ending in a bar are inhibitory, and those ending in an arrow are excitatory. Other abbreviations: BLA = basolateral amygdala; CeAl = lateral part of the central nucleus of the amygdala; CeAm = medial part of the central nucleus of the amygdala; CS = conditioned stimulus; ITC = intercalated nuclei; LA = lateral amygdala.

proposed to explain these results [20,69,645,995]. PL stimulation may enhance the CFR by activating BLA neurons that are part of an aversive BLA-to-CeAm-to-PAG circuit, while IL stimulation may depress the CFR by activating inhibitory ITC interneurons or the inhibitory neurons in CeAl, both of which would depress the output of CeAm projections to PAG.

Similar processes may also occur in primates as recent work indicates that *inactivation* of parts of the medial PFC in monkeys, which may be homologous to the rodent PL cortex, potentiated the long-term extinction of a CFR [496].

The involvement of the IL PFC in suppressing Pavlovian CFRs may also occur in humans. Using fMRI studies, it has been shown that as extinction learning proceeds, increases in the BOLD signal in the subgenual ACC (area 25, the presumed homolog of IL cortex) is correlated with BOLD decreases in the amygdala [394]. These authors have also reviewed the evidence that posttraumatic stress disorder, where patients exhibit a cue activated reexperience of a traumatic event when the cue no longer signals danger, may be related to a dysfunction of ventromedial PFC regulation of amygdala activity.

This research on different divisions of the medial PFC exerting opposing influences on fear-related consummatory processes may also be related to the therapeutic effectiveness of deep brain stimulation (DBS) for treating certain forms of clinical depression in human patients [421]. DBS of areas 32 (homologous to PL cortex) and 25 (homologous to IL cortex) has been associated with decreases in depressive symptoms [379]. Although it is usually argued that DBS is effective because it exerts inhibitory effects, there is also evidence that it might stimulate neural tissue near the site of the electrode [379]. Since some forms of severe depression are associated with intense anxiety states and since stress reactivity and fearfulness may be precipitating factors for depression [703], it is interesting to speculate that DBS may be therapeutic because it is ultimately modifying amygdala reactivity to stress and anxiety-inducing stimuli.

The importance of all of these findings is that they show that different neurons within the medial PFC cortex, through differential projections to the amygdala, can either increase or decrease the ability of stimuli to activate reflex-like conditioned fearfulness. Such findings open up the possibility that PFC projections to the amygdala might also be capable of downregulating or upregulating goal-directed avoidance or appetitive responses, depending on whether PFC neurons inhibit or activate positive or negative BLA neurons that project to the NA-VP circuit. How might the PFC be involved in promoting goal-directed avoidance responses? One possibility is that particular medial PFC efferents to the amygdala may depress the CFR to aversive stimuli so that the animal does not become immobile, while other PFC efferents act to facilitate goal-directed active avoidance responses to aversive stimuli by potentiating negatively valent amygdala neuron input to the NA-VP. Recent evidence has presented some support for such possibilities in rats [667], and hypothetical neural circuits mediating such effects are shown in Figure 1.17(A). Such an analysis suggests that PFC inputs to the amygdala may be involved in adaptive coping responses by inhibiting reflexive fear responses while promoting proactive responses that avoid primary aversive stimuli.

A recent report by Amemori and Graybiel [25] is also relevant to the role of the medial PFC in the modulation of goal-directed responses.

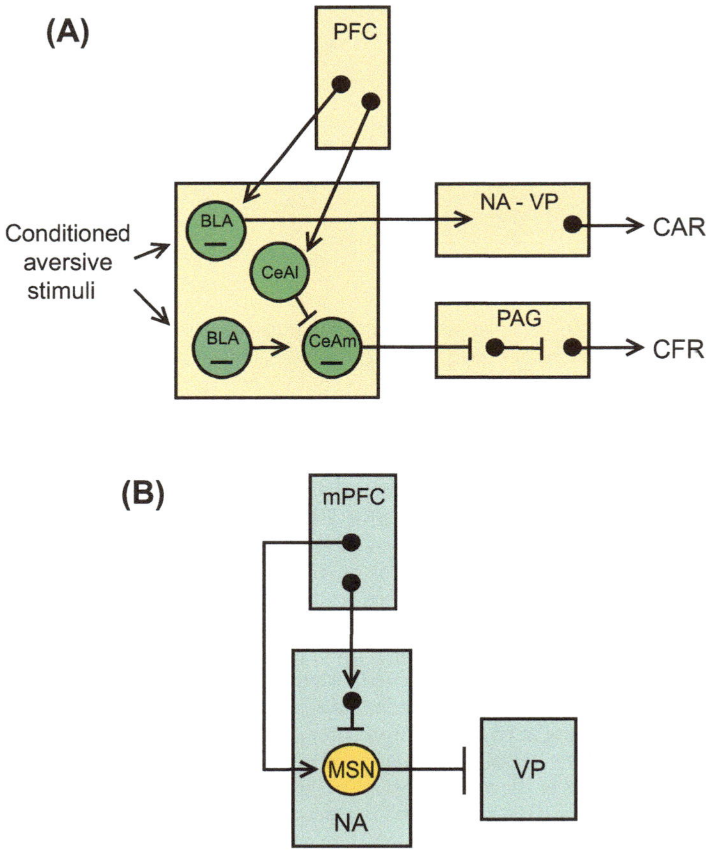

FIGURE 1.17 (A) Prefrontal cortex (PFC) connections with the amygdala can suppress the conditioned fear response (CFR) and facilitate the conditioned avoidance response (CAR) to an aversive stimulus by inhibiting CeAm output to the periaqueductal gray (PAG), while facilitating BLA projections to the nucleus accumbens-ventral pallidum circuit (NA-VP). Compare to Figure 1.6(B). Amygdala neurons that respond to aversive stimuli have a negative sign within their cell bodies. Other abbreviations: BLA = basolateral amygdala; CeAl = lateral part of the central nucleus of the amygdala; CeAm = medial part of the central nucleus of the amygdala. (B) Medial prefrontal cortex (mPFC) projections to the nucleus accumbens (NA) can either stimulate or depress medium spiny neuron (MSN) projections to the ventral pallidum (VP), depending on the particular neural circuits that are active. For both parts of this figure, axons ending in a bar are inhibitory, and those ending in an arrow are excitatory.

Rhesus monkeys were trained on an operant approach-avoidance task. A visual stimulus, composed of a yellow bar and a red bar, was presented on a screen. The lengths of the red and yellow bars corresponded to the amount of liquid food and the strength of an air puff, respectively, that the monkey would receive if it performed an approach response. If the monkey performed an avoidance response, it did not receive food or an air puff. An approach response consisted of moving a joystick in the direction of a plus sign, while an avoidance response consisted of moving the joystick in the direction of a square. Obviously, if the visual cue indicated a large food reward and a small air puff, the monkey would make an approach response, while if the predicted outcome was a small food reward and a strong air puff, an avoidance response would be appropriate. These response decisions, of course, were not absolute, but instead occurred across a continuum, so that as the relative sizes of the predicted food reward decreased and the air puff strength increased, the probability of approach responses to the visual signal decreased and the probability of an avoidance response increased. In one part of this study, the authors recorded from neurons in the dorsal anterior cingulate cortex, which is part of the primate medial PFC. Neurons could be categorized into two main types. For one group of neurons, increased neural spiking was correlated with approach responses, while neural activity in the other group was correlated with avoidance responses. Interestingly, in the dorsal parts of the dorsal ACC, these two types of neurons were intermixed, but in the ventral part of the dorsal ACC, which might be homologous to the rat's PL cortex, a predominance of avoidance neurons were detected. Significantly, microstimulation of this ventral part of the dorsal anterior cingulate region biased the monkeys' responses in the direction of goal-directed avoidance, which could be taken to indicate that such stimulation increased fearfulness, anxiety, or the ability of particular visual signals to stimulate amygdala circuits that promoted goal-directed avoidance responses.

The Amemori and Graybiel [25] study did not examine the brain regions that received inputs from the dorsal ACC. Since the medial PFC projects to both the ventral striatum and the amygdala, goal-directed responses regulated by the medial PFC could include influences on either the amygdala, striatum, or both regions. In this context, I want to present the results of another study that showed that medial PFC stimulation is capable of either increasing or decreasing the neural activity of NA medium spiny neurons (MSNs; the main output neurons of NA). Gruber, Powell, and O'Donnell [364] performed intracellular recordings from the NA of anesthetized rats while also electrically stimulating different parts of the medial PFC along the dorsal–ventral region comprising the PL and IL areas. Electrical stimulation at some sites increased the neural spiking of a particular MSN, while stimulation of neighboring sites (the two sites were about 0.5 mm apart) resulted in inhibitory effects. The authors suggested that different neurons in the rat medial PFC can either activate MSNs directly or inhibit MSNs indirectly through excitatory projections to inhibitory interneurons (the NA contains GABAergic and cholinergic inhibitory interneurons that synapse on MSNs). These possibilities are shown in Figure 1.17(B). This study lacked the anatomical detail needed to discern the exact regions of the medial PFC involved in these disparate effects. In support of these neurophysiological findings, a recent behavioral study has shown that activation of different parts of the medial network of the rat's PFC can either increase or decrease appetitive or aversive responses mediated by the NA-VP circuit [806].

In conclusion, different neurons in the medial PFC are capable of either increasing or decreasing the output of specific amygdala and nucleus accumbens circuits, allowing it to exert significant control over motivation and emotion. Since the medial PFC also projects to the hypothalamus, this would be another route for a PFC influence over appetitive and avoidance responses.

1.3.5.3 The PFC and Value-Based Decision Making

Within different contexts or situations, different voluntary goal-directed responses may vary in their positive or negative consequences, and such outcomes may also change over time. That is, situation-specific responses may vary in their resultant benefits and costs, and adaptive behavior should operate over time to maximize an individual's benefit/cost ratio. A growing body of literature indicates that the PFC is involved in such value-based decision making, and that different PFC circuits may influence different aspects of such decision making [842,846,1010]. The idea is that an individual's ongoing experiences are interpreted by PFC neural mechanisms, which then exert influences on the amygdala, NA-VP, or other regions to modify the way we act.

In the study by Amemori and Graybiel [25], under different stimulus conditions, a particular approach response resulted in different benefits (liquid reward magnitude) and costs (strength of an air puff). The monkey had to decide whether to make an approach or an avoidance response based on the predicted outcomes. Although this study examined the role of the medial PFC in these behavioral choices, the cognitive processes were complex and involved both stimulus-outcome associations and response-outcome associations. Recent work has provided evidence that the orbital network (central and lateral parts of the orbital PFC) is most concerned with learning and representing the relationship between stimuli and outcomes, while the medial network (medial PFC) is most concerned with learning and representing the relationship between particular responses and their outcomes. In the study by Rudebeck et al. [842], rhesus monkeys engaged in one of two tasks. In the first, the performance of one response resulted in a greater likelihood of reward than the performance of a different response. In the second task, while holding the type of response constant, performing the response in the context of one stimulus resulted in a greater probability of reward than performing the response to a second stimulus. Monkeys with lesions to the medial PFC, focused on area 24, exhibited deficits in maximizing their receipt of reward under task 1 (response-reward associations), while monkeys with lesions to the central part of the orbital PFC showed deficits in maximizing reward receipt during task 2 (stimulus-reward associations).

If one were to view the amygdala as assigning either a positive or negative motivational valence to a stimulus, while the NA-VP circuit might be concerned with the valence value of a response (regulating approach versus avoidance), it is interesting to speculate that orbital network projections to the amygdala might regulate complex decision making by updating stimulus-outcome associations that change over time, while medial PFC projections to ventral striatal circuits might regulate complex decision making by updating and selecting responses that have the most favorable outcomes. In other words, the executive or regulatory functions of the PFC, through connections with the amygdala and ventral striatum, might update or modify the value of particular stimuli and responses, respectively, with regard to their associated outcomes. Although there is some evidence for such proposals, the detailed circuitry mechanisms, such as those described for medial PFC control over the CFR, have not been described.

Orbital PFC output to the amygdala might modulate whether a particular stimulus activates positively or negatively valent amygdala neurons, and the degree to which such neurons are activated. In the section on the functional anatomy of the amygdala, experiments by Schoenbaum et al. [869], where rats learned a go–no go response to particular odors, were reviewed. Briefly, it was found that some neurons in the BLA responded to odors signaling reward, while others responded to odors signaling aversive consequences. More importantly, when reversal learning occurred, where a previously positive odor began to signal aversive consequences, the BLA neurons that had previously responded to the positive odor ceased to do so. Relevantly, when Saddoris, Gallagher, and Schoenbaum [848] lesioned the lateral orbital PFC on one side of the brain, the neural responses occurring in BLA during reversal training were slow to occur, suggesting that such changes were in part regulated by input from the orbital network [870].

With respect to medial PFC output to the striatum, this circuit might regulate the degree to which a particular stimulus activates the appetitive/approach or avoidance/rejection circuits that were described previously for the NA-VP regulation of behavior. It should be obvious that under natural conditions, most cases of value-based decision making are probably complex, with variations in the consequences associated with stimuli and responses occurring at the same time, so that the orbital and medial networks of the PFC would be working in concert [25]. However, careful experiments have begun to delineate the specific functional roles of each network. An important take-home message is that the PFC functions to modulate behavioral choices based on the predicted outcomes or consequences of particular acts within specific contexts.

1.3.5.4 Relevance of PFC Mechanisms to Social Behavior

Although much will be said about PFC projections to the amygdala, NA-VP circuit, and hypothalamus in subsequent chapters, it will be worthwhile to introduce the importance of such a PFC top-down regulation for social behavior at this point. If a stranger were to wink at your spouse, would it be appropriate to start a fight with that stranger? But if the stranger grabbed your spouse, what would you do? If a man passed an unfamiliar attractive woman on a deserted street, would it be appropriate for him to approach her and make sexual advances? As a very hungry visitor at a Thanksgiving dinner, if you were served first, would you take all of the white meat for yourself before passing on the serving plate to others? With respect to employment, would you call in sick often in order to engage in other activities that you deemed as immediately more rewarding, such as going to the movies? Finally, would you risk your family's life savings on a gambling deal that could have a super-large payoff but might also result in the loss of your entire savings? All of these examples exemplify how social context can influence behavioral choices, including social behavior, which might be mediated by interactions between PFC, amygdala, ventral striatum, and hypothalamus. Interestingly, the answers to most of the questions posed involve the classic view of PFC downregulation of subcortical and allocortical circuits that regulate basic motivational and emotion states.

The classic case of Phineas Gage is an early example of the process being described. Gage was a railroad worker whose brain was damaged during a construction accident in 1825. A recent analysis of Gage's skull has been used to reconstruct the most likely location of the brain damage, with the conclusion that the ventromedial PFC, probably involving aspects of both the medial and orbit networks, but sparing the lateral PFC, was bilaterally damaged [212]. Prior to the accident, Gage was intelligent, reliable, and socially adapted. He underwent a dramatic change in personality and social behavior after the accident. Although his language ability and intelligence remained normal, he became irresponsible and impulsive, he could not hold a steady job, and he disregarded social conventions. His physician described his symptoms as a break between his intellectual faculty and his animal propensities [212,319].

The effects of damage to the ventromedial PFC on human social behavior and decision making have been examined more recently by Damasio and colleagues [29,76,77,860]. The areas of the PFC that were damaged in these patients included medial, central, and lateral orbital regions, and areas 25 and 32 of the medial PFC. In a gambling task, patients with such lesions were compared to normal controls [77]. Using play money, subjects were instructed to select cards from one of two decks in order to maximize their profits. When cards were selected from one deck, the immediate reward was large, but periodically this reward would be accompanied by a large monetary penalty. The reverse was true for selecting cards from the other deck, where individual rewards were modest, but the periodic penalties were small. Over the long run, selections from the second deck would result in a monetary gain, while selections from the first deck would result in a monetary loss. Normal controls, over a long series of card selections, chose more cards from the second deck, while ventromedial PFC damage was associated with choosing more cards from the first deck. The authors suggested that PFC damage results in poor decision making because the patient's behavior is guided to a greater extent by the immediate consequences rather than by the long-term consequences of one's actions. Bechara et al. [76] similarly suggest that such patients fail to have anticipatory emotional experiences to the consequences of poor choices. Perhaps these patients cannot imagine the aversive nature of a strong punishment. It should be obvious how such a behavioral profile could have negative consequences for social decision making. It is intriguing to speculate that damage centered on area 32, which may be the homolog of the rodent prelimbic area, decreased anticipatory fear, as would be proposed based on the work of Milad and Quirk [645], who reported that prelimbic activation enhances the CFR in rats.

In relation to this interesting idea, in fMRI studies, Eisenberger [268] describes an increased BOLD response in the dorsal ACC, centered in areas 24 and 32, in response to social exclusion. She suggests that these regions are part of a neural network that mediates emotional pain or feelings of social distress associated with social rejection or with the dissolution of social bonds in humans. Since the rodent research shows that increased activation of the PL cortex (area 32 in humans) is associated with increases in fear-related processes, this proposal makes sense.

Interestingly, early onset damage to the PFC produces much more severe deficits in value-based decision making and social behavior than does adult onset damage. Individuals with adult damage usually do not harm others, and they have intact knowledge about appropriate social behavior (that they learned prior to the brain damage), but they make faulty decisions in the social and nonsocial realms because they appear to be guided by the immediate consequences of their actions. In contrast, individuals that received damage to the ventromedial PFC early in life do not have intact knowledge of appropriate social behavior, and they display an increased tendency toward stealing and exhibiting aggressive outbursts, risky sexual behavior, and poor parental behavior [29]. These results suggest that while the output of the ventromedial PFC to the amygdala and NA may regulate value-based social and nonsocial decision making, its outputs to other brain regions (the lateral PFC, for example) may be involved in the development of social knowledge and moral reasoning (perhaps as a result of the feedback the ventromedial PFC receives from the amygdala, which may be signaling the positive and negative consequences of one's actions: see Figure 1.15). When both of these functions are disrupted as a result of early brain damage, social behavior would be more severely affected.

Earlier in this chapter I reviewed the data showing that patients with Williams syndrome are hypersocial and have increased amygdala reactivity to positive facial expressions and decreased amygdala reactivity to negative facial expressions. A recent fMRI study suggests that these alterations in amygdala processing of social stimuli may in part be the result of a dysfunction of PFC control of the amygdala [672].

Finally, an interesting study on male rhesus monkeys provides evidence that medial PFC lesions decrease a monkey's responsiveness to social stimuli [843]. Normal male monkeys retrieve a desired food item with a short latency, but if they are concurrently shown a film of a sexually active female or another male, their food retrieval latencies increase, presumably because of their interest in the social stimuli. Such an increased latency was not observed in monkeys with lesions centered on areas 24 and 32 of the anterior cingulate cortex. Therefore, parts of the primate medial PFC seem to be important in regulating social responsiveness, perhaps through interactions with amygdala, NA-VP, and hypothalamic motivational and emotional networks.

To sum up, the PFC is obviously a complex neural region, and in humans, areas 24, 25, and 32 seem to be involved in functions that include the regulation of fear-related processes and social behavior. It is possible that the regulation of emotional processes is primary, which, in turn, affects social behavior. However, the medial and orbital PFC are composed of functionally heterogeneous populations of neurons that are likely to be involved in many other processes that also impact sociality.

1.4 CONCLUSIONS

This primer of functional neuroanatomy was meant to provide a comprehensive introduction to some of the most important neural functions that will be shown to influence social behavior. I have emphasized the roles of both the amygdala and the mesolimbic DA system in both goal-directed approach and avoidance responses and described the various nuclei of the hypothalamus as providing a significant activating force on VTA-DA neurons that project to NA. The modulatory role of different parts of the PFC is capable of either upregulating or downregulating activity in the amygdala, NA-VP circuit, and hypothalamus. I have also emphasized the importance of understanding the microcircuitry within neural regions, and how neural influences on this microcircuitry regulate the output of one neural region to another. Finally, the neural regulation of basic motivational and emotional processes in nonsocial settings may overlap with the regulation of social motivational and emotional processes.

CHAPTER

2

Basic Genetics and Epigenetics

2.1 INTRODUCTION

This chapter provides an introduction to the role of genes in neuron function and behavior. Genetic and environmental factors both influence behavior, including social behaviors, and these two factors often interact to mold the way in which brain and behavior develop. I will present the fundamentals of molecular biology, followed by concrete examples of how genes and environmental factors act together to affect developmental processes. These concepts will be important for a full understanding of the gene–social behavior relationships developed in future chapters.

Genes are composed of deoxyribonucleic acid (DNA) and are located within the chromosomes of a cell's nucleus. Figure 2.1 explains the terms *replication*, *transcription*, and *translation*. During meiosis and mitosis, the DNA in a particular cell is capable of replicating itself in order to be incorporated into newly formed cells. For example, during mitotic divisions, new cells are produced that contain the same DNA as the parent cell. During meiosis, a mature gamete (sperm or ova) acquires one member of each pair of chromosomes from a diploid germ cell. DNA replication provides the mechanism for genetic heredity, as copies of DNA are transferred to newly produced cells, which include gametes. But the term *heredity* cannot be fully appreciated without an understanding of what genes do. In each cell in our body, genes control the synthesis of ribonucleic acid (RNA), and this process is referred to as transcription. Several kinds of RNA exist, including messenger RNA (mRNA), ribosomal RNA, and transfer RNA (tRNA). My main concern will be mRNA, the form of RNA that contains the code that regulates protein synthesis. The process through which mRNA forms a template that controls the sequence of amino acids that make up a particular protein is referred to as translation. Therefore, with respect to protein synthesis, the following general sequence of events occurs: gene X (DNAx; a gene located at a particular locus on a particular chromosome) controls the synthesis of mRNAx, which, in turn, controls the synthesis of protein X. Similarly, gene Y (DNAy) would control the synthesis of protein Y, using mRNAy as an intermediary. Since the types of proteins synthesized in a cell influence the structure and function of the cell, one can see how genes regulate the structure and function of all cells, including neurons.

2.2 BASIC GENETICS

2.2.1 The Structure of Nucleotides

Several sources have provided the information in this section, and the reader is referred to these for a more comprehensive review of basic

http://dx.doi.org/10.1016/B978-0-12-416040-8.00002-X

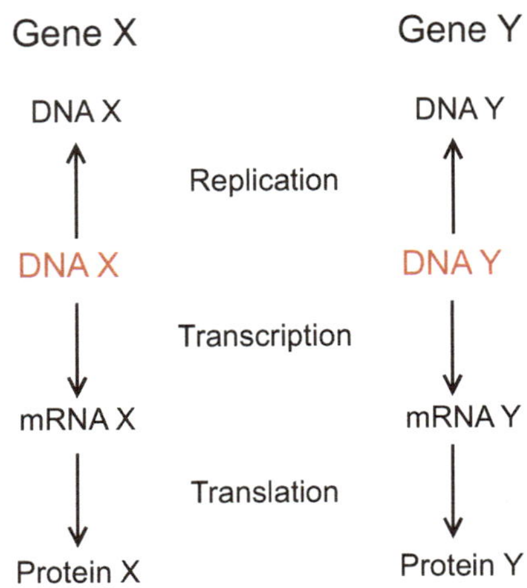

FIGURE 2.1 General depiction of the replication, transcription, and translation processes with respect to two separate genes, referred to as Gene X (DNA X in red) and Gene Y (DNA Y in red). Genes can replicate themselves during mitosis and meiosis (DNA-to-DNA). The transcription (DNA-to-mRNA) and translation (mRNA-to-protein) processes allow particular genes to regulate the synthesis of particular proteins.

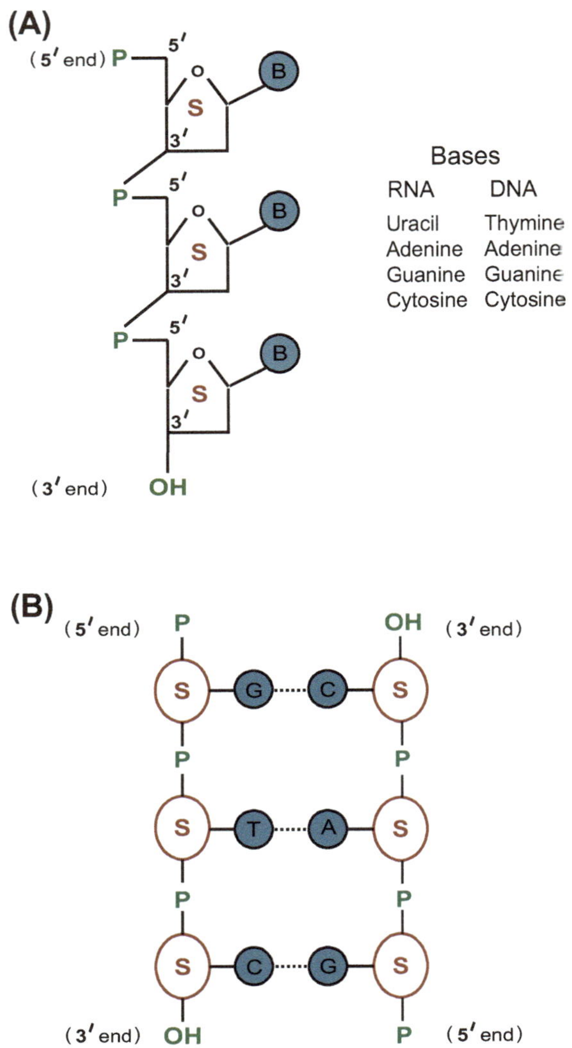

FIGURE 2.2 The general chemical structure of RNA and DNA. (A) RNA is composed of a single strand of nucleotides. Each nucleotide contains a pentose sugar (S = ribose), a base (B), which can be uracil, adenine, guanine, or cytosine, and a phosphate group (P). Nucleotides are linked together by covalent bonds between the phosphate group located at the 5′-carbon position on the sugar molecule of one nucleotide and the hydroxyl group (OH) located at the 3′-carbon position of the adjoining sugar molecule. The linkages that are formed are referred to as the sugar–phosphate backbone of the nucleotide sequence. O = oxygen. (B) DNA is composed of a double strand of nucleotides. The pentose sugar (S) is deoxyribose, and DNA contains the same bases as RNA, except that thymine is used instead of uracil.

genetic mechanisms: Alberts et al. [15]; Champe, Harvey, and Ferrier [175]; Hyman and Nestler [438]; Lewin [550]; Matthews, Freedland, and Miesfeld [613].

A single strand of DNA or RNA is composed of a sequence of nucleotides that are linked together by strong covalent bonds. A single nucleotide contains a nitrogenous base, a pentose (5 carbon) sugar, and a phosphate group. The strong covalent bond that links two nucleotides together within a stand of DNA or RNA occurs between the phosphate group ($PO_4 = P$) attached to the 5-carbon position on the sugar of one nucleotide (referred to as the 5′ position) and the hydroxyl (OH) group located at the 3-carbon position (referred to as the 3′ position) on the sugar of the next nucleotide in the sequence. This relationship is schematically illustrated in Figure 2.2(A). Notice how one end of the nucleotide sequence, referred to as the 5′ end, contains a free phosphate group, while the other end, referred to as the 3′ end, contains a free hydroxyl group. The different bases that are present in DNA include thymine (T), adenine (A), guanine (G), and cytosine (C). These same bases are present in RNA, except that uracil (U) is used in place of thymine. The pentose in DNA is deoxyribose, while that in RNA is ribose. A nucleotide sequence can be conceived as a series of bases protruding from a sugar–phosphate backbone, and a particular sequence might be written as 5′-pApTpTpG-3′.

While RNA is composed of a single strand of nucleotides, DNA contains two separate intertwining strands of nucleotides (forming a double helix), with the bases in one strand linked to the bases in the other strand by relatively weak hydrogen bonds (in comparison to the strong covalent bonds between the sugars in each separate strand). Double-stranded DNA is schematically shown in Figure 2.2(B). Note that particular bases are complementary to one another, which allows them to form hydrogen bonds: G is complementary to C, and A is complementary to T (A is also complementary to U, which is important for transcription, as will be described below). When viewing the two strands of nucleotides that compose DNA, note that one strand runs in the 5′-to-3′ direction, while its complementary strand runs in the 3′-to-5′ direction.

2.2.2 General Transcription and Translation Processes

During transcription, single-stranded RNA is synthesized from double-stranded DNA within the nucleus of a cell. In order for transcription to occur, the enzyme RNA polymerase must attach and bind to DNA. The DNA strands then unwind, and RNA polymerase moves along the strand that is referred to as the template strand in the 3′-to-5′ direction in order to catalyze the synthesis of an RNA strand that is organized in the 5′-to-3′ direction and is composed of bases that are complementary to those of the DNA template strand. This general process is shown in Figure 2.3(A). Because of this process, the base sequence in the synthesized RNA strand matches that of the 5′-to-3′ DNA strand, except that uracil replaces thymine. Since the nontemplate DNA strand matches the base sequence of the synthesized RNA, it is called the coding strand or the sense strand. By convention, when one refers to the nucleotide sequence of DNA that codes for a particular RNA, one uses the sequence shown by the 5′-to-3′ coding strand.

With respect to protein synthesis and translation, once a "mature" mRNA is synthesized within the nucleus of a cell, it leaves the nucleus and moves to the ribosomes within the cytoplasm. Triplet nucleotide sequences within mRNA, referred to as codons, code for particular amino acids. At the ribosomes, mRNA provides the information needed to link together the specific amino acids that compose a particular protein or peptide. As shown in Figure 2.3(A), the codons in the depicted mRNA provide the code for a peptide that contains the amino acid sequence tyrosine–proline–serine.

Through these basic mechanisms, it can be seen that when a particular gene is active, it regulates the synthesis of a particular protein. With respect to proteins within neurons, some neurotransmitters are peptides, and there are also different enzymes involved in the synthesis of neurotransmitters, different neurotransmitter receptors, and different ion channels. These are just a few of the proteins that influence neuron function, and different genes regulate the synthesis of each of these different proteins. Clearly, any process that affects the operation of such genes would influence brain and behavior mechanisms. Further, different alleles

The two nucleotide strands within DNA are linked together by hydrogen bonds (dashed lines) between complementary bases. Guanine (G) is complementary to cytosine (C), and thymine (T) is complementary to adenine (A). Compared to part (A), the sugar–phosphate backbone has been simplified, and deoxyribose had been drawn as a circle, although in reality its structure is similar to ribose (both are pentose sugars).

or variants of a particular neuron-relevant gene (different alleles of a gene have slightly different nucleotide sequences) can have important effects, including subtle effects on the amino acid sequence of a protein, which could have behavioral implications. For example, the affinity (binding strength) of a particular neurotransmitter for its receptor could be altered if the amino acid sequence of the receptor is modified.

2.2.3 Exons and Introns

In the previous section, I referred to mature mRNA leaving the nucleus en route to the ribosomes, where translation and protein synthesis occurs. What did I mean by mature mRNA? When a particular gene is transcribed within the nucleus, it initially produces what I will refer to as nuclear RNA. This nuclear RNA contains nucleotide sequences that are subsequently removed or spliced out in order to form the final (mature) mRNA that leaves the nucleus and controls protein synthesis. Therefore, a gene can be described as containing exons and introns. Exons are those nucleotide sequences that are present in the final mature mRNA. Introns are those nucleotide sequences that are removed from the nuclear RNA so that mRNA can be formed. Figure 2.3(B) shows a diagram of a generic 5′-to-3′ DNA coding strand, depicting exons and introns. In this example, the gene contains three exons and two introns. The nuclear RNA transcript would contain the entire nucleotide sequence present in the sense strand (with uracil replacing thymine). However, the mRNA that enters the cytoplasm would only contain the nucleotide sequence present in exons 1, 2, and 3, which would be joined together once the introns are spliced out. Note that in this example, the 5′ region of exon 1 contains an untranslated region (UTR), as does the 3′ region of exon 3. The UTR of an exon is composed of nucleotide sequences that are present in mRNA but are not translated into amino acids. Although UTRs within mRNA

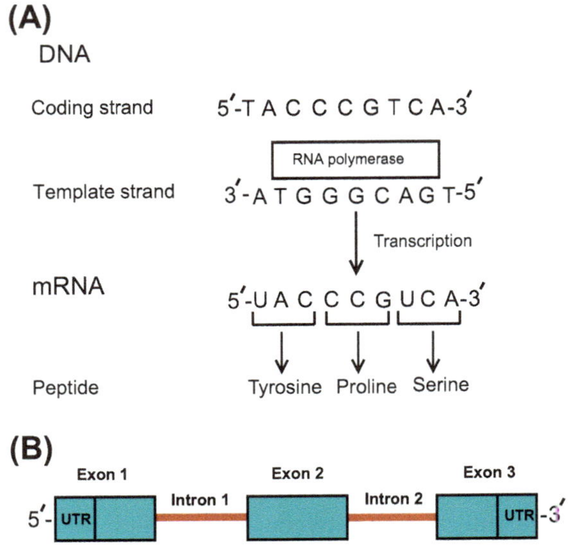

FIGURE 2.3 (A) A general overview of transcription and translation processes with respect to a particular gene. A part of one strand of DNA is referred to as the coding (sense) strand, while the other is the template strand. RNA polymerase moves along the template strand to catalyze the synthesis of single-stranded mRNA composed of bases that are complementary to those in the template. Therefore, the base sequence of the synthesized mRNA matches that in the DNA coding strand, except that uracil is substituted for thymine. During translation, mRNA leaves the nucleus and moves to the ribosomes, where protein synthesis occurs. Triplet nucleotide base sequences within the mRNA, referred to as codons, code for particular amino acids. Therefore, the nucleotide base sequence within the coding strand of a particular gene, which is transcribed to mRNA, reflects the amino acid sequence that makes up a particular protein or peptide. Abbreviations: A = adenine; C = cytosine; G = guanine; T = thymine; U = uracil. (B) Exons and introns within a DNA coding strand. The initial RNA strand that is synthesized during transcription is referred to as nuclear RNA. Certain segments of nuclear RNA are then spliced out of this initial transcript in order to produce mRNA. It is this mRNA that leaves the nucleus and travels to the ribosomes, where translation occurs. (Part (A) of this figure was a simplification that skipped the nuclear RNA to mRNA splicing step that occurs in the nucleus.) With this understanding, the coding region of DNA is defined as containing exons and introns. Nuclear RNA contains the same nucleotide base sequence as that contained within the exons and introns (with uracil substituting for thymine). In contrast, mRNA only contains the sequence represented by the joined exons because the intron sequences have been spliced out. Therefore, it is the exons that code for the synthesis of mRNA and

do not directly contribute to protein structure, they have been shown to influence the rate of translation, and therefore serve regulatory functions. For example, cytoplasmic factors can bind to nucleotide sequences within the UTRs in order to modulate the degree of translation (how much protein is actually produced: [519]).

Although introns do not directly code for the mRNA transcript and therefore do not directly affect protein structure, the following should be noted. A mutation (a modified nucleotide sequence) within an intron may alter whether the exons are spliced together properly. Second, regulatory factors (described below) have been found to bind to portions of DNA introns in order to influence the degree or rate of transcription [499].

2.2.4 The Regulation of Gene Transcription

One definition of a gene is that it is composed of a nucleotide sequence that is transcribed into RNA. However, there are regulatory regions, typically in the 5′ region upstream from where transcription begins, that control the degree or rate of transcription. These nucleotide sequences make up the regulatory region of a gene. A particular gene can be transcribed (and therefore translated) at different rates, allowing for the production of either small or large amounts of a particular protein, depending upon the presence or absence of transcription factors that act on the gene. Transcription factors refer to proteins that bind to particular nucleotide sequences in the gene's regulatory region in order to influence, in either a positive or negative manner, the rate of gene transcription. The current definition of a gene, therefore, includes both the coding region and the regulatory region, and this concept is shown in Figure 2.4. It is important to appreciate the difference between a mutation, or modification of the nucleotide sequence, within the coding region of a gene and a mutation within the regulatory region of that gene. The former is likely to affect the amino acid sequence within the protein that is synthesized and may therefore modify the nature of the protein produced. A mutation within the regulatory region is likely to affect whether a particular transcription factor can operate, which, in turn, will influence whether and to what degree a particular protein is produced within a cell.

The regulatory region of a gene contains nucleotide sequences that make up the promoter region (general regulatory region) and additional specific regulatory regions. The promoter region, which contains a nucleotide sequence referred to as the TATA box because it is rich in thymine and adenine, binds TATA-binding proteins and RNA polymerase. This region and its associated transcription factors are common to all genes and allow for a basal level of gene transcription. More specific transcription factors bind to nucleotide sequences that are typically upstream (on the 5′ end) from the promoter. Such specific nucleotide sequences are referred to as response elements, and some, but not all, genes contain particular response elements. In the example shown in Figure 2.4, two specific response elements are depicted: an estrogen response element (ERE) and a cAMP (cyclic adenosine monophosphate) response element (CRE). It will be worthwhile to describe the operation of these elements.

In order to present a specific example, it is known that the rat oxytocin receptor gene contains an ERE and a CRE in its 5′ flanking region upstream from both the TATA box and the transcription start site [52]. What this means is that the transcription rate of the gene and the amount of oxytocin receptor (OTR) protein that

for the amino acid sequence that is present in a particular protein. Note that an exon can contain an untranslated region (UTR), which means that although the particular base sequence in a UTR is present in mRNA, it is not translated into amino acids. In the example shown in this figure, the coding region of the gene contains three exons and two introns, and two of the exons contain UTRs.

is produced can be influenced by cellular factors that bind to ERE and CRE. The ovarian steroid estradiol increases the expression of OTRs in several brain regions (see Chapter 5). The classic mechanism of action of steroid hormones, such as estradiol, testosterone, and corticosterone, is referred to as a genomic mechanism of action. (See Stolzenberg and Numan [929] for an in-depth discussion of the various mechanisms through which estradiol can exert cellular effects.) Steroid hormones, being lipophilic, pass through the cell membrane and bind to specific receptors that are present in the cytoplasm of some cells. Estradiol (E) would bind to the estrogen receptor (ER), which changes the conformation of the receptor so that a region referred to as the ER's DNA-binding domain becomes exposed and active (capable of binding to an ERE). An activated E:ER complex forms a dimer with another E:ER complex, and this unit then translocates to the nucleus of the cell where the DNA-binding region of each ER binds to the regulatory regions of those specific genes that happen to contain an ERE (not all genes contain this response element). This mechanism, by which estradiol activates ERs so they can bind to EREs, allows estradiol to upregulate the transcription of the oxytocin receptor gene (as well as other genes with EREs in their regulatory regions) so more OTR protein is produced.

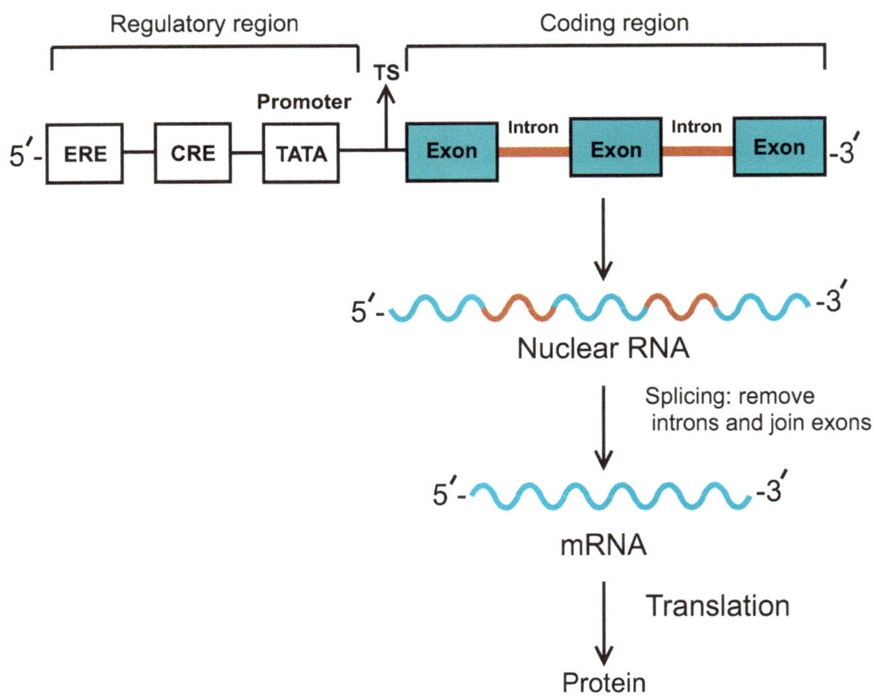

FIGURE 2.4 The regulatory and coding regions of a gene. The coding region of a gene contains the exons and introns that are transcribed into nuclear RNA, which, after the introns are removed, forms mRNA. The regulatory region of a gene contains those nucleotide sequences that are not transcribed but that control the degree or rate of transcription or the extent to which the coding region can produce RNA. The regulatory region contains the promoter (containing the TATA box), where RNA polymerase binds to DNA near the transcription start site (TS). Upstream from the TATA box are additional response elements that regulate transcription. The two that are shown are the cAMP response element (CRE) and the estrogen response element (ERE). See text for further details. Other abbreviations: A = adenine; T = thymine.

As will be described in subsequent chapters, oxytocin (OT) action in the brain plays a prominent role in a variety of social behaviors. The fact that E is capable of upregulating the expression of OTRs in those brain regions that contain ERs, such as the hypothalamus and medial amygdala, provides a mechanism through which an organism's internal physiological state—the amount of estradiol secretion—can influence those social behaviors that are affected by OT action in the brain.

It should be noted that OTRs are expressed at high levels in certain brain regions, such as the central nucleus of the amygdala (CeA), that do not contain ERs [51]. Obviously, there are alternate mechanisms to increase transcription of the OTR gene, and one of these involves activation of CRE. Bale et al. [51] have provided evidence that dopamine (DA) action on D1 receptors increases the expression of OTR in CeA by activating the CRE region of the OTR gene. In order to explain this effect, I need to first describe the intracellular effects that occur when DA binds to the D1 receptor. Recall from Chapter 1 that DA action on D1 receptors located on the output medium spiny neurons (MSNs) of the dorsal striatum exerts excitatory effects on those neurons. What are the mechanisms that allow DA action on D1 receptors to exert such excitatory effects, and how can this help us understand how DA action on D1 receptors in CeA increases the expression of OTRs? Figure 2.5 shows a very simplified analysis of the cellular effects that can be produced as a result of DA action on D1 receptors [158,331,401,438]. One type of neurotransmitter/neuromodulator receptor is

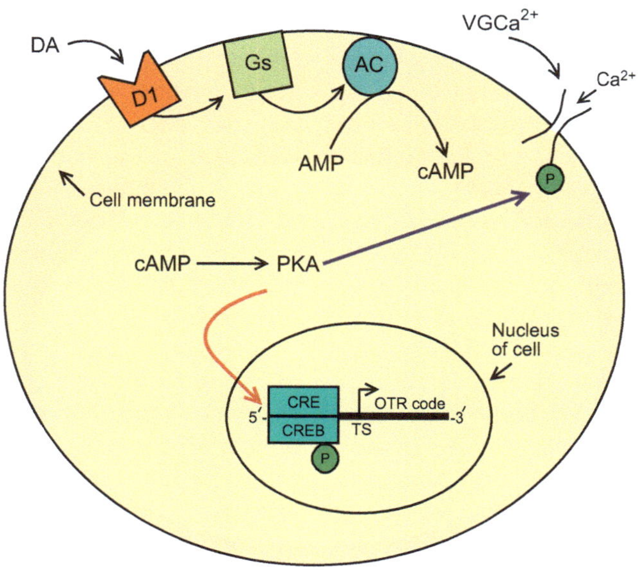

FIGURE 2.5 Dopamine (DA) action at the D1 dopamine receptor can exert genomic as well as nongenomic effects within neurons. When DA binds to the D1 receptor, it activates a stimulatory G protein (Gs), which stimulates adenylate cyclase (AC). AC catalyzes the conversion of AMP (adenosine monophosphate) to cAMP (cyclic AMP). In turn, cAMP activates protein kinase A (PKA). For a nongenomic effect (blue), PKA is shown as phosphorylating a voltage-gated calcium channel ($VGCa^{2+}$) in the cell membrane, which changes the functional activity of this channel. For an example of a genomic effect (red), PKA enters the nucleus of the neuron, where it phosphorylates the cAMP response element (CRE) binding protein (CREB) within the regulatory region of the oxytocin receptor (OTR) gene. Phosphorylation of CREB enhances the transcription of OTR RNA. Other abbreviations: P = phosphate group; TS = transcription start site. Nongenomic effects modify the function of proteins that already exist in a cell, while genomic effects result is the synthesis of new proteins.

referred to as a G-protein-coupled receptor, and the DA–D1 receptor is a member of that class. As shown in the figure, the intracellular component of the D1 receptor is coupled or connected to a stimulatory G protein (Gs). When DA binds to the D1 receptor, Gs is activated, and its subsequent conformational change activates the enzyme adenylate cyclase (AC). The activated AC then catalyzes the production of cAMP from AMP. In turn, cAMP activates protein kinase A (PKA). A kinase is an enzyme that phosphorylates (adds a phosphate group) to certain amino acids within a protein, which changes the function of the protein (this is referred to as a post-translational modification of protein function, since the protein has already been synthesized and is present in the cell). One protein that is likely to be phosphorylated by PKA is a certain type of voltage-gated Ca^{2+} channel (VG Ca^{2+} channel) within the cell membrane that allows Ca^{2+} to enter a neuron or other type of cell from the extracellular fluid, which would depolarize the cell. VG Ca^{2+} channels, which are located on the dendrites and cell bodies of MSNs, open in response to depolarization, and if such channels are phosphorylated, they stay open for a longer duration in response to depolarization. With this scenario, one can see how DA action on D1 receptors might enhance the excitability of MSNs. Let us assume that glutamate input to the striatum activates striatal MSNs by acting on glutamate receptors, with the result that Na^{+} enters the neuronal cell body to cause a depolarizing EPSP (excitatory postsynaptic potential). This EPSP, in turn, would cause the VG Ca^{2+} channels to open, resulting in calcium influx that would enhance the initial depolarizing EPSP caused by glutamate. Now, if DA were concurrently acting on D1 receptors in this same MSN, which would phosphorylate the VG Ca^{2+} channels, then they would stay open longer in response to the EPSP produced by glutamate, and the ultimate depolarization of the MSN would be greater, meaning that the likelihood of neuronal action potentials would be increased. As an aside, it should be noted that in this example glutamate would be considered a neurotransmitter because it directly causes ionic permeability changes within the neuron, while DA action on D1 receptors would serve a neuromodulatory function since it does not directly cause ionic permeability changes (phosphorylation of VG Ca^{2+} channels does not cause them to open at the resting membrane potential, but when membrane depolarization occurs they stay open longer), but instead enhances the depolarization caused by glutamate. In other words, DA–D1 action modulates, through amplification, the excitatory effects of glutamate.

Although DA–D1 effects that enhance the excitability of MSNs are considered nongenomic because the mechanism described involves modifications in the function of an already-synthesized protein, the stimulation of OTR expression by DA action on D1 receptors in CeA involves a genomic mechanism of action via an enhancement of the transcription of the OTR gene. A protein referred to as CREB (cyclic AMP response element binding protein) binds to the CRE unit in the regulatory region of the OTR gene. When enough PKA is produced as a result of cAMP activation, PKA can translocate to the nucleus in order to phosphorylate CREB, which then stimulates the transcription of genes that contain CRE (see Figure 2.5).

Important general principles are worth summarizing from these specific examples. First, a neurotransmitter or neuromodulator may not only affect the function of a neuron nongenomically but may also have genomic effects that allow new proteins to be produced within a neuron in order to modify its function. Second, since any particular gene has several response elements in its regulatory region, its expression can be induced in different neurons by different transcription factors. Further, the concept of combinatorial control is important. For a neuron that contains both cell membrane DA–D1 receptors and intracellular ERs, the following is possible. Either DA action on D1 receptors or E action on ERs would be capable of increasing OTR expression. However, such expression should be even higher when both the DA–D1 system and the E:ER system are activated concurrently over long periods of time. Finally, if a gene does not contain a particular response element, its transcription will not be affected by those intracellular events that act on that response element. For example, estrogen may bind to ERs in a particular neuron, but it will only affect the expression of those genes in the neuron that contain EREs.

Please note that many genes contain EREs and CREs and also that neurochemicals other than DA action of D1 receptors are capable of activating PKA. For example, norepinephrine (NE) action on the β-NE receptor is also coupled to a Gs protein that stimulates the cAMP–PKA pathway. The OTR gene was simply presented as a concrete example of how steroid hormones and neurotransmitters/neuromodulators are capable of modifying gene transcription mechanisms through actions within the regulatory control region of genes.

2.2.5 Genetic Polymorphisms

A particular gene that codes for the synthesis of an RNA and protein is located within a particular region of a chromosome. Different variants of a particular gene, referred to as different alleles, are the result of differences in the exact nucleotide sequence present in the gene. Two types of such polymorphisms that will be discussed in future chapters are single nucleotide polymorphisms (SNPs) and repetitive sequence polymorphisms. SNPs result from the presence of different nucleotides at a single site within the nucleotide sequence of a gene. As a specific example, when considering double-stranded DNA, some members of a population may exhibit a complementary A–T base pair at a particular site (with the A in the coding strand and the T in the template strand), while other members may exhibit a G–C base pair. When one draws or depicts the nucleotide sequence in a gene, the coding (5′-to-3′) strand is usually written on top (see Figure 2.3(A)). When referring to SNPs, it is probably best to refer to the nucleotide change in the 5′-to-3′ strand, and, therefore, based on the example above, one would say that some members of a population contain the A allele at the particular site being considered, while others contain the G allele. For example, one individual may show the sequence 5′-AAGG**A**C-3′, while another may exhibit 5′-AAGG**G**C-3′ (the SNP is shown in bold).

With respect to the functional consequences of SNPs, if a particular SNP were located in an exon, it could change an amino acid in a protein from one type to another. Using the above example, if the SNP were in an exon, note that AAG is the codon for lysine, GAC is the codon for aspartate, and GGC is the codon for glycine. Therefore, from these simple sequences, the translated peptide could be either lysine–aspartate or lysine–glycine. Although most SNPs are not located within exons that are translated into amino acids, they may still influence gene expression. An SNP within the UTR may affect the efficiency of translation, and an SNP within introns may affect the splicing process. In addition, SNPs within regulatory response elements, some of which may be contained within introns, could alter the degree of transcription. In many cases the exact functional significance of a particular SNP is not known, but a particular study may still find a correlation between a certain SNP and a behavioral or psychological trait.

Repetitive sequence polymorphisms involve nucleotide sequences, such as 5′-CAG-3′ that are repeated many times in tandem. Different alleles contain different numbers of repeats. For example, one individual may exhibit 5′-CAGCAGCAG-3′ (three repeats) while another may exhibit 5′-CAGCAGCAGCAG-3′ (four repeats). Repetitive sequence polymorphisms can be located within the 5′ flanking regulatory region of a particular gene and therefore may influence the rates of gene transcription (see Chapter 6).

2.2.6 Knockout Mutations

Transgenic mice, in which a native or wild-type gene has been modified, have provided valuable information with respect to the role of particular genes in behavior. The type of transgenic mouse I want to discuss is one that has been genetically engineered so that a native functioning gene has been replaced by a nonfunctional gene. Such a replacement would produce what is called either a knockout mutation or a null mutation. Let us say one wanted to create a mouse line with a knockout mutation of the OTR gene. First, one would manufacture a double-stranded DNA sequence that mirrored the sequence of the OTR gene and its surrounding nucleotide sequences, except that changes would be made to the gene to render it nonfunctional. This genetically engineered gene would then be injected into the nucleus of pluripotent mouse embryonic stem cells. Embryonic stem cells in which the transgene replaced the native OTR gene would then be harvested and inserted into very early mouse embryos (blastocysts). As these embryos develop and the embryonic stem cells divide and differentiate, some cells will contain the defective OTR gene while others will not (because the modified stem cells were inserted into a blastocyst that also contained normal cells). For those mice in which the defective OTR gene was incorporated into sperm and ova, males and females would then be bred together in order to produce offspring that are homozygous for the OTR gene knockout (OTR−/−). This defective gene would therefore be present in every cell of the transgenic offspring, including brain cells, preventing the synthesis of OTR protein in those cells in which it would normally be produced. For example, DA–D1 activation would not activate OTR in CeA and estradiol would not activate OTR expression in the hypothalamus and medial amygdala. In adulthood, one could study the social behavior of such mice. For example, as will be described in Chapter 5, the maternal behavior of OTR knockout mice (OTR−/−) has been compared to that of wild-type mice (OTR+/+) in order to determine whether the absence of OTR production influences maternal behavior.

2.3 BASIC EPIGENETICS

2.3.1 Introduction

The information in this section on epigenetics is based on several helpful reviews. In addition to Alberts et al. [15] and Lewin [550], the following were also very important: Caiafa and Zampieri [146]; Moore, Le, and Fan [662]; Roth [839]; Sweatt [954]; Tsankova, Renthal, Kumar, and Nestler [976]; and Zhang and Meaney [1073].

The control of transcription can be viewed as a hierarchical process. I have already described how the binding of general and specific transcription factors within the regulatory region of a gene allows transcription to proceed and also regulates the rate or efficiency of transcription. However, in order for these processes to occur, transcription factors and RNA polymerase must be able to bind to the DNA promoter and specific regulatory regions. If the organization or folding pattern of DNA were to be altered so that the nucleotide sequence remained the same but transcription factors and RNA polymerase could not gain access to the gene's regulatory region, then the gene would remain inactive or silent. When one considers embryology and cell differentiation, the fact that some, but not all, genes are active in any single cell makes sense. Cells, such as brain cells, blood cells, and skin cells, differ from one another in part because they produce different proteins, which means that different cells have different complements of active and inactive genes. More specifically, although all human somatic cells contain 23 pairs of chromosomes, and therefore every somatic cell contains the same genes, not all of these genes are transcribable in a particular cell, and different cells are distinguished by their particular population of active genes.

In general terms, epigenetics can be defined as those molecular processes that affect the organization of DNA so that transcription factors and RNA polymerase are either allowed or prevented access to a particular gene.

2.3.2 Heterochromatin and Euchromatin

Chromatin refers to the proteins and DNA that make up chromosomes. Importantly, DNA is coiled around a group of proteins called histones. Chromatin can exist in two major states: heterochromatin and euchromatin. In the heterochromatin state, DNA is wound tightly around histone proteins, which prevents the ability of transcription factors and RNA polymerase to gain access to a gene. Therefore, heterochromatin represents a closed and inactive gene state. Euchromatin is an open and active state, where DNA is separated from histones, which allows the exposed genes to be transcribed. Figure 2.6(A) depicts these different forms of chromatin. Epigenetic mechanisms, by acting on either histone proteins or DNA, can cause shifts between heterochromatin and euchromatin and

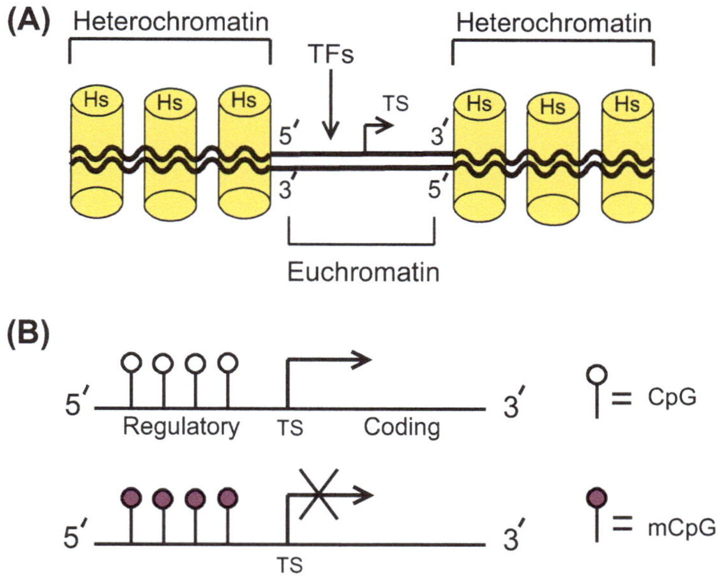

FIGURE 2.6 Epigenetic processes. (A) Heterochromatin and euchromatin. Chromatin refers to the DNA and proteins that make up chromosomes. In the heterochromatin state, DNA is tightly wound around a group of histone proteins (Hs), which prevents the binding of transcription factors (TFs) and RNA polymerase to the regulatory regions of those genes located in that wound DNA sequence. Therefore, heterochromatin represents a closed or inactive gene state. Euchromatin is an open or active gene state, where DNA is separated from histones, which allows the exposed genes to be transcribed. TS = transcription start site. (B) DNA methylation and the control of transcription. DNA methylation refers to the addition of a methyl group onto cytosine within DNA. Methylation is facilitated when a cytosine is followed by guanine (CpG site) in a particular nucleotide sequence. Therefore, one can distinguish unmethylated (CpG) from methylated (mCpG) cytosine bases. (Note that the "p" in CpG refers to the phosphate linkage between the two nucleotides within a single DNA strand. This nomenclature is used in order to differentiate CpG sites within a strand from the C–G complementary hydrogen bonding that occurs between DNA strands.) DNA methylation within the regulatory region of a gene can depress transcription in one of two ways: (1) by preventing transcription factor binding to particular response elements within the regulatory region of a gene; (2) by attracting histone deacetylases, which remove acetyl groups from histones, a process that favors a heterochromatin state. The X across the arrow at the transcription start site (TS) indicates that transcription of the gene is depressed.

therefore can either promote or interfere with gene transcription.

Histone proteins are positively charged and DNA is negatively charged, and the resultant electrostatic attraction contributes to the tight coiling of DNA around histones in the heterochromatin state. One factor that contributes to a shift to euchromatin involves a chemical modification of histones that decreases their positive charge. A class of enzymes called histone acetyltransferases (HATs) transfers an acetyl group ($COCH_3$) onto specific amino acids within histone proteins, which reduces the positive charge on the histones, allowing a particular segment of DNA to separate from histones, which then facilitates the transcription of genes located at that particular site. Another class of enzymes, histone deacetylases (HDAC), removes acetyl groups from histones. Therefore, HDACs favor a heterochromatin state and play a role in preventing the transcription of certain genes. Based on these mechanisms, one can see that if in a particular cell HDACs were dominant over HATs at a DNA site that contained gene 1, while HATs were dominant over HDACs at a DNA site that contained gene 2, then this cell would likely express the gene product of gene 2 but would not express the gene product of gene 1.

Another mechanism that alters the organization of chromatin involves DNA methylation, which is the addition of a methyl group (CH_3) onto cytosines in DNA. Not all cytosines are easily methylated, and the methylation process is facilitated if a particular cytosine is followed by guanine in a particular nucleotide sequence. This dinucleotide sequence is referred to as a CpG sequence. The p stands for the phosphate group within the sugar–phosphate backbone of single-stranded DNA. CpG is used to indicate that the C–G linkage is within a single strand of DNA rather than between the C–G hydrogen-bonded complementary base pair that links up two separate strands of DNA. DNA methylation is catalyzed by DNA methyltransferases (DNMTs), and the presence or absence of such methylation can be indicated as mCpG or CpG. DNA methylation, particularly within the regulatory region of a gene, is usually associated with the inactivation or suppression of gene transcription. First, a high density of methylated cytosines may oppose the ability of transcription factors to bind to their response elements. More importantly, mCpG sites attract methyl-DNA binding proteins, which in turn attract HDACs. As described previously, HDACs would favor heterochromatin formation and the silencing of transcription. Some of these processes are depicted in Figure 2.6(B).

Until recently, it was thought that epigenetic mechanisms were primarily involved in cell fate determination during embryological development and cell differentiation, where certain genes in a particular cell would be permanently silenced, while other genes would remain active, in this way determining the particular phenotype of a given cell [327]. However, recent research has indicated that early postnatal life experiences can also affect epigenetic processes that cause enduring effects on the transcription of genes within the central nervous system, which results in long-lasting effects on the behavior of the organism [839,1073]. Therefore, individual differences in early life experiences can result in individual differences in the expression of particular genes within the brain, which contributes to behavioral variation.

Although epigenetic mechanisms also contribute to gene expression in adults, and such processes have been implicated in learning and memory [954], it is the early postnatal developmental effects of epigenetic processes to which I will turn my attention.

2.4 EPIGENETIC EFFECTS ON PHYSIOLOGICAL AND BEHAVIORAL DEVELOPMENT

2.4.1 Introduction

In this section, I want to describe research that has shown that individual differences in early life experiences can give rise to individual differences in stress reactivity, and that epigenetic mechanisms contribute to such variation. The involvement of such epigenetic processes has been pioneered by Meaney's group [1073] and has dealt primarily with the physiological stress response (stress-induced activation of corticosterone release from the adrenal cortex), which is a neuroendocrine response. This research has shown how early life experiences can modify the way the brain responds to stressful stimuli, so that more or less corticosterone is released in response to stress. Since these modifications involve epigenetic changes that influence neuron function, such neuroendocrine variation can serve as a model for understanding how individual differences in early life experiences could also modify brain and behavior phenotypes via epigenetic processes.

2.4.2 The Physiological Stress Response

As described in Chapter 1, the physiological stress response involves the activation of the hypothalamic–pituitary–adrenal (HPA) axis by stressful and aversive stimuli [983]. Through a variety of neural pathways, physical and psychological stressors activate CRH neurons in the PVN of the hypothalamus. Subsequent CRH release into the hypothalamic–pituitary portal veins reaches the anterior pituitary to stimulate ACTH release into the blood. ACTH, in turn, stimulates corticosterone or cortisol release from the adrenal cortex. One function of these adrenal steroids is to promote gluconeogenesis by the liver, which raises blood glucose levels, providing a needed energy source. Since corticosterone and cortisol influence the synthesis of glucose, they are referred to as glucocorticoids.

Like other steroids, a dominant way in which glucocorticoids affect cell function is via a genomic mechanism. Similar to the way estradiol acts, glucocorticoids (GC) bind to intracellular glucocorticoid receptors (GR) that, in turn, bind to the glucocorticoid response element that is present in certain genes. Therefore, the genomic effects of glucocorticoids occur because the GC:GR complex serves as a transcription factor.

Once a stressful situation terminates, the physiological stress response also terminates. An acute stressful event does not cause the continuous release of glucocorticoids from the adrenal cortex because glucocorticoids exert a negative feedback effect on the HPA axis, and many of these effects are mediated by the action of glucocorticoids in the brain [983,1073]. Circulating blood levels of glucocorticoids are capable of entering the brain, and groups of neurons in certain parts of the brain, such as the amygdala, hippocampus, and hypothalamus, contain intracellular glucocorticoid receptors. Several studies have shown that glucocorticoid action on the hippocampus is one mechanism underlying the negative feedback effect of glucocorticoids on the HPA axis, and this process is depicted in Figure 2.7. The figure shows that glucocorticoids can reach the hippocampus, where they activate mechanisms that stimulate the output of those hippocampal neurons that contain GRs. Since the projection neurons of the hippocampus contain glutamate, the hippocampus is shown as inhibiting CRH neurons in the PVN indirectly, by first activating GABAergic neurons that then inhibit the CRH neurons [983]. As a result of this inhibition of CRH, ACTH and corticosterone/cortisol release subsequently decline.

The particular mechanism(s) through which glucocorticoid action on the hippocampus activates the output from this region, which then inhibits CRH neurons, is not known. If a genomic mechanism is involved, one possibility is that the transcriptional effects of glucocorticoids trigger the synthesis of proteins, such as neurotransmitter receptors, within the relevant hippocampal neurons, which allows these neurons to respond strongly to excitatory inputs from other neurons. Alternatively, genomic

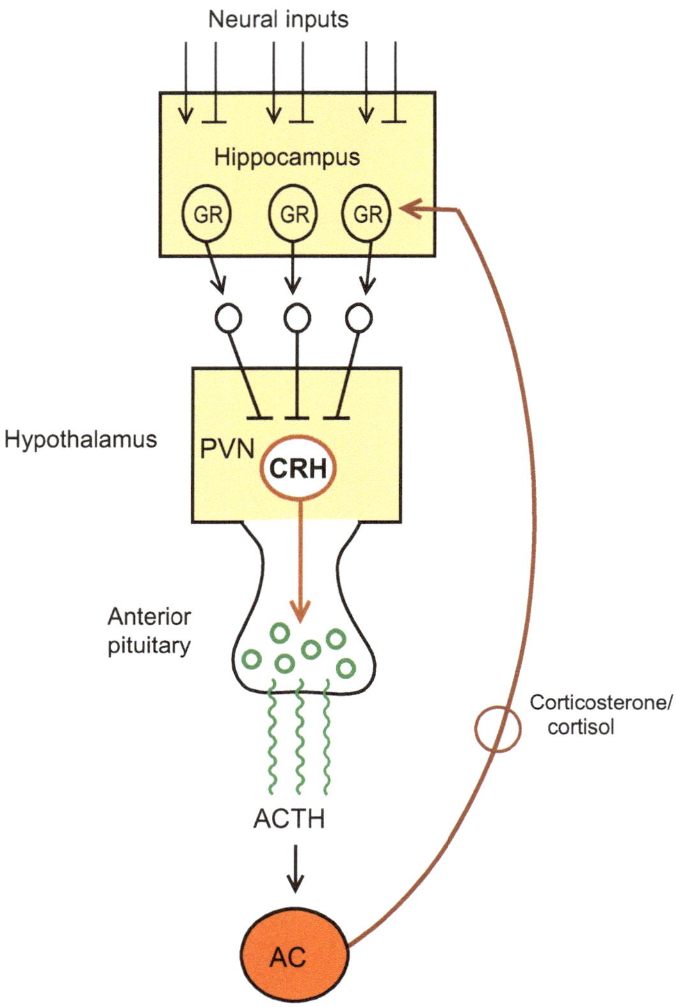

FIGURE 2.7 Glucocorticoid action on glucocorticoid receptors (GR) within certain hippocampal neurons exerts a negative feedback effect that depresses the hypothalamic–pituitary–adrenal axis. The action of corticosterone or cortisol on GRs increases the ability of particular hippocampal neurons to respond to excitatory inputs or decreases their ability to respond to inhibitory inputs. Such activated hippocampal neurons are shown as indirectly inhibiting corticotropin-releasing hormone (CRH)–containing neurons in the paraventricular nucleus (PVN) of the hypothalamus. A decrease in CRH release results in a decrease in adrenocorticotropin (ACTH) release from the anterior pituitary, which results in a decrease in the release of corticosterone/cortisol from the adrenal cortex (AC). Axons ending in a bar are inhibitory, and those ending in an arrow are excitatory.

effects may ultimately decrease the ability of the relevant hippocampal neurons to respond to inhibitory inputs.

Finally, please note that the hippocampus is involved in many functions, which includes important behavioral functions. The negative feedback system described above is referring to only a single facet of hippocampal function, one that involves the modulation of a neuroendocrine process.

2.4.3 Epigenetic Processes Mediate the Influence of Maternal Care on the Development of the Physiological Stress Response

2.4.3.1 Introduction

As I will review in many of the subsequent chapters in this book, the manner in which mothers treat their young has broad implications with respect to the development of social behavior, and this will be particularly evident in the chapters on the development of aggression and violence, and on the development of maternal responsiveness. In the current chapter on genetics, my focus will be on understanding the interactions between genes and environmental factors that influence the development of the physiological stress response. The principles developed here, however, will serve as an important model for understanding, in more general terms, how genetic and environmental factors interact to affect social development.

Liu et al. [557] have described stable individual differences in the level of maternal care that mother rats provide their young over the first 10 postnatal days of life (rat pups are typically weaned between 21 and 28 days of life). Some mothers show high levels of licking and grooming of their infants (High-LG), while other mothers show Low-LG. The amount of LG may signify the level of maternal attentiveness. These differences represent normal variations in maternal behavior, all offspring grow normally, and there is not a differential survival of offspring at weaning in these two groups.

As adults, the offspring of High-LG mothers show a reduced HPA stress response compared to offspring of Low-LG mothers. During a 20 min period of restraint stress, plasma corticosterone levels rise equally in both groups. However, once the stressful stimulus terminates, plasma corticosterone levels decline much more quickly in the High-LG group than in the Low-LG group [557]. Additional cross-fostering studies have shown that these differences in stress responsiveness are related to the maternal care received by the offspring rather than to the genetic inheritance of a stress-prone characteristic [1073]. When pups that were born to High-LG mothers were cross-fostered to Low-LG mothers (or vice versa), it was found that the rearing mother, not the biological mother, determined the stress response phenotype of the offspring in adulthood. In further support, given that licking and grooming provide tactile stimulation to the neonate, it has also been found that the experimental enhancement of tactile stimulation by stroking young pups with a brush also decreases the adult's HPA response to stress when compared to young that do not receive this extra tactile stimulation [1073].

One way to interpret these results is that adult rats that received less maternal care when they were young may have developed an enhanced responsiveness to stressful stimuli because of processes that resulted in an interference with the negative feedback effects of corticosterone on CRH release (see Figure 2.7). Importantly, in situ hybridization autoradiography was used by Liu et al. [557] to detect GR mRNA levels in the hippocampus, and it was found that GR mRNA levels were significantly lower in the Low-LG offspring when compared to the High-LG offspring. On the basis of these data, one can propose the following hypothesis: high or low levels of maternal care received by an infant rat pup either increase or decrease, respectively, the transcribability of the GR gene, with corresponding changes in the expression of GR mRNA and GR protein. Decreases in GR protein within the hippocampus of adults who received lower levels of maternal care decrease the effectiveness of one route through which corticosterone exerts its negative feedback effect on the HPA axis, resulting in an enhanced physiological stress response. Significantly, these results indicate that early life experiences exert enduring effects, lasting into adulthood, on stress responsiveness and gene expression.

Although these studies deal with the physiological stress response, they can serve as a model for understanding the mechanisms of brain plasticity that underpin the effects of early life experiences on the development of neural circuits that regulate social and emotional behaviors. Relevantly, the offspring of Low-LG mothers not only show increased physiological stress reactivity, but they are also more fearful in a variety of behavioral tests, such as exhibiting reduced exploratory activity in a novel open field when compared to High-LG offspring [150]. More generally, lower levels of maternal care during early postnatal development appear to increase the defensiveness, stress reactivity, and fearfulness of the affected offspring, and such effects persist into adulthood. With respect to sociality, it makes sense that one's level of anxiety and fearfulness would influence the degree and nature of one's social interactions with others. Therefore, this kind of data may provide a neurobiological understanding of maternal influences on personality development.

2.4.3.2 Epigenetic Events Mediate the Effects of Maternal Care on the Physiological Stress Response

The regulatory region of the GR gene contains the promoter region and several specific response elements. For hippocampal neurons, Weaver et al. [1022] have reported that offspring reared by either biological or foster mothers with a Low-LG phenotype exhibit increased DNA methylation at certain CpG sites within a specific regulatory region of the GR gene, while offspring reared by High-LG mothers show decreased DNA methylation at this site. Mechanisms, initiated by licking- and grooming-induced release of certain neurotransmitters into the hippocampus of rat pups, have been proposed to explain how high levels of maternal care decrease the degree of DNA methylation within this regulatory region of the GR gene (see [1073]).

Significantly, the higher levels of DNA cytosine methylation status that result from lower levels of maternal care are also associated with decreases in histone acetylation within the critical gene regulatory region. Recall that mCpG sites attract methyl-DNA binding proteins, which in turn attract HDACs. HDACs favor heterochromatin formation and the silencing of transcription. What these results suggest is that low levels of maternal care result in a partial silencing of the GR gene, causing a closed heterochromatin state within certain portions of the gene's regulatory regions, which decreases the ability of certain transcription factors to activate gene transcription. Since the GR gene has several specific response elements within its regulatory region, the process uncovered by Weaver et al. [1022] may decrease the expression of GR protein in certain cells, such has hippocampal neurons, without affecting its expression in other cells that rely on a different set of transcription factors and response elements.

The offspring of Low-LG mothers exhibited an increased methlyation of DNA and a decreased acetylation of histones within the GR gene of hippocampal neurons that persisted into adulthood, and this was associated with decreased expression of GR protein in the hippocampus and an increased HPA response to restraint stress. Importantly, when adult rats of Low-LG mothers were injected intracerebroventricularly over a seven-day period with an HDAC inhibitor, the stress reactivity of these rats became similar to that of rats that were raised by High-LG mothers. The rats injected with the HDAC inhibitor showed increased histone acetylation within the critical regulatory region, increased GR protein expression in the hippocampus, and a dampening in the duration of their corticosterone response to restraint stress. These results support the view that treatment with the HDAC inhibitor shifted the essential regulatory region of the GR gene from a heterochromatin to a euchromatin state, which then allowed transcription to proceed.

For an expanded view of the significance of these results, note that the nature of a mother's care of her young may influence the expression of genes other than the GR gene, and this process may similarly occur via an epigenetic process. Using a maternal deprivation paradigm, where rat pups were separated from their mothers for 4h/day over days 2–13 postpartum, Chen et al. [178] reported that the resultant decreased maternal care was associated with hypomethylation at CpG sites within the regulatory region of the CRH gene in the affected offspring, leading to an increased transcription of CRH within the PVN. This process would be another way to increase HPA responsiveness to stress in such maternally deprived rats. More broadly, Zhang et al. [1071]

found that maternal care influences the development of the GABA system within the hippocampus. Glutamic acid decarboxylase (GAD) is the enzyme that catalyzes the conversion of glutamate to GABA, and this enzyme is essential for GABA synthesis. The degree of expression of the GAD protein, in turn, is regulated by the transcription rate of the GAD gene. Zhang et al. reported that the adult offspring of High-LG mothers, when compared to the offspring of Low-LG mothers, exhibited higher GAD mRNA expression in the hippocampus, and this was associated with decreases in cytosine methylation and increases in histone acetylation within the regulatory region of the GAD gene. If a similar effect were to occur in the amygdala, it might possibly be related to the fact that the offspring of High-LG mothers are less fearful than those of Low-LG mothers. Recall from Chapter 1 that there are many GABAergic interneurons within the amygdala that function to restrain the output of the amygdala, and some of these output neurons regulate fear-related responses. Clearly, the nature and degree of maternal care has multiple effects on the brain development of infants, and decreases in maternal care may increase emotionality and stress reactivity via multiple pathways.

Do these results in rats have any relevance to primates, including humans? First, Coplan et al. [201] examined the development of bonnet macaque monkeys that were raised by their mothers under one of two different feeding regimens. In the low foraging demand (LFD) condition, food was freely available to the mother–infant dyad during postpartum months 3–6. In the variable foraging demand (VFD) condition, food was unpredictably available either in high or low amounts. The VFD condition resulted in a stressful mother–infant bond, with such mothers often breaking contact with their infants [835]. Importantly, in adulthood, the offspring reared under VFD conditions were more fearful, less gregarious, and had higher levels of CRH in cerebrospinal fluid (CSF) than did their LFD counterparts. Second, Lee, Gollan, Kasckow, Geracioti, and Coccaro [538] administered a parental bonding questionnaire to adult men and women in order to measure each subject's recollection of the quality of parental care that they received in childhood. Samples of CSF CRH levels were also obtained, and it was found that a significant negative correlation existed between the degree of remembered parental care and CRH levels—lower levels of recollected parental care were associated with greater CRH release. These results support the view that lower levels of parental care in humans can increase HPA responsiveness and CRH release. Lastly, McGowan et al. [624] performed a postmortem analysis of hippocampal tissue obtained from three groups of subjects. The subjects in Group 1 had a history of child abuse (sexual abuse, physical abuse, or parental neglect) and committed suicide as adults. Group 2 was composed of subjects who committed suicide as adults, but were not exposed to childhood abuse. Group 3 was composed of control subjects who did not experience childhood abuse and who did not commit suicide as adults, but died suddenly from other causes such as a heart attack. Analysis of GR mRNA levels in hippocampal tissue found that Group 1 had the lowest amount of mRNA, presumably indicative of lower GR protein expression and decreased negative feedback inhibition on the HPA axis. Significantly, when the DNA of the GR gene from hippocampal cells was analyzed, Group 1 exhibited increases in cytosine methylation and decreases in histone acetylation within a restricted regulatory region of the gene, and these epigenetic alterations presumably resulted in decreased transcription of the gene and reduced levels of GR mRNA. These combined results have important implications. Suicide is one of the outcomes of severe depression, and important characteristics of depression include elevated levels of CRH within the CSF and cortisol within blood plasma [418]. CRH not only acts as a releasing hormone for ACTH at the level of the anterior pituitary but is also present as a neurotransmitter/neuromodulator in diverse brain regions [323] such as the amygdala and dorsal raphe nucleus. Further, an enhancement of CRH release within the brain, which, as described, can result from the decreased negative feedback effects of glucocorticoids or from a hypomethylation of the CRH gene, has anxiogenic effects that may ultimately contribute to severe depression and other stress-related phenotypes [418,419]. One way in which CRH may have these effects is by modulating the activity of serotonin neurons in the dorsal raphe nucleus [490], and the central nucleus of the amygdala is an important source of CRH input to the dorsal raphe [353,762]. These results, therefore, indicate yet another way in which low levels of maternal care, through the mediation of epigenetic processes, can result in the development of an anxiety-prone phenotype in the affected offspring. I will have more to say about the relationship between CRH, serotonin, and behavioral hyperreactivity to stressful stimuli in the next chapter on aggression.

2.4.4 Genetic Polymorphisms Interact with Early Life Experiences to Influence the Development of Stress Reactivity and Psychiatric Disorders

The studies reviewed above on the effects of normal variations in maternal behavior on the development of the physiological stress response have described epigenetic processes that regulate variations in stress responsiveness *within the normal range*. However, the results of the McGowan et al. [624] study indicate that under circumstances that involve severe early life stress, epigenetic processes may contribute to pathological outcomes. As I will describe in more detail in Chapters 3 and 5, not all individuals who are exposed to early adverse life experiences, such as child abuse or neglect, go on to develop psychiatric disorders. Some individuals are resilient, while others are more susceptible to the deleterious effects of early life stressors. Gene–environment interactions may contribute to such differential effects of particular life experiences on the development of behavioral outcomes. Genetic polymorphisms within certain genes, giving rise to different alleles of that gene, may modulate the effects of environmental input on behavioral outcomes. In keeping with the current theme of stress responsiveness, I want to describe an example of a gene–environment interaction with respect to a gene involved in the physiological stress response and the associated behavioral dysfunctions that might be caused by the hypersecretion of CRH.

The FK506 binding protein 51, referred to as FKBP5, regulates the sensitivity of the GR to corticosterone/cortisol [634]. FKBP5 is a negative regulator of the GR, and when it binds to the receptor it decreases the affinity of corticosterone or cortisol for the GR. Therefore, higher levels of glucocorticoids would be needed to effectively activate the GR. What this means is that as the expression of FKBP5 protein increases, the negative feedback effects of glucocorticoids on the HPA axis and on CRH release would decrease.

In humans, different polymorphisms within the FKBP5 gene have been detected [499,634]. One SNP, which has been assigned the identifying number rs1360780 and is located within intron 2 of the gene, contains either an A nucleotide or a G nucleotide. Considering the diploid state, the following genotypes are possible: AA, AG, GG. The A allele has been referred to as a risk allele, because carriers of this allele (AA or AG genotypes) are more likely to develop psychiatric disorders if they are exposed to early life traumas. For example, Klengel et al. [499] have reported that about 80% of carriers of the risk allele who have also been exposed to childhood sexual or physical abuse develop posttraumatic stress disorder later in life, while this is true for only about 50% of individuals with the protective GG genotype. Importantly, Klengel et al. present data that show that the A allele is associated with enhanced transcription of the FKBP5 gene. Therefore, carriers of the risk allele would be expected to have a slightly enhanced HPA stress response due to an increase in a factor (FKBP5) that would decrease the sensitivity of the GR to cortisol and therefore dampen negative feedback mechanisms. What they suggest is that a genetic predisposition to respond to stress more strongly in the short term interacts with childhood abuse to increase the risk for the development of psychiatric disturbances in the long term. But what might be the mechanism through which early abuse interacts with this risk allele to increase the likelihood of pathology? Klengel et al. provide evidence that early life trauma further enhances the expression of the FKBP5 gene only in carriers of the risk allele, and this occurs via epigenetic mechanisms. They report that DNA demethylation (hypomethylation) at particular CpG sites within the regulatory region of the gene was detected only in those individuals that carried the risk allele and were also exposed to childhood abuse. DNA demethylation, of course, would promote transcription of the FKBP5 gene, which would further enhance the HPA stress response in the carriers of the risk allele.

In integrating this work with the research from Meaney's group, the following processes can be outlined:

1. Decreases in maternal behavior, via epigenetic mechanisms that act on the GR gene, can decrease the expression of the GR in the hippocampus, which decreases the negative feedback effects of glucocorticoids and enhances HPA stress responsiveness and CRH release within the brain.
2. The risk allele at rs1360780, coupled with adverse early life experiences, strongly enhances the expression of FKBP5, and this represents an interaction between a genetic polymorphism and the epigenetic effects induced by adverse early life events. The enhanced FKBP5 expression will further dampen the effectiveness of the GR.
3. These two effects, acting on two separate genes (GR gene and FKBP5 gene) and operating at the same time, should result in an amplified HPA stress reactivity: Fewer GRs will be synthesized, and those that are produced will bind less strongly to glucocorticoids. If these effects are severe enough, they may result in pathological consequences that may, in part, be due to hypersecretion of CRH within the brain throughout an organism's life span.

2.5 CONCLUSIONS

This chapter was meant to provide the basic genetic principles that will be needed to appreciate the gene–social behavior relationships that will be described in subsequent chapters. Indeed, Chapters 1 and 2 form an essential background unit for this book. Chapter 1 described those neural circuits that play important roles in social behavior. The current chapter on genetics provides an understanding of how genetic mechanisms can affect the operation of neural circuits. Proteins are key players in neuron function. Such proteins include: neurotransmitter/neuromodulator receptors; neuropeptide neurotransmitters/neuromodulators; ion channels; enzymes involved in the synthesis of neurotransmitters and neuromodulators; and steroid hormone receptors and other gene regulatory proteins. Clearly, gene modifications, either via genetic polymorphisms, epigenetic processes, or an interaction between the two, are in a position to exert powerful effects on how the neural systems related to social behavior operate.

In this chapter, I described how transcription factors regulate the expression of genes and how epigenetic processes determine whether transcription factors can gain access to particular genes. Environmental inputs affect both processes since neurotransmitters and hormones, for example, can activate transcription factors and affect epigenetic mechanisms. Indeed, interactions between genetic and environmental factors play dominant roles in the neural plasticity that underlies most forms of experience-induced behavioral changes [954].

With respect to the concrete examples presented in this chapter, I focused on the physiological stress response, with some expansion into behavioral control mechanisms. But my main goal was to present the genetic principles that could be applied to any gene–behavior relationship. For example, future chapters will emphasize the importance of oxytocin and its receptor (OTR) for social behavior. Genetic polymorphisms within the OTR gene will be related to individual differences in social behavior, and early life experiences will also be shown to influence the expression of the OTR gene via epigenetic mechanisms. The key take-home message is that genetic variation and environmental variation both contribute to individual variations in behavioral and psychological characteristics, and they often do so by interacting with one another.

CHAPTER

3

Aggressive Behavior

3.1 INTRODUCTION

Aggressive behavior, which is defined as behavior that has the potential of harming another individual, can occur when the interests of two or more individuals are in conflict [678]. A broader definition, which would be applicable to typical occurrences in humans, is that aggression can involve one individual obstructing the goal-directed responses of another individual. Such aggression could include verbal and psychological threats as well as physical assaults. A major emphasis of this chapter will be on the neurobiology of male competitive aggression in animals, wherein two conspecific males fight with one another to gain or retain access to an essential resource. Such aggression has also been referred to as rivalry aggression [970]. This analysis will then be related to certain aspects of aggression in humans.

Differential reproductive success is a major driving force behind evolution through natural selection—those individuals that produce the most offspring will transmit more of their genes to future generations. Those genes and the traits they regulate will then become common in the population. Under certain ecological and social contexts, survival and reproductive success require the acquisition and defense of vital resources, including food, shelter, territory, mates, or young. Since these essential resources can be in short supply relative to the number of individuals seeking them, individuals compete with one another to obtain these resources. Therefore, under these conditions, it is easy to understand how rivalry aggression can be adaptive [100] and why there should be genes that influence the neural circuitry underlying aggressive behavior.

But aggressive behavior can also be costly in terms of reproductive success [1043]. If an animal is wounded during a fight, or spends too much time and energy fighting, this will have consequences for the individual's long-term reproductive success. Aggressive behavior and conflict, therefore, are modulated by the social contexts in which they occur. Within social groups, conspecifics usually do not engage in lethal physical fights. Instead, a dominance hierarchy is established as a result of initial physical conflicts, and then threat displays and low-level physical aggression maintain the hierarchy because animals usually learn not to fight a battle they will lose. Therefore, although aggression has an obvious genetic basis, environmental factors and individual experience also play important roles in its expression and development.

In terms of classifying competitive aggression among males, one division is usually between offensive versus defensive aggression [4,100,670]. In the former, an individual initiates a threat or an attack on another individual

Neurobiology of Social Behavior
http://dx.doi.org/10.1016/B978-0-12-416040-8.00003-1

in order to maintain or obtain possession of a needed resource. In contrast, defensive aggression is usually preceded by other defensive responses such as escape, but if escape is not possible, aggression occurs as a protective response [722]. Both of these types of aggression are provoked by the actions of another, and both have a strong emotional component and can be referred to as affective aggression. Competitive aggression, however, is a dynamic process; during the establishment of a dominance hierarchy in a new social group, a particular individual may alternatively show offensive or defensive aggression, depending on whether he is winning or losing a particular encounter. (Because of the dynamic nature of intermale competitive aggression, investigators usually study "pure" defensive aggression and other defensive responses, such as freezing or escape responses, by exposing animals to predator-related, rather than conspecific, stimuli [153].)

In the analysis that follows, I will focus on the mechanisms underlying offensive aggression or the initiation of aggressive encounters. Importantly, Blanchard and Blanchard [100] have argued that anger-related aggression in humans is likely to be related to offensive aggression in nonhuman animals because the former usually functions to achieve or maintain (in response to a perceived threat) personal control or power over a particular situation, including an attempt to control crucial resources, and in this sense can be seen as similar to dominance and territoriality.

The understanding of the neural mechanisms of adaptive forms of aggression in nonhuman animals may help us understand the biology of pathological forms of aggression in humans. If we map out the neural circuits, neurochemicals, and genes that regulate aggression in appropriate situations, we may begin to understand how disruptions in these systems may lead to excessive and inappropriate violence. Therefore, the initial part of this chapter examines aggression in nonhuman animals, followed by a review of violence in humans. Pathological forms of human aggression include the following two types: intermittent explosive disorder (IED) and the aggression associated with psychopathy. IED is a pathological, reactive, and impulsive form of aggression with a strong emotional component, while the aggression of psychopathy typically lacks emotion and is related to a deficit in empathy [194,377,943]. In this chapter, I am concerned with affective aggression, and I will delay a discussion of psychopathy until the final chapter of this book, which will deal with a broad overview of human sociality, including a review of the neurobiology of empathy.

3.1.1 Aggression in Rodents

Much of the experimental work on the neurobiology of offensive aggression in nonhuman mammals has been conducted on rodents, with some additional work on primates. It will be worthwhile to briefly characterize aggression in these species (see [677,678]).

Under natural conditions, mice live in small reproductive units consisting of a single dominant male who sires virtually all litters, and one or several breeding females with their offspring. Offensive aggression is shown by the dominant male in order to prevent new males from entering his territory and also to expel maturing male offspring from the breeding group once they reach puberty.

Rats, in contrast, are more colonial in nature than mice, and form more elaborate dominant–subordinate relationships. Mature male offspring may remain in the group, where a dominant or alpha male typically prevails in conflicts with rival beta males, which in turn prevail over even more subordinate males. Dominance status results in priority access to resources, including sexually active females. Aggressive behavior is a basic part of life in a colony of rats, and it is particularly intense during the early stages of colony formation, before a stable dominance hierarchy is achieved.

The most important provocation or trigger for offensive aggressive behavior in rats and mice is the intrusion of an unfamiliar adult male into their home territory. Male aggression in these rodents is a tool for excluding other males from the breeding unit. The adaptive significance of such behavior should be obvious. This understanding has given rise to the widely used resident–intruder test of offensive aggression in experimental laboratory research [642]. In this test, a mature male rodent is isolated in a cage for several days or weeks. This procedure results in the cage being established as the male's territory, and the territory owner is referred to as the resident. Subsequently, an unfamiliar male, the intruder, is placed in the cage, and aggressive encounters between the individuals are measured. The resident typically shows offensive aggression and initiates attacks, while the intruder shows escape responses and defensive aggression when attacked. As will be shown, experimental manipulation of the resident is employed to examine the neural mechanisms that operate to either enhance or depress offensive aggression.

3.1.2 Aggression in Old World Monkeys

Rhesus monkeys will serve as an example [597]; these monkeys reside in large troops composed of several female-headed families, each spanning three or more generations of kin, plus numerous immigrant adult males. Females spend their entire lives in the troop in which they were born, whereas virtually all males emigrate from their natal troop around the time of puberty and eventually join a different troop. When males emigrate, they initially join other emigrant males to form bachelor groups before they join a group that contains females. Entrance into a new female family group is a difficult process, and it involves a high rate of aggression with other males already in that group who try to prevent the entry of the novel males.

Within established troops, aggression among monkeys occurs frequently during the course of attaining a higher social status as younger males mature. Disputed access to food and sexual partners underlies most instances of male conflict. Once dominance is established along with a stable hierarchy, rates of physical aggression markedly decrease, being replaced by threat displays, avoidance, and submissive postures. However, the degree to which a monkey displays ritualized low intensity or physical high intensity aggression differs greatly among individuals. Even in a long-term stable group with an established hierarchy, some monkeys seem unable to appropriately modulate or control their aggression—they escalate otherwise low intensity agonistic encounters into fights of excessive severity with heightened risk of injury. An understanding of the neurobiology of such escalated aggression in monkeys may be relevant to the causes of pathological forms of affective aggression in humans.

3.2 NEURAL SYSTEMS OF OFFENSIVE AGGRESSION IN NONHUMAN MAMMALS

3.2.1 The Hypothalamic Aggression Area in Rodents

3.2.1.1 Introduction

Kruk [515] has reviewed the research on what has come to be called the hypothalamic aggression area (HAA). Electrical stimulation of this region in male rats elicits offensive attacks against other rats. In the absence of a rat-like object (mice will be attacked), attack does not occur. The HAA is in the hypothalamic region extending from the anterior hypothalamic nucleus (AHN) through the ventromedial nucleus of the hypothalamus (VMN), with effective electrode sites located in the more lateral parts of these regions. As shown in Figure 3.1, this hypothalamic region lies ventral to the fornix and includes the lateral parts of the AHN,

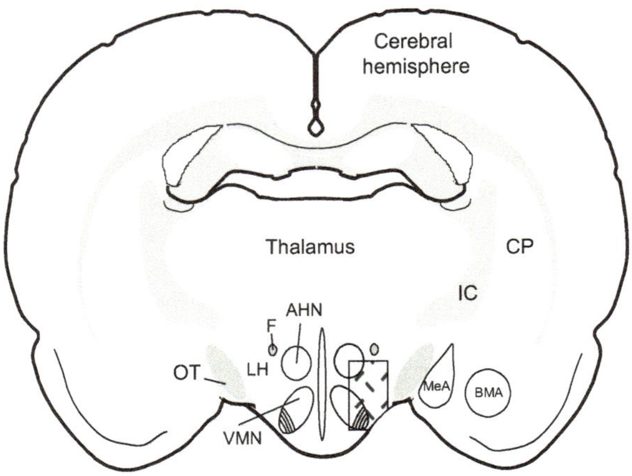

FIGURE 3.1 Based, in part, on research findings reviewed in Kruk [515], this frontal section through the rat brain shows the approximate location of the hypothalamic aggression area, which is outlined by a stipple-filled box. The ventrolateral part of the ventromedial nucleus of the hypothalamus (VMN) is shown with striped lines. Other abbreviations: AHN = anterior hypothalamic nucleus; BMA = basomedial amygdala; CP = caudate/putamen; F = fornix; IC = internal capsule; LH = lateral hypothalamus; MeA = medial nucleus of the amygdala; OT = optic tract. *The brain section shown has been modified from Swanson LW. Brain maps: structure of the rat brain. 2nd ed. Amsterdam: Elsevier; 1998/1999.*

the ventrolateral part of the VMN (VMNvl), and ventral parts of the lateral hypothalamus. Since the effective sites where stimulation leads to aggression occupy a relatively large area, and since electrical stimulation affects both neuronal cell bodies as well as axons of passage, the exact location of the critical "aggression" neurons cannot be ascertained. Kruk [515] points out that the HAA overlaps to some extent with the medial hypothalamic region to which the amygdala projects, and the importance of this anatomy will soon become clear. Importantly, lesions of the HAA have been found to reduce offensive aggression in rats [515,722].

Under natural conditions, brain mechanisms assess the appropriateness of attacking a particular individual in a given situation. Electrical stimulation of the HAA overrides or bypasses this modulatory influence of context, which has given rise to the view that activation of the HAA stimulates the final common motor pathway that produces offensive aggression in the presence of another rat. The following evidence supports this view [515]:

1. When two male rats are placed in an unfamiliar neutral setting outside their home cages, aggression usually does not occur. Stimulation of the HAA induces aggression in this situation.
2. Using a resident–intruder test, the resident male will attack an intruder male without stimulation. However, if the intruder is a sexually receptive female rat, the resident usually engages in sexual behavior with the female and does not attack her. But when the HAA of the resident is electrically stimulated, he will attack the female.

Point (2) is very interesting. If one were to view sexual behavior as a prosocial response and offensive aggression as an antisocial response, one can conclude that under natural conditions social stimuli are differentially processed. Intruder male stimuli promote rejection and aggression, while female stimuli activate acceptance and sexual behavior. Direct stimulation of the HAA overrides prosocial behavior and activates aggression. Deeper insights into some of the relevant mechanisms have been provided by research on mice.

3.2.1.2 VMNvl and Offensive Aggression in Male Mice

Lin et al. [556] have emphasized the importance of one subregion of the HAA for offensive aggression in male mice, the ventrolateral part of the VMN of the hypothalamus. The VMN is a complex region, and in addition to its ventrolateral part, there is a dorsomedial nucleus (VMNdm). While the VMNvl, as part of the HAA, has been associated with offensive aggression, VMNdm has been associated with defensive responses, including freezing and escape, particularly in response to predator cues [153,363,924]. Using the resident–intruder model, Lin et al. [556] reported that selective optogenetic stimulation of VMNvl neurons in residents elicited a rapid attack against intact and castrated male intruders and against sexually receptive females. Without optogenetic stimulation, castrated males were not attacked, and mating behavior was directed toward receptive females. Optogenetic stimulation of VMNdm of residents resulted in escape and freezing defensive responses, which underlines the anatomical specificity of the observed effects. This study also reported that selective inhibition of VMNvl neurons greatly reduced the display of offensive aggression directed at an intact male intruder. Finally, and significantly, when recording from single neurons within VMNvl of resident males, those neurons that were activated by intruder male stimuli during offensive aggression were inhibited by intruder female stimuli during sexual encounters.

It will be worthwhile to explore the ramifications of this study. Social stimuli appear to be differentially routed within the brain so that either prosocial or antisocial responses occur. An intact intruder male activates the VMNvl of the resident mouse, resulting in offensive aggression, while female stimuli result in a depression of this aggression circuitry and presumably activate other neural circuits that promote mating behavior in the resident male. However, optical stimulation of VMNvl, if strong enough, can override the inhibition induced by female stimuli. In rodents, olfactory signals from conspecifics play an important role in social behaviors, including male sexual behavior [432] and male offensive aggression [266,678]. As indicated in Chapter 1, the medial amygdala (MeA) is an important recipient of olfactory input [952], and MeA, in turn, projects to various hypothalamic nuclei via the stria terminalis [154]. Not surprisingly, therefore, MeA output has been shown to promote both male sexual behavior [432] and male offensive aggression [1000]. In a complex anatomical study, Choi et al. [184] have shown that different MeA neurons in male mice respond to different odors and that the neurons that respond to the odor of a predator are separate from those that respond to the odors of a sexually receptive female. In a similar fashion, separate and distinct MeA neurons appear to respond to the odors of an intact male versus a sexually receptive female [761,852]. With this understanding, Figure 3.2 shows a hypothetical neural model of the differential routing of social stimuli that, under normal conditions, results in a resident male reacting differently to a female versus a male intruder, and is based on the facts that MeA neurons project to the medial preoptic area (MPOA), an area important for male sexual behavior [432], and to the VMNvl. The model proposes that intruder male stimuli activate one group of MeA neurons that project to VMNvl within HAA to trigger offensive attack. In contrast, female stimuli activate two populations of MeA neurons, one projecting to the MPOA (stimuli related to the female's sexual receptivity) and the other projecting to VMNvl (stimuli that the female shares with males). MPOA efferents are shown to play a dual role, with one output promoting male sexual behavior, while the other output inhibits VMNvl responses to female odors. Interestingly, my own personal observations of male sexual behavior in rats show that when a sexually *inexperienced* male is presented with a sexually receptive female, his initial responses alternate between aggression and mating. Mating eventually wins out, and once the male is sexually experienced and has copulated, aggression is rarely observed toward a receptive female. Based on the model shown in Figure 3.2, as a result of sexual experience, a learning mechanism may result in the MeA-to-MPOA projection becoming dominant over the MeA-to-VMNvl projection when a male is exposed to female stimuli. However, if the VMNvl were to be optically stimulated, that stimulation would override any inhibition of VMNvl by MPOA, and the male would then attack the sexually receptive female.

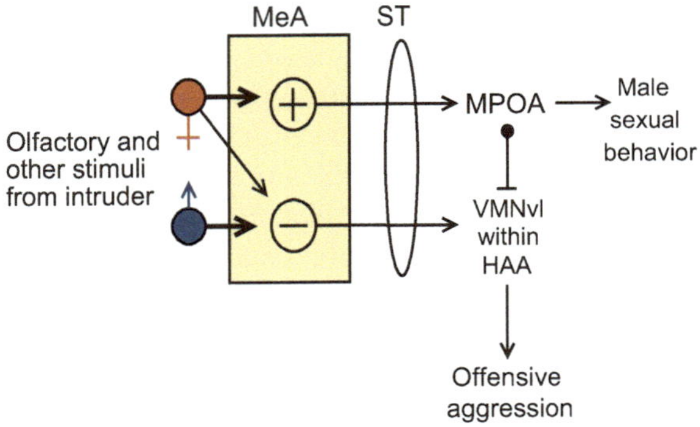

FIGURE 3.2 For rodents, a neural model showing how stimuli from male and female intruders can be differentially processed in the brain of a resident male during a resident–intruder test so that intruder males are attacked while sexually receptive intruder females evoke male sexual responses. Intruder male stimuli activate medial amygdala (MeA) neurons in the resident that project to the hypothalamic aggression area (HAA). Stimuli from sexually receptive females primarily activate a MeA-to-medial preoptic area (MPOA) circuit, which promotes sexual behavior in the resident male. Since some female stimuli might also weakly activate the HAA, the model also proposes that male sexual responses are increased by an inhibitory connection from MPOA to the ventrolateral part of the ventromedial nucleus of the hypothalamus (VMNvl) within the HAA. MeA neurons that respond to stimuli that evoke prosocial responses from the male are shown with a positive valence, while MeA neurons that respond to stimuli that evoke antisocial responses are shown with a negative valence. Axons ending in a bar are inhibitory, and those ending in an arrow are excitatory. The thickness of the connections from male and female stimuli to MeA neurons represents the strength of the connections. ST = stria terminalis.

Some interesting work on prairie voles conforms to aspects of the model shown in Figure 3.2 [343]. Prairie voles are a monogamous vole species, and when a male mates with a particular female, he forms a preference for her. Once a male is attached to his mate, he will show aggressive responses toward intruder males (referred to as mate guarding) and toward intruder novel females, while affiliative responses (huddling) are shown toward the familiar mate. Aggressive responses to strange females are presumably involved in the maintenance of monogamy. Therefore, a complex form of social recognition occurs. Gobrogge et al. [343] used Fos immunohistochemistry to uncover the neural systems that are activated when a bonded male is either exposed to an intruder male, his familiar female partner, a strange female, or no social stimulus at all. The expression of the Fos protein within specific neuron populations is used as an indicator of the neural systems that are active during particular stimulus conditions and/or behaviors. Box 3.1 presents a more detailed discussion of the use of Fos expression in neurobiological research. Gobrogge et al. [343] found that Fos was expressed in MeA neurons when males were exposed to each social stimulus, and there were no significant differences in the number of Fos-expressing cells. Such immunohistochemical data, however, do not inform us with respect to whether the population of MeA neurons activated by familiar female stimuli is the same or different from the population activated by an intruder male or strange female. Some support for the possibility that different MeA neurons, with different projections to the hypothalamus, were activated by stimuli that resulted in aggression versus stimuli that promoted affiliation comes from the analysis of Fos expression in the anterior hypothalamus (presumably part of the HAA) of male voles after exposure to one of the various stimuli. Intruder males and strange females, social stimuli causing aggression, activated a significantly greater number of anterior hypothalamic neurons in the resident male, as measured by Fos expression, than did reintroduction of the familiar female into the resident male's cage.

In Chapter 1, I presented evidence that reward-related (appetitive) stimuli activate positively valent amygdala neurons, while aversive stimuli activate negatively valent amygdala neurons, and that these different amygdala neurons were presumed to give rise to distinct neural circuits controlling either approach or avoidance, respectively. But how should one view the valence of MeA neurons in the context of offensive aggression, where a male shows an

BOX 3.1

FOS AS A MARKER OF NEURON ACTIVITY

The detection of the Fos protein, a product of the *cfos* gene, within neurons is a common procedure used in behavioral neuroscience research to identify neurons and circuits within the brain that are active during certain behavioral situations. This method is based on the assumption that Fos expression is positively related to neuronal depolarization. In an example from aggression research, Haller, Toth, and Halasz [378] analyzed Fos expression in neurons of resident male rats during aggressive encounters. Three groups of rats were exposed to one of the following conditions for 20 min: resident rats that were not exposed to an intruder (baseline control group); resident rats exposed to auditory, visual, and olfactory cues from an intruder that was located behind a perforated Plexiglas barrier, which prevented overt aggression (psychosocial control group); resident rats directly exposed to an intruder, allowing for the occurrence of offensive aggression (aggression group). Approximately 2 h after the initiation of these exposure conditions, the rats were sacrificed and the brains were immunohistochemically processed for the detection of Fos. Since Fos is expressed in the nucleus of a neuron, its detection provides for an excellent spatial resolution of the activated neurons. To the extent that Fos expression is linked to neuron depolarization, the Fos-expressing neurons detected in the rats that engaged in overt aggression represent those neural systems that have become active during aggression. Such research, like the use of fMRI or PET scans in humans, is correlational in nature and simply identifies neurons that may be involved in a particular behavior. Subsequent experimental research, using the Fos data, could then specifically examine whether activity within the identified neurons is actually necessary for the particular behavior under investigation.

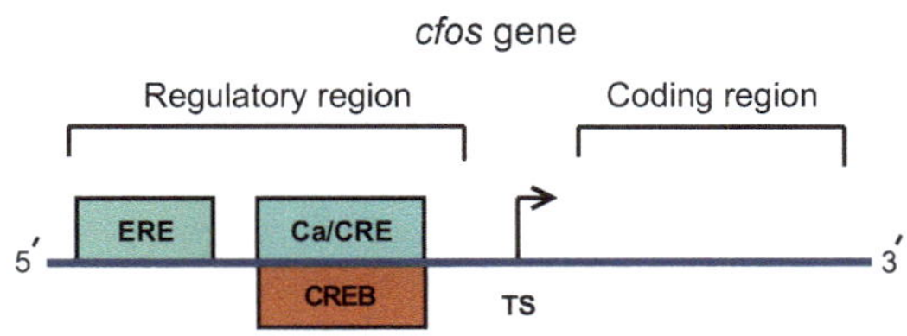

An advantage of the Fos procedure is that it can be combined with other neuroanatomical methods, such as neural tract tracing or immunohistochemical detection of neurotransmitters. With such combined techniques, one could determine whether neurons that express Fos during a certain behavior also contain a particular neurotransmitter and/or also project to a specific brain region. Such an analysis contributes toward understanding the chemical neuroanatomy that underpins a particular behavior.

What are the mechanisms that regulate Fos expression within neurons? Basically, extracellular signals, such as neurotransmitters, neuromodulators, and hormones, produce intracellular biochemical changes that result in transcription factor action within the regulatory region of the *cfos* gene, which stimulates transcription of the gene and Fos protein expression [665,892,1024]. A simplified diagram of the *cfos* gene is shown in the figure above. Although the regulatory region of the gene contains many response elements (DNA response elements are shown in blue), only two are shown, the estrogen response element (ERE) and the calcium and cAMP response element (Ca/CRE). The cAMP response element binding protein (CREB, which is shown in red) is bound to Ca/CRE. As reviewed in Chapter 2, these response elements

Continued

BOX 3.1 *(cont'd)*

would allow Fos expression to be activated by estradiol or by any neurotransmitter or neuromodulator that increases intracellular levels of calcium or cAMP and protein kinase A (PKA; in the figure, TS = transcription start site in the DNA coding region). Clearly, many *excitatory* neurotransmitters increase intracellular calcium levels, which in turn would activate CaMK (calcium modulated kinase). CaMK, by phosphorylating CREB, could then activate Fos expression via Ca/CRE stimulation. This relationship between calcium influx, induced by glutamate activation of the NMDA receptor, for example, and Fos expression has given rise to the view that Fos expression within a neuron represents neuronal depolarization and excitation, which in turn relates to the idea that Fos expression labels or marks neurons that become active during certain behaviors. However, the fact that estradiol can also activate Fos expression, as a result of the E-ER complex being bound to the ERE, indicates that Fos expression can also occur without neuronal depolarization. In addition, many neuromodulators may increase cAMP and PKA within neurons without causing neuronal depolarization, and yet such neuromodulators would be able to increase Fos expression via CREB phosphorylation and CRE activation [414]. Therefore, although Fos expression is typically understood as an indication of neuronal activation, this may not always be the case, and it is certainly possible that the intracellular processes that increase Fos expression may also be part of a cascade of events that actually decrease the neuronal excitability of the neuron within which Fos is expressed.

In relation to this last point, it should be noted that Fos itself serves as a transcription factor. Once Fos is expressed within the nucleus of a neuron, it binds to a response element referred to as an AP1 site, which is located within the regulatory region of certain genes. Such Fos binding then affects the expression of these other genes. Therefore, Fos is sometimes referred to as the product of an immediate early gene (*cfos*), and as a transcription factor it then activates late-responding structural genes that produce proteins that alter the function of the Fos-expressing neuron. Since it would seem to be just as likely for Fos-induced late responding genes to decrease as to increase neuron excitability in the affected neurons, it is probably best to conceive of Fos expression as marking neurons that have become transcriptionally active and modified in some way during a particular behavioral event. Even with this broad definition that is undefined with respect to excitation versus inhibition, Fos expression still marks neurons that have changed in some way during particular behaviors.

approach response and *initiates* an attack that is provoked by an intruder? In Chapter 1, offensive aggression was considered a rejection response mediated by a motivational system that regulates aversion. Indeed, during offensive aggression, the intruder is repelled. In relation to social behavior, and based in part on Figure 3.2, I will try to develop the argument throughout this book that there are distinct prosocial and antisocial circuits, with the former promoting continued contact between conspecifics, while the latter promotes rejection, separation, and the elimination of continued contact between conspecifics. In this broad context, it can be suggested that social stimuli that activate prosocial behaviors stimulate positively valent neural circuits, including those that involve the amygdala, while social stimuli that activate antisocial responses, which include offensive aggression, activate negatively valent neural circuits.

3.2.1.3 *Efferent and Afferent Neural Connections of the HAA in Rats*

Two studies have anatomically explored the larger neural circuitry within which the rat HAA is embedded. Roelling et al. [818] injected an anterograde tracer (PHAL) into the HAA to explore its efferent connections, and Toth, Fuzesi, Halasz, Tulogdi, and Haller [968] injected a retrograde tracer (cholera toxin subunit beta) into this region to uncover its afferent connections. Figure 3.3 presents some of the connections that were uncovered. The HAA receives significant afferent inputs from the amygdala, including MeA and basomedial amygdala (BMA). Such input would allow olfactory, somatic sensory, and gustatory stimuli to reach the HAA [154,760]. Significant inputs also arise from the medial PFC (primarily infralimbic cortex; the orbital PFC did not project to the HAA), lateral septum (LS), MPOA, and the bed nucleus of the stria terminalis (BST). The BST is a telencephalic nucleus that lies dorsal to the MPOA and anterior hypothalamus (see Figures 4.1(A) and 5.6 for the anatomical locations of MPOA, BST, and septal area). Amygdala efferents traveling in the stria terminalis both pass through and synapse in the BST, and BST projections to the hypothalamus, including the HAA, are primarily GABAergic [255,256]. Three important outputs of the HAA were to the brainstem, including the periaqueductal gray (PAG), the mediodorsal thalamus (MDT), and the LS. Therefore, there was a reciprocal relationship between the HAA and LS and between the medial PFC and HAA (although the HAA does not project directly to medial PFC, it could affect the medial PFC indirectly through its projections to the MDT: see Chapter 1). Based on this anatomy, one might propose that amygdala inputs to HAA drive aggressive behavior in response to certain social stimuli and that the output of the HAA to the PAG and other lower brainstem sites provides the route to cranial and spinal motor neurons that control consummatory aggressive responses. The reciprocal relationships between the HAA and the medial PFC and LS suggest a modulatory role of medial PFC and LS on HAA output; it is also possible that medial PFC and LS are sources of processed (polymodal) sensory inputs to the HAA (see Chapter 1). It is significant to note that the major output neurons of the LS are GABAergic and that LS lesions enhance aggression, suggesting that LS may restrain aggression by inhibiting the HAA [887,889]. Because the output of the BST to the HAA is also GABAergic, it might also serve to restrain HAA activity and depress aggression.

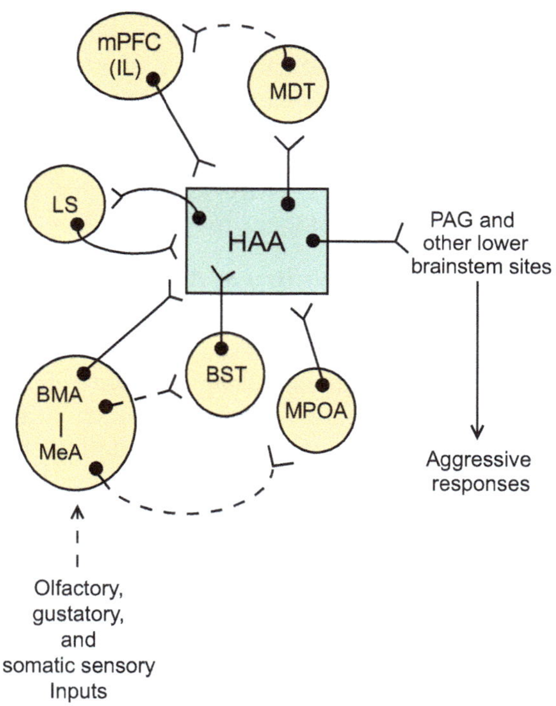

FIGURE 3.3 Some of the efferent and afferent neural connections of the hypothalamic aggression area (HAA) in rats. The connections shown in solid lines were uncovered from the neuroanatomical research of Roelling et al. [818] and Toth et al. [968]. Connections shown in dashed lines are known from the findings of other studies. See text for details. Other abbreviations: BMA = basomedial amygdala; BST = bed nucleus of the stria terminalis; IL = infralimbic part of the medial prefrontal cortex (mPFC); LS = lateral septum; MDT = mediodorsal nucleus of the thalamus; MeA = medial nucleus of the amygdala; MPOA = medial preoptic area; PAG = periaqueductal gray.

Finally, the fact that MPOA projects strongly to the HAA fits with the neural model that was proposed in Figure 3.2.

Similar anatomical work has explored the neural connections of the hypothalamic aggression region in male golden (Syrian) hamsters; this research has also emphasized a core circuitry that includes a MeA-to-HAA-to-PAG circuit [235].

Anatomy, however, does not by itself inform us about function. Additional experiments are needed to provide information on the exact role the described projections play in offensive aggression. As one example, based on the dominant view of PFC function described in Chapter 1, Toth et al. [968] have proposed that medial PFC input to the HAA may restrain or downregulate aggression. They note that the output of medial PFC is composed of glutamate neurons and they suggest that such projections to the HAA might synapse on local GABAergic inhibitory neurons within the HAA, which would then depress the output of the HAA to the PAG. However, this proposal is only conjecture, and it is equally possible that medial PFC neurons synapse directly onto HAA output neurons, providing polymodal sensory inputs that drive their activity and promote aggression (see section on the PFC). In conclusion, experimental behavioral neuroscience research is needed to uncover the exact function, if any, of the anatomically described connections with respect to offensive aggression.

3.2.2 The Neurobiology of Aggression in Cats

Siegel and his coworkers [357,895] have experimentally examined the neural circuitry of affective aggression in cats. Affective aggression includes pronounced sympathetic and behavioral responses such as increased heart rate and blood pressure, growling, hissing, arched back, piloerection, and an aggressive paw strike at nearby objects. A cat may show these responses when provoked by a strange cat or dog that enters its territory. Siegel et al. [895] argue that, unlike rats, there is no clear distinction between offensive and defensive aggression in cats, and the term *affective aggression* is used to refer to aggressive episodes that are a mixture of both responses. In Siegel's laboratory experiments, affective aggression is produced by electrical stimulation of either the anterior medial hypothalamus (AMH) and adjoining VMN or the PAG (rostral dorsolateral part), while the cat is in a cage with an anesthetized rat (aggression is not shown to an anesthetized rat without such stimulation). The threshold electrical current (in mA) required to produce hissing, or the latency of the hiss response at a fixed level of current, is used as a measure of the intensity of the stimulation-induced aggressive behavior. The fact that affective aggression can be elicited by electrical stimulation of AMH/VMN suggests that this region may be homologous to the HAA in rats.

3.2.2.1 Neural Circuitry of Affective Aggression in Cats

Figure 3.4 shows the neural circuits that the work of Siegel and his colleagues have defined as underlying affective aggression. When affective aggression occurs after stimulating the PAG, that stimulation presumably excites PAG neurons with descending projections to lower brainstem–spinal cord circuits that cause the somatic motor and autonomic aspects of affective aggression. The projection from the AMH/VMN to PAG appears to be glutamatergic, which would allow the AMH/VMN to excite PAG affective aggression neurons. Affective aggression elicited by stimulation of the AMH/VMN is blocked by lesions of the PAG or by microinjection of glutamate receptor antagonists into the PAG. Further, after the injection of a retrograde tracer into the affective aggression region of the PAG, many neurons double-labeled with the retrograde tracer and glutamate are detected in the AMH/VMN. Corresponding supportive evidence in rats shows that most of the neurons in the HAA are glutamatergic [426].

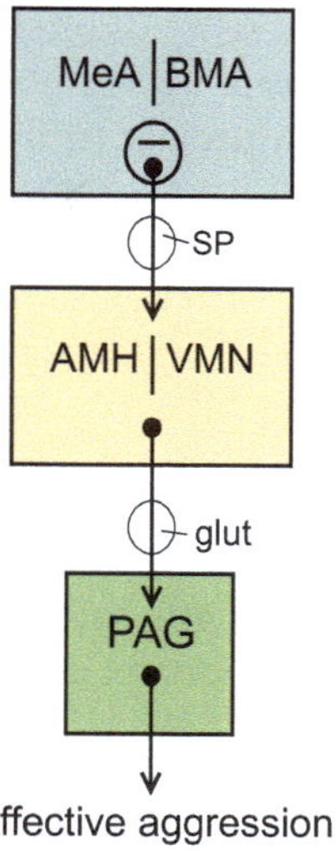

FIGURE 3.4 Based on the research of Siegel and his colleagues [357], the figure shows a description of the neural circuitry that influences affective aggression in cats. Substance P (SP)-containing neurons in the medial and basomedial amygdala (MeA/BMA) activate the anterior medial hypothalamus (AMH) and the ventromedial nucleus of the hypothalamus (VMN). AMH/VMN glutamatergic (glut) projections to the periaqueductal gray (PAG) stimulate circuits that promote affective aggression. MeA/BMA neurons that enhance aggression are shown with a negative valence.

Additional work shows that neurons in the amygdala synapse in the AMH/VMN to modulate their activity and the intensity of the aggressive response. Concurrent stimulation of the medial and BMA facilitates affective aggression elicited by AMH stimulation. An experiment by Shaikh, Steinberg, and Siegel [885] indicates that the peptide neurotransmitter substance P (SP) is present in the MeA-to-AMH projection that modulates affective aggression. They implanted stimulating electrodes into both the MeA and AMH. In separate trials, the AMH was stimulated alone or the two regions were stimulated together (MeA–AMH), and the latency for the hissing response from the time of stimulation was recorded. The latency to hiss was significantly lower after concurrent stimulation of MeA and AMH when compared to the latency to the aggressive response after AMH stimulation alone. Systemic administration of an SP receptor antagonist reversed the facilitation of AMH induced aggression by concurrent stimulation of the MeA. More importantly, when the antagonist was injected directly into AMH, it caused an 80% decrease in MeA facilitation of affective aggression caused by AMH stimulation. In a final anatomical study, they showed that when a retrograde tracer, fluorogold (FG), was injected into the AMH aggression site, many SP+FG double-labeled cells were found in the MeA.

Siegel's experimental work on cats conforms to some of the anatomy in Figure 3.3 that outlines potential neural circuits for offensive aggression in rats. Although the effective stimulation sites in cats appear to be slightly medial to the HAA of rats, the work on both species points to the involvement of a core circuit including amygdala-to-anterior hypothalamus/VMN-to-PAG.

Research on aggression in both cats and rats indicates the importance of the PAG as a premotor region with descending projections that activate aggressive responses. In Chapter 1, I reviewed the role of the PAG in the behavioral immobility associated with the conditioned fear response. As I will subsequently demonstrate, the PAG is a heterogeneous region, and different PAG projections give rise to different types of responses. Further, while nonsocial stimuli can be routed directly from the amygdala (CeAm) to PAG, social stimuli, such as pheromones, are typically routed from the amygdala (MeA and BMA) to the hypothalamus, which then projects to PAG [899].

3.2.3 SP and Offensive Aggression in Rats

In addition to the research on cats, two studies have explored the role of SP in offensive aggression in rats. In the Halasz et al. [374] study, resident males were exposed to an intruder male for 20 min. During the aggression tests, hard bites were defined as bites that included kicking and that startled the intruder, while soft bites were not associated with kicking and did not startle the intruder. One hour later, the resident males were sacrificed and their brains were immunohistochemically processed and analyzed to detect the presence of SP receptors (NK1-type receptor) and Fos. One aspect of this study, therefore, asked whether neurons that are active during offensive aggression contain SP receptors, which would imply that SP might be activating these neurons.

Aggressive encounters increased activation of NK1 receptor-containing cells, as measured by Fos expression, in both the MeA and the HAA. In another part of this study, when resident rats were systemically injected with an NK1 receptor antagonist 20 min before the resident–intruder encounter, the number of hard bites was decreased and the latency to initiate hard bites was increased, but the number of soft bites was not affected. Therefore, the NK1 antagonist did not eliminate offensive aggression but did decrease its intensity.

In the second study, Halasz et al. [375] selectively destroyed NK1 receptor-containing neurons within HAA by microinfusion of SP-conjugated saporin into HAA. Saporin is a ribosome inactivating protein, and when it is conjugated with SP it will selectively lesion neuron populations that express SP receptors. Violent attacks (hard bites with kicking) were dramatically reduced by this selective lesion, while milder forms of aggression (soft bites and offensive threats) remained unaltered. These investigators also reported that only about 5% of neurons in the HAA contained NK1 receptors and that these neurons tended to be located on dendrites in the ventral lateral hypothalamus, just lateral to VMNvl. It is interesting to speculate that some of these NK1 receptors may have been located on the dendrites of VMNvl neurons that extend into the adjoining part of the ventrolateral hypothalamus [650].

To conclude, a population of SP neurons within the MeA that projects to the HAA operates to increase the intensity of offensive aggression. Other MeA neurons that contain either GABA or glutamate also project to the hypothalamus [95]. One possibility is that MeA glutamate input to the HAA is activated by intruder male stimuli and stimulates a baseline level of offensive aggression in the resident male, and that the added activation of SP MeA neurons then serves to enhance the intensity of the offensive attack. As will be reviewed below, the MeA also contains a population of vasopressin neurons that may be involved in aggression. Finally, since the MeA is involved in prosocial as well as antisocial responses, intruder males are presumed to activate negatively valent MeA neurons that give rise to aggression through projections that include the HAA.

3.2.4 Consummatory versus Goal-Directed Mechanisms

As reviewed in Chapter 1, basic motivated behaviors are usually categorized into goal-directed versus consummatory components. One can conceive of consummatory responses as being regulated by brainstem motor mechanisms, while goal-directed responses, which are more complex and variable, are probably integrated and controlled by telencephalic motor systems (see Figure 1.3). With respect to aggression, the MeA-to-HAA-to-PAG pathway probably regulates the reflexive aspects of aggressive behavior via projections to the brainstem and spinal cord, while the voluntary and goal-directed aspects of aggression are likely to be regulated by neural pathways that involve the telencephalon. Examples of goal-directed aspects of offensive aggression would include a resident male approaching a potential intruder from a distance, and a resident male patrolling those areas of his territory where he previously encountered intruders.

Since the PAG not only projects to brainstem motor control systems, but also projects to the VTA (see Chapter 1), it is likely that the mesolimbic dopamine (DA) system may be involved in the proactive or voluntary, rather than the reflexive, aspects of aggression. Also note that the MeA projects to BLA/BMA [154], and that some neurons in these latter regions project to the nucleus accumbens–ventral pallidum circuit (NA–VP) circuit. As previously described, it is possible that DA release into NA allows the VP to respond to aversively labeled stimuli, which then lead to proactive and voluntary forms of offensive aggression. In support of this view, Ferrari, Van Erp, Tornatzky, and Miczek [293] have reported, using in vivo microdialysis recordings, that DA is released into the NA shell of resident rats during aggressive encounters. Further, such DA release was shown to occur in anticipation of a regularly occurring encounter with an intruder. Importantly, Beiderbeck, Reber, Havasi, Bredewold, Veenema, and Neumann [83], using the resident–intruder model, have shown that intense levels of male offensive aggression are mediated by DA release into the NA: microinjection of a DA D2 receptor antagonist in the NA disrupted offensive aggression. Finally, the offensive aggressive responses shown by bonded male prairie voles toward novel intruder females are also dependent upon DA action in the NA [36]. During resident–intruder tests, just which elements of offensive aggression can be conceived as being either consummatory or goal directed remains to be determined.

3.2.5 Vasopressin, the Anterior Hypothalamus, and Aggression

Two important neuropeptides produced by hypothalamic neurons are oxytocin (OT) and vasopressin (AVP for arginine vasopressin), and throughout this book they will be shown to play central roles in the regulation of a variety of social behaviors [991]. These peptides serve as hormones in the periphery and as neurotransmitters within the brain [140,151]. As described in Chapter 1, both peptides are synthesized by neurons in the paraventricular nucleus (PVN) and supraoptic nucleus (SON) of the hypothalamus, and many of these neurons send their axons to the posterior pituitary (neural lobe) where the peptides are released into the blood supply to exert peripheral effects as hormones. Since OT and AVP do not cross the blood brain barrier in significant amounts, peripherally released OT and AVP are unlikely to directly affect brain mechanisms. However, additional neuronal cell bodies in the PVN contain either AVP or OT, and these neurons project to targets within the brain, allowing OT and AVP to act as neurotransmitters or neuromodulators. Although the primary source of OT projections within the brain is derived from the PVN, in many species the sources of AVP projections within the brain are derived not only from the PVN but also from other brain nuclei, including the anterior hypothalamus, MeA, and BST [240,241,824]. AVP will be the focus of this section as it has been shown to be involved in the regulation of offensive aggression.

Another important point to note about AVP is that it is also present in a population of PVN neurons that influence the anterior pituitary (rather than the posterior pituitary), and in this hormonal capacity it potentiates the ability of CRH to stimulate adrenocorticotropic hormone (ACTH) release [419,926]. AVP release into the anterior pituitary, therefore, is able to promote ACTH-induced corticosteroid release from the adrenal cortex. Corticosteroids can enter the brain where they can affect neural circuits regulating behavior. Therefore, keep in mind that AVP may affect behavior indirectly via its influence over the hypothalamic–pituitary–adrenal axis, where is synergizes with CRH to stimulate ACTH release.

There are three types of receptors for AVP, all of which are G-protein-coupled receptors [151]: the V1a, V1b, and V2 receptors. The V1a receptor has a widespread localization in the brain, and the V2 receptor is located in the kidney (where AVP exerts antidiuretic effects). The V1b receptor is strongly expressed in the anterior pituitary where AVP coacts with CRH to stimulate ACTH release. The V1b receptor also has a restricted distribution in the brain [926]. Most studies on the role of AVP in aggression have emphasized its action on V1a receptors within the brain, where AVP exerts excitatory effects on neurons through its action on this receptor type [792].

Ferris's laboratory has presented research on male golden (Syrian) hamsters that shows that AVP action on the V1a receptor in the anterior hypothalamus (presumably homologous to the HAA of rats) stimulates offensive aggression [297,298,300]. In anatomical work it was shown that V1a receptors were located in the anterior hypothalamus. Then, using the resident–intruder model, the effects of injecting AVP agonists or antagonists into the anterior hypothalamus of the resident on his offensive aggressive responses toward an intruder were examined. Microinjection of a V1a receptor antagonist into the anterior hypothalamus suppressed offensive aggression, while direct injection of AVP into the anterior hypothalamus increased the aggressive behavior of the resident. Aggressive behavior was measured by the latency to initiate aggression against the intruder and the number of bites exhibited during a time-limited test, typically 10 min.

Given that AVP acts on the anterior hypothalamus to promote offensive aggression in male golden hamsters, what is the neural source of this vasopressin? Following the model presented for SP, it would seem likely that a MeA-to-anterior hypothalamus AVP pathway might stimulate aggression. However, the distribution of AVP neurons in the golden hamster brain is different from that in other rodents [296,649]—the hamster MeA does not contain AVP neuronal cell bodies, which suggests the possibility that the source of AVP is actually derived from those AVP neurons that reside directly within the anterior hypothalamus [297].

Aggression in males is often linked to testosterone, and castration of males in many species decreases aggression [161,677]. Importantly, the synthesis of V1a receptors in certain brain regions is under the control of gonadal hormones. Using the resident–intruder test, Delville, Mansour, and Ferris [236] showed that castration eliminates offensive aggression in male golden hamsters and that aggression is reinstated with testosterone injections. The binding of radioactive AVP to the ventrolateral hypothalamus, adjoining the VMNvl aggression region, was reduced by castration and reinstated with testosterone treatment. Importantly, AVP injections into this hypothalamic region increased aggression in intact or testosterone-treated males but not in castrated males. In some species, testosterone also influences AVP synthesis. In rats, testosterone increases, and castration decreases, the synthesis of AVP in MeA [151]. Therefore, it is possible for testosterone to affect aggressiveness by influencing the synthesis of both AVP and its V1a receptor.

There appears to be an important sex difference in hamsters with respect to AVP regulation of aggression. In the resident–intruder test, resident female golden hamsters are more aggressive than males, and Gutzler, Karom, Erwin, and Albers [371] have reported that AVP microinjections into anterior hypothalamus decrease, and V1a receptor antagonist injections increase, offensive aggression in females. These findings have given rise to an important question: Does AVP action on the hypothalamus directly affect aggression neural circuitry, or is some other system, such as one involved in anxiety or fear-related responses, being directly affected [14,151]? An alternative, more concrete, interpretation of this sex difference is that different neural circuits are activated by AVP in males and females. Within the anterior hypothalamic region, the areas that are involved in offensive aggression and defensive aggression overlap, although the defensive region tends to be medial to the offensive region [153,426]. Perhaps AVP activates offensive aggression in males and defensive behaviors (escape, freezing) and defensive aggression in females. There may be important sex differences in AVP systems in hamsters [241], and research needs to be aimed at uncovering the details of the neural network on which AVP acts in hamsters and other species. On which neurons in the anterior hypothalamus is AVP acting in male and female hamsters, and what are the inputs and outputs of the relevant AVP-sensitive neurons in each sex?

Similar to the research on male hamsters, research on pair-bonded male prairie voles indicates that AVP acts on V1a receptors in the anterior hypothalamus to promote offensive aggression toward intruder males (mate guarding) and novel intruder females [344]:

1. In a microdialysis study, AVP release into anterior hypothalamus was greater on exposure to a novel female when compared to a familiar female partner. This increased release was positively correlated with the amount of aggression. These results suggest that stimuli that induce antisocial responses cause the release of AVP into the anterior hypothalamus.
2. Intra-anterior hypothalamic injections of AVP caused a naïve (nonbonded) male to show aggression toward a novel female compared to the low aggression shown by control males. In contrast, when a V1a receptor antagonist was injected into the anterior hypothalamus of a bonded male, he showed less aggression toward a novel female than did males that received control injections of cerebrospinal fluid (CSF).
3. Paired-bonded males showed greater numbers of V1a receptors in the anterior hypothalamus compared to naïve nonbonded males, suggesting that once a male forms a preference for a particular female, his anterior hypothalamus becomes more sensitive to AVP.

In these experiments by Gobrogge et al. [344], the source of AVP input to the hypothalamus was not determined, although the investigators suggested that the AVP arose from neurons within the anterior hypothalamus itself [343]. However, since MeA in voles contains AVP neurons [241], this source cannot be ruled out.

In rats, the anterior hypothalamus and the ventrolateral hypothalamus have been shown to contain V1a receptors [460,510]. However, I am not aware of any study that has directly tested whether AVP action on the HAA of rats can promote aggression. Recall that rats not only have AVP-producing neurons in the hypothalamus but that such neuron populations also exist in MeA and BST. In addition, there is an intense innervation of the LS by AVP axon terminals that originate from MeA and BST neuronal cell bodies [240,241]. Veenema, Beiderbeck, Lukas, and Neumann [989] have provided preliminary evidence that AVP release into the LS increases, while AVP release into the BST decreases, male offensive aggression. This finding is important because it shows that vasopressin is not an exclusively antisocial neuropeptide. As will be shown in greater detail in Chapter 6 on Social Attachment, depending on the neural system involved and its site of action, AVP can exert either prosocial or antisocial effects. The Veenema et al. [989] findings coincide with this view by suggesting that AVP action on the BST promotes prosocial interactions by depressing the aggression neurocircuitry, while its putative antisocial action on the LS may activate aggression.

Recall that the outputs of both the LS and BST to the HAA are GABAergic. Allaman-Exertier, Reymond-Marron, Tribollet, and Raggenbass [19] have suggested that AVP action on V1a receptors in the LS excites inhibitory *interneurons* that in turn depress the activity of the GABAergic projection neurons. Such an action on the LS may allow AVP to promote aggression by disinhibiting the HAA. Perhaps vasopressin action on BST V1a receptors directly activates its GABAergic projection neurons, which would dampen aggressive responses. Figure 3.5 provides a summary diagram of some of the chemical neuroanatomy that might underpin male offensive aggression

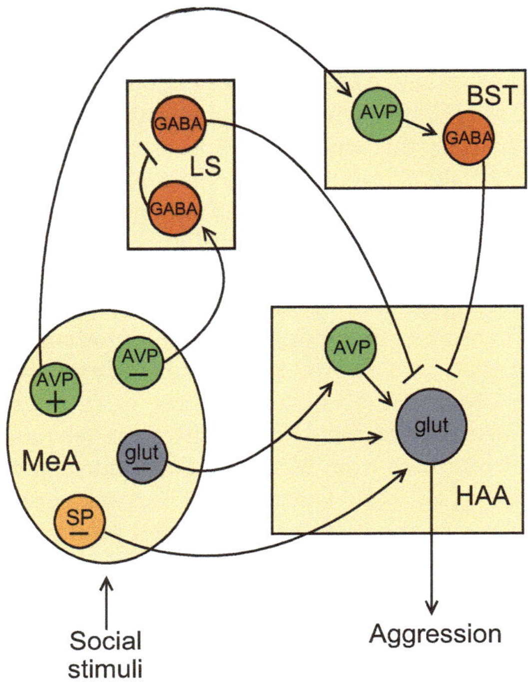

FIGURE 3.5 Chemical neuroanatomy of some of the limbic (amygdala, lateral septum (LS), bed nucleus of the stria terminalis (BST)) and hypothalamic circuits that regulate offensive aggression in male rats. Within the medial amygdala (MeA), neurons with a negative valence respond to social stimuli by promoting aggression, while those with a positive valence are involved in restraining aggression. As shown in the figure, research suggests that vasopressin (AVP)-containing neurons can exert either proaggressive or antiaggressive effects, depending on the particular neural circuits that are active during a social interaction. See text for additional details. Axons ending in a bar are inhibitory, and those ending in an arrow are excitatory. Other abbreviations: glut = glutamate; HAA = hypothalamic aggression area; SP = substance P.

in rats. Note that the AVP circuit comprising some neurons in MeA and BST is proposed to be a prosocial circuit, and one of its functions is to depress aggression. Such a view coincides with research from Goodson's laboratory [349,413]. Although the circuits shown in Figure 3.5 are speculative and based on preliminary evidence, they emphasize the point that a complete understanding of the neural circuits regulating social behavior necessitates an understanding of the microcircuitry within the critical brain nuclei, and a delineation of the sites within this circuitry at which specific neurochemicals act.

Although most research on the stimulatory effects of AVP on aggression have suggested an action on V1a receptors in the hypothalamus, there is some work that indicates that AVP action of the V1b receptor may also stimulate aggression [926]. In particular, in resident–intruder aggression tests, male mice with a knockout mutation of the V1b receptor (V1bR −/−) display fewer attacks and have longer attack latencies than V1b wild-type (+/+) controls. Perhaps AVP action on the V1b receptor promotes aggression because of an action on the anterior pituitary where AVP acts with CRH to stimulate ACTH release. In support, research shows that under certain situations corticosteroid action on the brain potentiates male offensive aggression in rats [516,644].

Finally, although vasopressin action on the brain has prominent stimulatory effects on male aggression, there is evidence that OT action on the oxytocin receptor (OTR) has prosocial effects that depress male offensive aggression—male mice with a knockout mutation of the OTR show increased levels of offensive aggression in resident–intruder tests [247,957], and male rats that are highly aggressive synthesize less OT in PVN than do less aggressive rats [148].

3.2.6 The Prefrontal Cortex and Aggression

Prefrontal cortex (PFC) projections to the amygdala are often described as being involved in a top–down regulation of a variety of basic appetitive and aversive motivational processes, which would include aggression, and a dominant proposal is that PFC output to the amygdala might act to suppress amygdala–hypthothalamic–brainstem circuits that control aggressive responses, providing the basis for a cognitive or executive control of aggressive impulses based on rational decision making (see Chapter 1). More broadly, however, the PFC is a complex region and is composed of the medial PFC and orbital PFC, and each of these major areas is composed of subareas. The PFC not only interconnects with the amygdala but also with the nucleus accumbens and the hypothalamus. This diverse connectivity may allow the PFC to influence aggression in a variety of ways, depending on the particular circuits that are active.

Two studies on rats have examined the role of the PFC in offensive aggression; they suggest different roles for the orbital and medial PFC. In an early study, De Bruin, Van Oyen, and Van De Poll [226] described the effects of electrical lesions of the orbital PFC on intermale aggression in rats. Two male rats were put in a large cage separated by a partition for 24 h. The partition was then removed and their social interactions were recorded. Control or orbital PFC lesioned rats were paired with normal males. This type of aggressive situation is somewhat similar to the resident–intruder model, except that each rat could be considered both a resident and an intruder. The observed aggressive encounters included Lateral Threat, which is a broadside or lateral approach toward an opponent and includes a sideways pushing response. Lateral Threat is not overly aggressive and is considered a threat posture. More intense overt aggression included the Keep Down response, where one male keeps the other down with its front paws while the other is lying on its back, and the Chase response, where one animal pursues the other fleeing animal. The results indicated that the orbital PFC lesioned rats showed more Keep Down and Chase responses than did the controls. Also, following a mutual boxing encounter, where both animals stand upright with their paws on the opponent, the controls tended to follow this act with a Lateral Threat, whereas orbitally lesioned rats were more likely to exhibit the overt aggressive responses of Chasing or Keeping Down. The orbital lesions were centered in the lateral orbital (LO) and anterior insular regions (see Chapter 1). These data point to a dissociation between threat and aggression as a result of lesioning the orbital PFC, with the balance shifting toward overt aggression in the lesioned rats, which suggests an intensification of aggression. These findings have been confirmed more recently with excitotoxic amino acid lesions of the orbital PFC, although the resultant increase in aggression was not as pronounced, presumably because the rats were not tested in a territorial setting [844]. With respect to the potential restraining role of orbital PFC output on aggression, there is also evidence that this is the case in rhesus monkeys [583].

Halasz, Toth, Kallo, Liposits, and Haller [373] also examined the involvement of the PFC in the offensive aggressive responses of rats. Using the resident–intruder model, male rats were allowed to engage in aggressive behavior for 20 min. The resident was subsequently sacrificed, and the expression of Fos within various parts of the prefrontal cortex was quantified and compared to that of control rats that did not engage in aggressive behavior.

There are two major types of neurons in the PFC: glutamatergic pyramidal neurons that provide the major efferent projections to brain regions outside the PFC, and local circuit neurons that remain within the PFC. Most of these local circuit neurons are GABAergic and restrain the output of the pyramidal neurons. An important aspect of the immunohistochemcial analysis performed by these investigators was to process the brain tissue in order to colocalize Fos expression with protein markers for pyramidal cells or GABAergic cells. Their major findings can be summarized as follows:

1. Aggressive encounters increased Fos expression above that seen in controls within the following prefrontal cortical regions: infralimbic (IL), medial orbital (MO), ventral orbital (VO), and lateral orbital (LO).
2. When comparing the IL PFC with the LO PFC during aggressive encounters, it was

found that most of the IL cells that contained Fos were pyramidal cells, while in the LO cortex, about half of the Fos-activated cells were pyramidal cells, while the remainder were GABAergic.

The authors concluded that although it is generally believed that pathological or intense aggression is caused by prefrontal deficits, their findings show that most Fos-expressing cells during aggression are prefrontal pyramidal cells, suggesting that aggression is associated with strong output from the PFC. Correlational studies like this one, however, are difficult to interpret. Perhaps the activation of pyramidal cells in the PFC actually served to restrain aggression, and that if such activation had been suppressed, aggression might have been more intense than that which was actually observed.

A fuller appreciation of these results also requires a consideration of anatomical details, and this study did not investigate the neural structures to which the Fos-expressing pyramidal neurons projected. I have already reviewed the data, which shows that the medial PFC (IL) has strong projections to the HAA in rats, while the VO and LO PFC do not project to the HAA (see Figure 3.3 and [968]). In contrast, both the orbital PFC and the medial PFC project to the amygdala. Given this background, the following scenario is possible: some proportion of the activated neurons in the medial PFC may have provided polymodal sensory inputs to the HAA in order to stimulate aggression, which coincides with the suggestion of Halasz et al. [373]. In contrast, the LO cortex, perhaps via projections to the amygdala, might depress overt aggressive responses. During intense aggression, this effect of the LO pyramidal projection neurons might be partially restrained by the activation of GABAergic interneurons within the LO cortex. This kind of analysis also fits with the results of the lesion study by De Bruin et al. [226], where it was found that LO lesions increased aggression in rats. However, such hypothesizing about the results of the Halasz et al. [373] paper is pure conjecture, and real answers will require more direct neurobiological investigation.

3.2.7 Serotonin and Aggression

Serotonin (5-HT), a monoamine neurotransmitter, is significantly involved in the regulation of male offensive aggression, although its exact role is controversial. A dominant view is the "5-HT deficiency hypothesis" of heightened aggression, which proposes that intense levels of aggressive behavior are associated with low 5-HT neural transmission within the forebrain [723]. According to this view, 5-HT normally acts to restrain aggression. In contrast, others have argued that this hypothesis is obsolete and that 5-HT action in the forebrain actually stimulates aggression [224,520]. In a meta-analytical review of studies that examined the effects of *systemic* pharmacological manipulation of 5-HT levels, most studies found that drugs that increased serotonin neural transmission decreased aggression, but a minority of studies did find a positive relationship between 5-HT and aggression [162].

The serotonin neural system is a complex one, and before a critical evaluation of its role in aggression can be undertaken, a review of its neuroanatomy and neurochemistry is necessary. 5-HT neuronal cell bodies are located in the midbrain, pons, and medulla [59,326,376,544]. These include the dorsal and median raphe nuclei located along the midline of the midbrain and pons (see Figure 1.1), and the caudal medullary raphe neurons. While the medullary raphe neurons have major projections downstream to the spinal cord, the other 5-HT cell groups have widespread ascending projections that include those to the hypothalamus, PFC, and amygdala, areas that are critically involved in aggression. The issue to be resolved is whether 5-HT input to the forebrain decreases aggression, increases aggression, or has both influences, depending on the details of the neural circuitry that is engaged. The latter is possible since different

5-HT neurons have different inputs and outputs [59,376], and because 5-HT can have different neuronal effects within a particular target region depending on the type of receptor it acts upon and whether such action is exerted on projection neurons or interneurons within the target region.

The widespread forebrain projections of 5-HT neurons are obviously related to the diverse functions that they influence, and it is likely that distinct serotonin circuits regulate specific functions. Highly relevant are the fact that in addition to being involved in aggression, 5-HT neural systems influence arousal, attention, and sleep–wake cycles [661], stress reactivity, and fear-related processes [416]. Therefore, when the effects of 5-HT manipulations on aggression are examined, one can ask whether the observed effects are the result of a direct influence on aggression neural circuits or an indirect effect. For example, lowering 5-HT transmission within certain circuits may decrease aggression (suggesting a positive relation between aggression and 5-HT) not because aggression circuits are directly affected but because the animal is sedated. Similarly, to the extent that offensive aggression decreases when fearfulness increases [101], and since 5-HT can have anxiolytic or anxiogenic effects depending on its site and mechanism of action [551,574], 5-HT may indirectly influence aggression by affecting fearfulness. Given this complexity, this section will focus on serotonin action within the aggression neural circuits that were previously described, with the goal of uncovering potentially direct effects of 5-HT on aggression.

The synthesis, storage, and inactivation mechanisms regulating 5-HT are shown in Figure 3.6 [137,416,493]. The amino acid precursor for 5-HT is tryptophan, which is converted to 5-hydroxytryptophan (5-HTP) by the rate-limiting enzyme tryptophan hydroxylase (TPH). 5-HTP is then converted to 5-HT by decarboxylation. Monoamine oxidase A (MAOA) breaks down 5-HT into an inactive metabolite, 5-hydroxyindole acetic acid (5-HIAA), both within the axon terminal for any synthesized 5-HT that has not yet been stored in synaptic vesicles and also in the synaptic cleft for 5-HT that has been released. The former process regulates the amount of 5-HT incorporated within synaptic vesicles and the latter process terminates the action of synaptically released 5-HT. Terminating the action of synaptically released

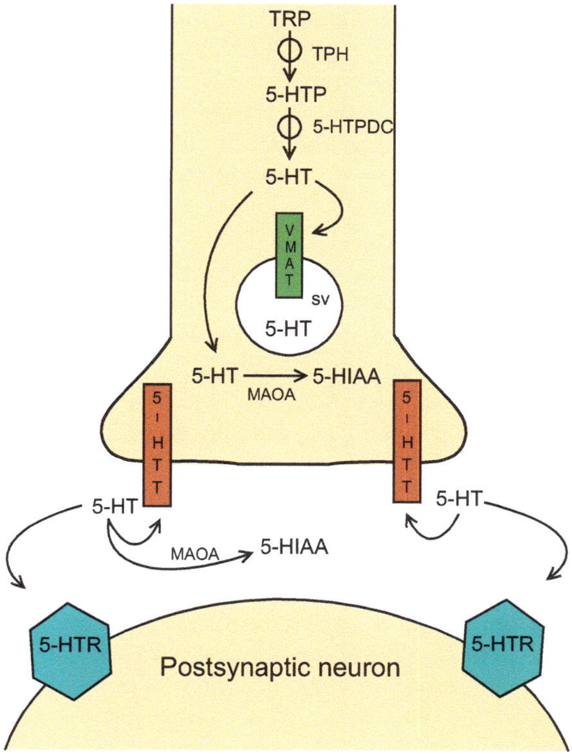

FIGURE 3.6 The serotonin (5-HT) synapse. 5-HT is synthesized from the amino acid precursor tryptophan (TRP) and then stored in synaptic vesicles (SV) within the axon terminals of a 5-HT-containing neuron. Monoamine oxidase A (MAOA) breaks down 5-HT into an inactive metabolite, 5-hydroxyindole acetic acid (5-HIAA), both within the axon terminal for any 5-HT that has not been stored in SVs and within the synaptic cleft for 5-HT that has been released. The termination of the action of synaptically released 5-HT also occurs via reuptake through the serotonin transporter pump (5-HTT). Other abbreviations: 5-HTP = 5-hydroxytryptophan; 5-HTPDC = 5-hydroxytryptophan decarboxylase; 5-HTR = 5-HT receptor; TPH = tryptophan hydoxylase; VMAT = vesicular monoamine transporter.

5-HT also occurs via reuptake through the serotonin transporter pump (SERT), also referred to as the 5-HT transporter (5-HTT).

There are at least 15 different 5-HT receptors, complicating the study of 5-HT effects on behavior due to the resultant diversity of physiological effects that 5-HT action can have [384,416,493,648,653,856]. These receptors have been divided into seven major classes (5-HT1-7), and most are G-protein coupled receptors. With respect to aggression research, most attention has been devoted to the 5-HT1A, 5-HT1B, and 5-HT2A receptors, so their mechanisms of action will be described in detail (see Figure 3.7). 5-HT1A receptors are localized on the cell bodies and dendrites (somatodendritically) of 5-HT neurons in the brainstem where they serve as inhibitory autoreceptors. 5-HT1A receptors are also located on the dendrites and cell bodies on many nonserotonergic neurons. The action of 5-HT on the 1A receptor causes an inhibitory hyperpolarization as a result of G-protein activation of K^+ efflux. Therefore, activation of 5-HT1A autoreceptors on raphe 5-HT neurons inhibits those neurons and suppresses 5-HT release, while 5-HT action on 5-HT1A receptors on nonserotonergic neurons, such as those in the prefrontal cortex, for example, would inhibit neurons located there [789]. 5-HT1B receptors can be categorized as either presynaptic autoreceptors or presynaptic heteroreceptors. The former are located on the axon terminals of serotonin neurons, while the latter are located on the axon terminals of nonserotonin neurons. The action of 5-HT on the 1B receptor depresses the activity of voltage-gated Ca^{2+} channels, causing a presynaptic inhibition that decreases exocytosis and neurotransmitter release. Therefore, as a presynaptic autoreceptor, 5-HT1B activation decreases 5-HT release, while as a presynaptic heteroreceptor, its activation blocks the release of some other neurotransmitter. Finally, the 5-HT2A receptor is localized on the cell bodies and dendrites of non-5-HT neurons, where 5-HT exerts excitatory depolarizing effects.

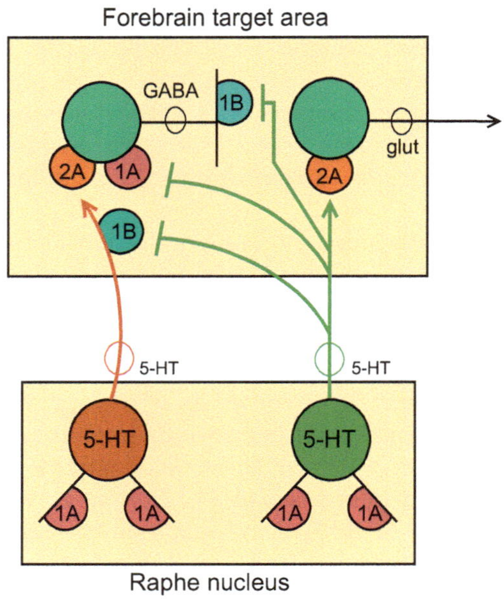

FIGURE 3.7 The disparate effects of serotonin (5-HT) action on 5-HT1A (1A), 5-HT1B (1B), and 5-HT2A (2A) receptors. The 5-HT2A receptor is typically located on the soma or dendrites of non-5-HT neurons, where 5-HT exerts excitatory effects. 5-HT1A receptors can be localized on the soma or dendrites of either 5-HT neurons (in the raphe nucleus), or non-5-HT neurons. 5-HT exerts inhibitory effects at the 1A receptor. The 5-HT1B receptor is a presynaptic receptor that can be located on the axon terminals of either 5-HT or non-5-HT neurons. 5-HT exerts presynaptic inhibition (depresses neurotransmitter release) at the 1B receptor. The figure shows two 5-HT neurons in the raphe nucleus that project to a forebrain target area. The forebrain target is shown as containing a glutamatergic (glut) projection neuron and an inhibitory GABAergic interneuron. The 5-HT neuron shown in red depresses the output of the glutamatergic projection neuron because of its stimulatory effect on the 2A receptor located on the cell body of the GABAergic inhibitory interneuron. The 5-HT neuron shown in green promotes the output of the projection neuron because it: (1) directly activates the projection neuron as a result of 5-HT release onto the 2A receptor; (2) inhibits the release of 5-HT from the red serotonin neuron and GABA release from the GABA interneuron via an action on the 1B presynaptic receptors; (3) inhibits the GABAergic interneuron via 5-HT's action on the somatodendritic 1A receptor. Axons ending in a bar are inhibitory, and those ending in an arrow are excitatory.

In order to critically evaluate the role of 5-HT in aggression, correlational studies will be reviewed first, followed by experimental research. Rhesus monkeys have been examined in natural social groups, and the following correlation has been found [597]: among male rhesus monkeys, the number of unrestrained high intensity aggressive acts covaries inversely with CSF 5-HIAA concentrations. (Low intensity aggression [ritualized threats involved in establishing and maintaining dominance hierarchies] does not correlate with 5-HIAA levels.) Similar findings have been reported for other mammalian species [642]. The negative correlation between 5-HIAA levels in CSF and intense aggression has been interpreted to mean that low central 5-HT neural transmission is associated with severe and unrestrained aggression and is based on the view that low levels of 5-HIAA are a reflection of low levels of synaptic release of serotonin. An alternative mechanism is possible, however: a low functioning MAOA enzyme would result in higher 5-HT levels and lower 5-HIAA levels, and if this were the process regulating the negative correlation between aggression and 5-HIAA, it would support a positive relationship between 5-HT neural transmission and aggression. Therefore, a clearer picture would emerge if one correlated brain 5-HT levels with aggression. This approach was taken by Van Erp and Miczek [985], who used in vivo microdialysis to measure changes in extracellular 5-HT levels in the medial PFC (prelimbic cortex (PL) and IL areas) of resident rats during aggressive encounters with an intruder. 5-HT levels were found to decrease during the aggressive encounters, supporting the view that a decrease in medial PFC 5-HT transmission promotes the occurrence of aggressive behavior in male rats. In related findings in mice, highly aggressive strains were found to have lower levels of 5-HT in PFC than less aggressive strains [156,157]. Recall that Halasz et al. [372] reported increases in Fos expression in medial PFC pyramidal neurons during offensive aggression, and it was suggested that such neurons might provide polymodal sensory inputs to the HAA to promote aggression. Since medial PFC pyramidal neurons contain inhibitory 5-HT1A receptors [789], it is possible that 5-HT depresses these neurons and therefore depresses aggression, while a fall in 5-HT would enhance aggression.

An interesting study by Haller, Toth, and Halasz [378] examined the expression of Fos within TPH raphe neurons of resident male rats during aggressive encounters. Three groups of rats were employed: resident rats that were not exposed to an intruder (baseline control group); resident rats exposed to auditory, visual, and olfactory cues from an intruder who was located behind a perforated Plexiglas barrier, which prevented overt aggression (psychosocial control group); resident rats directly exposed to an intruder, allowing for the occurrence of offensive aggression (aggression group). For rats in the aggression group, a significant negative correlation was observed between the number of Fos-expressing TPH neurons and the number of resident attacks against the intruder during the 20 min test. This finding would be consistent with the 5-HT deficiency hypothesis of intense aggression.

Turning to experimental research, the systemic administration of 5-HT1A and 1B receptor agonists has long been known to depress offensive aggression in resident male rodents exposed to an intruder [224,723]. These systemic administration studies are extremely difficult to interpret because 5-HT1A and 1B receptors serve as both autoreceptors on 5-HT neurons and as receptors on nonserotonergic neurons. Since 5-HT action on each receptor exerts inhibitory effects (inhibitory postsynaptic potentials at the 1A receptor or presynaptic inhibition at the 1B receptor), the question that needs to be resolved is whether aggression decreases because of a dominant autoreceptor effect or because of an inhibitory effect on nonserotonergic neurons. The former outcome would conflict with, while the latter would support, the 5-HT deficiency hypothesis. Even when relatively selective agonists for the 5-HT1A *autoreceptor* [224] are administered systemically, the resultant effects on aggression are hard to interpret. Within the raphe nuclei, there are 5-HT projection neurons that project to the forebrain, and there are also 5-HT interneurons that act to inhibit these projection neurons [450]. Since both sets of neurons contain 1A autoreceptors, if inhibition of interneurons were the dominant effect of a particular drug, any influences on aggression would be associated with increased release of 5-HT into the forebrain. In contrast, if the dominant action of the 5-HT1A autoreceptor agonist were to inhibit the raphe projection neurons, any effects on aggression would be associated with decreased release of 5-HT into the forebrain.

There is some evidence that 5-HT1A autoreceptors exhibit a heightened sensitivity to 5-HT in highly aggressive stains of mice when compared to less aggressive strains [157]. This heightened sensitivity was associated with a net decrease in the output of 5-HT projection neurons to the forebrain, since 5-HT levels in PFC were lower in the highly aggressive strains, supporting the 5-HT deficiency hypothesis of intense aggression.

There are several mouse knockout mutation models (null mutations) that, for the most part, support an inhibitory role for 5-HT in the control of aggression. *Pet-1* is a gene that regulates the synthesis of the Pet-1 transcription factor, which is important for the development of 5-HT neurons [559]. Hendricks et al. [402] investigated the function of *Pet-1* by generating *Pet-1* deficient mice. Genetic deletion of *Pet-1* disrupted the development of the central 5-HT system, and adult mice with this null mutation were hyperaggressive. Counts of 5-HT-immunoreactive cell bodies indicated a 70% deficiency in the rostral and caudal parts of the raphe brainstem system in adult null mice. No effect on other monoaminergic systems was evident. Further, confocal microscopic imaging showed that the density of 5-HT-immunoreactive axons in the forebrain was greatly diminished in the null mice. Using the standard resident–intruder test for intermale aggression, the null mice, when compared to controls, attacked the intruder with a very short latency, initiated more attacks, and produced significantly more bite wounds. These results are supported by research that shows that a null mutation of the *Tph2* gene, which eliminates the synthesis of TPH, and therefore serotonin, within the brain, results in increased offensive aggression in the mutant mice [543]. Support for the view that 5-HT action on the 5-HT1B receptor suppresses aggression comes from the work of Saudou et al. [859] and Bouwknecht, van der Gugten, Hijzen, Maes, Hen, and Olivier [115], who reported that a null mutation of the 5-HT1B receptor gene also resulted in an increase in offensive aggression in male mice.

These genetic studies are important, but because serotonin's effects on so many neural systems are affected by these genetic deletions, the potential sites and mechanisms through which serotonin influences aggression cannot be determined. Also, since the null mutations existed throughout the lifespan of the genetically manipulated mice, some of the observed effects may have been the result of influences on brain development rather than due to modifying the action of 5-HT on adult neural circuits [544]. Recent important research by Audero et al. [45] has eliminated some of these difficulties. These investigators created a transgenic mouse strain that expressed 5-HT1A receptors *exclusively* within raphe serotonin neurons. The administration of a 5-HT1A receptor agonist to these male mice in adulthood was found to decrease the neuronal firing of raphe serotonin neurons. In addition, such agonist treatment also increased the offensive aggression of these transgenic mice in comparison to control mice. Therefore, when stimulation of 5-HT1A receptors can *only* act on autoreceptors to inhibit 5-HT release into the forebrain, aggression is increased, which is consistent with the deficiency hypothesis.

More direct approaches for evaluating the 5-HT deficiency hypothesis would be to examine the effects on aggressive behavior of neuropharmacological manipulation of 5-HT transmission within selected neural sites known to be involved in aggression. In particular, if 5-HT1A agonist microinjection into a forebrain site known to be involved in aggression were to cause a decrease in offensive aggression, this would provide good evidence for the 5-HT deficiency hypothesis of heightened aggression: since 5-HT1A autoreceptors are only located on 5-HT neurons in the brainstem, such an effect would be the result of inhibiting nonserotonergic neurons in the forebrain.

The available rodent studies have provided evidence that 5-HT action on the 1A receptor in the AHN or on either the 1A or 1B receptor in the PFC can decrease aggression, and that the action at the level of the AHN antagonizes the proaggressive effects of vasopressin. Using radioactive ligands, Ferris, Stolberg, and Delville [300] reported that vasopressin V1a, 5-HT1A, and 5-HT1B binding sites were present in the AHN of hamsters. Since 5-HT1A receptors cannot be autoreceptors in this location, their presence in the AHN is most likely on AHN cell bodies or dendrites. They found that AVP microinjection into the AHN promoted enhanced aggression in resident hamsters, and this aggression was blocked by coinjection of a 5-HT1A receptor agonist into AHN. These results suggest that serotonin action on the 5-HT1A receptor suppressed aggression by inhibiting HAA projection neurons that are stimulated by vasopressin (see Figure 3.8).

In the review of the role of the PFC in aggression, one suggestion was that the output of the orbital PFC suppresses aggression by depressing amygdala input to the HAA, while the output of the medial (infralimbic) PFC might promote aggression via a direct stimulatory effect in the HAA. Several studies in mice have shown that the microinjection of 5-HT1A or 1B receptor agonists into the orbital PFC suppresses male offensive aggression, and that injection of selective 1B agonists into the infralimbic medial PFC has a similar suppressive effect [168,223,278]. The inhibitory effects of 5-HT1A agonists in the orbital PFC on aggression are the most consistent with the serotonin deficiency hypothesis, since these results indicate that an increase in 5-HT suppresses aggression. 5-HT1A receptors are located on both glutamatergic pyramidal projection neurons and GABAergic interneurons in the PFC [31,789,790]. Given that the orbital PFC projects strongly to the amygdala [622], one possibility is that specific 5-HT circuits act on the 1A receptor in the orbital PFC to suppress inhibitory interneurons that synapse on pyramidal neurons. Through this process of disinhibition, these active pyramidal glutamatergic neurons may then project to and stimulate GABAergic inhibitory interneurons in the amygdala, which would result in an inhibition of amygdala input to the HAA (see Figure 3.8). With respect to the medial PFC, although it has not been specifically tested whether activation of 5-HT1A receptors at that site suppresses aggression, since the pyramidal output neurons in the medial PFC contain inhibitory 5-HT1A receptors [789], it is possible that 5-HT also acts there to inhibit aggression. In other words, 5-HT may act in orbital PFC to promote its output and in infralimbic medial PFC to depress its output, with both actions favoring a decrease in aggression. These proposals are depicted in Figure 3.8.

Although research on the role of 5-HT in aggression is still controversial, the overall evidence seems to support the idea that lowered levels of forebrain 5-HT, with a resultant decrease in 5-HT1A and 1B receptor stimulation, are associated with an increase in intense, unrestrained aggression, providing good support for the 5-HT deficiency hypothesis. However, there is also some evidence that increases in serotonin may increase aggression, and this may be due, in part, to an action on 5-HT2A receptors. Shih et al. [893] and Juarez et al. [464] have reported that systemic treatment with a 5-HT2A receptor antagonist decreased aggression in a highly aggressive mouse strain. It is worth emphasizing that brainstem raphe 5-HT neurons have diverse inputs and outputs, and that particular 5-HT forebrain projections can act on a variety of receptors. It appears that selective afferent excitation of particular raphe 5-HT neurons that project to either AHN or PFC to act on the inhibitory 1A or 1B receptors depresses aggression, while activation of other raphe 5-HT

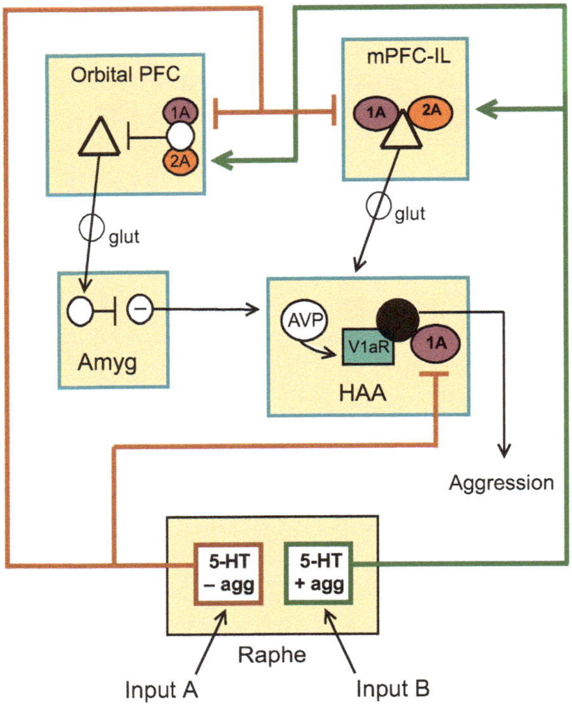

FIGURE 3.8 The potential inhibitory and excitatory effects of serotonin (5-HT) on the neural circuitry that regulates aggressive behavior. Amygdala (Amyg) projection neurons with a negative valence promote aggression by stimulating the hypothalamic aggression area (HAA). The infralimbic cortex within the medial prefrontal cortex (mPFC-IL) is also shown as stimulating aggression via a glutamatergic (glut) projection to HAA. The orbital PFC is shown as depressing aggression via glutamatergic projections to inhibitory interneurons within the amygdala. Within the raphe nucleus, two 5-HT neurons with distinct neural connections are shown as either depressing (5-HT – agg; red) or enhancing (5-HT + agg; green) aggression. The 5-HT neuron shown in red is driven by input A and depresses aggression by: (1) inhibiting HAA and mPFC-IL projection neurons via an action on 5-HT1A receptors (1A); (2) inhibiting inhibitory interneurons within the orbital PFC via an action on the 1A receptor. Such an effect would enhance the output of orbital PFC projections to the amygdala. The 5-HT neuron shown in green is driven by input B and enhances aggression by stimulating mPFC-IL projection neurons and orbital PFC inhibitory interneurons via an action on 5-HT2A receptors (2A). Note that a shift from a dominant effect of 5-HT on the 1A receptor to a dominant effect on the 2A receptor would cause a shift from a hypoaggressive to a hyperaggressive state. Axons ending in a bar are inhibitory and those ending in an arrow are excitatory. Other abbreviations: AVP = vasopressin; V1aR = vasopressin 1a receptor.

neurons by a separate set of neural inputs stimulates forebrain projections that synapse on excitatory 5-HT2A receptors to promote aggression. Since pyramidal cells in medial PFC not only contain 5-HT1A receptors but also 2A receptors [789], to the extent that these pyramidal neurons project to and stimulate the HAA, one can envision how 5-HT could either depress or promote aggression, depending on the circuits activated and the receptors affected (see Figure3.8). Perhaps certain social situations are associated with an activation of 5-HT input to PFC 5-HT1A receptors, while other circumstances result in a stimulation of serotonin input to 5-HT2A PFC circuits (see [85]), with a resultant shift in the effects of 5-HT from antiaggressive to proaggressive. Furthermore, GABAergic inhibitory neurons in PFC also contain 5-HT2A receptors [31,790], making it possible for serotonin action on such neurons in the orbital PFC to promote aggression (see Figure 3.8).

A final point to consider is that 5-HT's inhibitory effects on aggression could be viewed as part of a neural network that indirectly promotes prosocial behavior, since conspecifics would be less likely to provoke antisocial responses. Significantly, in a double-labeling immunohistochemical analysis of Fos expression in dorsal raphe (DR) neurons that projected to the AHN, Gobrogge and Wang [345] have reported higher activation in this circuit when long-term bonded male prairie voles interacted with their mated partner and weak activation when these males interacted with a novel female. The familiar partner provoked affiliative responses, and the strange female provoked aggressive responses from the male.

3.3 NEUROBIOLOGY OF IMPULSIVE OR AFFECTIVE AGGRESSION IN HUMANS

3.3.1 Introduction

Aggression neural circuits, and the associated genetic machinery that underpins the development of these circuits, evolved because under certain ecological conditions aggressive behavior is adaptive and enhances reproductive success. Therefore, neural circuits for aggression are present in all animals, including humans. In this section, research will be reviewed that ties together the nonhuman animal literature on aggression with the human literature, based on the understanding that similar neural mechanisms regulate aggression across mammalian species, and asks what happens when these circuits become dysfunctional. The focus will be on extreme and unrestrained forms of violent aggression in humans, aggression that is inappropriate, not constrained by context, and can be considered pathological—in other words, out-of-control anger that results in aggression that is impulsive and maladaptive. Such aggression implies a dysregulation of the neural mechanisms that constrain affective aggression, which could include prefrontal and serotonergic systems.

An early study by Raine, Meloy, Bihrle, Stoddard, LaCasse, and Buchsbaum [796] will set the stage for the discussion to follow. In a positron emission tomography (PET) scan study that measured glucose utilization, brain activity was compared between three groups of human subjects: normal controls, convicted murderers who exhibited impulsive or affective aggression, and convicted murderers who exhibited planned (premeditated) aggression. Affective/impulsive murderers, in response to threats or provocation, committed their crimes with violence that was uncontrolled, not planned, and emotionally charged. Murderers exhibiting planned or premeditated aggression committed crimes that lacked emotion and were planned ahead of time in a purposeful way. (Please do not equate the goal-directed aspects of aggression exclusively with premeditated murder, as impulsive or affective aggression can have both goal-directed and consummatory components. Just as I have described for male offensive aggression in animals, in the heat of an affective aggressive episode in humans, both voluntary [proactive and goal-directed] and reflexive responses occur.)

All the subjects were injected with a radioactively labeled glucose compound and then engaged in a cognitive task that is known to increase metabolic activity in the PFC. After 32 min, the subjects were placed in a PET scanner, and images of glucose uptake as a proxy for brain activity were acquired. The findings were as follows:

1. Affective murderers had significantly lower PFC activity, as measured by glucose uptake, than did controls, while premeditated murderers did not. The PFC areas affected included the lateral and ventromedial PFC.
2. Both affective and premeditated murderers had higher activity in the amygdala, thalamus/hypothalamus, and midbrain when compared to controls.

One way to interpret these data is that increases in amygdala, hypothalamus, and midbrain activity are related to increases in all types of aggression. For individuals with affective aggression traits, decreases in PFC activity may contribute to this baseline hyperactivity in amygdala–hypothalamus–midbrain aggression circuits, while other factors may cause such hyperactivity in premeditated murderers. For affective aggression, a lack of top–down regulation by the PFC of the amygdala and other sites may result in a hyperresponsiveness to aversive and threatening social stimuli, leading to impulsive violence. In contrast, relatively normal activity within certain regions of the PFC in murderers that planned their aggression might be related to their ability to organize their crimes ahead of time. Please realize that both groups of individuals were not overtly aggressive at the time of the scans, and, therefore, what is being measured is brain activity during a cognitive task that may be related to heightened aggressive responses that could occur in other contexts.

Two major types of pathological aggression in humans that map onto aspects of affective and premeditated aggression, respectively, are IED and the planned, cold-blooded, and callous aggression associated with psychopathy, although it should be noted that premeditated violence and aggression can be committed by individuals who are not diagnosed as psychopaths [99,194,794,897]. This chapter will focus on the neurobiology of IED, which is an example of inappropriate affective aggression in humans. This disorder is characterized by chronic elevated levels of anger and frequent acts (twice per week) of impulsive, emotionally driven physical or verbal aggression that are not justified by the context in which they occur [193]. Chapter 7 will review the neurobiology of psychopathy and the aggressive behavior that is associated with this disorder. In that chapter, recent studies using more refined and sensitive methods than that used by Raine et al. [796] will show that dysfunctions within the PFC also contribute to psychopathy. Given the complexity of the PFC, it should not be surprising that its dysfunction should contribute to both IED and to psychopathy.

With respect to the role of the hypothalamus in pathological impulsive aggression in men, there are case study reports that have measured abnormal neural activity in the hypothalamus of aggressive patients and that have also shown that deep brain stimulation of the hypothalamus, which may inhibit the HAA, significantly decreased the incidence of uncontrollable and impulsive aggression [826].

3.3.2 PFC Function and Affective Aggression in Humans

Many researchers view PFC control mechanisms as downregulating or restraining the amygdala's response to aversive or threatening stimuli, exerting a cognitive or rational control over emotional responsiveness. In Chapter 1, I reviewed evidence that PFC mechanisms might receive feedback from the negative consequences of one's actions, allowing it to inhibit future responses that would result in punishment. In this chapter, we reviewed evidence in nonhuman animals that the orbital PFC, in particular, appears to downregulate or suppress offensive aggressive behavior. Putting these and other findings together, Davidson et al. [218] predicted that a dysregulation of particular PFC circuits would lead to inappropriate affective aggression. This proposal is supported by the earlier work of Raine et al. [796], and there is now additional supportive evidence.

To begin, an interesting study shows that during "appropriate" imagined affective aggression in humans, PFC activity decreases when imagined aggression increases, which would presumably allow amygdala–hypothalamic–midbrain circuits to operate with less restraint. Using healthy subjects in a PET scan study, Pietrini, Guazelli, Basso, Jaffe, and Grafman [774] asked subjects to imagine one of four scenarios: (1) neutral: subjects were in an elevator with their mother and two other men; (2) cognitive restraint: same as (1), except that the two men were violent against the subject's mother, but the subject did nothing; (3) physical restraint: the subject tried to protect the mother, but was restrained by one of the men; (4) unrestrained aggression: the subject was able to attack the men that were assaulting his mother, either killing or seriously injuring them. In each of the three imagined aggressive episodes, subjects showed decreased activity in the orbital and ventromedial PFC, with the decrease being largest during the unrestrained aggression condition.

Coccaro et al. [193] examined the brain activity of patients diagnosed with IED. Using functional magnetic resonance imaging (fMRI) while subjects viewed emotionally salient facial expressions, the brain responses of IED subjects were compared to those of healthy controls (HC). The IED subjects were also administered a Lifetime History of Aggression (LHA) scale, which measured the frequency of their aggressive outbursts during their lifetimes. While in the scanner, subjects were shown images of the following facial expressions: anger, disgust, fear, happiness, neutral, or surprised. The results showed: (1) IED subjects showed greater amygdala activation specifically to angry faces than did HCs.

Differences in amygdala activation to other emotional expressions were not detected between the two groups; (2) in the HC group, a negative correlation was detected between the BOLD signal in the orbital/ventromedial PFC and the amygdala. That is, when the orbital and ventromedial PFC BOLD signal went up, the amygdala BOLD signal went down. Significantly, this correlation did not exist in the IED subjects. No correlations between PFC activity and amygdala activity were observed in IED subjects (also see [682]); (3) subjects in the IED group with greater LHA scores had greater activation of the amygdala in response to angry faces.

These findings show that there is a specific threat-related amygdala hyperactivity in IED subjects that does not generalize across other types of emotional stimuli. These findings also suggest that for impulsively aggressive subjects, the orbital and ventromedial PFC are not engaged during the processing of angry faces to exert control over amygdala reactivity, and that an uncoupling of PFC–amygdala circuitry may contribute to the neurobiology of affective aggressive disorders in humans.

Similarly, an fMRI study compared the brain responses of men who had physically abused their wives (batterers) with normal control males while the subjects viewed either aggressive or neutral words. During exposure to aggressive words, the PFC BOLD response was lower and the amygdala response was higher when batterers were compared to the controls [533].

Note that each of these studies suggests that both the orbital and ventromedial PFC may be involved in restraining affective aggression in humans. I will have more to say about the ventromedial parts of the PFC later.

3.3.3 Serotonin and Affective Aggression in Humans

As reviewed by Manuck et al. [597], research on the role of serotonin in affective aggression in humans is consistent with the main thrust of the nonhuman animal research. Most research indicates a negative relationship between central 5-HT neural activity and aggressiveness, although, as in the animal research, 5-HT action on the 2A receptor may enhance aggression. In addition, a dysregulation of 5-HT action on the orbital and ventromedial PFC has been implicated in impulsive aggression. I will selectively review studies that support these conclusions.

When healthy human subjects are compared to those who committed violent crimes that were classified as either impulsive or planned, the levels of 5-HIAA in the CSF of the impulsive violent offenders were significantly lower than controls ([597,998] and cf. [819]). Although some meta-analytic reviews have concluded that the negative correlation between aggression and CSF 5-HIAA levels may not be as strong as that proposed by Manuck et al. [597], these reviews lumped together research on all sorts of aggression, including impulsive and nonimpulsive forms of aggression [259]. Overall, it seems clear that high levels of impulsive aggression are associated with low CSF levels of 5-HIAA.

The above conclusion conforms with clinical research, which indicates that one of the pharmacological therapies that is effective in treating individuals with IED is the administration of selective serotonin reuptake inhibitors, such as fluoxetine or paroxetine, treatments that would increase the synaptic availability of serotonin by decreasing the function of SERT [681]. An interesting laboratory experiment strongly supports this clinical evidence [88]. Healthy human subjects were administered a LHA scale and on the basis of their scores were separated into high and low aggression groups. Half of the subjects in each group were then treated systemically with paroxetine 4h before engaging in a computer game that measured their degree of aggressiveness in response to a provocation or threat by a fictitious opponent. Subjects with high scores on the LHA scale showed high levels of provocation-induced aggression, and this aggression was reduced to low levels, comparable to that exhibited by subjects with low LHA scores, by prior paroxetine treatment. An interpretation of these results [88] is that the subjects with low aggression scores on the LHA scale had normal brain 5-HT levels, which served to restrain their aggression during the computer game. In contrast, the subjects with high LHA scores may have had deficient 5-HT transmission, and paroxetine functioned to raise 5-HT synaptic levels in these subjects, which decreased their aggression in response to provocation.

The research on nonhuman animals indicated that serotonin action on 5-HT1A receptors in either the PFC or the hypothalamus can decrease offensive aggression. In this context, there is evidence for a decrease in the functional activity of 5-HT1A receptors in highly aggressive humans. As background, and in relation to the diversity of neural systems affected by serotonin, in addition to its antiaggressive effects, 5-HT action on the hypothalamus also exerts neuroendocrine effects that result in the enhanced release of prolactin and ACTH from the anterior pituitary [191,597]. In a study by Cleare and Bond [191], healthy men filled out several self-rating questionnaires that measured anger, hostility, impulsivity, and aggression. The subjects then received systemic administration of either ipsapirone or a placebo, and neuroendocrine responses to these treatments were measured. Ipsapirone is a selective 5-HT1A receptor agonist, and it causes increases in blood levels of ACTH and prolactin. ACTH, of course, increases cortisol release from the adrenal cortex. The main finding was a significant negative correlation between ipsapirone-induced increases in cortisol/prolactin and aggression scores, suggesting that highly aggressive individuals have decreased functional activity at 5-HT1A receptors. The implication is that a decrease in the function of 5-HT1A receptors at some neural sites decreases neuroendocrine events, but at other sites, such as the PFC or HAA, increases aggression. In this regard, there is support for a decrease in 5-HT1A receptors within the PFC of aggressive humans. Parsey et al. [738] performed a PET scan study on healthy men and women. All the subjects were administered a questionnaire that measured a lifetime aggression score. Subjects were then injected with a radioactive ligand for the 5-HT1A receptor, and brain scans were taken over a period of 2h. There was a significant negative correlation between severity of lifetime aggression and the binding potential of the 5-HT1A ligand in several brain regions that included the dorsal raphe, amygdala, medial PFC, and orbital PFC. (Binding potential [BP]: the amount of radioactive ligand binding, which can result from either a higher density of 5-HT1A receptors or a greater affinity of the ligand for these receptors, or both.) Since 5-HT1A receptors are postsynaptic receptors on non-5-HT neurons in the PFC and amygdala, these results support the proposal that decreased postsynaptic activity at 5-HT1A receptors might contribute to heightened aggression. Note, however, that this is a correlational study and that other interpretations are possible. Decreased 5-HT1A binding at raphe *autoreceptors* could have caused increased 5-HT release, which then resulted in a compensatory downregulation of the postsynaptic receptor. Alternatively, high endogenous 5-HT levels might compete with the radioligand, in this way decreasing BP (cf. [740]). Therefore, some could interpret this data as supporting the view that an overactive 5-HT system promotes aggression (see [1041]). However, the decreased BP in PFC along with the finding of a blunted neuroendocrine response to ipsapirone, suggest that a postsynaptic, rather than an autoreceptor deficit, may be more closely related to heightened aggression. With respect to a likely effect on 5-HT1A receptors in the orbital PFC, New et al. [681] treated human patients diagnosed with IED with fluoxetine over a 12-week period. The fluoxetine not only decreased the patients' aggression, but in a PET study it was found that glucose uptake in the orbital PFC was increased, suggesting increased orbital PFC activity. Based on Figure 3.8 and the results reviewed in the previous section on the role of the PFC in human aggression, an increase in the output of the orbital PFC to the amygdala (presumably to downregulate amygdala responsiveness to presumed threat) would be predicted if long-term fluoxetine treatment increased 5-HT action on the 1A receptor located on inhibitory interneurons in orbital PFC. Additional data support this point of view [682].

While 5-HT action on the 1A receptor acts to suppress aggression, the nonhuman animal research suggested that serotonin action on the 2A receptor increases aggression, and some research on humans is consistent with this finding. Rosell et al. [827] injected subjects with a radioactive ligand that binds to 5-HT2A receptors. A positive correlation was observed between the amount of ligand binding to the orbital PFC and the degree of *current* aggressiveness. They compared three groups: healthy, current IED, and previous IED (in remission) subjects. The subjects with current IED symptoms had significantly greater binding of the 5-HT2A receptor ligand in the orbital PFC than the other two groups (cf. [208]).

Relatedly, in healthy subjects, 5-HT action on 5-HT2A receptors tends to promote risk taking in a gambling task: systemic pharmacological treatment of subjects with a 5-HT2A receptor antagonist increased the number of *risk-averse* responses [586].

In combining the nonhuman animal literature with the human literature, it appears that our current state of knowledge supports much of the circuitry shown in Figure 3.8. Although 5-HT has an antiaggressive effect via actions on the 1A receptor, it may stimulate aggression via an action on the 2A receptor, and a major site of action may be in the PFC. Overall, IED appears to be associated with a shift from 5-HT1A to 5-HT2A receptor dominance. Importantly, the treatment of IED patients with quetiapine, a drug with 5-HT1A agonistic effects and 5-HT2A antagonistic effects, is effective in decreasing impulsive aggression [494].

3.3.4 Vasopressin, Oxytocin, and Aggression in Humans

In the nonhuman animal research, AVP action on the V1a receptor, particularly in the anterior hypothalamus, was shown to exert proaggressive effects. There is also evidence in humans for a positive relationship between aggression and vasopressin. Coccaro et al. [192] reported a highly significant positive correlation between LHA in personality-disordered male subjects and CSF levels of vasopressin. In an experimental study, Thompson et al. [959] treated healthy men with intranasal AVP or saline. Intranasal application of AVP produces elevation of this peptide in CSF within 10 min, and AVP levels remain elevated for at least 80 min [110]. Fifteen minutes after intranasal treatment, the subjects viewed neutral, happy, or angry faces and rated each face on a scale of approachable/friendly-to-unapproachable/threatening. For happy faces, men injected with AVP rated the faces as significantly less approachable and more threatening than did men that were injected with saline, suggesting that the AVP-treated subjects were biased toward antisocial responsiveness. Significantly, some, but not all, fMRI research has found that intranasal administration of AVP to healthy men is associated with an increased amygdala BOLD response to socially aversive images [136,1077]. Such intranasal AVP treatment is also associated with a disruption in the functional connectivity between the PFC and amygdala [136]. This latter result could be interpreted to mean that under natural situations, if aggression were to occur because of an endogenous increase in AVP, the ability of the PFC to regulate amygdala responses to aversive social stimuli might be decreased, perhaps resulting in a more sustained and less restrained aggressive response to a social threat. Where AVP acts to exert these potential effects remains to be determined. Although animal studies have emphasized the anterior hypothalamus as a site of action, since V1a receptors are located on neurons in the PFC of rhesus monkeys [1064] and on neurons in the basal amygdala of humans [571], it could be that AVP directly affects PFC and/or amygdala function to influence aggressiveness in humans.

One study has explored the effects of repetitive sequence polymorphisms within the human V1a receptor gene (AVPR1A gene) on the amygdala's response to threatening facial expressions [639]. The AVPR1A gene is located on chromosome 12, and its coding region contains two exons and one intron [637]. Within the 5′ regulatory region of the gene there are several sites where particular nucleotide sequences are repeated several times. One of these repetitive sequence sites, RS3, will be my main focus. This site is polymorphic because the number of repetitive sequences can vary, giving rise to different alleles. RS3 has at least 16 different alleles. For example, allele 334 at the RS3 site would consist of repetitive sequences containing a total of 334 nucleotides, while allele 320 would contain 14 fewer nucleotides. Meyer-Lindenberg et al. [639] have reported that when human subjects view angry or fearful faces while in an fMRI scanner, the BOLD response in the amygdala, as expected, is greater than when they view geometric forms. Most importantly, carriers of allele 334 at RS3 had the largest amygdala BOLD response to aversive social stimuli when compared to subjects who carried other alleles. Just how allele 334 might affect amygdala reactivity is not known. The fact that the RS3 polymorphism is located in the regulatory region of the gene suggests that variations in gene transcription might be affected, and there is some evidence that allele 334 might be associated with an increased production of V1a receptors in the brain [500]. Since the V1a receptor is located in the basal amygdala of primates, including humans [571,1064], amygdala responsiveness might have been directly affected by the degree of expression of the V1a receptor on particular amygdala neurons. Relevantly, research on rats has shown that AVP acts on the V1a receptor to stimulate amygdala projection neurons that respond to aversive stimuli [431]. Also, recall that a population of AVP neurons is located in the amygdala of some species [240].

Combining these results, one might predict that when an individual is threatened or provoked under natural conditions, those individuals who carry allele 334 and who also respond to this social situation with elevated brain levels of endogenous AVP should show the highest levels of affective aggression because negatively valent amygdala neurons would be highly activated and less restrained by PFC mechanisms.

Although these studies are emphasizing a positive role for vasopressin on aggression, in subsequent chapters it will be shown that certain vasopressin neural systems can exert prosocial effects. When one measures CSF levels of AVP, or studies genetic polymorphisms that might globally increase V1a receptor expression, or examines the effects of increasing CSF levels of AVP via intranasal administration, AVP could be affecting multiple neural systems, and what is being detected is AVP's dominant effect under very specific testing conditions. A more focused analysis on specific neural systems will likely show, for example, that AVP action on the anterior hypothalamus or on certain amygdala neurons might promote antisocial aggressive responses, while AVP action at other sites, for example, within the BST (see Figure 3.5), or NA–VP circuit, might promote affiliative behaviors.

Finally, several studies have indicated that OT neural systems exert antiaggressive effects. In personality-disordered subjects, a significant negative correlation has been observed between LHA scores and CSF OT levels [537]. In a computer game performed by healthy subjects who had to choose between an angry or happy face in order to obtain a monetary reward, control subjects tended to pick happy faces more than angry faces, even when higher rewards were associated with angry faces, while subjects that were administered intranasal OT were more likely to choose the angry face under this high reward condition [274]. These results suggest that central OT decreases aversion to angry faces. In support of this proposal, several fMRI studies on healthy subjects have shown that intranasal OT administration decreases the amygdala BOLD response to angry or fearful faces [251,324,492], while increasing the amygdala BOLD response to happy faces ([324] and cf. [251]). Interestingly, Kirsch et al. [492] showed that intranasal OT not only decreased the amygdala BOLD response to angry faces but also decreased the functional connectivity between the amygdala and the midbrain (the correlation between the BOLD response in these two regions decreased compared to control subjects receiving intranasal saline). Based on our review of the neural circuitry underlying animal aggression, these results suggest that intranasal OT might decrease the activation of an amygdala–hypothalamus–PAG circuit in response to an angry face.

Several single nucleotide polymorphisms (SNPs) within the OTR gene, resulting in different alleles, have been associated with changes in aggressive behavior in humans. These findings should be understood in the context of the animal work that showed that OTR knockout mice are hyperaggressive. In humans, carriers of certain alleles exhibit heightened aggressiveness [593]. The OTR gene is located on chromosome 3, and its coding region contains four exons and three introns [263,637,1048]. One particular SNP, referred to as rs1042778, is located in the 3′ untranslated region (UTR) of exon 4. The UTR of deoxyribonucleic acid (DNA) is transcribed into the final messenger ribonucleic acid (mRNA) that is used as a template for the amino acid sequence of a protein. Although the nucleotides in the UTR do not code for amino acids, they do regulate the translation of the mRNA into the protein, in this case the OTR (see Chapter 2). Therefore, alterations in the nucleotide makeup of the UTR of DNA can ultimately influence OTR expression. In regard to this particular SNP, Malik et al. [593] genotyped 160 cases of children (mean age = 11.5 years) displaying extreme and persistent aggressive behavior and compared them to healthy controls. They found that the cytosine (C) allele at site rs1042778 was overrepresented in male aggressive cases when compared to controls, who were more likely to carry another nucleotide base at this site. Of course, not all aggressive subjects carried the C allele, which demonstrates that to the extent that this SNP variant contributed to the development of aggression, other genetic and experiential factors are also involved: the development of abnormal and pathological aggression is determined by multiple interacting factors. With respect to decreases in OT being associated with increases in aggression, it would be important to determine the manner in which this particular OTR SNP affects OTR expression in the brain.

If an increase in OT neural activity decreases aggression, what general mechanisms might be involved? One possibility is that OT simply activates prosocial neural circuits, and these circuits, in turn, inhibit antisocial circuits, which would include aggression circuits. It is also possible that OT directly upregulates prosocial circuits and directly downregulates antisocial circuits. As reviewed in Chapter 1 [431], animal studies clearly show that OT action can directly excite amygdala inhibitory neurons that depress the response of amygdala projection neurons to aversive stimuli, showing that OT can directly downregulate circuits that respond to aversive stimuli.

Our full understanding of the role of OT and AVP in human social behavior is constrained by a paucity of information with respect to the location of the receptors for these neuropeptides in the human brain [103,440,571,572]. Once appropriate radiolabeled nonpeptide ligands are fully developed, which would be capable of crossing the blood-brain barrier after systemic administration, then they could be combined with PET scanning to achieve an accurate measure of the spatial location and density of oxytocin and vasopressin receptors in the human brain. Even with the absence of such important information, because of the evidence indicating the opposing roles of vasopressin and OT in aggression control, it has been proposed that therapies that modulate OT and AVP transmission in the brain might be used clinically to treat pathological forms of aggression, such as IED [637,923].

3.4 EARLY ADVERSE LIFE EXPERIENCES, GENES, AND THE DEVELOPMENT OF AGGRESSION

3.4.1 Introduction

Early adverse social experiences, represented by decreases in the quality and/or quantity of social interactions early in the life of an organism, have profound and enduring effects on the development of social behaviors [466,703,988]. In Chapter 2, I described how low levels of parental care act to increase HPA responsiveness and stress reactivity in the affected offspring. As will be demonstrated in this and subsequent chapters, negative early life social experiences also cause a dysregulation in the development of neural circuits that influence social behaviors, resulting in decreases in prosocial and increases in antisocial behaviors. In this chapter, I will describe how early social stress leads to the development of heightened affective aggression in adulthood, and I will concentrate on the evidence that indicates that a dysregulation in the development of serotonergic neural systems contributes to this heightened aggression. It will also be shown that there is interindividual variability in the response to early social stress and that gene–environment interactions contribute to such variability. That is, certain genes increase an organism's vulnerability to early social stressors, while other gene types exert protective effects.

3.4.2 Early Adverse Social Experiences and the Development of Aggressive Behavior

Correlational studies in humans have clearly shown a positive relationship between parental neglect/abuse of children and the development of impulsive aggression in the affected offspring [176,312]. Experimental studies on nonhuman animals have shown that this is a cause–effect relationship. Beginning with the early studies of Harlow [390,391], it has been known that early social deprivation in rhesus monkeys results in increases in affective/impulsive aggression. Careful studies by Suomi's group [942] have emphasized the role of maternal separation, which can be viewed as a model of maternal neglect, on the development of intense affective aggression in rhesus monkeys. Under natural conditions, rhesus monkey infants are virtually in exclusive contact with their mothers throughout the first month of life, and only after this period do they begin to interact with other members of their social group. Weaning occurs at between 3 and 4 months of age, indicating that the mother–infant bond remains intense over that period of time; puberty occurs between 3 and 4 years of age.

In exploring the effects of maternal separation on the development of aggression in an experimental setting, mother-reared (MR) rhesus monkeys have been compared to peer-reared (PR) monkeys [942]. MR monkeys are raised in social groups composed of several mother–infant pairs. Such rearing conditions allow the developing infant to interact with its mother during early development and with its mother and peers during later periods. In contrast, PR monkeys are separated from their mother near the time of birth and hand-reared by humans in a nursery for several weeks. At about 40 days of age, such monkeys are placed in a group with three other PR monkeys until 6 months of age, after which they are placed in a larger social group. In comparison to MR monkeys, PR juvenile and adult monkeys show higher levels of aggressive responses toward unfamiliar conspecifics, do not integrate well into the larger social group, and exhibit increases in fear-related responses, as indicated by their tendency to avoid novel or unfamiliar situations [410,873,942]. Therefore, one explanation for the heightened aggression is that PR monkeys perceive a novel conspecific as overly threatening, leading to a strong impulsive aggressive response.

In male rats, maternal separation also increases offensive aggression as measured by the resident–intruder paradigm [990]. Two groups of rats were formed: maternal separation (MS) rats were separated from their mothers for 3 h/day on postnatal days 1–14; controls were left undisturbed. All rats were weaned at 21 days of age and housed in groups of four similarly treated subjects until the time of the aggression test. In adulthood (3–4 months of age), using the resident–intruder model of offensive aggression, it was found that resident MS rats were more aggressive that resident control males when exposed to an unfamiliar intruder.

Proper social development not only involves a normal mother–infant relationship but also involves normal peer–peer relationships during the juvenile period. Indeed, one can conceive of a continuum where an effective mother–infant relationship facilitates the transition to appropriate juvenile peer–peer interactions, and that the total pattern of social interactions during early development, in turn, influences the development of adult social behaviors [390]. Within this context, research has shown that a disruption of juvenile peer–peer interactions also leads to heightened aggression.

Kempes et al. [476] compared the social behavior of two groups of subadult rhesus monkeys. One group was mother-only reared during the first year of life and then housed with peers, while the second group was composed of monkeys raised with both a mother and peers during the first year of life. The monkeys who were peer deprived during their first year showed inept social behavior when placed in the larger social group of peers, reacting with aggressive or submissive responses to nonthreatening social situations, such as being groomed. This socially incompetent behavior exhibited a mixture of offensive and defensive responses that appeared to be related to an increase in fear-related responses toward other group members, since the peer-deprived monkeys tended to avoid rather than approach group members.

Research on male rats has shown that postweaning social isolation results in an intense and severe increase in aggression during resident–intruder tests [969]. Male rats were weaned at 19 days of age and housed either individually (social isolation) or in groups (controls) for 7 weeks. Subsequently, all rats were housed individually for 3 days and then tested for aggression by being exposed to an unfamiliar intruder male for 20 min. The rats that experienced postweaning social isolation exhibited a mixture of intense offensive aggression coupled with some defensive responses, such as running away from the intruder. This type of aggression was therefore somewhat ambiguous and may have been influenced by an increase in fearfulness. With respect to the intense offensive aggression, the socially deprived resident males showed a dramatic increase in violent attacks (hard bites aimed at vulnerable parts of the intruder's body) that were not preceded by threat signals. In a related Fos immunocytochemical study [970], the socially deprived hyperaggressive rats showed increased Fos expression in the MeA and BMA and in the HAA, indicating an increased activation of the aggression circuit. The authors propose that this postweaning social isolation–induced hyperaggression is a good animal model of human IED or extreme affective aggression, where there is a mismatch between the provocation and the aggressive response.

3.4.3 A Dysregulation of Serotonin Neural Systems Mediates Some of the Effects of Early Social Deprivation on the Development of a Hyperaggressive State

In accordance with the serotonin deficiency hypothesis of heightened impulsive aggression, several studies in rodents and primates have indicated that early social deprivation lowers serotonergic neural system function and may also produce a shift in the PFC from 5-HT1A receptor to 5-HT2A receptor dominance. As reviewed in the sections on serotonin and aggression, such changes should lead to increases in affective aggressive responses to relatively nonthreatening social stimuli.

In rhesus monkeys, CSF levels of 5-HIAA have been measured in MR and PR individuals. Although there is substantial variability in both groups, 5-HIAA levels are significantly lower in the hyperaggressive PR monkey group [410,886]. These results have suggested that early maternal deprivation decreases 5-HT synaptic release, which, according to the 5-HT deficiency hypothesis, should increase affective aggression. Similar support comes from Veenema et al. [990], who found that early maternal separation in male rats not only increased offensive aggression, as described above, but was also associated with a decrease in 5-HT immunoreactive axon terminals in the anterior hypothalamus.

Whitaker-Azmitia, Zhou, Hobin, and Borella [1030] examined the effects of postweaning social isolation on 5-HT neural circuit development and found a decrease in 5-HT axon terminal input to the forebrain in socially deprived rats. A problem with this study is that the density of 5-HT terminals was measured by using an antibody to SERT (5-HTT) as a proxy. However, low levels of SERT do not prove low levels of 5-HT-positive terminals, and it would have been more appropriate to use an antibody against 5-HT itself, as was done in the Veenema et al. [990] study. In the context of lower 5-HTT forebrain labeling in the socially deprived rats, a PET study in humans that employed a radiotracer that binds to the 5-HTT [315] found a significantly lower level of radiotracer binding in the anterior cingulate cortex (ACC: a part of the medial PFC; see Chapter1) of subjects with IED when compared to healthy controls. The authors suggested that pathological impulsive aggression might be associated with a decrease in serotonin innervation of the ACC. It would have been interesting to know about the early life histories of the IED patients. I will have more to say about the ACC below.

Previously, I reviewed the work by Caramaschi et al. [157], which suggested that the 5-HT1A *autoreceptor* exhibited a heightened sensitivity to 5-HT in highly aggressive strains of mice when compared to less aggressive strains and that this heightened sensitivity was associated with a decrease in 5-HT release into the PFC. Significantly, Arborelius et al. [38] reported that dorsal raphe neurons of adult rats that were subjected to periods of maternal separation for 3h/day over postnatal days 2–14 showed an enhanced response to the inhibitory effects of 5-HT when compared to controls. They suggested that early social deprivation might increase the affinity of the 5-HT1A autoreceptor in the raphe nucleus to 5-HT, in this way enhancing 5-HT negative feedback, which would decrease 5-HT synthesis and 5-HT release into target forebrain sites.

With respect to forebrain 5-HT receptors, in a PET study on rhesus monkeys that used a radiotracer that binds to 5-HT1A receptors, Spinelli et al. [919] have reported decreased binding in several forebrain regions, including the ACC, when PR monkeys were compared to MR controls. To the extent that 5-HT exerts antiaggressive effects by binding to postsynaptic 5-HT1A receptors, these results add additional support to the view that social deprivation–induced decreases in 5-HT activity at specific receptors in select forebrain regions, such as the medial PFC, may result in heightened aggression. This decrease in 5-HT1A binding may also contribute to a shift toward the dominance of 5-HT2A receptors.

In my review of the animal literature on aggression, I emphasized medial PFC input to the HAA and orbital PFC input to the amygdala (see Figure 3.8). However, the medial PFC, including the ACC, also projects to the amygdala, and this may be another route through which PFC mechanisms modulate aggression. I have already reviewed the evidence suggesting that the projection of IL part of the medial PFC to the HAA might promote aggression, but the role of another part of the medial PFC, the PL, was not discussed in detail. There is some evidence that PL input to the amygdala might regulate aggression and that such input might also be affected by early life experiences. Although pyramidal cells in the prelimbic part of the medial PFC of rats contain both 5-HT1A and 5-HT2A receptors, the dominant effect of 5-HT release into the PL cortex is inhibitory via an action on 1A receptors [789]. Interestingly, Benekareedy et al. [85] have reported a switch in receptor dominance in rats exposed to 3h/day of maternal separation between postnatal days 2–14. While normally reared rats conformed with the findings of Puig et al. [789], the dominant effect of 5-HT on pyramidal neurons in the prelimbic cortex of rats that were subjected to early maternal separation was excitatory, and this effect was blocked by a 5-HT2A receptor antagonist. As already reviewed, since 5-HT action on the 2A receptor appears to have proaggressive effects, the heightened PL response to 5-HT via the 2A receptor might be another avenue through which early social deprivation enhances aggressive responsiveness. Recall from Chapter 1 that the PL area in rats is likely to be homologous to area 32 of the human medial PFC, and that some have included area 32 as part of the dorsal ACC along with area 24 [268]. In my review of the rodent literature (see Figure 1.16), evidence was presented that the output of the PL cortex to the basal amygdala *increases* fear-related responses by activating basolateral amygdala (BLA) output neurons that have a negative valence. One can expand on this idea by suggesting that PL stimulation of BMA output neurons that have a negative valence might enhance responsiveness to mildly threatening social stimuli, in this way promoting an impulsive affective aggressive response through the activation of a BMA-to-HAA circuit. Figure 3.9 shows how a switch from 5-HT1A to 5-HT2A dominance might enhance affective aggression. The only problem with this interpretation is that PL neural connections to BMA, in contrast to its projections to BLA, are weak [994]. Therefore, it is also possible that 5-HT2A activation of PL input to the BLA (rather than BMA) results in increased fearfulness, which then contributes, by enhancing defensive responses, to the mixture of heightened offensive *and* defensive aggression shown by socially deprived organisms. Please note that if this switch from 5-HT1A to 5-HT2A receptor dominance caused by early social deprivation also occurs on IL pyramidal neurons that project to HAA, such a switch would also be expected to enhance offensive aggression.

In conclusion, early social deprivation is likely to increase adult aggressiveness by resulting in a general decrease in the release of 5-HT into the forebrain and by causing a switch in the dominance of 5-HT receptors in the medial PFC, and perhaps the orbital PFC, from the 1A to the 2A type. This combined action should serve to potentiate affective aggression. But what might be the mechanism through which early social deprivation causes a dysregulation in the development of serotonin neural circuits? An answer to this question might be related to the fact that 5-HT can influence the development of serotonin and non-serotonin neural systems during the perinatal period of brain development [108,326,544,905]. Most important for our purposes is the evidence that high levels of 5-HT *during brain development* exert a negative feedback effect on the development of 5-HT neural circuits [1027]. Supranormal amounts of perinatal 5-HT release are compensated for by a 5-HT-induced downregulation of 5-HT neural circuit formation and/or function. Therefore, the following hypothesis can be offered: the stressful effects of early

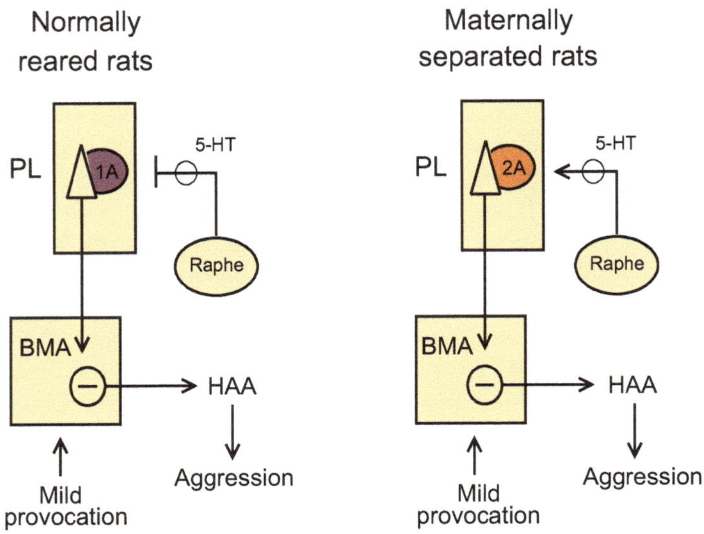

FIGURE 3.9 In addition to the infralimbic prefrontal cortex and orbital prefrontal cortex, the output of the prelimbic part of the medial prefrontal cortex (PL) may also contribute to the regulation of aggression. This figure presents a neural model involving PL in order to explain how early social deprivation might operate to enhance the probability of an aggressive response to a mild provocation. Projection neurons (pyramidal neurons; cell bodies shown as triangles) from PL are shown as stimulating the output of those basomedial amygdala (BMA) neurons with a negative valence that project to and stimulate the hypothalamic aggression area (HAA). Serotonin (5-HT)-containing raphe neurons are shown as synapsing on these PL neurons that project to BMA. In normally reared rats, the dominant effect of 5-HT on PL pyramidal neurons is inhibitory through an action on 5-HT1A receptors (1A). Such an effect would depress aggressive responses. In adult rats that had been exposed to early maternal separations, the dominant effect of 5-HT on PL pyramidal neurons is shifted to a stimulatory effect via an action on 5-HT2A receptors (2A). Such an effect would enhance aggressive responses. See text for additional details. Axons ending in a bar are inhibitory, and those ending in an arrow are excitatory.

social deprivation activate 5-HT release during the postnatal period of brain development, perhaps as a coping mechanism to help an organism deal with *acute* stressors [450]. However, if the stress is prolonged, as occurs in most early social deprivation paradigms, then the long-term enhancement of 5-HT release induced by chronic stress may feed back to suppress the full development of serotonin neural systems, leading to an enhancement of emotionality and affective aggression in adulthood. In support of this hypothesis, I will present the evidence that sustained perinatal increases in 5-HT activity result in a negative feedback effect on the development of 5-HT neural circuits.

In an early in vitro study, Whitaker-Azmitia and Azmitia [1028] cocultured rat fetal raphe neurons with fetal hippocampal neurons. One dependent variable was the growth of 5-HT raphe axons into the hippocampal target site. The addition of a nonselective 5-HT receptor agonist to the culture medium decreased 5-HT raphe input to the hippocampus. As will become important later, the addition of pargyline, an MAO inhibitor, to the culture had similar effects. Since MAO action is one of the ways to terminate 5-HT activity, pargyline would have increased endogenous 5-HT levels in the developing raphe neurons and in the extracellular environment. Such enhanced 5-HT levels were proposed to have a negative feedback effect on raphe neuron outgrowth. Similarly, Whitaker-Azmitia, Zhang, and Clark [1029] reported that the systemic treatment of fetal and neonatal rats in vivo with MAOA inhibitors decreased 5-HT innervation of the cortex, as measured at postnatal day 30. Even when drug administration is restricted to the postnatal period, treatment with drugs that increase endogenous 5-HT levels has a negative feedback effect on the full development of 5-HT neural circuits, and it is these effects that are most relevant to the effects of early social deprivation. Maciag et al. [584] have reported that chronic systemic treatment of neonatal rats from postnatal days 8–21 with a selective serotonin reuptake inhibitor (SSRI), which antagonizes the action of the 5-HTT, resulted in reductions in the expression of both TPH in dorsal raphe neurons and the serotonin transporter in cortex. These effects persisted into adulthood, well beyond the period of SSRI treatment, and the 5-HTT immunoreactivity decrease was taken as evidence for decreased cortical innervation by 5-HT axons. Finally, Martin, Liu, and Wang [609] systemically treated male prairie voles with a 5-HT receptor agonist beginning during the fetal period and continuing through postnatal day 20. On postnatal day 80, immunohistochemical detection of 5-HT axons, using an antibody directed against 5-HT, found a decreased serotonin innervation of the amygdala in the 5-HT agonist–treated voles in comparison to controls.

These results, taken together, suggest that early social deprivation may lead to sustained 5-HT release, which then exerts a negative feedback effect on the development of the 5-HT neural system, and that this dysregulation persists into adulthood, with the potential of causing enhanced emotionality and aggression.

3.4.4 The Potential Role of CRH as a Mediator of the Effects of Early Social Deprivation on the Development of 5-HT Neural Systems

In Chapter 2, I described how low levels of parental care and maternal deprivation promote CRH release in the brain as part of the stress hyperactivity that develops in the affected offspring. Since CRH, functioning as a neurotransmitter/neuromodulator, has been shown to act on the DR to affect 5-HT release [314,490,581], it is possible that early life stress, by promoting CRH release into the raphe, modulates 5-HT release in such a way as to result in the persistent dysregulation of sertonergic function that ultimately leads to an aggressive phenotype.

There are two types of CRH receptors, the CRH1 (CRHR1) and the CRH2 receptor (CRHR2), and both types of receptor are located in DR [490]. CRH has a higher affinity for CRHR1 than for CRHR2, which means that low levels of CRH release selectively act on CRHR1, while higher levels of CRH release also affect CRHR2 [581]. Importantly, CRH action at the CRHR2 on DR 5-HT neurons depolarizes these 5-HT neurons and results in the release of 5-HT into the forebrain, which includes release into the medial PFC [314,490]. A source of CRH release into the DR derives from CRH neurons located in the central nucleus of the amygdala (CeA) [353], and electrical stimulation of CeA activates 5-HT release into medial PFC, an effect that is mediated by activation of CRHR2 in DR [314]. Forster et al. [314] suggest that when *adult* organisms are exposed to high stress and fear-inducing stimuli, the forebrain release of 5-HT by CRH hyperactivity may serve as a coping mechanism for the affected organism in that 5-HT may act to offset or decrease intense fear, anxiety, and stress responses.

Given this data, the following proposal can be offered. Exposing *young* organisms to early adverse social stress, such as maternal neglect, maternal abuse, or general social deprivation, may also activate a CRHR2-induced release of 5-HT into the medial PFC and other forebrain sites as a coping mechanism. However, if the early life social stress is severe and prolonged, the chronic activation of 5-HT neural systems in the developing brain results in a dysregulation of these systems so they become hyporesponsive in adulthood, particularly at the 5-HT1A postsynaptic receptor, leading to heightened anxiety, stress reactivity, and increased likelihood for the development of an impulsive aggressive phenotype.

Kolber et al. [506] developed a transgenic mouse line where an enhancement of CRH synthesis in the brain could be both spatially and temporally limited. When CRH synthesis (and presumably release) was enhanced in the forebrain (including CeA) between embryonic day 15 and postnatal day 21 (the time of weaning), in adulthood these mice displayed heightened anxiety and stress reactivity (decreased exploration of a novel open field area). Therefore, early life environmental events that trigger the chronic release of CRH may facilitate the development of a phenotype of heightened reactivity to mild stressors, which may promote affective aggressive responses to mildly threatening stimuli.

With respect to the antiaggressive effects of OT, since OTRs are located in CeA [431], it would be important to determine whether OT action in the amygdala might restrain CRH release from CeA neurons. Perhaps exposing young organisms to social stressors, such as maternal neglect, not only affects CRH and 5-HT systems, but also depresses OT function in the amygdala and other sites (see Chapter 5).

3.4.5 Gene–Environment Interactions and the Development of Aggressiveness

3.4.5.1 Introduction

Several studies have indicated that genetic polymorphisms or genetic null mutations that increase 5-HT levels during brain development are associated with an increase in aggression in adult animals and humans. Many of these studies also emphasize the involvement of gene–environment interactions, such that the aggression-enhancing effects of early adverse social experiences are potentiated in individuals with certain genotypes. One way of viewing this interaction is as follows: some genotypes, but not others, may mildly or moderately increase 5-HT levels during the perinatal period of brain development; when this genetic effect is coupled with a chronic social stress–induced 5-HT release during the perinatal period, the resultant extended action of supranormal levels of 5-HT release might then have a large negative feedback effect on the full development of 5-HT neural systems, with this depression of 5-HT function enduring into adulthood. Lower endogenous serotonin levels and lower synaptic release of 5-HT in adulthood, perhaps coupled with a switch to 5-HT2A receptor dominance over 5-HT1A dominance, would then promote increases in affective aggression in response to relatively mild forms of provocation in adulthood.

3.4.5.2 MAOA and Aggression

Monoamine oxidase enzymes catalyze the metabolic degradation of 5-HT, norepinephrine (NE), and DA. There are two forms of MAO, the A and B forms, each produced by a separate gene, and several studies have linked MAOA and its gene to aggressive behavior in animals and humans [137]. Interestingly, each of these genes is located on the X chromosome, and therefore males have only one copy of each gene. If a null mutation were to occur in the MAOA gene located on a single X chromosome, it would cause an effective knockout of this gene in males but not in females.

The MAO enzymes are located in the presynaptic terminals of NE, 5-HT, and DA neurons, where they regulate the amount of monoamines that are stored in synaptic vesicles. MAO is also produced by extrasynaptic elements, such as astrocytes, where it functions to degrade monoamines that have been released synaptically ([137]; also see Figure 3.6). Given this relationship, one would conclude that as MAO activity increases, monoamine activity would decrease, and as MAO activity decreases, monoamine activity would increase. For example, in the latter case, more monoamines would be stored in synaptic vesicles and therefore more would be released by action potentials, and once released, the monoamines would act for a longer time because one of the termination mechanisms (metabolic degradation) would be reduced. (Recall that the other termination mechanism is via reuptake through transporter pumps, such as the 5-HTT pump in the case of serotonin; I will return to the importance of the 5-HTT and its gene later.)

Evidence suggests that the primary targets for MAOA are 5-HT and NE, while the primary target for monoamine oxidase B (MAOB) is DA. However, each enzyme is capable of degrading each of the monoamines so that if MAOA were not present, the B form could substitute and operate on NE and 5-HT. However, early in development, including the early postnatal period, during the time of brain development, MAOA is the dominant enzyme, and very little MAOB is produced [137,165].

The first indication of the involvement of the MAOA gene in aggression came from Bruner et al. [135], who reported on a single-point mutation of the MAOA gene in an extended Dutch family. The mutation caused an effective knockout of this gene throughout development in male carriers: a functional MAOA enzyme was no longer produced. The mutation of a single nucleotide (C to T mutation) changed glutamine (CAG) to a termination codon (TAG), resulting in the termination of translation before the complete synthesis of MAOA. In females, with this mutation on only one X chromosome, normal amounts of MAOA were produced.

Selective MAOA deficiency in the men of this family resulted in a marked disturbance of monoamine metabolism and decreased levels of 5-HIAA (the 5-HT degradation product) were found in their urine. The behavioral phenotype of the males in this family was characterized by borderline mental retardation and a tendency toward aggressive outbursts, often in response to anger, fear, or frustration. Other types of impulsive behavior in this family included arson, attempted rape, and exhibitionism.

The authors noted that in humans, as we have already reviewed, impulsive aggression has been linked to low 5-HIAA CSF levels. They also note that this finding has usually been interpreted to mean that a reduction in central 5-HT function is linked to impulsive aggression. In light of their results, they suggested an alternative explanation. The reduced 5-HIAA levels in the CSF of aggressive individuals may be caused by low MAOA activity, which would actually result in increased central 5-HT function. In other words, they argued that increased serotonergic activity in adulthood might be causing heightened aggression. However, the treatment of normal adult animals and humans with MAO inhibitors does not cause aggression [112,632]. Therefore, MAOA deficiencies early in development may have different effects than a decrease in function that is restricted to adulthood. As described in the introduction to this section, increased levels of 5-HT during the perinatal period may dysregulate 5-HT neural circuit development, leading to an increase in adult aggression, and the low levels of CSF 5-HIAA in aggressive individuals may indeed be indicative of 5-HT hypofunction.

Animal models have confirmed the human results reported above. Cases et al. [165] performed a knockout mutation of the MAOA gene in mice and studied aggression using the resident–intruder model, and it was found that the mutant mice showed enhanced offensive aggression (also see [112]). In 7-day-old MAOA mutants, 5-HT immunoreactivity was enhanced in various parts of the brain, including the cortex. Importantly, the brain content of 5-HT was greatly elevated throughout the early postnatal period through day 25 and then slowly normalized thereafter, presumably because MAOB began to compensate for the loss of MAOA. The authors concluded that the increased aggression of adults might have been related to persistent structural and functional alterations induced during development by abnormal exposure to high levels of monoamines (5-HT and NE) early in life.

Importantly, Shih et al. [893] have reported that the hyperaggressiveness observed in adult MAOA knockout mice is reduced by treating such mice with a 5-HT2A receptor antagonist. Therefore, exposure to high levels of 5-HT (and perhaps NE) during development may not only downregulate adult 5-HT function via a negative feedback mechanism but may also cause the 5-HT2A receptor to become dominant in certain brain regions.

3.4.5.3 Genetic Polymorphisms in the MAOA Gene

A total mutation of the MAOA gene is a rare occurrence. If one were looking for genetic contributions to extreme or pathological aggression, which occurs relatively frequently in human societies, one would not expect that the Dutch mutation would be the cause. However, the research on the Dutch extended family does suggest that variations in the function of the MAOA gene may contribute to aggression in animals and humans. Perhaps different alleles of the MAOA gene would have differential influences on an individual's aggressive phenotype.

There are genetic polymorphisms in the MAOA gene that occur at high frequencies in the human population [137]. Variations in the nucleotide sequence within the regulatory region of the MAOA gene have given rise to two important alleles: MAOA-L and MAOA-H. The L allele is low functioning or less effective in degrading 5-HT and NE, while the H allele is more effective, as determined by in vitro studies. This genetic polymorphism is the result of a variable-number tandem repeat (VNTR) sequence in the upstream promoter region (MAOA u-VNTR). For example, a 30-base pair sequence can be repeated three or four times, giving rise to two different alleles. The 3-repeat sequence has been associated with lower expression of MAOA (MAOA-L), while the 4-repeat sequence is associated with higher expression (MAOA-H). As you might expect, the MAOA-L allele, which would presumably result in increased levels of 5-HT (and NE) during brain development, has been associated with increased aggressive tendencies in human males.

In a groundbreaking study, Caspi et al. [166] reported on a significant gene–environment interaction in the development of pathological and severe aggression in human males. A large human population of about one thousand children was genotyped, and these findings were related to other data that was available for this population, which included the occurrence of early childhood abuse and the display of aggressive criminal behavior by the children, as assessed at ages 3 through 21 years. Between 3 and 11 years of age, 8% of the children experienced definite maltreatment, 28% experienced probable maltreatment, and 64% were not maltreated. These different treatment groups did not differ in MAOA genotype proportions. Under the condition of definite childhood abuse, the composite antisocial behavior score was significantly greater for MAOA-L genotype boys, while the effect of childhood maltreatment on antisocial behavior was much weaker in MAOA-H males. Low levels of antisocial behavior were observed for both genotypes under the nonmaltreated condition. Although individuals having the combination of low activity MAOA and maltreatment were only 12% of the entire male population, they accounted for 44% of the violent convictions. These basic findings have been replicated [143,311,1023], indicating that the MAOA-L genotype predisposes human males to become aggressive under the influence of early life social adversity (for example, physical, emotional, or sexual abuse by one's parents), while the MAOA-H genotype may protect the individual from the effects of early life adversity. These kinds of data give some insight into why early life adversity can lead to psychopathology in some, but not all, individuals. The results also indicate that having the MAOA-L gene does not destine one to be severely aggressive. One's genetic background interacts with one's environment in determining a particular phenotype.

How can one present a neurobiological model to explain this interaction, based on the research and hypotheses previously reviewed? One possibility is that early life stress causes increased release of 5-HT and NE into cortical and limbic areas. While these monoamines may be restrained and degraded promptly in MAOA-H individuals, in individuals with the MAOA-L genotype the brain may be exposed to supernormal levels of 5-HT (and NE), which may lead to long-term changes in 5-HT neurocircuitry, promoting the development of an aggressive phenotype.

In human populations that are composed of healthy males who are not violent criminals, one can still detect personality differences in the degree of a person's anger and aggressiveness, even if hostility is covert rather than overt, or represented verbally rather than physically. Therefore, an important question is whether one's latent aggressiveness is related to one's MAOA genotype. Meyer-Lindenberg et al. [636] showed that even when early childhood experiences are not taken into account, there are significant neurobiological differences between MAOA-L and MAOA-H genotype males. This study examined a large sample of healthy nonviolent volunteers, and fMRI analyses examined their brains' response to angry and fearful faces that are known to activate the amygdala. In response to emotionally salient visual faces, fMRI BOLD activity was increased in the amygdala, and this increase was greater in the MAOA-L individuals. Although this study did not measure aggressive tendencies, it does suggest that MAOA-L individuals have the potential to overreact to aversive stimuli (also see [539]).

A study by Eisenberger et al. [269] did relate MAOA genotype to anger and to brain reactivity under a condition of social exclusion. Prior to the fMRI neuroimaging tasks, healthy human subjects completed several self-report questionnaires that measured anger, hostility, and hypersensitivity within social settings, giving rise to two separate composite scores of trait aggression and interpersonal hypersensitivity. While in the fMRI scanner, each subject played a computer game with two other fictitious subjects. During the game, the experimental subjects were first allowed to play the game (social inclusion), but were subsequently excluded from playing (social exclusion).

Statistical analyses showed that on the self-report measures, MAOA-L individuals had higher trait aggression and higher interpersonal hypersensitivity scores than did carriers of the MAOA-H genotype. The authors suggest that this combination of personality characteristics might result in MAOA-L individuals being more prone to reactive or affective aggression. In response to social exclusion, the fMRI measure indicated an increased BOLD response in the dorsal ACC of the MAOA-L subjects. To the extent that this area of the cingulate cortex included area 32, which may be homologous to the PL cortex of rodents, the following argument can be made, based on the circuitry shown in Figure 3.9. In the MAOA-L subjects, increased activity in the dorsal ACC may have caused increased activity in amygdala neurons that respond to aversive social stimuli, and this may lead to a propensity to respond aggressively. In support of some of these findings, it has been reported that MAOA-L healthy adults, in comparison to their MAOA-H counterparts, are actually more aggressive toward fictitious opponents when they are provoked during a competitive computer game [518].

Interestingly, and in relation to Figure 3.9, in a positron emission tomography (PET) study that used a radiotracer ligand for the 5-HT1A receptor, Mickey et al. [641] report that brain 5-HT1A binding levels are lower in women with the homozygous MAOA-L genotype compared to those carrying the MAOA-H allele on each X chromosome, suggesting that the dorsal ACC, and other brain regions, would be under decreased 5-HT inhibition in women with two copies of the low-functioning MAOA allele.

There has been some concern that the in vitro studies characterizing the effects of MAOA genetic polymorphisms on MAOA expression might not be an accurate reflection of MAOA expression in vivo [18]. Therefore, Alia-Klein et al. [18] set out to measure MAOA expression in the brains of healthy adult males who had filled out a personality questionnaire that measured trait aggression. In a PET study that used a radioligand that would bind to MAOA, they found a strong negative correlation between brain MAOA expression and aggression such that lower MAOA expression (as measured by radioligand binding) in cortical and subcortical regions was related to higher self-reported aggression. For example, the correlation between aggression and MAOA expression in the medial PFC was −0.61 ($p=0.001$).

All of these results point to the conclusion that the MAOA-L genotype and lower levels of MAOA expression during brain development render one prone to increases in affective aggression, and when this genotype is coupled with severe early life social stresses, an increased probability for the development of pathological violence and criminality occurs. Based on my proposals, this gene–environment interaction is conceived as affecting the development of serotonin neural circuits, which in turn leads to heightened amygdala responsiveness to socially aversive stimuli in adulthood, presumably due in part to a dysregulation of PFC control over amygdala reactivity. But MAOA not only degrades 5-HT it also degrades NE. Can additional evidence be provided that the MAOA-L genotype and the MAOA mutation effect on the development of aggression is primarily the result of an increase in 5-HT levels rather than NE levels?

3.4.5.4 5-HTT Polymorphisms and the Development of Aggression

In both humans [422] and rhesus monkeys [86], there is a genetic polymorphism within the regulatory region of the serotonin transporter gene (located on chromosome 17) that has been referred to as the serotonin transporter-linked polymorphic region (5-HTTLPR). Although the exact location of this polymorphism differs between the two species, the similarities are that the polymorphism consists of a nucleotide length variation within the regulatory region, giving rise to either a short (s) allele or a long allele (l), and the short allele is associated with decreased transcription of 5-HTT mRNA and decreased expression of 5-HTT protein. Therefore, during the perinatal developmental period, the brains of carriers of one (s/l) or two (s/s) copies of the short allele should have increased exposure to extracellular serotonin, due to decreased reuptake, compared to those with the l/l genotype. In line with the argument I have been making, there is evidence that carriers of the short allele are more susceptible to the aggression-promoting effects of early adverse life events.

In rhesus monkeys, PR monkeys are hyperaggressive and have lower levels of 5-HIAA in CSF when compared to their MR counterparts only if they also carry at least one short allele of the 5-HTTLPR [64,86,873], while those PR monkeys with the l/l genotype are protected from these effects of peer rearing and maternal separation. Importantly, carriers of the s allele are not hyperaggressive if they are MR. These results indicate an important gene–environment interaction in the development of heightened aggressiveness. These findings also fit with my proposal that early chronic social stress may increase the release of 5-HT from raphe neurons, and if 5-HT is not either quickly sequestered back into the serotonin axon terminal via an efficient 5-HTT, or quickly degraded via an efficient MAOA enzyme, then the higher-than-normal synaptic levels of 5-HT may have deleterious effects on serotonin neural system development, leading to decreases in 5-HT release and increases in emotionality and hyperresponsiveness to aversive stimuli later in development (not all findings are consistent with my interpretation; [417]). Please note that this developmental argument is emphasizing that the low CSF levels of 5-HIAA in *adulthood* are due to decreased 5-HT release because of the negative feedback effects of high synaptic levels of 5-HT during early development.

Results consistent with the monkey findings have been reported for humans. Reif et al. [801] genotyped a population of male criminals and also categorized them into groups of violent and nonviolent offenders. A questionnaire was administered to each subject that, through retrospective recollection, measured each subject's early life history, such as instances of physical abuse and instances of parental separation or loss. High levels of criminal violence were associated with early childhood adversity only in carriers of the s allele of the 5-HTTLPR. Carriers of the s allele who did not experience early adversity were much less likely to be violent offenders. More recent research has replicated this important finding [198].

In a laboratory experiment with healthy humans playing a computer game, Verona, Joiner, Johnson, and Bender [993] reported that male subjects who were stressed by air blasts were more likely to administer shocks to an opponent (the subjects thought they were shocking an opponent, but actual shocks were not administered) if they were homozygous for the s allele of the 5-HTTLPR (s/s) compared to males carrying the l/l or s/l genotype. Women did not show this effect, although it has been argued that the s allele may make women prone to anger but not necessarily physical violence [198]. Under nonstressed conditions, the number of shocks administered was low for all subjects. Although this study did not measure the early life experiences of the subjects, it does suggest that a double dose of the s allele makes men more prone to aggressive behavior while they are currently experiencing a stressful event.

Several fMRI studies have indicated a positive association between the 5-HTTLPR short allele and the amygdala BOLD response to aversive facial expressions [671], making it likely that carriers of the short allele would be hyperresponsive to aversive social stimuli, and that this hyperresponsiveness might lead to aggression under stressful life conditions. In line with our review of MAOA polymorphisms, the interaction between early life stressors and the short allele of the 5-HTT may lead to a dysregulation of PFC control of amygdala reactivity, leading to heightened emotionality, anxiety, and aggression.

Finally, PET scan studies on rhesus monkeys [188] and humans [216], using a radioligand that binds to 5-HT1A receptors, have reported that carriers of the s allele of the 5-HTT have reduced 5-HT1A binding levels in cortical brain regions. David et al. [216] suggested that the lower transcriptional efficiency of the s allele leads to an increase in 5-HT postsynaptic activity that, in turn, leads to a compensatory desensitization or downregulation of 5-HT1A receptors.

3.5 CONCLUSIONS

This has been a long and complex chapter, and many controversies exist, particularly with respect to the involvement of 5-HT in aggression. In addition, while research on humans has emphasized a disruption in orbital and medial PFC function as contributing to aggression, not enough research has been done on the role of the PFC in the aggression of nonhuman animals. In my view, increased animal research on the neurobiological mechanisms through which the PFC interacts with the amygdala, hypothalamus, and other brain regions will be necessary to fully appreciate how PFC dysfunction relates to aggression in humans. Further, very little animal research has been done on the role of the NA–VP circuit in aggression, with most research concentrating on amygdala projections to a hypothalamic–midbrain system, projections that are likely to regulate consummatory, rather than goal-directed, responses. An investigation of the connections between the amygdala and the NA–VP circuit, and the effects of the hypothalamus and PFC on such connections, might provide insights into the more voluntary aspects of offensive aggression. Finally, although this chapter emphasized the neurobiology of offensive aggression, pathological aggression in animals and humans appears to be ambiguous, presenting a mix of anger and fearfulness, or more objectively, a mix of offensive and defensive aggression. More research needs to be aimed at uncovering the similarities and differences underlying the neural bases of these two forms of aggression.

The content of this chapter can be summarized with the following main points:

1. A core neural circuit comprising a MeA/BMA-to-HAA-to-PAG pathway promotes aggression in a variety of species.
2. Within this core circuit, substance P and vasopressin act as proaggressive neuropeptides. However, vasopressin is not an exclusive antisocial peptide, since it can exert prosocial effects within other neural circuits. Oxytocin, which has prosocial effects, antagonizes aggressive tendencies.
3. The view has been presented that social stimuli can be processed over separate and distinct neural circuits, one reacting to social stimuli having a positive valence, while the other reacts to social stimuli with a negative valence. These distinct circuits give rise, respectively, to prosocial or antisocial (including aggressive) responses.
4. The PFC exerts a modulatory influence over the core neural circuit for aggression. While medial PFC (infralimbic) projections to the HAA may promote aggression, projections of the orbital PFC to the amygdala may decrease aggression. The medial PFC also projects to the amygdala, and such projections may have diverse effects on aggression, depending on the particular medial PFC area involved.
5. The role of 5-HT neural systems in aggression is complex and controversial. Most, but not all evidence, is consistent with the serotonin deficiency hypothesis of impulsive affective aggression. Evidence suggests that there are distinct projection circuits from 5-HT brainstem neurons to the forebrain, and that activation of certain circuits depresses aggression, while activation of other circuits may increase aggression. 5-HT action on 5-HT1A receptors in the PFC and HAA suppresses offensive aggression. In contrast, 5-HT action on 5-HT2A receptors, perhaps in the PFC, may stimulate aggression. The effects of 5-HT on aggression may be due to its direct effects on circuits that regulate aggression, or may be due, more generally, to its effects on circuits that mediate an organism's responsiveness to many types of stressful stimuli, which would include social threats. A disruption of PFC regulatory functions by abnormal 5-HT neural transmission may contribute to such hyperreactivity to stressful events. Another mechanism by which 5-HT circuits could influence aggression is through a direct effect on neural systems that regulate fearfulness and anxiety. All of these processes are interrelated, and careful research will be needed to fully appreciate the specific mechanisms through which 5-HT can either decrease or increase aggression within particular contexts.
6. Genetic polymorphisms influence aggression: certain variants of the AVP1AR, OTR, MAOA, and 5-HTT genes have been related to increased aggressive tendencies in humans.
7. Early adverse social experiences, such as maternal neglect or social isolation, increase reactive aggression in the affected offspring. This aggression presents itself as a mixture of offensive and defensive responses. Early social deprivation may increase aggression by ultimately causing a decrease in adult

5-HT neural transmission and by causing a shift within the medial PFC from 5-HT1A to 5-HT2A receptor dominance. As to the underlying mechanisms, early adverse social experiences may initially cause increased 5-HT release, but during the early postnatal period of brain development, large increases in endogenous 5-HT release may exert a negative feedback effect that blocks the full and complete development of 5-HT neural circuits, resulting in 5-HT hypofunction in adulthood and the functional dominance of the 5-HT2A receptor over the 5-HT1A receptor.

8. Genetic effects interact with early adverse social experiences to influence the development of aggression. Carriers of the MAOA-L allele or the short allele of the 5-HTT gene react more strongly to the aggression-promoting effects of early life social stress. This interactive effect was interpreted within the context of the proposal that heightened 5-HT release during early brain development operates to suppress the normal development of 5-HT neural circuits. This 5-HT neural system dysfunction then appears to lead to a hyperresponsiveness of the amygdala to aversive social stimuli, due in part to faulty functional connectivity between the PFC and amygdala. A hyperresponsiveness of the hypothalamus to aversive inputs is an additional possibility.

CHAPTER

4

Sexual Behaviors and Sexual Differentiation

4.1 INTRODUCTION

Male and female sexual behaviors are the most elementary forms of prosocial behaviors. During sexual encounters, males and females are attracted to each other and engage in various types of courtship responses that ultimately culminate in copulation. As described in Chapter 1, the phase of sexual behavior during which males and females are attracted to one another, seek each other out, and engage in courtship responses is referred to as the appetitive (goal-directed) phase, while the behaviors that comprise copulation are referred to as the consummatory phase. The performance of the consummatory phase of motivated behaviors is generally considered to be a rewarding experience.

Chapters 4, 5, and 6 review the core prosocial behaviors that occur in mammals. Sexual behavior occurs in both sexes of all mammals, the mating system of most mammals is polygamous or promiscuous, and males and females typically leave each other after mating (Chapter 4). Therefore, although sexual behaviors are prosocial or positive social interactions, in most mammals they represent short-lasting encounters that do not involve the formation of enduring bonds between mating partners. Following mating, in the typical mammal, the next major prosocial behavior is maternal behavior (Chapter 5), which does represent an enduring social bond between a mother and her infant(s). Although most mammalian mating partners leave each other after mating, about 5% of mammals are monogamous, and the mating partners do form an enduring bond and stay together after mating is completed (Chapter 6). In such species, both maternal and paternal behaviors are exhibited. In the chapters that follow, the neurobiology of each of these forms of prosociality will be reviewed.

In this chapter, we are interested in two broad issues: (1) The regulation of male and female sexual behavior in adults. What are the underlying neural mechanisms that control the appetitive and consummatory aspects of male and female sexual behaviors? Are the mechanisms that regulate appetitive sexual responses distinct from those that regulate consummatory sexual responses? (2) Sexual differentiation or the developmental processes that cause the sexes to become different. What are the mechanisms that cause the typical male mammal to be sexually attracted to females and the typical female mammal to be sexually attracted to males? Although heterosexuality is the norm in mammals, homosexual interactions do occur. Can an

Neurobiology of Social Behavior
http://dx.doi.org/10.1016/B978-0-12-416040-8.00004-3

understanding of the developmental neurobiological processes that give rise to heterosexuality shed light on the neurobiological underpinnings of homosexuality in humans?

4.2 THE HORMONAL BASIS OF SEXUAL BEHAVIOR

This section will briefly review the role of steroid hormones in the sexual behavior of male and female mammals [677]. Testosterone, secreted from the testes, is essential for male sexual behavior. Castration leads to a decline in sexual behavior, which can be reinstated with exogenous testosterone administration. For female mammals, estradiol, secreted by the ovaries, is the primary hormone that stimulates female sexual behavior, although in some species the subsequent secretion of progesterone from the ovary potentiates the stimulatory effect of estradiol. Steroid hormones secreted from the gonads into the systemic blood circulation are capable of entering the brain, and that is where they exert their most important effects on sexual behavior.

In many mammals, breeding seasons occur, while in other mammalian species, sexual activity can occur throughout the year [677]. In the typical mammal, when breeding occurs, the testes and ovaries are functional, gonadal steroid hormones are secreted, and gametes are produced. During nonbreeding seasons, when males and females are not sexually interested in each other, the gonads are dormant and steroids and gametes are not produced. When mammals are in reproductive condition, males typically exhibit relatively continuous sexual interest in receptive females, and testosterone levels in the blood remain elevated throughout the breeding season. In contrast, the typical female mammal exhibits cyclical sexual activity during the breeding season, which is referred to as an estrous cycle. During the estrous cycle, females show sexual behavior only around the time of ovulation, and such behavior is typically preceded by a surge in estradiol secretion from the ovaries. Therefore, for the typical (nonprimate) mammalian male and female, nonreproductive sexual behavior does not occur. Males mate when testosterone levels are high and sperm is being produced, and females mate only near the time of ovulation. Importantly, for most nonprimate mammals, steroid hormones are essential for both the appetitive and consummatory aspects of sexual behavior.

Concerning the role of testosterone in the sexual behavior of male primates, the following research is instructive. Rhesus monkeys are seasonal breeders, strongly suggesting that gonadal steroids influence both male and female sexual behavior in this primate species. Wallen, Eisler, Tannenbaum, Nagell, and Mann [1007] studied the sexual behavior of group-living male rhesus monkeys in an outdoor enclosure during the breeding season. Systemic administration of a long-acting gonadotropin-releasing hormone (GnRH) antagonist caused blood levels of testosterone to decline (see Chapter 1). Male sexual behavior declined significantly by 1 week posttreatment and was almost completely eliminated at 4 weeks posttreatment. In conformance with these findings, other studies have shown that castration causes a decline in the sexual behavior of male rhesus monkeys and that this decline can be reversed with testosterone treatment [773]. Kwan, Greenleaf, Mann, Crapo, and Davidson [524] studied a population of hypogonadal men whose blood levels of testosterone were very low. Based on self-reports of sexual activity and sexual feelings, after these men were treated with testosterone the frequencies of these sexual responses increased. Bagatell, Heiman, Rivier, and Bremmer [47] administered a GnRH antagonist to healthy men. This treatment caused testosterone levels to decline, as did the reported frequency of sexual desire (libido), sexual fantasies, and intercourse. These effects were reversed by concurrent testosterone treatment.

Wallen [1005] has been a strong proponent of the view that testosterone primarily regulates the appetitive aspects of sexual behavior in men (sexual desire, interest, and libido), rather than the consummatory aspects of the behavior (the ability to copulate), since hypogonadal men exhibit erectile responses to erotic films. However, other studies indicate that testosterone does influence erectile function in men, although the threshold level of testosterone needed to stimulate erection appears to be much lower than that needed to stimulate sexual desire [643].

The case for an appetitive-consummatory distinction in the hormonal regulation of sexual behavior is stronger for female primates, including women. Most nonprimate female mammals will only engage in sexual activity around the time of ovulation, but female primates are capable of copulating throughout their ovulatory cycles [1004]. Although ovariectomy depresses sexual activity in macaque monkeys, it does not eliminate it [195,1004]. Wallen has made the distinction between steroid effects on desire and ability, and he has presented evidence that the periovulatory surge of estradiol primarily modulates the female primate's sexual attraction toward and pursuit of males [1004,1005,1009]. Studies on women are consistent with these findings on monkeys; although women can engage in sexual intercourse throughout their menstrual cycles, their sexual desire or libido peaks near the time of ovulation [1005]. As one example, Haselton and Gangestad [395] studied normally cycling women who provided daily reports of their sexual interests and feelings for 35 days. Near ovulation, the women reported feeling more physically attractive and had a greater interest in attending social situations where they might meet men [791,823].

In my review of the neurobiology of sexual behavior in nonprimate mammals, steroid effects on hypothalamic and limbic circuits will be shown to influence both the appetitive and consummatory aspects of sexual behavior. The fact that human females can engage in sexual intercourse at any time and for multiple reasons emphasizes the importance of cognitive control over consummatory sexual activity in women [354], and such control undoubtedly also occurs in men. What is clear, however, is that in all mammals, human and nonhuman, gonadal hormones influence the appetitive aspects of sexual behavior: attraction to a sex partner, sexual desire, and libido. Further, the fact that steroid hormones affect libido more than they affect the ability to copulate in primates suggests that the neural mechanisms regulating appetitive sexual activity are at least partly distinct from those that regulate consummatory sexual responses. Understanding the neurobiology of the appetitive aspects of sexual behavior in mammals may be particularly relevant to understanding the neurobiological dysfunctions that underlie pathological sexual attractions, such as pedophilia.

4.3 THE NEUROBIOLOGY OF MALE SEXUAL BEHAVIOR IN NONPRIMATE ANIMALS

4.3.1 Introduction

Most of the experimental research on the neurobiology of male sexual behavior has been conducted on rats [432]. The appetitive phase of male sexual behavior in rats has also been referred to as the anticipatory, preparatory, or precopulatory phase. During this phase, the male engages in behaviors to seek out and approach a sexually receptive female, also referred to as an estrous female. Once the male is in proximal contact with an estrous female, he engages in the consummatory acts of copulation, which include mounting the female from the rear with accompanying pelvic thrusts, intromissions (penile insertions), and ejaculation. During sexual behavior in rats, the male mounts and dismounts a female several times, with some of the mounts resulting in a successful intromission; the male typically requires several intromissions before ejaculation occurs. Following an ejaculation, a male becomes temporarily refractory to female stimuli, and during this postejaculatory interval (PEI) the male does not engage in sexual behavior. Following the PEI, a new bout of mounts, intromissions, and ejaculation resumes. A male may ejaculate up to eight times with a receptive female before he reaches a state of sexual satiety [432].

The consummatory aspects of male sexual behavior are studied by measuring copulatory responses during a time-limited test, such as the frequency of mounts, intromissions, and ejaculations that occur when a male interacts with an estrous female. Several experimental paradigms have been created to selectively examine the appetitive aspects of male sexual behavior, and some of these will be briefly described. Everitt [275] has used operant conditioning procedures to study appetitive male sexual behavior in rats: during initial conditioning trials, male rats mate with a receptive female in the presence of a flashing light, which becomes a conditioned reinforcer. In subsequent 15 min operant sessions, the males are trained to press a lever for the light, and at the end of the 15 min session the male is presented with an estrous female and allowed to copulate with her (the primary reward). Pressing the operant lever to gain access to a receptive female is used as a measure of appetitive sexual motivation.

Pfaus and Phillips [769] have used a bilevel cage to study appetitive male sexual behavior. On each of several training trials, a male is placed in this cage for 5 min and then a receptive female is introduced into the cage for another 20 min. During this 20 min interval, the male seeks out and mates with the female. Once a male is experienced in this situation, the measure of appetitive sexual motivation is the number of level changes that the male exhibits in his search for the receptive female during the 5 min period prior to the actual placement of the female in the cage.

The procedures just described require a neutral stimulus (flashing light or cage area) to acquire attractive properties as a result of being paired with sexual activity. Other researchers have studied appetitive sexual motivation in male rats by using a preference test to examine the attractive or incentive motivational properties of estrous (sexually receptive) female stimuli. In one example of such a preference test, a test male rat is placed in the central compartment of a three-compartment cage, with an estrous female in one of the outside compartments and a nonestrous female or a male in the other. The test male is allowed to freely move around the cage, but he is separated from the stimulus rats by a wire mesh barrier, which allows for auditory, olfactory, and visual stimulation but prevents physical interaction and copulation. In a time-limited test, the amount of time the test male spends in the compartment with the estrous female is used as a measure of appetitive male sexual motivation [437].

4.3.2 The Medial Preoptic Area and Male Sexual Behavior

Lesions or damage to the medial preoptic area (MPOA), located in the rostral hypothalamus, have been shown to eliminate copulation (mounts, intromissions, and ejaculation) in all species that have been examined, including nonhuman primates [393,432]. Figure 4.1 shows frontal and sagittal sections of the rat brain, indicating the location of the MPOA as well as other important structures. In addition to the MPOA, note the locations of the nucleus accumbens (NA) and ventral pallidum (VP) in the telencephalon, the ventromedial nucleus (VMN) in the hypothalamus, the ventral tegmental area (VTA) and periaqueductal gray (PAG) in the midbrain, and the nucleus paragigantocellularis (nPGi) and gigantocellularis (nGi) in the reticular formation of the ventral medulla.

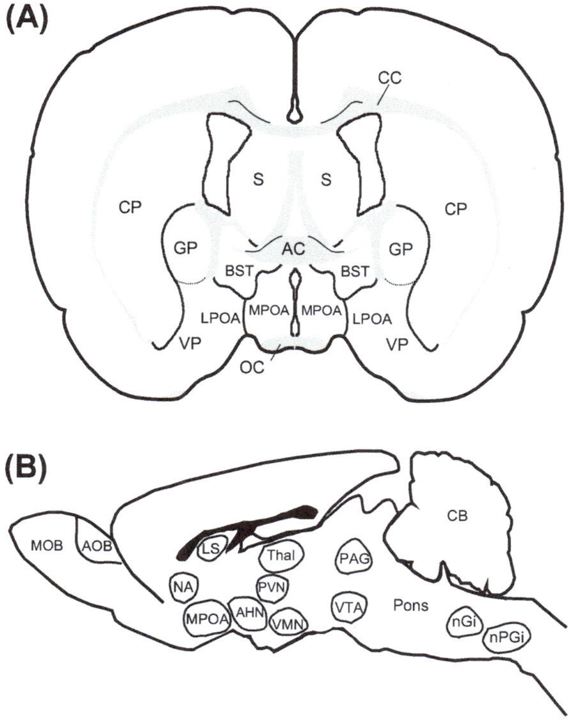

FIGURE 4.1 Frontal (A) and sagittal (B) sections through the rat brain at the level of the medial preoptic area. Note the locations of the following brain regions: AOB = accessory olfactory bulb; AHN = anterior hypothalamic nucleus; BST = bed nucleus of the stria terminalis; LPOA = lateral preoptic area; MOB = main olfactory bulb; MPOA = medial preoptic area; NA = nucleus accumbens; nGi = nucleus gigantocellularis; nPGi = nucleus paragigantocellularis; PAG = periaqueductal gray; PVN = paraventricular nucleus of the hypothalamus; Thal = thalamus; VMN = ventromedial nucleus of the hypothalamus; VP = ventral pallidum; VTA = ventral tegmental area. Other abbreviations: AC = anterior commissure; CB = cerebellum; CC = corpus callosum; CP = caudate/putamen; GP = globus pallidus; LS = lateral septum; OC = optic chiasm; S = septal area (the lateral septum lies laterally and the medial septum is toward the middle of the septal area). *The brain sections have been modified from Swanson LW. Brain maps: structure of the rat brain. 2nd ed. Amsterdam: Elsevier; 1998/1999.*

Not only do MPOA lesions disrupt male mating behavior but the MPOA is also a site where testosterone acts to stimulate male sexual behavior, which is consistent with the fact that MPOA neurons contain high densities of intracellular androgen receptors [432,901]. In castrated male rats, testosterone implants into the MPOA stimulate mating behavior [217], while implants of an antiandrogen into this region disrupt copulation in males that have been treated systemically with testosterone [623]. As reviewed in Stolzenberg and Numan [929], the aromatase enzyme, which converts testosterone to estradiol, is present in the MPOA. Therefore, the stimulatory effects of testosterone on male sexual behavior may be due to the action of testosterone on androgen receptors and also to an action of estradiol on estrogen receptors in MPOA. Indeed, implants of estradiol into MPOA also stimulate male sexual behavior in castrated rats [432].

Two sources of sensory input to the MPOA that appear to drive output circuits relevant to male sexual behavior are olfactory inputs relayed through the medial amygdala and genitosensory tactile inputs ascending from the spinal cord with a relay in the midbrain central tegmental field [73,154,252,626,821].

In most of the studies cited above, the role of the MPOA in consummatory sexual responses was studied. That is, the effects of lesions of the MPOA or testosterone/estradiol stimulation of the MPOA on mounts, intromissions, and ejaculations were studied. Clearly, the MPOA is critically involved in the regulation of the consummatory aspects of male sexual behavior, but is it also involved in the appetitive aspects? In an influential series of studies, Everitt and colleagues [275,277] proposed that the MPOA only regulates the consummatory aspects of male sexual behavior (the execution of copulatory responses), while the mesolimbic dopamine (DA) system, acting independently from the MPOA, regulates the appetitive aspects of male sexual behavior. Testosterone, however, regulates both appetitive and consummatory male sexual responses. The findings supporting these views were that excitotoxic MPOA lesions, which would produce neuron-specific lesions without damaging axons of passage through the MPOA, eliminated consummatory mounting, intromission, and ejaculation responses in the presence of a receptive female but did not depress an operant lever pressing response, which was reinforced by a conditioned reinforcer (flashing light) that was used to ultimately gain access to a receptive female. Preoperatively, sexually experienced male rats learned the operant response. After MPOA lesions, during the first 15-min postoperative session, male rats pressed the lever at the same high rate as they did preoperatively, but when the female was presented, they could not mate with her. In contrast, manipulations of the mesolimbic DA system influenced appetitive sexual responses without affecting consummatory copulatory responses. Finally, castration was found to decrease operant responding and copulatory responses, both of which could be reinstated with systemic testosterone injections.

With respect to the appetitive operant response, the following is worth noting: although lever pressing for the flashing light conditioned reinforcer was not disrupted during the first 15-min postoperative session, over subsequent sessions the operant response rate did decline, and this was related to the fact that mating behavior and the rewarding properties of intromission and ejaculation could not occur; the absence of primary reinforcement caused a loss of the secondary reinforcing properties of the flashing light [275]. However, since the MPOA is clearly involved in regulating intromissions and ejaculations, could some of its neural circuits also be involved in allowing the flashing light to acquire reinforcing properties? I will return to this issue later.

Everitt [275] argues that the mesolimbic DA system (the dopamine circuit originating in the VTA and terminating in the NA) regulates appetitive male sexual motivation independent of the MPOA. However, such a view conflicts with neuroanatomical tracing studies showing that the MPOA provides a massive neural input to the VTA ([716,929]; also see Chapter 1). Based on this anatomy, the MPOA is certainly positioned to affect the appetitive aspects of male sexual behavior.

4.3.3 A Neurobehavioral Model

Figure 4.2, relying heavily on the research reviewed in Stolzenberg and Numan [929] and in Chapter 1 of this book, presents a neural model of how the MPOA might regulate *both* the

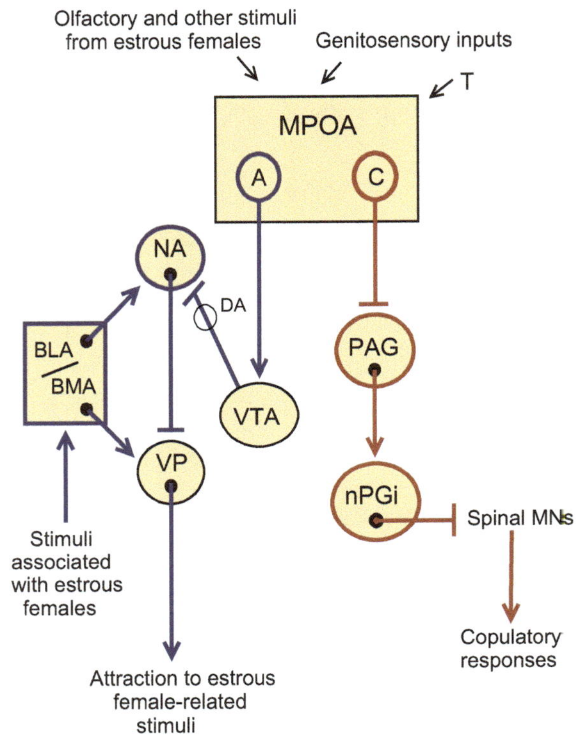

FIGURE 4.2 A neural model showing the circuits that may regulate the appetitive (blue) and consummatory (red) components of male sexual behavior in rats. The medial preoptic area (MPOA) is proposed to contain two separate populations of neurons controlling either appetitive (A) or consummatory (C) male sexual responses. MPOA neurons that regulate consummatory responses do so, in part, through inhibitory projections to the periaqueductal gray (PAG). MPOA neurons that regulate appetitive sexual responses in males do so via excitatory projections to ventral tegmental area (VTA) dopamine (DA) neurons. See text for additional details. Axons ending in a bar are inhibitory, and those ending in an arrow are excitatory. Other abbreviations: BLA = basolateral amygdala; BMA = basomedial amygdala; MNs = motor neurons; NA = nucleus accumbens; nPGi = nucleus paragigantocellularis; T = testosterone; VP = ventral pallidum.

appetitive and consummatory aspects of male sexual behavior. After describing the model, I will present the evidence for its support.

The figure shows two separate populations of MPOA neurons, one regulating appetitive male sexual behaviors and the other regulating consummatory male sexual responses. The MPOA is shown as being stimulated by testosterone, olfactory stimuli from receptive females, and genitosensory inputs that result from intromissions. The consummatory output from the MPOA is shown as inhibiting the PAG, which is shown as exciting the nPGi. The nPGi in the reticular formation of the ventral medulla provides a major inhibition over sexual reflexes and other copulatory responses by inhibiting motor neuron pools in the lumbar and sacral spinal cord that control sexual responses. The MPOA, by inhibiting the PAG, removes a major source of excitation from nPGi, in this way facilitating the activity of spinal motor neurons that regulate consummatory copulatory responses.

The interaction of a separate population of MPOA neurons with the mesolimbic DA is proposed to regulate appetitive male sexual behavior; this is conceived as an interaction between a specific and a nonspecific motivational system. Note that in this model, sensory stimuli associated with an estrous female reach both the MPOA and the nucleus accumbens–ventral pallidum (NA-VP) circuit. Recall from Chapter 1 that the mesolimbic DA system regulates a variety of goal-directed responses: DA release into NA can increase the motivational salience of a variety of biologically significant stimuli. To the extent that different hypothalamic neurons regulate distinct motivational states, and given that these hypothalamic neurons project to VTA, the release of DA into NA can be selectively activated by neural output from distinct hypothalamic nuclei, which, in turn, respond to their own suite of sensory and physiological inputs. Such an interaction can result in motivational specificity by restricting the types of stimuli that activate DA release into NA at a particular point in time, which, in turn, regulates the types of stimuli that can be processed through the NA-VP circuit. In the presence of a sexually receptive female, if the MPOA did not activate DA release into NA, as would occur in castrated males (or in males outside the breeding season), then appetitive sexual responses should not occur. The model shows that when the MPOA is properly primed with testosterone, estrous female stimuli activate MPOA projections to VTA. DA release into the NA acts to disinhibit the VP, allowing the VP to respond to stimuli that are associated with receptive females and that reach the VP via projections from the basolateral amygdala/basomedial amygdala (BLA/BMA) nuclei in the amygdala (see Figure 1.10 in Chapter 1). In contrast, testosterone action on MPOA would not be expected to increase hunger or thirst, which are regulated by other neuron populations within the hypothalamus that respond to different external stimuli (food- or water-related) and need to be primed by other internal factors and stimuli, such as decreased glucose levels or dehydration (see Chapter 1).

When discussing the general characteristics of appetitive motivational processes, one is referring to an organism's attraction to a rewarding stimulus, and, therefore, appetitive behavior can also be called reward-seeking behavior. A rewarding stimulus, be it a sex partner, food, infants, etc., has several properties [91]: (1) it is attractive and activates appetitive approach behaviors; (2) it has reinforcing properties so than once the reward is obtained, the occurrence of consummatory responses toward the rewarding stimulus reinforces or strengthens those particular appetitive responses that successfully obtained the desired goal; (3) at least for humans, the acquisition of the reward and the subsequent consummatory responses that occur result in pleasurable hedonic sensations. With respect to the mesolimbic DA system, Berridge [91] has been a strong advocate of the position that DA release into NA primarily regulates reward-seeking (appetitive) responses and is not critically involved in the reinforcing and hedonic aspects of reward. However, in this and subsequent chapters, I will review the evidence that DA release into NA is not only essential for a variety of appetitive responses to biologically significant stimuli but that the DA release that occurs during consummatory behavior also has reinforcing effects, strengthening those particular appetitive responses that were successful in obtaining the primary reward.

4.3.4 MPOA Interactions with the Mesolimbic DA System and Appetitive Male Sexual Behavior

4.3.4.1 The MPOA and Appetitive Male Sexual Behavior

Unlike the findings from Everitt's group, other research shows that the MPOA is importantly involved in appetitive male sexual behavior. Hurtazo et al. [437] tested sexually experienced male rats in a three-compartment cage, with a stimulus estrous female in one compartment and a stimulus male rat in the other. During a 10-min test, the time that the test male spent near each stimulus rat was recorded. Wire mesh partitions prevented physical contact between animals. Control test males spent most of their time in the vicinity of the estrous female, while test males that received lidocaine injections into MPOA showed a dramatic decrease in their preference for the receptive female. Lidocaine blocks voltage-gated Na^+ channels, and therefore would block action potentials and neural activity within MPOA neurons (and axons of passage).

With respect to DA systems in the brain, in Chapter 1 I described the mesolimbic DA system and the nigrostriatal system. However, there are several other DA neural pathways in the brain, and one of these is referred to as the incertohypothalamic DA pathway. Dopamine neurons within the periventricular hypothalamus of this system (see Figure 1.2) send their axons into the MPOA. DA action on MPOA neurons has been shown to affect both appetitive and consummatory male sexual behavior in rats [432]. With respect to appetitive sexual activity, Pfaus and Phillips [769], using the bilevel cage test, found that haloperidol (a mixed D1/D2 receptor antagonist) microinjections into the MPOA dramatically reduced the number of female-searching level changes during the 5-min period prior to the introduction of the estrous female into the apparatus. Others have reported similar findings [668].

Although most experimental research on the neurobiology of male sexual behavior has been done on rats, a substantial body of research has also been performed on a bird species, the Japanese quail [57]. These researchers used a social proximity score as a measure of appetitive male sexual behavior in quails, and copulatory responses as a measure of consummatory male sexual behavior. Male birds were placed in a two-compartment cage with the male on one side of a partition and a receptive female on the other side. There was a window in the partition that permitted the male to view the sexually receptive female. Sexually experienced males were placed in this apparatus, and after a fixed period of time the partition was removed and the birds were allowed to mate. Preoperatively, the males spent most of their time viewing the female through the window (high social proximity score), and when the partition was removed, they copulated with the female. Postoperatively, males with lesions in the rostral MPOA no longer were interested in viewing the female through the window (elimination of appetitive motivation), but when the partition was removed, they were capable of mating. In contrast, males with lesions located in the caudal MPOA continued to spend most of their time viewing the female through the window, but when the partition was removed, they failed to mate (elimination of consummatory sexual responses). These findings support the view that different populations of MPOA neurons are involved in the regulation of appetitive and consummatory male sexual behavior in birds.

These results in quail are supported by a very interesting study done on rats [623]. Consummatory copulatory behavior of sexually experienced male rats was examined during mating tests with a receptive female, and appetitive responses were examined during a preference test. In the preference test, the male was placed in a three-compartment cage, with an estrous female in one compartment and a nonestrous female in the other. Although the male could physically interact with each female, intromissions were prevented because masking tape was placed over the female's vaginal opening. Intact males showed normal copulatory behavior during the mating tests and preferred the estrous female during the preference test. Castration abolished both copulation and the preference for an estrous female. Following these castration results, males were treated systemically with testosterone, which restored both appetitive and consummatory sexual responses. Importantly, if testosterone-treated males were injected with an antiandrogen in the dorsal MPOA, they showed a decrease in their preference for an estrous female, but their copulatory behavior was normal. In contrast, antiandrogen injection into the ventral MPOA did not interfere with the preference for the estrous female but significantly decreased copulatory responses. These results suggest that testosterone acts on different populations of MPOA neurons to influence either the appetitive or consummatory aspects of male sexual behavior in rats.

Given these results, how can we interpret the original findings of Everitt and Stacey [277]? A likely possibility is that their MPOA lesions, which disrupted copulation but did not depress an operant response used to gain access to an estrous female, may have been located in an MPOA region that only contained consummatory MPOA neurons, and that larger lesions may have been needed to disrupt both phases of male sexual behavior. Additionally, perhaps under their particular test conditions, the MPOA may have been only *temporarily* nonessential for appetitive sexual responses over the first few 15-min postoperative operant tests. In reference to Figure 4.2, perhaps MPOA-directed DA release into NA during the *preoperative* sexual behavior sessions not only regulated appetitive sexual motivation, but also promoted the acquisition of the sexually reinforced operant response. The latter *reinforcing* effects of DA release during sexual activity may have resulted in synaptic strengthening within the NA-VP circuit so that the flashing light was able to *temporarily* activate VP output circuits in the absence of MPOA activity during the initial postoperative tests. Therefore, MPOA interaction with the mesolimbic DA system during sexual behavior with the estrous female may have enabled the flashing light to temporarily acquire secondary reinforcing properties (cf. [570]). As will be shown in subsequent chapters, such potential reinforcing effects of DA release into NA appear to play a role in the formation of an enduring mother–infant bond (Chapter 5) and in the formation of pair bonds between mates in monogamous vole species (Chapter 6). In other words, MPOA-directed DA release into NA may have both appetitive and reinforcing effects with respect to male sexual behavior.

4.3.4.2 The Mesolimbic DA System and Appetitive Male Sexual Behavior

If DA release into NA were involved in the appetitive aspects of male sexual behavior, then one would expect extracellular DA levels to increase in the NA during the anticipatory precopulatory period. Further, if DA release into NA also influenced the reinforcing properties of sexual behavior, allowing neutral stimuli that are associated with sexual activity to become secondary reinforcers and therefore conditioned attractive stimuli, then one would also expect DA to be released into NA during the consummatory phase of sexual behavior when copulatory responses are actually occurring. Research does indeed indicate that DA is released into NA during both phases of male sexual behavior [213,303,1026]. In these microdialysis studies, it was found that extracellular DA levels increased significantly in NA of male rats when a sexually receptive female was presented behind a screen, and DA levels increased even further once the screen was removed and copulation was permitted. In addition, Wenkstern et al. [1026] reported that DA levels increased in the NA of naïve male rats on their first exposure to a sexually receptive female but did not increase significantly when such males were exposed to a nonestrous female.

Employing systemic injections of a DA receptor antagonist (haloperidol), Ettenberg and colleagues have done some interesting work on the role of DA neural systems in both the appetitive aspects of male sexual behavior and in the reinforcing effects of engaging in consummatory sexual responses. Concerning the latter, Lopez and Ettenberg [568] studied the running speed of male rats down an alleyway to reach a female in a goal box. First, the males were placed in the goal box with the female; the female was behind a partition that allowed olfactory, visual, and auditory contact but did not allow physical contact and mating. On some occasions, an estrous female was in the goal box, and on other occasions a nonestrous female was in the goal. After exposure to the female, the male was placed in the start box and his running speed to the goal was measured. Naïve sexually inexperienced males ran faster for the estrous than the nonestrous female, indicating the innate attractiveness of certain estrous female stimuli. These inexperienced males were then allowed to mate with an estrous female in a new environment outside the runway apparatus. Some of these males were treated systemically with haloperidol during the mating experience, while others served as controls. At the doses used, haloperidol treatment did not affect any measure of consummatory sexual behavior. Two days following this sexual experience, when the animals were drug free, they were again tested in the alleyway, and their postsexual experience run times were compared to their presexual experience run times. For the control males, sexual experience decreased their run time to a goal box that contained an estrous female (they ran faster). In contrast, sexual experience did not have this effect on the males that had been treated with haloperidol; they did not increase their speed of running for an estrous female. These researchers suggested that sexual experience enhanced the incentive value of estrous female cues above the level observed in naïve males through associative (learning) mechanisms and that DA release into the brain during consummatory sexual behavior was involved in promoting this effect.

Concerning the role of DA neural systems in the attraction of a male rat to an estrous female (appetitive phase of sexual behavior), Lopez and Ettenberg [569] used the runway procedure describe above and examined how fast sexually naïve males would run from the start box to the goal box to gain access to an estrous female. Systemic injections of low doses of haloperidol decreased the running speed for the estrous female without affecting the time to gain access to either a nonestrous female or an empty goal box. These results suggest that depression of DA neural activity decreased the male's attraction to estrous female stimuli without affecting general locomotor ability.

These studies used systemic injections of DA receptor antagonists, and therefore, given that there are multiple DA systems within the brain, the neural site regulating the behaviors and processes described above cannot be ascertained. However, other studies have clearly shown that the NA is a likely site where DA acts to promote the appetitive aspects of male sexual behavior. Everitt [275] microinjected amphetamine, which stimulates the release of DA from axon terminals, into NA, and this treatment increased a male rat's operant bar press response rate to gain access to an estrous female. Pfaus and Phillips [769], using the bilevel cage procedure, found that haloperidol microinjections into the NA depressed the number of anticipatory level changes that were shown by male rats as they searched for an estrous female. However, once the female was introduced into the apparatus and was encountered by the male, the performance of consummatory sexual responses (mounts, intromissions, ejaculations) was not affected. These results certainly fit with the view that the mesolimbic DA system regulates appetitive male sexual responses, while another system regulations consummatory sexual behavior (see Figure 4.2). Finally, Bialy, Kalata, Nikolaev-Diak, and Nikolaev [94] examined the effects of injecting a DA-D1 receptor agonist into the NA on the initiation of copulation in sexually inexperienced male rats. In a 15-min test, only 40% of control males initiated copulation with an estrous female, while 90% of the D1 agonist–treated males did so. One interpretation of these results is that the activation of DA-D1 receptors in NA increases the male's attraction to estrous female stimuli so that the naïve male transitioned more quickly from precopulatory to copulatory activity. In Chapter 3 on aggression, I indicated that when a male is first exposed to an estrous female he alternates between rejection responses and sexual responses. In reference to Figures 3.2 and 4.2, one can understand how D1 receptor stimulation within NA might add to MPOA-directed DA release into NA in order to promote appetitive approach responses that ultimately lead to copulation.

4.3.4.3 MPOA Interaction with the Mesolimbic DA System and Appetitive Male Sexual Behavior

The above studies indicate that both the MPOA and the mesolimbic DA system are involved in the appetitive aspects of male sexual behavior. Since MPOA neurons project to and terminate in the VTA, this data would suggest that MPOA interaction with the mesolimbic DA system is involved in regulating the appetitive aspects of male sexual behavior. However, currently, there is no direct experimental evidence that proves this. An interesting study that has not been performed, but should be, would involve exposing a male rat with a unilateral MPOA lesion to an estrous female while also measuring DA release into NA. As will be described in more detail in Chapter 5 on Parental Behavior, the projections of the MPOA to the VTA and the projections of the VTA to the NA are primarily ipsilateral and uncrossed, which means that the right MPOA projects to the right VTA, which projects in turn to the right NA. Similarly, the MPOA on the left side of the brain would connect to the left VTA, which in turn would connect to the left NA. Since unilateral MPOA lesions do not disrupt sexual behavior (bilateral lesions are necessary; [432]), if MPOA projections to VTA direct DA release into NA during both the precopulatory and copulatory stages of male sexual behavior, microdialysis data should show decreased DA release into the NA ipsilateral to the MPOA lesion, but not in the NA on the contralateral side of the brain. In other words, during a sexual encounter, a male with an MPOA lesion on the right side of the brain should show a lower level of DA into the right NA in comparison to the left NA.

In examining Figure 4.2, it can be seen that separate populations of MPOA neurons are conceived as regulating appetitive and consummatory male sexual behavior. It is not surprising, therefore, to find that VTA lesions do not interfere with consummatory male sexual responses [118]. However, other midbrain lesions have been found to interfere with both the appetitive and consummatory aspects of male sexual behavior. Figure 4.3 shows a sagittal section of the rat brain with a representation of the neural pathways through which the MPOA reaches both the VTA (appetitive) and the PAG (consummatory). The green line in the section shows that a properly placed lesion in the midbrain should be able to interfere with both pathways [704,903]. Indeed, research has shown that electrical lesions at such a midbrain site, referred to

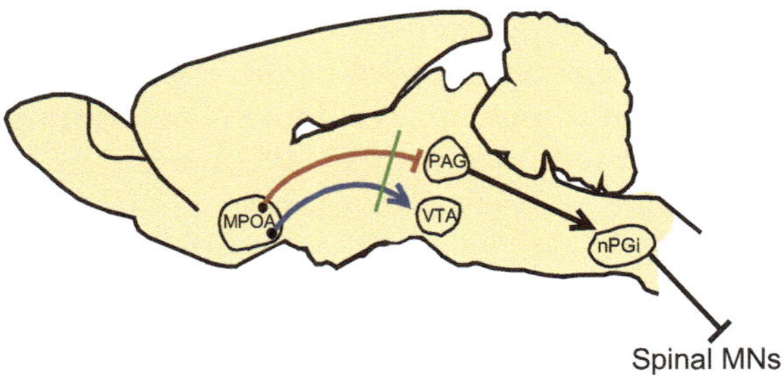

FIGURE 4.3 A sagittal section through the rat brain showing distinct medial preoptic area (MPOA) consummatory (red) and appetitive (blue) circuits. An appropriately placed lesion in the midbrain that is capable of destroying axons (shown in green), such as an electrical lesion, would be able to interfere with both circuits. Axons ending in a bar are inhibitory, and those ending in an arrow are excitatory. Other abbreviations: MNs = motor neurons; nPGI = nucleus paragigantocellularis; PAG = periaqueductal gray; VTA = ventral tegmental area.

as the dorsolateral tegmentum (DLT), disrupt both the preference of a male rat for an estrous female (appetitive) and his copulatory responses (consummatory) with such females [117,265]. Since these electrical lesions would damage both neurons and axons of passage through this part of the midbrain, it is possible that they damaged MPOA axons descending to both the VTA and to the PAG. Supporting the conclusion that these electrical lesions disrupted both types of sexual responses because axons of passage coursing through DLT were damaged, Romero-Carbente, Camacho, and Paredes [820] have reported that neuron (cell body)-specific excitotoxic amino acid lesions of the DLT do not disrupt the appetitive and consummatory aspects of male sexual behavior.

More research is certainly needed to support the proposal that MPOA activation of the mesolimbic DA system controls the appetitive aspects of male sexual behavior. Based on the findings of McGinnis et al. [623] and Pfaus and Phillips [769], I would predict that the injection of an antiandrogen into the dorsal MPOA on one side of the brain and a DA receptor antagonist into the NA on the opposite side of the brain should depress appetitive male sexual behavior (level changes in a bilevel cage or preference for an estrous over a nonestrous female), while leaving consummatory copulatory responses relatively intact. This procedure, referred to as an asymmetrical neural intervention, which will be described in more detail in Chapter 5 on Parental Behavior, would selectively interfere with appetitive male sexual neural systems on both sides of the brain, while presumably leaving consummatory neural systems intact. That is, MPOA-directed DA release into NA would be disrupted on both sides of the brain, while MPOA projections to PAG would remain functional.

Finally, in further support of the model shown in Figure 4.2, excitotoxic lesions of the BLA have been shown to disrupt a male rat's operant lever pressing response to gain access to a sexually receptive female without subsequently affecting his ability to copulate with her [276,626].

4.3.5 The MPOA and Consummatory Aspects of Male Sexual Behavior

Although evidence clearly exists that MPOA descending projections to the brainstem are involved in the regulation of the consummatory aspects of male sexual behavior, the exact site of termination of such descending projections has not been determined, and it is highly likely that MPOA efferents terminate in more than one brainstem site to influence the execution of copulatory behavior in response to the proximal cues from an estrous female [117,302,820]. In Figures 4.2 and 4.3, I suggested that the PAG is one of the sites where MPOA projects to influence consummatory sexual behavior. Recall from Chapter 1 that the PAG can be considered a premotor nucleus and that different parts of the PAG project to different parts of the medullary reticular formation, whose outputs form the reticulospinal tract. For consummatory male sexual behavior, the following mechanism is proposed: consummatory MPOA neurons project to and inhibit the output of a select group of neurons in the PAG that project to and excite the nPGi. Since it is also proposed that nPGi provides a descending inhibition over motor neuron pools in the lumbar and sacral spinal cord that regulate copulatory responses and genital reflexes (erection and ejaculation), MPOA output would reduce excitation of nPGi and facilitate copulation through a process of disinhibition.

The evidence for this proposal can be enumerated as follows: (1) anatomical research shows that MPOA neurons that contain androgen receptors contribute to an MPOA-to-PAG-to-nPGi-to-spinal cord circuit [606,673,690]; (2) with respect to genital reflexes, erection and ejaculatory response are spinal reflexes: in anesthetized male rats with spinal transections (the connections between the brain and the spinal cord are severed), genital stimulation, such as distension of the urethra, can elicit erections and ejaculation responses that are controlled by spinal motor neurons [199,606]. Such genital reflexes cannot be stimulated in rats in which the spinal cord has not been transected, indicating that the spinal mechanisms that control certain genital reflexes are under a descending inhibition from the brain [606]. Marson and McKenna [606] identified the nPGi as a major source of this tonic descending inhibition over genital reflexes regulated by the spinal cord: in anesthetized male rats whose spinal cord was not transected, genital stimulation elicited erection and ejaculation responses in males with excitotoxic amino acid lesions of nPGi. Importantly, several studies have shown that stimulation of MPOA efferents is capable of facilitating genital reflexes in intact male rats [432,607,751]. Putting these results together, it appears that in intact rats, MPOA output can reduce nPGi inhibition over spinal genital reflexes; (3) the above studies examined erectile and ejaculatory responses to genital stimulation in anesthetized or retrained rats. What roles do PAG and nPGi play in the regulation of consummatory male sexual responses in freely behaving rats? Electrical or neuron-specific excitotoxic amino acid lesions of nPGi have been shown to facilitate copulation [691,1053]. In comparison to controls, sexually experienced males with nPGi lesions exhibit an increased number of ejaculations before reaching sexual satiety. In sexually naïve male rats, on their first exposure to an estrous female during a 30-min test, males with nPGi lesions were more likely to copulate to ejaculation (95%) than were control males (50%), and the lesioned males also demonstrated a decrease in the number of mounts and intromissions that preceded ejaculation. One interpretation of these results is that nPGi lesions, by removing descending inhibition over spinal mechanisms, decreased that amount of genitosensory tactile stimulation needed to achieve ejaculation [1053].

Importantly, electrical and neuron-specific excitotoxic amino acid lesions of the PAG have also been shown to facilitate copulation in male rats [118,387]. For example, in a 30-min test, males with PAG lesions ejaculated 11 times, while controls ejaculated 6 times [118]. Further, males with excitotoxic lesions of the ventrolateral part of the PAG show shorter postejaculatory intervals, which may be another mechanism that contributed to an increased number of ejaculatory responses in a time-limited test.

Overall, these results are consistent with the proposed neural circuit shown in Figures 4.2 and 4.3 for the regulation of consummatory male sexual behavior, although not all findings are in agreement [605]. It is very likely that other MPOA circuits contribute to the consummatory aspects of male sexual behavior, but these remain to be determined. The goal of these last two major sections, however, was to show that the MPOA is involved in both appetitive and consummatory male sexual responses, and that these dual influences are exerted over two separate neural circuits.

4.3.6 The Prefrontal Cortex and Male Sexual Behavior

Only a few studies have explored the involvement of the prefrontal cortex (PFC) in male sexual behavior, but these studies are instructive. Recall that the output of the PFC can modulate the activity of amygdala, NA-VP, and hypothalamic circuits that are involved in motivated behaviors, and, for rats, as one moves in a dorsoventral direction through the medial PFC the following regions are observed (see Figure 1.14): anterior cingulate cortex (ACC), prelimbic cortex (PL), and infralimbic cortex (IL).

Agmo, Villalpando, Picker, and Fernandez [11] examined the effects of thermal lesions of the medial prefrontal cortex on the sexual behavior of sexually experienced male rats. The lesions destroyed the ACC and the dorsal parts of the PL area. However, since the lesions were produced by thermocoagulation, both cell bodies and axons of passage would be destroyed. These lesions appeared to interfere with the appetitive aspects of male sexual behavior, since at 1 week postlesion the affected males had extremely long latencies (greater than 60 min) to initiate copulation (as measured by mount and intromission latencies) when compared to control males (less than 2 min). However, once copulation was initiated, relatively normal consummatory sexual behavior ensued. The authors suggested that these lesions reduced males' sensitivity to estrous female stimuli, making activation of the behavior difficult. In relation to Figure 4.2, since the destroyed medial PFC region is known to project to the BLA [880] and could be a route through which some female stimuli are relayed from the cortex to the BLA, perhaps such lesions interfered with proper BLA input to the VP that would be needed for appetitive male sexual responses to female stimuli. Recall that BLA lesions also interfere with a male's operant responding to gain access to a receptive female without interfering with consummatory copulatory responses [276,626]. However, BLA-lesioned males do not show such extremely long mount and intromission latencies as observed after ACC + PL lesions. It is likely, therefore, that these medial PFC lesions interfered with additional processes necessary for the initiation of sexual behavior. Relevantly, the medial PFC also projects to the MPOA and to the NA-VP circuit [55].

In an interesting study, Davis et al. [219] allowed male rats to mate through ejaculation and then immediately injected them with either saline or lithium chloride (LiCl). LiCl produces gastrointestinal upset and is an aversive stimulus. In normal males, such an aversive consequence after mating inhibits subsequent mating behavior (only 20% of males continue to mate and ejaculate). However, males with excitotoxic amino acid lesions of the PL and IL portions of the medial PFC not only showed normal mating behavior (mounts, intromissions, ejaculations) but also continued to mate normally after the LiCl treatment. Davis et al. [219] suggested that such medial PFC-lesioned males lacked the ability to suppress seeking a sexual reward (ejaculation) in the face of the aversive consequences associated with mating.

In comparing the Davis et al. [219] study with Agmo et al. [11], in the latter study sexual behavior was depressed, while in the former study it was not. This difference is likely the result of two factors: Davis et al. produced neuron specific lesions that did not damage the ACC but did include IL. In contrast, Agmo et al. produced nonspecific lesions that destroyed both the cell bodies in the affected regions and axons of passage through the regions, and their lesions damaged ACC and PL, but did not directly affect the IL area.

How might one explain the continuance of sexual behavior in the face of aversive consequences in male rats with excitotoxic lesions of PL and IL? In Chapters 1 and 3, I emphasized the neural projections of the medial PFC to the amygdala, NA-VP circuit, and the hypothalamus in the regulation of motivated behaviors.

However, the medial PFC also has a dominant projection to the VTA [55,159]. Importantly, with respect to mesolimbic DA system function, PL and IL inputs to the VTA appear to terminate on GABAergic interneurons that inhibit VTA-DA projections to the NA [79,563]. Therefore, it is possible that in normal male rats exposed to aversive consequences following mating, some medial PFC neurons register these aversive consequences, and then through various projections, act to depress the initiation of sexual behavior. This action could involve depression of the mesolimbic DA system, BLA/BMA input to the NA-VP circuit, and/or depression of MPOA output (see Chapters 1 and 3).

The research of Davis et al. [219] conforms to the typical top-down PFC executive regulation of limbic, striatal, and hypothalamic processes, where the PFC is conceived as downregulating or restraining basic motivational processes [397]. However, the research by Agmo et al. [11] demonstrated that the medial PFC also plays a stimulatory role in basic motivational processes. Therefore, as previously described in detail within Chapters 1 and 3, different populations of neurons within the PFC may either downregulate or upregulate motivational processes.

4.4 THE NEUROBIOLOGY OF FEMALE SEXUAL BEHAVIOR IN NONPRIMATE ANIMALS

4.4.1 Introduction

Most of the experimental research on the neurobiology of female sexual behavior has been done on rats and some other rodents [766]. The estrous cycle of female rats is typically 4 days in length, which means that the female ovulates once every 4 days, with a period of sexual activity that brackets the time of ovulation. A female is said to be in behavioral estrus when she is sexually active, and if she mates with a male and pregnancy ensues, then estrous cycles are suspended during the 22 days of pregnancy. Upon giving birth, female rats exhibit a postpartum estrus, where mating can occur on the first postpartum day [332]. If mating occurs during the postpartum estrus, then while the female is lactating and caring for one litter she is also pregnant, and she gives birth to this second litter once the young from the first litter are weaned. Except for the postpartum estrus, estrous cycles do not occur during lactation.

Behavioral estrus is dependent upon the ovaries: ovariectomy eliminates sexual activity in rats and most other mammals, and in ovariectomized rats, adequate levels of behavioral estrus can be reinstated by estradiol alone, but when estradiol is followed 24–48 h later by the administration of progesterone, then full behavioral estrus becomes evident [677,766]. Significantly, during the normal estrous cycle, behavioral estrus and ovulation are preceded by sequential peaks in the secretion of estradiol and progesterone from the ovaries.

The occurrence of female sexual behavior near the time of ovulation is an excellent example of how the physiological state of an organism can influence the neural pathways over which social stimuli are processed. Nonestrous females actively reject the advances of sexually active male rats, indicating that male stimuli are processed over antisocial pathways. In contrast, estrous females actively seek out males and accept their advances, indicating that crucial steroid hormone effects on brain mechanisms enable male stimuli to be processed over prosocial acceptance neural pathways.

When a female rat is in behavioral estrus, sexual activity can be divided into appetitive and consummatory components [929]. The consummatory component, referred to as the lordosis posture, has been the most studied aspect of sexual behavior in female rats [763]. In response to mounts by the male, an estrous female remains stationary while arching her back with her head and rump elevated, in order to expose her vaginal region, which permits intromission. If lordosis does not occur in response to male mounting stimulation, then intromission is prevented. One measure of the intensity of a female's sexual receptivity is the lordosis quotient (LQ), which is simply the number of lordosis responses divided by the number of mounts that the female receives. For example, if a male mounts a female 10 times, but the female only shows a lordosis response to five of these mounts, then her LQ is 0.5 or 50%.

The female rat is an important initiator of sexual interactions; she either approaches the male during estrus, which allows him to mount her, or she avoids or escapes from him. Surprisingly, although such escape responses are dominant when a female is not receptive (nonestrous females), they also occur during estrus. Pacing behavior is the term used to refer to the manner in which the female rat regulates copulatory bouts during estrus [273], and such behavior is perhaps the best measure of the appetitive phase of female sexual behavior in rats. When an estrous female is placed in a partitioned cage with a sexually active male, which allows the female to escape and return to the male through an opening in the partition that is too small to allow the male to pass from one side to the other, then the active role of the female in pacing copulatory interactions becomes obvious. Two measures of paced mating behavior include the percent of exits from the male's compartment following a particular copulatory stimulus and the contact–return latency after a particular copulatory stimulus [273]. Females are more likely to escape from a male following an intromission than following a mount, and the contact–return latency (the amount of time that the female remains away from the male following mounts, intromissions, and ejaculations before reentering the male's chamber) is greater in duration following ejaculations and intromissions than following mounts. With respect to appetitive sexual motivation, Erskine et al. [273] suggest that the contact–return latency is a measure of the female's motivation to reinitiate mating.

Importantly, paced mating behavior has adaptive significance for the female [273]. A sufficient amount and an appropriate timing of vaginocervical stimulations are needed to effectively activate the corpora lutea in the ovaries through neuroendocrine mechanisms, which results in the enhanced progesterone secretion that is necessary for the maintenance of pregnancy. Paced mating behavior has been shown to increase the likelihood of a successful pregnancy. If a female receives five nonpaced intromissions, including one ejaculation (she is in a single cage with a male that does not permit her to escape and return to the male, resulting in an interintromission interval of about 30 s), the probability of pregnancy is about 10%, but if she is in a partitioned cage that allows her to pace copulatory bouts so that intromissions occur every 1–2 min, then about 70% of females become pregnant after five intromissions, including one ejaculation. Given the adaptive advantage of paced mating, one can assume that appetitive brain mechanisms have evolved in the female that cause her to become more sexually attracted to males following a certain critical latency after the last intromission stimulus, which in turn promotes a reinitiation of mating by the female.

In conclusion, for the female rat, paced mating, where the female actively initiates and terminates copulatory bouts, is the primary component of appetitive or goal-directed female sexual behavior, while the copulatory act, composed of lordosis, is the primary consummatory component. Importantly, when given a choice, females prefer to mate with males under conditions that allow pacing to occur, rather than to mate with a male in a confined area where the female cannot regulate copulatory bouts [202,452]. Therefore, paced mating presumably also has reinforcing or rewarding properties, allowing a female to learn to choose a male or a situation where paced mating previously occurred.

4.4.2 The Neural Circuitry Regulating Lordosis

4.4.2.1 Introduction

Figure 4.4 shows elements of the neural circuitry that regulate the lordosis response in rats [360,640,763,766]. This response to male mounting stimuli is considered a supraspinal reflex that requires hormonal priming, particularly at the level of the hypothalamus. The primary components of this system are the neurons in the ventrolateral division of the hypothalamic ventromedial nucleus (VMNvl), whose axons descend to and *stimulate* the output of the PAG. The PAG, in turn, projects to and stimulates a nucleus in the medullary reticular formation referred to as the nucleus reticularis gigantocellularis (nGi). nGi projects to and *excites* motor neuron pools in the lumbar spinal cord that promote contractions in those muscles that regulate lordosis. At the spinal cord level, these multisynaptic descending influences originating from the VMNvl summate with somatosensory inputs from male copulatory stimuli in order the fully activate the critical spinal motor neurons that trigger the lordosis response. Without the hormonally controlled descending stimulation from the hypothalamus, male copulatory stimuli alone are not sufficient to fully activate the lumbar motor neurons that control lordosis. Estradiol and progesterone are shown as acting at the level of VMNvl to stimulate lordosis. VMNvl dendrites that extend into the region lateral to the VMN contain oxytocin receptors (OTRs), and the action of OT, presumably derived from axons originating from the PVN, on these VMNvl dendrites is also stimulatory for lordosis. Further, the MPOA is shown as inhibiting lordosis, although the exact neural mechanisms underlying this effect are not known. Finally, a β-endorphin pathway originating from neurons in the arcuate nucleus of the hypothalamus, which lies ventromedial to the VMN, is shown as depressing the inhibitory MPOA neurons so that the lordosis response can be fully activated.

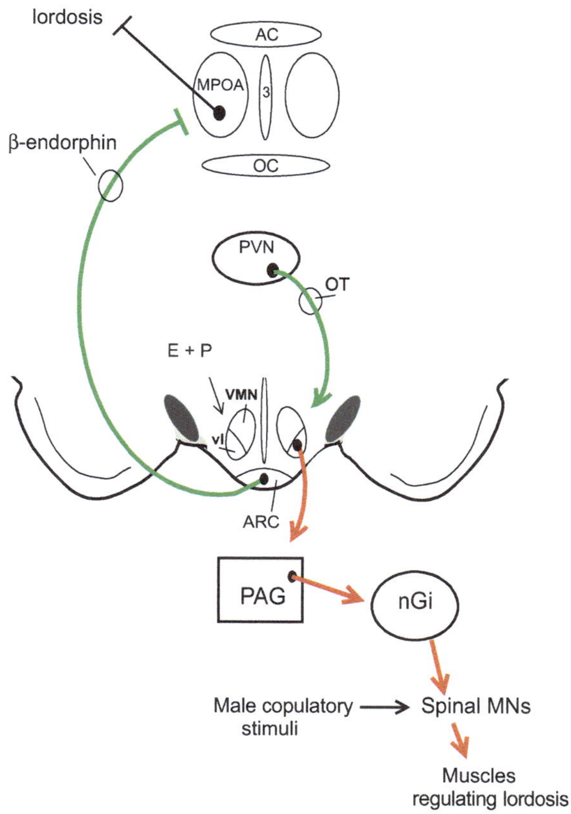

FIGURE 4.4 The neural circuitry regulating the lordosis response in female rats. Three rostral-to-caudal partial frontal sections through the hypothalamus are shown, indicating the medial preoptic area (MPOA), paraventricular nucleus (PVN), and ventromedial nucleus (VMN). The primary circuit for lordosis (shown in red) involves neurons in the ventrolateral (vl) part of the VMN that send excitatory projections to the periaqueductal gray (PAG). PAG output to the nucleus gigantocellularis (nGi) in the medulla activates nGi neurons that stimulate those spinal motor neurons (MNs) that regulate the muscle contractions of lordosis. Estradiol (E) and progesterone (P) act on VMNvl neurons to stimulate lordosis, and the functional effectiveness of VMNvl is enhanced by oxytocin (OT) input from the PVN. The MPOA is shown as suppressing lordosis, although the specific neural pathways involved in this effect are not shown. β-endorphin-containing neurons in the arcuate nucleus (ARC) contribute to the facilitation of lordosis by inhibiting those MPOA neurons that depress lordosis. Axons ending in a bar are inhibitory, and those ending in an arrow are excitatory. Other abbreviations: 3 = third ventricle; AC = anterior commissure; OC = optic chiasm.

In comparing Figure 4.3 with Figure 4.4, it is interesting to note some differences in the regulation of consummatory male and female sexual responses. In particular, the PAG is shown as inhibiting male consummatory responses, while it excites lordosis. This difference, in part, may be related to the fact that PAG projections to different nuclei within the medullary reticular formation appear to be important for male (nPGi) and female (nGi) responses. As already indicated, the PAG is a complex nucleus that regulates a variety of reflex-like stereotyped responses such as those involved in escape and withdrawal, aggression, male copulation, and lordosis. In the next chapter, evidence will be presented that indicates that certain PAG neurons also regulate nursing behavior. The best way to conceive of this diversity is to understand that different PAG neurons, with different projections to the medullary reticular formation, ultimately influence different cranial and spinal motor neurons, resulting in different kinds of responses.

4.4.2.2 *Evidence for the Neural Model*

Some of the evidence in support of the model shown in Figure 4.4 can be outlined as follows:

1. Daniels, Miselis, and Flanagan-Cato [214] injected a transsynaptic retrograde neuroanatomical tracer (pseudorabies virus) into the lordosis-producing back muscles of a female rat. This tracer was taken up by axons terminals and then transsynaptically transported to neurons in the central nervous system that had connections to these back muscles. In addition to labeling neuronal cell bodies of lumbar spinal motor neurons, labeled cell bodies were also detected in the nGi, PAG, and VMNvl.
2. Lesions of the VMN or PAG disrupt lordosis in hormone-primed female rats [765,850], as does interference with the neural connections between these two critical brain regions [404]. Lesions of nGi also disrupt the lordosis response [1070].
3. VMNvl neurons contain both estrogen receptors (ERs) and progesterone receptors (PRs), and estradiol action on ERs in VMNvl induces the expression of PRs [766,847]. Estradiol and progesterone act locally within the VMN to stimulate lordosis in ovariectomized rats [508,841]. Although some of these lordosis-promoting effects of estradiol and progesterone on VMNvl neurons occur by the binding of these steroids to intracellular receptors and the subsequent transcriptional effects of the hormone–receptor complex on the genome (see Chapter 2), additional effects of these steroids also occur by the binding of these steroids to membrane receptors with the subsequent activation of second messenger systems (see [508]).
4. Estradiol induces the expression of OTRs within the VMNvl [234,239,360], and blocking the action of OT at the level of the VMN depresses the lordosis response in steroid-primed rats [618,745].
   As I will indicate throughout this book, OT action on OTRs has potentiating effects on a variety of prosocial behaviors, such as lordosis, maternal behavior, and pair bonding in monogamous species. The neural site where OT acts influences the kind of response it promotes, and in this case, its action on VMNvl promotes lordosis, presumably as a result of stimulating the VMNvl-to-PAG projection.
5. Hoshina, Takeo, Nakano, Sato, and Sakuma [425] reported that neuron-specific excitotoxic lesions of the MPOA enhanced lordosis in estrogen-primed ovariectomized rats (also see [1032]). Control rats that received a 1 μg suboptimal dose of estradiol exhibited an LQ of 40%, while the MPOA-lesioned females had an LQ of 90%. These results indicate that certain MPOA neurons exert a depressing effect on the consummatory lordosis response. As I will outline below, this depressing effect

may allow for the occurrence of pacing behavior.

The mechanism through which MPOA neurons depress lordosis remains to be determined. Since the MPOA projects to both the VMN and the PAG [704,903], it is possible for certain MPOA neurons to suppress or restrain the activity of lordosis-promoting VMNvl and PAG neurons.

6. Micevych and his colleagues have provided evidence that the full activation of lordosis may require estradiol action on membrane receptors within the arcuate nucleus of the hypothalamus, which ultimately activates a β-endorphin opioid pathway to the MPOA that inhibits those MPOA circuits that restrain lordosis [187,640]. In support of the proposal that β-endorphin action on μ opioid receptors (MORs) in MPOA facilitates lordosis, the following evidence is offered: a mouse strain with a knockout mutation of the MOR gene exhibits deficits in lordosis after hormone priming [906]. While wild-type control mice had an LQ of about 90%, the mutant strain exhibited an LQ of 60%. Therefore, opioid action on MORs is not essential for lordosis to occur but appears to enhance lordosis. Further, β-endorphin infusions into the third ventricle enhance, while infusions of β-endorphin antiserum into the third ventricle depress, lordosis in steroid hormone-primed rats [964,965].

4.4.2.3 *Sex, Aggression, and the VMN*

In the context of the research on the hypothalamic aggression area that was reviewed in Chapter 3, research was presented that neurons in the VMNvl promote male offensive aggression. It may seem surprising that the VMNvl also plays an important role in the consummatory aspects of female sexual behavior in rats. However, as suggested by the work of Lin et al. [556], it is highly likely that different VMNvl neurons are involved in antisocial aggressive responses and prosocial sexual responses. Some VMNvl neurons with projections to certain parts of the PAG may promote aggression, while other VMNvl neurons that presumably project to other parts of the PAG promote lordosis. Importantly, when female rats are not in behavioral estrus, they reject mount attempts by males. Perhaps male stimuli activate VMN aversion circuits in nonestrous females, while the same male stimuli activate the lordosis circuitry in estrous females. As previously suggested, hormone-dependent differential neural processing of social stimuli can influence behavioral outcomes.

While intense offensive aggression is common in male rats, such aggression is typically low or absent in females under most situations. However, female offensive aggression toward male rats and other conspecifics, as tested in the resident–intruder test, is very high during the early postpartum period. Such aggression is referred to as maternal aggression, and it likely serves to protect vulnerable offspring from dangerous conspecifics, such as male rats, that might be infanticidal [703]. Hansen [385] has shown that maternal aggression depends on the VMN and other parts of the hypothalamic aggression area, since excitotoxic lesions of this region disrupted aggression without interfering with other aspects of maternal behavior. Confusingly, however, a postpartum estrus occurs on day 1 postpartum. Therefore, the same male stimulus should be capable of activating both maternal aggression and sexual behavior in normal female rats on day 1 postpartum [13]. A male rat would be able to present stimuli that, on the one hand, could activate VMNvl neurons promoting aggression, while on the other could activate other VMNvl cells that stimulate lordosis. Such an ambivalent situation does not appear to be adaptive or efficient. The following is important, however: the occurrence of maternal aggression is highly dependent upon the presence of the young [703]. Under natural or seminatural conditions, during the postpartum estrus, a female rat typically leaves her burrow and nest site to go above ground to mate. Under these conditions, when she is away from her offspring, male stimuli do not evoke aggression but instead stimulate exclusive sexual responsiveness [332]. Clearly, both physiological state and the social context influence whether male stimuli will promote agonistic or acceptance responses in female rats.

4.4.3 The MPOA and Appetitive Female Sexual Behavior

Several studies have indicated that the MPOA plays an important role in the appetitive aspects of female sexual behavior in rats, regulating the female's attraction toward a sexually active male. Indeed, the MPOA may play a primary role in regulating goal-directed pacing behavior: following an intromission, after a preferred time interval, the MPOA may regulate attraction, approach, and return to the male, which results in the initiation of a new copulatory bout.

Whitney [1032] placed steroid-primed female rats in a cage that contained three stimulus arenas, one with sexually active males, a second with sexually inactive males, and a third containing ovariectomized females. The steroid-primed female had access to each arena, while the stimulus rats were confined to their particular arena. Steroid-primed control females (estradiol plus progesterone) spent significantly more time with the sexually active males than did females that received electrical lesions of the MPOA. Employing neuron-specific excitotoxic amino acid lesions of the MPOA, similar results have been reported by Guarraci and Clark [366]. In both of these studies, when given a choice that presumably reflects the motivational preference of the female, the amount of time that hormone-primed females with MPOA lesions chose to spend with a sexually active male dropped to close to zero. Importantly, if such MPOA-lesioned females were confined to a single cage from which they could not escape, with a sexually active male, then consummatory female sexual behavior was enhanced, as reflected by a higher LQ following MPOA lesions [1032]. These findings support a dual role for MPOA in the control of female sexuality: suppression of consummatory and enhancement of appetitive female sexual behavior. It is likely that separate MPOA neuron populations are involved in these two functional effects (cf. [849]).

Using a paced mating paradigm, similar results have been reported by others. Neuron-specific excitotoxic lesions of MPOA, while not disrupting lordosis, disrupted pacing behavior: the lesioned females spent less time with the males, the exit frequency following an intromission was increased, as was the contact–return latency [368,1050].

Estradiol acts on the VMNvl to promote lordosis. Is it possible the estradiol acts on the MPOA to promote the appetitive aspects of female sexual behavior in rats? There is some evidence that this is the case [920]. A strand of RNA capable of binding to and inhibiting the mRNA strand that codes for one of the major types of ER (ERα) was microinjected into MPOA. This procedure dramatically reduced the expression of the ERα protein in the MPOA, which would therefore interfere with estradiol action on MPOA. When given a choice between spending time near a sexually active male or a castrated male (direct physical contact was prevented), estradiol-primed control females chose the sexually active male, while experimental females in which the expression of ERα in MPOA was disrupted did not show a preference. Importantly, when these experimental females were confined to a single cage with a sexually active male, their LQ was significantly higher than controls. These results suggest that estradiol action on ERα-containing MPOA neurons enhances the female's attraction and approach toward sexually active males while also suppressing the consummatory female sexual lordosis response. These two effects presumably occur via estrogen action on different neuronal populations in MPOA (cf. [849]).

4.4.4 The Mesolimbic DA System and Appetitive Female Sexual Behavior

Given the role of the mesolimbic DA system in the regulation of a variety of goal-directed appetitive responses, it makes sense that this system would also be engaged during the appetitive aspects of female sexual behavior. In support of this view, microdialysis studies have shown that DA is released into NA when an estrous female rat interacts with a sexually active male [768]. Becker and her colleagues have done important research relating DA levels in the NA to paced mating behavior [78,452]. These studies compared female rats that engaged in paced mating behavior with females who were confined with males in a single cage (nonpaced mating). Samples of DA in the extracellular fluid within the NA were obtained at 1 min intervals. For females that engaged in nonpaced mating, DA did not increase in NA. For females that paced their copulatory bouts at their preferred interval, DA increased in the NA in the period just before an intromission occurred and during the period immediately after the occurrence of an intromission. These correlational findings have several implications. The increase in DA that occurs prior to an intromission may be involved in regulating appetitive reward-seeking responses that are necessary for the reinitiation of a copulatory bout. The increase in DA levels after an intromission may exert two effects: one effect may promote goal-directed avoidance or exit responses. However, such increases may also exert reinforcing or rewarding effects that result from a female receiving coital stimulation at her preferred rate. Therefore, as described for the involvement of the mesolimbic DA system in male sexual behavior, for females, DA may also exert two major effects. DA release into NA may regulate goal-directed responses toward (and away from) a sexually active male, and the release of DA that occurs during or immediately after consummatory behavior and the receipt of coital stimulation may also have reinforcing properties, strengthening the subsequent appetitive responsiveness of the female to those stimuli that were associated with DA release into NA.

Not much experimental work has been done on the involvement of the mesolimbic DA system in the appetitive aspects of female sexual behavior, and the research that has been done is equivocal. Guarraci, Megroz, and Clark [367] reported that excitotoxic amino acid lesions of NA had no effect on paced mating behavior in estrous female rats. Recall, however, that there is some controversy concerning the role of DA action at the level of NA with respect to the regulation of goal-directed responses. Based on Chapter 1 and on Figure 4.2 in this chapter, one view that has received support is that DA may act to depress NA activity, which releases VP from inhibition, and it is increased VP activity that regulates reward-seeking and other goal-directed responses. If that model is appropriate for paced mating, then VP lesions, rather than NA lesions, should disrupt appetitive female sexual behavior. Interestingly, Jenkins and Becker [451] found that while small excitotoxic lesions of NA did not disrupt paced mating behavior, large lesions, which may have encroached on the VP, resulted in hormone-primed females who would not approach a sexually active male at all. Such females never even initiated a single copulatory bout.

As far as I am aware, no one has explored the effects of DA antagonist injection directly into NA on paced mating behavior. I would predict that such injections would disrupt paced mating, causing a major increase in contact–return latencies. However, such injections would not be expected to interfere with consummatory lordosis responses if the male and female were confined to a single cage that prevented paced mating. Similarly, the injection of DA receptor agonists into NA would be predicted to enhance the female's appetitive responses.

Ellingsen and Agmo [270] have examined the implication from Becker's correlational research that DA release into NA is important for appetitive female sexual motivation through the use of systemic injections of flupenthixol, which is a mixed D1/D2 antagonist. Estrous female rats were tested in two situations: a paced mating chamber and a sexual preference test. Females injected with either 0.25 or 0.50 mg/kg of flupenthixol did not differ from control females (0 mg/kg) in their paced mating behavior. In particular, the contact–return latencies did not distinguish the various groups. In addition, all females, irrespective of treatment, spent more time in the vicinity of a cage that contained a sexually active male. Finally, the doses of flupenthixol used in this study were found in previous studies not to have an effect on general locomotor activity. The authors conclude that central DA systems do not play a major role in appetitive female sexual motivation.

I tend to disagree with the conclusions of Ellingsen and Agmo [270]. The results from studies using systemic injections of neuroactive drugs are difficult to interpret, and the use of a drug that blocks both D1 and D2 receptors complicates matters even further. Since the MPOA, VMN, and NA may all be involved in female sexual behavior, and since the incertohypothalamic DA system provides DA input to both MPOA and VMN, systemic injections of DA antagonists are likely to affect all three targets. Importantly, the effects of DA on VMN may oppose the effects of DA on the NA. Mani [594] has shown that DA action on D1 receptors in VMN promotes lordosis. The promotion of lordosis (consummatory female sexual activity defined by behavioral immobility) may suppress active appetitive sexual responses. With such an understanding, one can conceive that while the inhibition of DA action on NA may depress paced mating, the inhibition of DA action on VMN may promote paced mating. If that were the case, the two effects may cancel each other out, so that systemic treatment with relatively low doses of DA receptor antagonists would not produce a major effect on appetitive female sexual motivation.

Further, the fact that doses of flupenthixol were used that did not have an effect on general motor activity suggests that the function of the nigrostriatal DA system was not affected. If the nigrostriatal DA system remained functional, why would one assume that the function of the mesolimbic DA system was strongly depressed? Of course, if one disrupted movement through the use of higher doses of flupenthixol, and such doses also affected sexual behavior, then the specificity of the results would be called into question. The only way out of these dilemmas is through the use of site-specific injections of selective DA receptor antagonists into the NA to explore the necessity of mesolimbic activity for appetitive female sexual behavior. In conclusion, although the correlational microdialysis research supports a role for the mesolimbic DA system in appetitive female sexual motivation, appropriately designed experimental research will be needed to prove that this is the case.

In examining the microdialysis findings in rats, I noted that DA release into NA occurs as the female reinitiates a copulatory bout and that there is a second DA peak that occurs following an intromission by the male. It was suggested that the DA peak that follows intromission might have reinforcing properties. There is some research that supports the reinforcing effects of copulation-induced DA release into NA. Meisel and colleagues have studied the role of DA in the sexual behavior of female hamsters [631]. Copulatory bouts in hamsters are different from what has been described for rats. Estrous female hamsters do not engage in active pacing behavior and they are relatively immobile during sexual interactions; in fact, they may hold the lordosis posture for several minutes while the male mounts at his own pace [631]. However, during lordosis, female hamsters do make perineal movements in response to the perivaginal stimulation that occurs during male mounting, and such movements facilitate the occurrence of intromission [631].

DA is released into NA during copulation in female hamsters, and the stimuli associated with intromissions are involved in triggering such release [504]. In some interesting research, it has been shown that DA release into NA during copulation in female hamsters may modify the nature of the female's consummatory lordosis response via a learning-type mechanism. When sexually inexperienced estrous female hamsters mate with sexually inexperienced male hamsters, the proportion of mounts that result in an intromission (the hit rate) is about 50% [119]. However, the hit rate increases to about 70% when sexually experienced females mate with inexperienced males. It has been suggested that the DA release that occurs in females following an intromission serves to reinforce or strengthen those particular perineal movements that facilitated intromission [119,631]. In support of this proposal, sexually experienced female hamsters did not develop a higher hit rate when mating with sexually inexperienced males if they had received 6-hydroxydopamine (6-HD) lesions of NA just prior to the time they were gaining sexual experience [119]. 6-HD is a neurotoxin that destroys DA neurons and, therefore, would lesion VTA-DA axons that terminate in NA, which in turn would disrupt the DA release that occurs in NA during copulation. The authors also presented evidence that 6-HD lesions of NA did not disrupt lordosis duration (a major measure of consummatory sexual behavior in hamsters). The only critique I have of this study is that the authors did not measure whether active perineal movements occurred after 6-HD lesions. In other words, it is possible that the 6-HD lesions simply prevented the occurrence of perineal movements during copulation, rather than preventing DA from reinforcing those perineal movements that resulted in intromission.

Using the conditioned place preference (CPP) paradigm, Meisel, Joppa, and Rowe [630] provided additional evidence that engaging in sexual activity has reinforcing properties for estrous female hamsters and that such reinforcing properties are dependent upon DA action on D2 receptors. The CPP apparatus consisted of three compartments, a center neutral compartment and two distinct side compartments, one of which was gray and the other white. During conditioning trials, estrous female hamsters were placed in one of the distinct side compartments and allowed to mate with a male, and then they were placed alone in the other compartment. During the test trial, females were placed in the center chamber and were allowed to move freely through the apparatus. During this test trial, males were not present in the apparatus, and the amount of time the female spent in each compartment was recorded. Normal females spent significantly more time in the compartment that was associated with mating behavior. These results were interpreted to mean that copulatory activity had reinforcing properties and that stimuli that were associated with the primary reward of copulation (the particular place or compartment) became secondary reinforcers or rewards that served to attract the female to that particular compartment. Such a learning mechanism, of course, under natural conditions, would serve to attract sexually active females to those environments that in the past had contained sexually active males. The following evidence was presented to support the proposal that DA neural systems supported CPP learning: if female hamsters were systemically treated with a DA D2 receptor antagonist during the conditioning trials where mating occurred in a particular compartment (lordosis was not affected on these trials), then a subsequent CPP for that compartment was not formed (females were drug free during these test trial sessions).

Results from research on rats also suggest that engaging in copulation is reinforcing for estrous females, and that such reinforcing effects may be dependent upon DA release into NA. Using the CPP paradigm, Jenkins and Becker [453] allowed female rats to mate in two distinct compartments during conditioning trials. In one compartment, paced mating was permitted, while in the other nonpaced mating occurred. The females formed a preference for the compartment that was associated with paced mating. Since paced mating, but not nonpaced mating, results in DA release into NA, such release may have reinforced the learning required for the formation of the CPP. It was specifically suggested that the release of DA into NA following coital intromission stimuli might be the event that is reinforcing. In support of the view that the events that occur during and after vaginocervical stimulation are rewarding, Meerts and Clark [629] have reported that estrous female rats will form a CPP for a compartment in which they were placed immediately after receiving artificial vaginocervical stimulation via a rubber-tipped probe.

4.4.5 MPOA Interactions with the Mesolimbic DA System and the Appetitive and Rewarding Aspects of Female Sexual Behavior

In a manner similar to my arguments for the neural control of the appetitive aspects of male sexual behavior, I am going to suggest that MPOA interactions with the mesolimbic DA system may also be involved in the appetitive aspects of female sexual behavior. This proposal is based on the following facts that have already been presented: the MPOA has strong anatomical connections to the VTA, and there is evidence that both the MPOA and the mesolimbic DA system are involved in the appetitive aspects of female sexual behavior. Although this proposal is attractive, there is currently no neurobehavioral evidence that provides direct support. However, there is some indirect evidence. As I indicated previously, when given a choice, an estrous female rat prefers to mate with a male under paced mating conditions rather than under nonpaced mating conditions [202,203]. Coria-Avila and Pfaus [203] allowed estrous female rats to mate with males that were scented with an almond extract. Some of the females mated with males under paced mating conditions, while other females engaged in nonpaced mating. In one part of this study, it was determined that females were attracted to and chose to mate with almond-scented males only if the almond scent had previously been associated with paced mating. Please note that the females did not form a partner preference to a particular male, as would occur in monogamous species after mating (see Chapter 6). The females in this study would mate with any almond-scented male if that odor had previously been associated with paced mating.

In another part of this study, the females were subsequently exposed to the almond odor alone for 1 h, and then the brains were immunohistochemically processed for the detection of Fos-labeled neurons. The almond odor that was associated with paced copulation produced significantly more Fos-labeled cells in the MPOA and VTA, relative to the same odor associated with nonpaced copulation. One interpretation of these data is that the attraction of an estrous female to male stimulus characteristics that are associated with paced mating is the result of MPOA activation of the mesolimbic DA system.

The MPOA may direct DA release into NA so that appetitive female sexual behavior occurs and the subsequent MPOA-directed release of DA into NA following an intromission might also have reinforcing effects. Is there any evidence to support the latter point of this hypothesis? Meerts and Clark [628] have reported that excitotoxic amino acid lesions of the MPOA prevent estrous female rats from forming a CPP for a cage compartment that was associated with vaginocervical stimulation. This result is particularly dramatic since other studies show that such MPOA lesions potentiate the consummatory aspects of female sexual behavior by increasing the lordosis quotient.

4.4.6 A Working Neural Model

Although the results are far from conclusive, Figure 4.5 presents a working neural model of the regulation of female sexual behavior in rodents. The figure shows the consummatory pathways

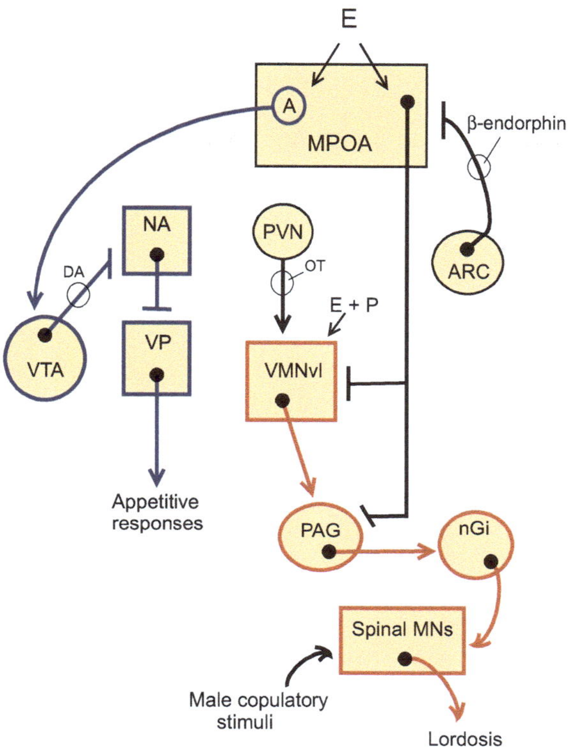

FIGURE 4.5 A neural model of the circuits that mediate female sexual behavior in rodents. The proposed appetitive circuits are shown in blue and include appetitive (A) medial preoptic area (MPOA) neurons interacting with the mesolimbic dopamine (DA) system. The consummatory circuits that trigger lordosis originate from the ventrolateral part of the ventromedial nucleus of the hypothalamus (VMNvl), and the full circuit is shown in red. Modulatory circuits are shown in black, and include: (1) MPOA suppression of lordosis via proposed inhibitory connections to VMNvl and periaqueductal gray (PAG); (2) oxytocin (OT) enhancement of lordosis via paraventricular nucleus (PVN) projections to VMNvl; (3) an arcuate nucleus (ARC) β-endorphin projection that inhibits those MPOA neurons that suppress lordosis. Axons ending in a bar are inhibitory, and those ending in an arrow are excitatory. Other abbreviations: E = estradiol; MNs = motor neurons; NA = nucleus accumbens; nGi = nucleus gigantocellularis; P = progesterone; VP = ventral pallidum; VTA = ventral tegmental area.

indicated in Figure 4.4 combined with my tentative view of the neural regulation of appetitive female sexual behavior. The MPOA is shown as regulating active appetitive sexual responses associated with paced mating via its connections with the mesolimbic DA system. Such MPOA-directed DA release into NA may also underlie the reinforcing effects of sexual stimulation, although the exact neural mechanisms involved are yet to be defined. VMNvl output to the PAG is the major controller of lordosis, which involves both motor inhibition as well as lordosis (the female needs to stand still while the male mounts). In analyzing this model, one can conceive of how a female switches between appetitive and consummatory sexual activity during mating. When the MPOA is the dominant system, active appetitive sexual responses occur and lordosis is suppressed. When the VMNvl and ARC are most active, the MPOA is inhibited and descending lordosis-promoting projections to the PAG are activated. The exact stimuli and other mechanisms that cause the toggling back and forth between these two basic neural processes that create the basis of female copulatory bouts remain to be determined.

In comparing Figure 4.5 with Figure 4.2, which describes a neural model for male sexual behavior, some interesting differences are apparent. For male sexual behavior, evidence was presented for two major MPOA neural populations, one stimulating appetitive and the other consummatory male sexual behavior. For females, the VMNvl contains the main stimulatory population for female consummatory responses, while the MPOA contains neurons stimulatory for female appetitive behavior and inhibitory for female consummatory responses. Interestingly, the PAG appears to be a major brainstem relay point for both male and female consummatory sexual responses, although functional sex differences occur, with the PAG exerting either an inhibitory or an excitatory role for male and female consummatory responses, respectively.

Finally, in comparing these two neural models, it is proposed that MPOA interactions with the mesolimbic DA system may be involved in the appetitive aspects of both male and female sexual behavior. In the next chapter, much stronger evidence will be presented that shows that MPOA interactions with the mesolimbic DA system are also involved in the appetitive aspects of maternal behavior. In Chapter 5, I will discuss the important issue of the specificity of the MPOA for these different types of reproductive behaviors. Are different MPOA neurons involved in the appetitive aspects of maternal behavior and male and female sexual behaviors, or is a common general population of MPOA neurons involved in all three behaviors? As I will show, we currently don't have an answer to this question. But whatever the evidence ultimately concludes, what is clear is that the MPOA plays a very important role in appetitive motivation related to reproduction, and this role may also extend to other social behaviors (see Chapters 6 and 7).

4.5 THE NEUROBIOLOGY OF SEXUAL BEHAVIOR IN PRIMATES

4.5.1 Introduction

Compared to research on nonprimate mammals, much less experimental research has been done on the neural mechanisms controlling sexual behavior in nonhuman primates. In humans, a modest brain imaging literature has provided correlational evidence with respect to the neural basis of sexual behavior in men and women, and most of this research has examined sexual behavior in men. Importantly, however, this research shows that there is good overlap between the subcortical mechanisms that regulate sexual behavior in rodents and primates.

The emphasis of this section will be on the brain mechanisms in primates that appear to regulate the appetitive goal-directed phase of sexual behavior, with less attention being given to the consummatory aspects of sexual behavior. The reasons for this focus are twofold: in men, and particularly women, steroid hormones have their primary influence on sexual desire and libido (appetitive sexual motivation); poorly regulated and excessive sexual desire may lead to aberrant, inappropriate, and pathological sexual behavior. Understanding the brain mechanisms of normal appetitive sexual motivation may provide insights into neural dysfunctions that could cause faulty sexual motivation.

4.5.2 Sexual Behavior in Monkeys

4.5.2.1 Males

As in rodents, electrical lesions of the MPOA and adjoining anterior hypothalamus severely disrupt male copulatory behavior in rhesus [910] and marmoset [561] monkeys without interfering with pituitary regulation of testosterone secretion from the testes. Lloyd and Dixson [561] suggested that the lesions interfered with the sexual arousal and interest necessary for the initiation of copulatory behavior since the lesioned male marmosets showed no sexual interest in receptive females, as indicated by low levels of anticipatory erections and anogenital investigation of females. These lesions also eliminated or dramatically decreased the occurrence of mounts, intromissions, and ejaculations in both the marmoset and rhesus males. It is likely that the large lesions that were produced in these studies interfered with both appetitive and consummatory mechanisms. Interestingly, Slimp et al. [910] cite earlier clinical findings on human males, where damage to the rostral hypothalamus and MPOA was associated with a decrease in sexual interest/libido.

Oomura, Yoshimatsu, and Aou [727] recorded from single neurons in MPOA during various aspects of male sexual behavior in macaque monkeys. MPOA neurons were detected that increased their neural activity during the commencement of sexual activity with a receptive female. Importantly, in an operant response paradigm, MPOA neurons demonstrated an increase in action potential frequency when a male was pressing a lever to gain access to a receptive female, but not when the male pressed the lever to obtain a food reward.

Male marmoset monkeys prefer the perivaginal pheromonal scents of ovulatory females to those from nonovulatory females. Using this knowledge, Ferris et al. [299] performed a functional magnetic resonance imaging (fMRI) study on fully conscious male marmosets that were exposed to scents from ovulatory and nonovulatory females. The blood-oxygen-level dependent (BOLD) response in the MPOA was significantly greater when the males were exposed to the ovulatory scents, which would normally attract males to females that emit such odors.

4.5.2.2 Females

Kendrick and Dixson [477] examined the effects of electrical lesions of the hypothalamus on the sexual behavior of female marmoset monkeys. Normal female marmosets are receptive to males through their 28-day reproductive cycle, in the sense that they will allow the male to mount them and will not terminate male mounting by withdrawing from the male. However, female proceptive behaviors that invite the male's attention, which include rhythmic tongue protrusions that solicit male copulation, are restricted to the periovulatory period, suggesting that the periovulatory increase in estradiol promotes appetitive motivation. In this study, all females were ovariectomized and treated with estradiol. Therefore, control females showed both tongue-flicking and receptivity (permitting mounts by males). Females with lesions of the MPOA and rostral anterior hypothalamus showed an almost complete elimination of tongue flicking, although they permitted male mounting. In two females with more caudal lesions that infringed upon the VMN, there was some evidence that receptivity was affected in that these females either refused or terminated male mounts. These results are consistent with the proposal derived from the rodent research on females, where MPOA was viewed as being involved in appetitive sexual motivation, while the VMN was concerned with the consummatory lordosis response.

Aou, Oomura, and Yoshimatsu [34] recorded from single neurons within MPOA and VMN of sexually active female macaque monkeys during the breeding season. When engaging in sexual activity, the female presents her rump area to the male and stands still, which allows the male to mount. MPOA neurons decreased and VMN neurons increased their firing rate during the presenting posture. However, immediately before the initiation of the presenting posture, some MPOA neurons increased their firing rates. One interpretation of these data is that one population of MPOA neurons (the ones that increased firing prior to presenting) is involved in the appetitive initiation of sexual behavior. In contrast, the actual present posture may represent the consummatory response, which, as in rats, may require an inhibition of MPOA neurons, which enhances VMN activation (see Figure 4.5).

4.5.2.3 Conclusions

Although the existing nonhuman primate research has been restricted to an examination of the hypothalamus, and a larger circuitry analysis has not been performed, these results are roughly comparable to the findings from the rodent research. It appears that a population of MPOA neurons is involved in the appetitive aspects of male and female sexual behavior in monkeys, with another population of MPOA neurons being involved in the consummatory aspects of male sexual behavior; more caudal hypothalamic regions that include the VMN influence consummatory responses in females.

4.5.3 Sexual Behavior in Humans

4.5.3.1 Heterosexual Males

Most studies on the potential neural mechanisms underlying the appetitive aspects of sexual behavior have been conducted on young healthy heterosexual men (for a recent review, see [330]). In a typical study, men are placed in an fMRI scanner and brain BOLD responses are measured in various regions of interest (ROI) while the participants view either erotic (either single photos of nude women or sexually explicit video clips of heterosexual couples engaged in sexual activity) or nonerotic visual stimuli. In some studies, the men indicate their degree of subjective sexual interest and desire by applying varying degrees of hand pressure to a pressure sensitive gauge, while in other studies they rate their degree of sexual arousal after the completion of the scans. Finally, some studies correlate the occurrence of erections with the subject's reported degree of experienced sexual arousal and desire.

These studies are probably focusing on the anticipatory or appetitive phase of male sexual activity since it is presumed that viewing erotic visual stimuli activates the desire to engage in sexual activity and because actual copulatory behavior is not occurring. However, interpreting such data may not be that simple. When viewing erotic video clips, the subject may imagine himself engaging in copulatory behavior, and, therefore, the BOLD responses associated with viewing such images may represent a combination of both appetitive and consummatory neural processes.

A few caveats are worth considering with respect to interpreting these imaging studies (see [330]). First, the temporal resolution of fMRI scans is not a faithful reflection of neural activity, since changes in neural activity occur over milliseconds, while the temporal resolution capability of fMRI scans is measured in seconds. Second, the spatial resolution of most fMRI scans usually does not distinguish separate nuclei within a larger ROI. For example, BOLD responses are typically measured throughout the entire amygdala or hypothalamus, and it is also not obvious whether BOLD responses in the ventral striatum are clearly separated from those in the ventral pallidum. Third, earlier studies tended to use a 1.5 T scanner, while more recent studies have used a 3 T scanner. The latter creates a stronger magnetic field, which may result in the more recent studies detecting significant BOLD responses that were missed in earlier studies. Finally, in some cases the anatomical demarcations used by certain investigators require greater accuracy. For example, in a study by Brunetti et al. [134], the reported activation of the hypothalamus that occurred when men viewed erotic images appears to include both the lateral preoptic area of the hypothalamus and the adjoining ventral pallidum in the telencephalon, with a central focus within the latter structure (see their Figure 2).

Most studies agree on a common group of structures that show a greater BOLD response when men view erotic visual stimuli in comparison to control stimuli. In evaluating these structures, please refer to the part of Figure 4.2 that describes the neural circuits regulating appetitive sexual responses in male rodents. The following brain regions are among those that show increased BOLD responses: hypothalamus, midbrain, amygdala, NA-VP circuit, medial PFC (including areas 24, 25, and 32 of the anterior cingulate cortex: see Chapter 1), and the insular cortex [1,75,134,180,295,380,471,717,742,782,1011]. These findings are certainly consistent with a view that the neural circuitry underlying the appetitive aspects of sexual behavior in men is likely to include hypothalamic (MPOA) projections to the mesolimbic DA system, which in turn increase the responsiveness of the NA-VP circuit to the neural input that this circuit receives from the amygdala. In fact, the output of amygdala neurons with a positive valence may be one of the routes over which visual erotic stimuli reach the NA-VP circuit. The output of the NA-VP may give rise to sexual interest, desire, and imagined goal-directed sexual activity. Hypothalamic output to the brainstem may also influence the occurrence of erections. An important take-home message is that the limbic and subcortical mechanisms of appetitive male sexual motivation show a strong overlap between men and male rats. I have no doubt that particular dysfunctions within this system could lead to either hypersexuality or hyposexuality. A study by Abler et al. [1] is relevant. Healthy male subjects received subchronic treatment with either paroxetine or a control treatment (placebo). Paroxetine is a selective serotonin reuptake inhibitor (SSRI), and one of its side effects is a decrease in sexual libido. While watching erotic videos in an fMRI scanner, the control subjects showed the expected increase in BOLD responses in amygdala, ACC, hypothalamus, midbrain, and ventral striatum. The paroxetine-treated men, who reported a drug-related decrease in sexual desire, demonstrated significant decreases in the BOLD responses in anterior cingulate cortex (area 25), ventral striatum, and midbrain while watching the erotic videos. Perhaps increased 5-HT action within the ACC (via an action in 5-HT1A receptors) inhibited ACC input to the NA-VP region.

With respect to the involvement of the medial PFC, recall that my review on rodents indicated that damage to some parts of the medial PFC disrupts the initiation of male sexual behavior in rats, while damage to other parts of this region actually disinhibits sexual behavior. As reviewed in Chapter 1, areas 24, 25, and 32 of the medial PFC in humans represent a heterogeneous population of neurons (see [25]) with projections that can either upregulate or downregulate neural activity within the ventral striatum, amygdala, and hypothalamus. Medial PFC activation that occurs while men are watching erotic videos may be the result of increased neural activity within certain medial PFC neurons that relay processed visual inputs to the amygdala, NA-VP circuit, and hypothalamus (see [786]), with such inputs playing a role in stimulating male erotic interests and appetitive motivation.

The anterior part of the insular cortex is involved in a variety of subjective emotional experiences [204]. In the primate brain, unlike that of the rodent, all parts of the insular cortex are located deep within the lateral fissure of the cerebral cortex and are covered by parts of the frontal, parietal, and temporal lobes (compare Figure 7.1 with Figure 1.14). The anterior agranular part of the insular cortex forms a border with, and is considered to be part of, the very posterior aspects of the lateral orbital prefrontal cortex [787]. A more detailed description of the anatomical location of the insular cortex will be presented in Chapter 7. The insula is strongly and reciprocally interconnected with the amygdala and with area 24 of the anterior cingulate cortex within the medial PFC network [685]. Thalamic and cortical inputs to the insula relay a variety of sensory inputs, including viscerosensory inputs from internal organs. Craig [204] notes that fMRI studies show that the anterior insular cortex and the anterior cingulate cortex often co-activate together during the viewing of emotional visual scenes. He suggests that the insula may be involved in subjective emotional experiences while the anterior cingulate cortex may modulate the initiation of motivated behaviors, which might occur through its projections to ventral striatum (see [262]).

Although I am delaying a full discussion of empathy for Chapter 7, it should be noted that the anterior insula and anterior cingulate cortex are considered to be part of the brain's empathy system [271]. Empathy has usually been studied in the context of one person observing another person who is exposed to a painful or aversive situation, and it is defined as a process through which the observer generates an affective state that is similar to what is believed to be occurring in the individual that is being exposed to a negative situation. However, a broader definition of empathy would simply be the process through which one individual imagines the emotional experiences, whether negative or positive, that another individual is having. With this understanding, activation of the anterior insula and anterior cingulate cortex when men watch erotic videos may also be tapping into the process whereby the viewer is reconstructing and experiencing the sexual acts that he is observing.

As the preceding discussion suggests, many of the brain regions that are activated while men are viewing erotic videos may also be active during motivational states other than sex. Indeed, in Chapter 1 I referred to the mesolimbic DA system as a nonspecific motivational system that is involved in regulating an organism's goal-directed responses to a variety of biologically significant stimuli. It is likely that activation of the mesolimbic DA system by certain hypothalamic nuclei, particularly the MPOA, may be involved in providing behavioral specificity by activating overt or covert (imagined) reward-seeking responses toward sex-related stimuli. Some evidence exists to support this proposal for human males. Childress et al. [180] studied brain activation in response to cocaine-related photos (such as drug paraphernalia) or erotic photos while male cocaine addicts were in an fMRI scanner. The cocaine-related and erotic stimuli were subliminally presented for only 33 ms, which was followed by a 467 ms presentation of a neutral photograph. Postscan tests indicated that the subjects had no recognition memory for the subliminally presented stimuli. The results indicated that "unseen" cocaine-related cues resulted in heightened BOLD responses in the amygdala, NA-VP circuit, insula, and caudal orbital frontal cortex. Significantly, the "unseen" sexual cues not only activated these regions, as indicated by an increased BOLD response, but also activated the hypothalamus and midbrain tegmentum. It is interesting to speculate that cocaine-related cues, which would be secondary reinforcers for cocaine addicts, may have been able to directly activate the NA-VP circuit, without the need for a major VTA activation, because of previous learning (see my earlier discussion of Everitt's work on operant responding in male rats for a conditioned reinforcer that had been associated with sexual activity). In contrast, erotic photos of women may be primary incentives for males and may stimulate MPOA activation of the mesolimbic DA system.

In a related study, Oei et al. [717] reported that subliminally presented erotic photos of women activated the nucleus accumbens (presumably representing increased DA input to this region) and the anterior cingulate cortex of men. Treatment of subjects with haloperidol (a dopamine receptor antagonist) decreased these BOLD responses, while treatment with levodopa (a precursor to dopamine that acts as a dopamine agonist) increased the BOLD response to erotic photos in these areas.

Although activity in the anterior cingulate cortex appears to be clearly related to increases in appetitive sexual motivation in men, the medial PFC is a heterogeneous region and some of its neurons are also involved in emotional and motivational downregulation. Related to these issues, a study by Beauregard et al. [75] showed that when men watched an erotic video, the amygdala and hypothalamus were activated, as measured by an enhanced BOLD response in comparison to watching a neutral video. However, when the men were instructed to inhibit their emotional and sexual responses to the erotic visual stimuli, their reported sexual arousal decreased, and this was associated with a decrease in amygdala and hypothalamic activation, along with a significant increase in the BOLD response measured in both the rostral part of the dorsolateral PFC and in area 32 (homologous to the rat's prelimbic cortex) of the anterior cingulate cortex. As described in Chapter 1, this potential interaction between the "cognitive" lateral PFC and the medial PFC (area 32) may be one route over which rational control of an individual's motivation and emotion is exerted (see Figure 1.15). Interestingly, some studies have reported that the orbital PFC is active, as indicated by an increased BOLD response, when men view erotic visual stimuli [180,471,800,864]. Since the orbital PFC is one of the links through which the lateral PFC can influence the medial PFC (lateral PFC-to-orbital PFC-to-medial PFC; see Chapter 1), perhaps in some studies where orbital PFC activity was detected, the subjects were attempting to suppress their sexual arousal even without specific instructions to do so. (It is also possible that increased activity in the orbital PFC (along with anterior insula activity) may mediate sexual feeling states. That is, different parts of the orbital PFC may be involved in either emotional experience or emotional regulation.)

These overall results on the involvement of the medial PFC in sexual motivation in men fit nicely with the work by Agmo et al. [11] and Davis et al. [219] on the positive and negative involvement of different parts of the medial PFC in the sexual behavior of male rats. These results also have larger implications with respect to irrational sexual activity in humans: dysfunctions within certain parts of the PFC system that interfere with the ability of the PFC to downregulate limbic, striatal, and hypothalamic circuits could result in inappropriate and risky sexual activities.

4.5.3.2 *Heterosexual Females*

Although far fewer imaging studies have been performed on heterosexual women while they watch visually presented sexual stimuli of nude males or heterosexual couples engaged in sexual activity, the results obtained with respect to the brain regions that are active, as measured by increased BOLD responses, are roughly similar to that which has been described for men [330,336,1011]. The activated regions include the hypothalamus, NA-VP circuit, amygdala, anterior cingulate cortex, and insula. These results are not surprising when one compares Figures 4.2 and 4.5, where MPOA interaction with the mesolimbic DA system is proposed to underlie the appetitive aspects of both male and female sexual behavior in rats.

A few studies have presented results that argue for a sex difference, with hypothalamic and amygdala activation being greater in men than in women when each sex watches erotic videos [380,471]. However, these studies did not control for the stage of the female's menstrual cycle. Recall that sexual interest and libido peak during the periovulatory period in women. Gizewsky et al. [336] have shown that visually erotic stimuli clearly activate the amygdala and hypothalamus when imaging is performed near the time of ovulation.

4.5.3.3 *Homosexual Males and Females*

When homosexual men and women view erotic photos or videos, stimuli appropriate to their sexual orientation increase the BOLD response in the hypothalamus and ventral striatum, while inappropriate sexual stimuli do not [742,782]. For example, while male genital stimuli activate the ventral striatum in homosexual men, female genital stimuli do not. In heterosexual men, of course, just the opposite is observed [782]. These results make complete sense. Homosexual men are attracted to other homosexual men, just as heterosexual men are attracted to heterosexual women. To the extent that MPOA activation of the mesolimbic DA system regulates appetitive sexual motivation, one would expect to see increased activity within the hypothalamus and NA-VP circuit when an individual is attracted to a preferred sexual incentive. These results also suggest that it is the types of erotic stimuli that gain access to hypothalamic appetitive circuits that influence one's sexual orientation or sexual partner preference. In other words, differential neural processing of sex-specific stimuli may participate in the regulation of individual differences in sexual preference and orientation.

4.5.4 Pedophilia

In this section, I will briefly review what we know about the neurobiology of pedophilia in order to provide some evidence that dysfunctions in the neural circuitry that participates in the regulation of appetitive male sexual behavior can give rise to pathological/abnormal forms of male sexual behavior.

Pedophilia is defined as a persistent sexual interest in prepubescent children as evidenced by an adult's sexual fantasies, urges, and goal-directed sexual behavior [882]. In its purest form, the pedophile's sexual preference is restricted to children, and there is no sexual interest in adults. Pedophilia occurs primarily in males (estimated to be about 5% of the male population), and the preferred sex of the child may be girls, boys, or both sexes [882]. Multiple assessment measures are used to identify pedophilic individuals, and these include the individual's admitted sexual urges, his erectile responses to erotic photos of children versus adults, and a history of committing sexual offenses that includes possessing child pornography and sexually molesting children. According to Seto [882], a significant proportion of pedophiles possess child pornography but have never committed a physical sexual offense. Seto suggests that an interaction between sexual interest in children and a criminal personality may promote the occurrence of actual physical sexual contact between a pedophile and a child. In this context, most studies on the neurobiology of pedophilia have examined male pedophiles that have committed a physical sexual offense with a child and are in prison hospitals. Some studies have compared such pedophiles with healthy adult males, while other studies have used nonsexual offenders/prisoners as controls. The latter comparison is used to subtract out the element of criminality, presumably giving a more specific picture of the potential neural mechanisms that influence pedophilia [155].

The etiology of pedophilia is largely unknown. Although there is some evidence that the experience of sexual child abuse may cause the abused child to develop pedophilic tendencies, this evidence is not strong in that a large majority of pedophiles have not been sexually abused when they were children [318,454]. In an interesting retrospective study that looked at the development of erotic preferences, Freund and Kuban [318] asked adult male pedophiles and healthy adult males to recall their curiosity about seeing nude children or nude adults when they were children. Pedophiles recalled that they were more interested in seeing nude children than nude adults and that this interest began between 7 and 11 years of age. In contrast, healthy adult males recalled being more interested in seeing nude adults than nude children during this same age span. These results indicate that erotic preferences develop in juveniles prior to the onset of puberty. Importantly, however, a large proportion of healthy adult males did recall an interest in seeing nude children, but they also remembered that this interest disappeared at about the time of puberty. Freund and Kuban suggest that in healthy males there is an active devaluation of erotic stimuli involving children that occurs around the time of puberty and that this process does not occur in pedophiles.

Sexual behavior in humans is under greater and more complex regulation than it is in nonhuman animals. Context, culture, and morality, for example, influence the display of sexual behavior in humans. A male rat will attempt to copulate with any novel female that he encounters, while this is clearly not the case in humans. What this suggests is that neocortical control mechanisms exert a major top–down cognitive/rational regulation of goal-directed sexual behavior in humans. For example, as already indicated, lateral PFC-to-orbital PFC-to-medial PFC circuits may exert strong control over hypothalamic–amygdala–ventral striatal circuits that control appetitive sexual behavior in humans, appropriately constraining the circumstances under which an individual's sexual urges are acted upon. Indeed, a dominant theory of pedophilia is that pedophiles exhibit a major dysfunction in PFC executive control mechanisms [155]. Relevantly, the IQ of pedophiles is significantly lower than that of the general population.

In an fMRI study, Walter et al. [1012] presented *adult* erotic photos and neutral photos to a group of male pedophiles and healthy male controls.

As measured by an increased BOLD response to erotic stimuli, the healthy male subjects, but not the pedophiles, demonstrated activation of the hypothalamus. This result, of course, accords with the sexual preferences of the tested subjects, with controls, but not pedophiles, exhibiting a sexual preference for adult erotica.

Sartorius et al. [857] presented pedophiles and healthy male adults with sexually nonexplicit photos of women, men, boys, and girls dressed in either swimsuits or underwear while examining the amygdala BOLD responses to these stimuli. The amygdala BOLD response to adult photos did not distinguish the two groups. However, when the photos of children were presented, the amygdala BOLD response decreased in the control group, while in pedophiles the amygdala BOLD response increased. These results are consistent with the hypothesis that there is an active devaluation of the sexual attractiveness of children in the normal population but not in pedophiles.

A few studies have shown that when pedophiles view their preferred sexual stimuli (for example, photos of nude prepubescent girls for heterosexual pedophiles), the brain activation patterns that occur are similar to those shown by heterosexual adult males that view nude women [776,783,864]. These activated regions include the amygdala, NA-VP area, and ACC. Somewhat surprisingly, I am not aware of any study that has reported hypothalamic activation when adult male pedophiles view their sexually preferred stimuli. Perhaps this is related to the nature of the stimuli that were presented in these studies, which were photos rather than video clips. That is, perhaps the intensity of sexual arousal is lower under conditions where only photos are presented. In line with the idea that similar mechanisms regulate libido in pedophiles and normal men when they view their preferred stimuli, treatments that are known to decrease libido in normal men also decrease pedophilic urges in sex offenders. Such treatments include antiandrogen administration (cyproterone acetate, which antagonizes the action of androgens at the androgen receptor) and SSRIs, which increase extracellular 5-HT levels [356,462,882].

A noteworthy difference between heterosexual pedophiles and healthy heterosexual men in the manner in which the brain responds to sexually explicit photos was detected by Schiffer et al. [864] for the orbital PFC. When compared to neutral photos, the orbital PFC was activated by photos of nude women and girls in the control group, while this brain region (area 47) was not activated by any sexual stimulus in the pedophile group. This result is consistent with a dysfunction in prefrontal cortex processing of sexual stimuli in pedophiles.

In addition to fMRI studies, some studies have examined differences in brain structure between pedophiles and control subjects using MRI voxel-based morphometry measures of gray matter and white matter volume. Poeppl et al. [777] compared the brains of male pedophile offenders with male nonsexual criminals and reported decreased gray matter volumes in the amygdala and dorsolateral PFC of the pedophiles (also see [865]). Similarly, Cantor et al. [155] examined differences in white matter (axon pathways or neural tracts) between male pedophiles and control men with a history of nonsexual crimes. A major finding was a decreased volume of a major neural tract in the pedophile group: the superior fronto-occipital tract. This neural pathway connects the lateral PFC with the visual association cortex (cf. [167]). The visual association cortex includes the extrastriate cortex in the occipital lobe. It is possible that such a pathway influences how particular visual stimuli are processed and whether such stimuli can gain access to limbic regions such as the amygdala. In this regard, most fMRI studies show that the occipitotemporal regions of the neocortex are activated, in terms of an increased BOLD response, when normal subjects view emotional visual stimuli, which include sexual stimuli, in comparison to when they view neutral visual stimuli. This increased BOLD response probably represents enhanced attention to and processing of emotional visual stimuli. Also note that the occipital visual association cortex projects to the inferior temporal cortex (ventral part of temporal neocortex, which is also considered to be part of the visual association cortex), where processes related to visual form perception occur. In regard to this functional neuroanatomy, additional findings from the Beauregard et al. [75] study are very relevant. Recall that when healthy male subjects viewed adult erotic videos, the amygdala and the hypothalamus were activated. However, when such men were instructed to suppress their sexual arousal to these erotic stimuli, hypothalamic and amygdala activation was decreased while lateral PFC and ACC (area 32) activity was increased. Importantly, when not suppressing their sexual arousal, the occipitotemporal regions of the visual association cortex was also more active in subjects when viewing adult sexual stimuli than when viewing neutral stimuli. Under the condition of emotional regulation, where the subjects suppressed their sexual arousal, an enhanced BOLD response in the inferotemporal cortex was no longer detected. Perhaps a pathway from the lateral PFC to the occipital lobe depressed the relay of sexually explicit visual stimuli to the inferotemporal cortex, which, in turn, prevented such stimuli from being relayed to the amygdala (see [621]). If this were the case, and if the fronto-occipital tract were involved in this type of emotional regulation, then one would predict that pedophiles might be less likely to suppress the processing of visual sexual stimuli of children and the sexual arousal that they produce.

The fMRI functional studies and the MRI anatomical studies, of course, are correlational in nature, and therefore cannot prove cause–effect relationships. However, based on an integration of the animal studies with the human studies, Figure 4.6 presents a rudimentary model

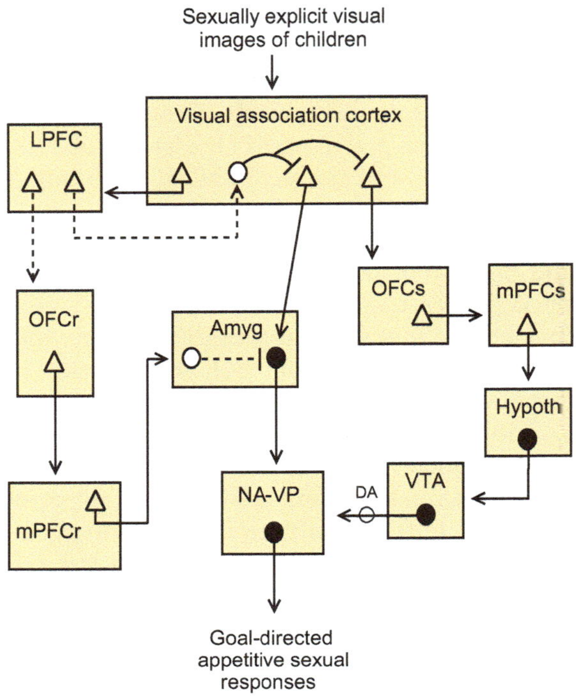

FIGURE 4.6 A tentative neural model that proposes some of the neural dysfunctions that may underlie pedophilia in men. These neural dysfunctions allow sexually explicit visual images of children to gain access to those amygdala and hypothalamic neural systems that promote goal-directed (appetitive) male sexual behavior. There is evidence that the neural connections shown in dashed lines may not be fully functional in male pedophiles. The first of these is the projection from the lateral prefrontal cortex (LPFC) to the visual association cortex, a projection that would normally inhibit the transfer of sexually explicit images to the amygdala (Amyg), and to those part of the orbital and medial prefrontal cortices (OFCs; mPFCs) that provide stimulatory sensory inputs to the hypothalamus (Hypoth). Another dysfunctional circuit may be between the LPFC and the regulatory parts of the OFC and mPFC (OFCr; mPFCr). These regulatory prefrontal regions would normally act to depress amygdala output by stimulating inhibitory interneurons in the amygdala. Such inhibitory amygdala interneurons may also be dysfunctional in pedophiles. As a result of these neural dysfunctions, sexually explicit images of children become potent activators of those hypothalamic-to-mesolimbic dopamine (DA) circuits that underpin the male pedophile's strong sexual attraction toward children. Axons ending in a bar are inhibitory, and those ending in an arrow are excitatory. Other abbreviations: NA-VP = nucleus accumbens–ventral pallidum circuit (DA input to NA-VP facilitates VP output); VTA = ventral tegmental area.

of some of the dysfunctions that may underlie pedophilia. This model is also based on the anatomy presented in Chapter 1 (see Figure 1.15, and also see [61,621]). As described in the figure caption, those pathways indicated by dashed lines may be dysfunctional in pedophiles, which then results in appetitive male sexual responses toward children. Dysfunctions in the connections of the lateral PFC with the orbital prefrontal cortex (OFC) may prevent PFC inhibition of amygdala responses to sexual stimuli from children. Further, dysfunctions in the connections of the lateral PFC with the visual association cortex may allow sexually explicit visual stimuli from children to be processed by the visual association cortex so that such stimuli can ultimately be relayed to the hypothalamus and amygdala where they exert excitatory effects. It is interesting to speculate that the decreased gray matter volume of the lateral PFC in pedophiles may represent a decreased number of excitatory output pyramidal neurons, while the decreased gray matter volume of the amygdala that is observed in pedophiles may represent a decreased number of inhibitory interneurons.

This figure represents an attempt to explain why an active devaluation of sexual stimuli emanating from children does not occur in pedophiles near the time of puberty, while such devaluations do occur in healthy controls. Dysfunctions in the connections of the lateral PFC may interfere with the ability of socialization processes to suppress sexual responses to children by disrupting a top–down cognitive control of sexual impulses. The proposed figure, however, is very far from explaining the neurobiology of pedophilia. Primary among the mechanisms that must ultimately be explained is the initial propensity, occurring at a very early age and prior to puberty (between 7 and 11 years), for pedophiles to express a greater interest in viewing nude children than nude adults, while the reverse developmental trend exists in normal boys. The etiology of pedophilia is undoubtedly very complex, and the potential involvement of experiential and genetic factors is yet to be explored and uncovered.

Figure 4.6 also emphasizes that in normal individuals, inappropriate sexual responses are restrained by PFC regulatory mechanisms. For healthy adult males, the occurrence of sexual behavior is usually consensual and appropriate, and can be defined as a prosocial response. However, when the regulatory functions of the PFC are disrupted, abnormal appetitive male sexual behavior can occur because inappropriate stimuli gain access to prosocial circuits, leading to functionally antisocial behaviors such as pedophilia.

4.6 SEXUAL DIFFERENTIATION AND THE DEVELOPMENT OF SEXUAL BEHAVIOR

4.6.1 Introduction

The purpose of this section is to describe the neurobiological processes that result in the development of sexual behavior and sexual preferences. Male mammals typically show male copulatory responses directed toward female conspecifics, while female mammals typically exhibit female copulatory responses directed toward male conspecifics. How does this important sexual difference in the display of sexual behavior toward a partner of the opposite sex come about? Is it because circulating levels of steroids in *adult* mammals differ by sex, with males secreting more testosterone and females more estradiol in adulthood? If that were the case, then one would conclude that fully functional neural circuits for male and female sexual behavior and motivation reside in the brains of both sexes and that if an adult male were given estradiol he would be attracted to other males and exhibit female copulatory responses, while if an adult female were given testosterone, she would be attracted to other females and would display male copulatory responses.

Alternatively, perhaps as a result of developmental processes, the neural circuits regulating sexual behavior actually differ by sex. Perhaps the particular circuits regulating male sexual behavior are fully functional only in males, while different circuits regulating female sexual behavior are fully functional only in females. As I will show, this latter suggestion approximates reality in that there are sex differences in the neural circuits that regulate sexual behavior.

Although typical forms of sexual behavior, motivation, and partner preference are heterosexual, homosexuality also exists. By understanding the nature of the developmental neurobiological processes that regulate the ontogeny of sexual behavior and motivation, insights into the development of both heterosexuality and homosexuality are provided. With respect to human sexual orientation, early views espoused the primary role of socialization processes and proposed that the way a child was raised by its parents affected the child's sexual development [660]. In the review that follows, which integrates animal and human studies, it will become clear that biological factors, rather than socialization processes, exert the most significant effects on how the brain develops with respect to sexual behavior and sexual preferences.

Table 4.1 shows some typical sex differences that occur in mammals. How do these sex differences come about? With respect to the embryological development of the gonads and reproductive tract, the following sequence of events determines sexual development [737,944]: the sex-determining region of the Y chromosome (SRY gene, which is only present in males) regulates the development of the undifferentiated fetal gonadal tissue into testes. If SRY is not present, as in XX females, then the undifferentiated fetal gonadal tissue develops into ovaries. Therefore, fetal gonadal tissue is bipotential and develops as testes or ovaries depending on the presence or absence of the SRY gene on the Y chromosome. During the remainder of fetal development, with respect to hormone secretions, the fetal ovaries are quiescent, while the testes secrete the steroid testosterone and a peptide referred to as Mullerian inhibiting substance (MIS). It is the secretion of these two testicular hormones during the fetal period that directs the subsequent development of the reproductive tract, resulting in masculinization (the development of a male internal reproductive tract and male external genitalia) and defeminization (suppression of a female reproductive tact and female external genitalia development). In the absence of testicular secretions, feminization occurs (the development of a female internal reproductive tract and female external genitalia).

TABLE 4.1 Some Typical Sex Differences in Mammals

Type	Male	Female
Chromosomes	XY	XX
Gonads	Testes	Ovaries
Internal reproductive tract	Vas deferens and seminal vesicles	Uterus and fallopian tubes
External genitalia	Penis and scrotum	Clitoris and vaginal labia
Sexual behavior	Male copulatory behavior directed toward females	Female copulatory behavior directed toward males

Based on the pioneering work of Jost [463], Figure 4.7 provides a summary of reproductive tract development. With respect to this figure, a few points are worth considering. First, depending upon the enzymes present in target tissues, testosterone (T) can be converted into other steroids, which then act on the target tissue along with testosterone. The enzyme 5α reductase converts testosterone into dihydrotestosterone (DHT), and aromatase enzymes convert testosterone into estradiol (E). Second, with respect to the development of the internal reproductive tract, there are two separate primordial (undifferentiated) systems, the Mullerian (female) system and

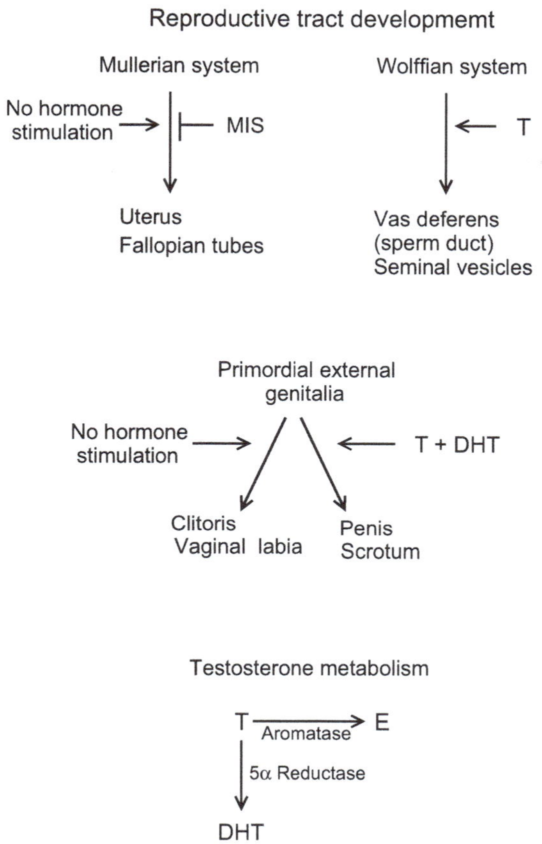

FIGURE 4.7 Reproductive tract development and testosterone (T) metabolism. For reproductive tract development in the embryo, T secretion from the fetal male's testes stimulates the Wolffian system to develop into a vas deferens and seminal vesicles, and stimulates the primordial external genitalia to develop into a penis and scrotum (dihydrotestosterone [DHT] co-acts with T to produce this effect on external genitalia development). Mullerian inhibiting substance (MIS) secretion from the fetal testes inhibits the development of the Mullerian system. In the absence of T, DHT, and MIS effects, as would occur in the female fetus, the Wolffian system does not develop, the Mullerian system develops into uterus and fallopian tubes, and the primordial external genitalia form the clitoris and vaginal labia. The bottom panel shows that the steroid hormone testosterone can also serve as a precursor for other steroids, depending on the particular enzymes that are present within the cells of a target tissue. The aromatase enzyme system catalyzes the conversion of T to estradiol (E), and 5 alpha reductase converts T to DHT. With respect to the hormonal effects shown in the top and middle panels, lines ending in a bar represent inhibitory effects, and lines ending in an arrow signify stimulatory effects.

the Wolffian (male) system. In the absence of testicular secretions, the Mullerian system develops while the Wolffian system does not. In the presence of testicular secretions, MIS suppresses Mullerian development (defeminization) while testosterone stimulates Wolffian development (masculinization). In contrast, the external genitalia develop from a single bipotential primordial tissue: in the absence of steroids, feminization occurs, and in the presence of T and DHT masculinization occurs. Third, once the gonads have developed on the basis of genetic differences between males and females (XY versus XX), the future development of the reproductive tract is determined by the fetal hormonal environment rather than by sex chromosomes: if a functional fetal rabbit testes in transplanted into a fetal XX female rabbit prior to reproductive tract development, then when this XX rabbit is born it will not possess a uterus and fallopian tubes but will have a penis, scrotum, seminal vesicles, and vas deferens; if a fetal XY rabbit is castrated in utero prior to reproductive tract development, this XY rabbit will be born with a female internal reproductive tract and female external genitalia [463]. The reader is referred to Glickman, Short, and Renfree [341] for interesting exceptions to these general mammalian rules concerning the sexual differentiation of the peripheral reproductive system.

The important issue for us is whether sex differences in sexual behavior and motivation develop in a manner similar to the development of the peripheral reproductive system. Do secretions from the fetal testes masculinize and defeminize the brain, promoting the development of functional male sexual circuits and suppressing the development of functional female sexual circuits, so that in adulthood male copulatory responses are directed toward females, while in the absence of testicular secretions, do feminized neural circuits develop so that in adulthood female copulatory responses are directed toward males? Further, which model of reproductive system development might the brain follow: is there a single primordial neural system that can develop in either a male or female direction, or are there two separate primordial neural systems, one with the potential to regulate male behavior and the other with the potential to regulate female sexual behavior?

4.6.2 Sexual Differentiation of Brain and Behavior in Nonprimate Mammals

4.6.2.1 *Introduction*

Several excellent and extensive reviews have been written on the effects of fetal and early postnatal testicular secretions on the masculinization and defeminization of the brain and sexual behavior [17,40,49,242,541,617,666,779,900]. A simplification of the overall data is shown in Figure 4.8. The figure shows that gonadal *steroids* can have organizational and activational effects on the brain mechanisms controlling sexual behavior [40,772]. Organizational effects refer to developmental effects: testicular steroids secreted during the perinatal period (prenatal and early postnatal period) masculinize and defeminize brain circuits so that in adulthood male copulatory responses are directed toward females. In the absence of steroid secretion during the critical perinatal period of brain development, feminization of the brain occurs, resulting in female copulatory responses being directed toward males in adulthood. Activational effects of steroids refer to the action of steroids on the neural circuits that have already been developed, with such action affecting the functional activity of those circuits in postpubertal (sexually mature) organisms. For example, in adults,

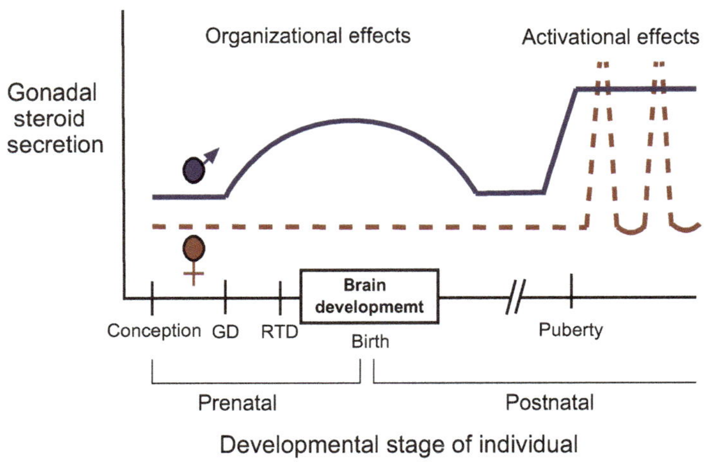

FIGURE 4.8 A representation of gonadal hormone secretions from males (solid blue line) and females (dashed red line) during the prenatal and postnatal periods. Hormonal secretions from the ovaries of females are relatively quiescent until puberty, at which time the cyclic production of ovarian steroids (primarily estradiol) activates discrete periods of sexual motivation and behavior in most female mammals. Organizational effects refer to developmental effects. For the female, the absence of gonadal hormone secretions during the prenatal and early postnatal period results in the development of a female reproductive tract and in a feminization of those brain circuits that control sexual behavior. For males, the testes secrete hormones during the prenatal and early postnatal periods, and the resultant organizational effects of these secretions promote the development of a male reproductive tract and a masculinization and defeminization of those brain circuits that control sexual behavior. After the early postnatal period in males, testicular secretions decline until puberty. At puberty, relatively constant amounts of steroids (primarily testosterone and dihydrotestosterone) begin to be secreted from the testes, which results in the activation of relatively constant levels of male sexual motivation and behavior. Abbreviations: GD = gonad development; RTD = reproductive tract development.

testosterone acts on the fully developed male circuitry to affect the appetitive and consummatory aspects of male sexual behavior, while estradiol acts on the fully developed female circuitry to affect the appetitive and consummatory aspects of female sexual behavior. In adulthood, female sexual behavior is cyclic (estrous cycles) because estradiol secretion is cyclic, while male sexual behavior is constant (during the breeding season) because testosterone secretion is relatively constant.

Figure. 4.8 also shows that the development of different structures occurs in stages: the reproductive tract differentiates before the brain, through mechanisms that I have already described. Therefore, a perturbation in testicular steroid secretions during the period of brain development could result in feminized sexual behavior in an XY individual that looks like a male in terms of peripheral reproductive system development.

In rodents, the masculinizing and defeminizing effects of testosterone are primarily, but not exclusively, due to the conversion of testosterone to estradiol in brain cells that contain aromatase enzymes [617]. However, in primates, testosterone appears to be the major steroid that causes the masculinization and defeminization of brain development with respect to sexual behavior [1006].

Although the presence or absence of perinatal testicular steroid secretions is the primary factor that affects the development of sex differences in sexual behavior, there is some evidence that the genes located on sex chromosomes can also influence sex differences in behavior independent of perinatal gonadal secretions, although such behavioral differences between the sexes primarily involve nonsexual behaviors [40]. For example, genes on the Y chromosome (other than SRY) have been shown to influence aggressive behavior [41].

4.6.2.2 Behavioral Analysis

Most experimental studies on the development of sexual behavior have been done on rats. A substantial part of brain differentiation in rats occurs over the first 10 postnatal days, so experiments can be performed on neonates rather than on fetuses. A summary of some of these results is shown in Table 4.2 [49,617]. If male rats are castrated on day 1 of life, when they are injected with testosterone in adulthood, they show low levels of mounting behavior when exposed to a sexually receptive female. When such males are injected with ovarian hormones in adulthood (estradiol or estradiol followed by progesterone), they show high levels of lordosis behavior when mounted by a sexually active male. The sexual behavior of these XY males has been feminized. Importantly, if neonatal males are castrated later in life, on day 15 instead of day 1, they behave just like typical males in adulthood: when they are injected with testosterone they show high levels of mounting, but when they are injected with ovarian steroids they do not show lordosis when exposed to a sexually active male. In other words, exposure to the organizing effects of testicular steroids during a critical period over

TABLE 4.2 Sexual Differentiation of Sexual Behavior in Rats

Organizational Neonatal Treatment	Activational Adult Treatment (70–90 days old)
1. Male rat castrated on day 1 of life	E + P: high lordosis T: low mounting behavior
2. Male rat castrated on day 15 of life	E + P: low lordosis T: high mounting behavior
3. Female rat injected with testosterone (or estradiol) on day 1 of life (and ovariectomized in adulthood)	E + P: low lordosis T: high mounting behavior
4. Female rat injected with testosterone (or estradiol) on day 15 of life (and ovariectomized in adulthood)	E + P: high lordosis T: low mounting behavior

Abbreviations: E = estradiol; P = progesterone; T = testosterone.

the first few postnatal days masculinizes and defeminizes their sexual behavior. With respect to the XX female rat (with a uterus and vagina), if they are injected with either testosterone or estradiol on day 1 of life, then in adulthood they show low lordosis after being injected with ovarian steroids and high mounting behavior when injected with testosterone. These females have been masculinized and defeminized by the steroid injection on day 1 of life. However, if the testosterone (or estradiol) injection is delayed until day 15 of life (after the critical period of brain development), then such females are feminized just like normal females, showing high lordosis behavior (when treated with estradiol or estradiol plus progesterone and exposed to a sexually active male) and low mounting behavior (when treated with testosterone and exposed to a sexually receptive female) in adulthood.

The above results show that there is a postnatal critical period over which exposure to either testosterone (which can be converted to estradiol in the brain by aromatase) or estradiol masculinizes and defeminizes the rodent brain. In the rat, after approximately postnatal day 10, the brain circuits underlying sexual behavior appear to be fixed and can no longer be changed by hormone treatment. One caveat should be mentioned. The analysis presented so far suggests that feminization of the brain is the default process: in the absence of testicular steroid exposure during the first postnatal week, the brain develops in a female direction with respect to sexual behavior. However, there is some evidence that estradiol secretion from the ovaries of females during the third postnatal week (after the perinatal period of testosterone/estradiol-induced masculinization and defeminization) potentiates the brain feminization process [49,129]. Therefore, one conclusion is that testicular secretions during the first postnatal week masculinize and defeminize the brain, while ovarian secretions during the third postnatal week fully feminize the brain.

A deeper analysis of these results shows that it is incorrect to assume that normal (control) male rats will never show lordosis and that normal (control) female rats will never show mounting behavior [243,874,962]. Female rats that are not exposed to perinatal testosterone or estradiol will show some mounting toward an estrous female if they are injected with testosterone in adulthood, but the level of such mounting behavior (number of mounts in a 20-min test, for example) is much lower than that shown by normal males or by females that had been injected with either testosterone or estradiol during the perinatal critical period. Similarly, normal male rats are capable of showing some lordosis responses when injected with high levels of estradiol in adulthood, but such lordosis responses are much less frequent than that shown by normal females and by males that have been castrated during the perinatal critical period. It is probably best to assume for the typical male and female rat, under normal developmental conditions, that the neural circuits for both male and female sexual behavior are present in some form in the brain of both sexes, but that the male circuits are more fully developed and functional in males while the female circuits are suppressed, with the opposite being true for females.

McCarthy [617] has asked why the defeminization process that results in the suppression of brain circuits controlling female sexual behavior is necessary. In other words, although it is understandable that testosterone/estradiol masculinizes the brain during the critical period, why should defeminization occur, since males do not have ovaries and would not be exposed to the activational effects of estradiol in adulthood? An answer to this question involves the fact that aromatase is present in certain neurons and therefore testosterone could be converted to estradiol in the adult brain. Since estradiol alone can trigger female sexual behavior, the defeminization process ensures that under normal physiological conditions the typical adult male will not show female sexual behavior, presumably because the full development of the female circuits that could respond to estradiol would have been suppressed.

The studies reviewed so far have examined the effects of perinatal steroids on the sexual differentiation of copulatory (consummatory) responses. However, when one discusses the concept of sexual orientation or preference, one is referring to the sex of the partner that an individual animal chooses to engage in sexual behavior with, rather than the particular copulatory responses that are shown. In a T-maze situation, for example, the typical adult male rat, when given a choice to mate with either an estrous female or a sexually active male, will choose the female, and the typical adult female rat in estrus will choose the sexually active male rat [5,72,403]. Therefore, under the activational effects of testosterone, the typical male will choose to mate with a female and will also exhibit male copulatory responses. Similarly, under the activational effects of estradiol or estradiol plus progesterone, the typical female will choose to mate with a sexually active male, and she will show lordosis in response to his mounts. Therefore, sexual partner preferences and copulatory responses are concordant in the typical rat. In addition, like copulatory responses, the development of sexual partner preferences is also under the control of the organization effects of perinatal testicular secretions [403]: the presence of testicular steroid secretions during the prenatal and postnatal critical periods masculinizes and defeminizes mate choice, which causes the typical male to choose a female and not a male as a mating partner.

In interesting research performed on sexual partner preferences in male domestic sheep [828], it has been determined, as expected, that most rams choose to mate with an estrous ewe (female-oriented rams). However, in a natural population, approximately 8% of rams are attracted to and choose to mate with another male (male-oriented rams). Importantly, such male-oriented rams have normal male genitalia and show male copulatory responses, such as mounting, toward other males, and they do not display female-typical receptive behaviors. These results indicate that the sexual differentiation of sexual partner preference may occur independently from the sexual differentiation of copulatory responses. More specifically, there may be slightly different critical periods and different underlying cellular processes through which perinatal steroids influence the development of each of the following: masculinization of copulatory behavior, masculinization of partner preference, defeminization of copulatory behavior, and defeminization of partner preference. The male-oriented rams appear to have masculinized and defeminized copulatory responses, but their mate choice or partner preference appears to be feminized. It is important to understand these distinctions when one considers the various phenotypes that could result from selective and specific disturbances of the effects of perinatal steroid secretions. For example, if perinatal steroid-induced defeminization of copulatory responses and partner preferences were to be selectively disrupted without interfering with steroid-induced masculinization of these processes, the result would be a bisexual individual.

4.6.2.3 *Brain Mechanisms: MPOA*

In examining the brains of typical adult male and female rodents, clear sex differences exist in many brain regions, with some brain regions containing more neurons in males and other brain regions containing more neurons in females. Several important reviews have been written on this subject, and the reader is referred to De Vries and Simerly [242] for an excellent and comprehensive summary and analysis of much of this data. The importance of detecting sexual dimorphisms in the brain is that some of these differences may contribute to sex differences in sexual behavior. Importantly, as described by De Vries and Simerly, many of the sexual dimorphisms in brain anatomy are driven by the presence or absence of testicular secretions during the perinatal critical period of brain development. The obvious implication of these findings is that perinatal testicular steroids cause the male brain to develop differently from the female brain, resulting in the masculinization and defeminization of neural circuits that control the appetitive and consummatory aspects of sexual behavior.

Because of the positive involvement of the MPOA in the appetitive and consummatory aspects of male sexual behavior and in the appetitive aspects of female sexual behavior, in this section I will describe sex differences in MPOA anatomy, differences that result from the presence or absence of perinatal testicular secretions, which have been correlated with differences in the sexual behavior of male and female rats. In a classic study, Gorski, Harlan, Jacobson, Shryne, and Southam [351] described the sexually dimorphic nucleus of the MPOA (SDN-POA). This nucleus has also been referred to as the central subdivision of the medial preoptic nucleus (MPNc; [903]). In adult rats, the volume of this nucleus is much larger and also contains many more neurons in males than in females. This morphological sex difference is shown in Figure 4.9(A).

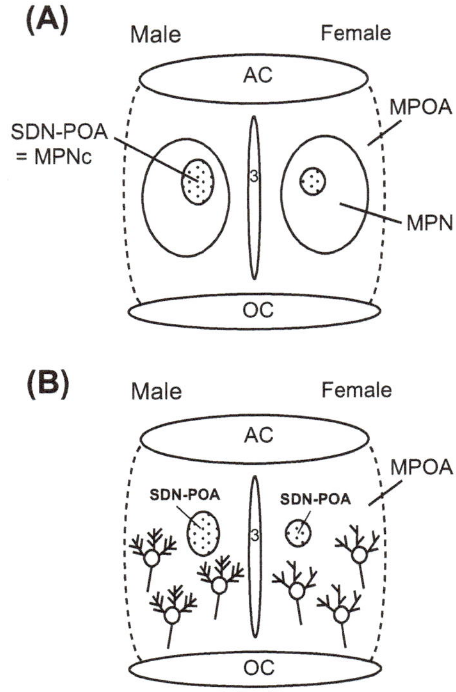

FIGURE 4.9 Sexual dimorphisms within the medial preoptic area (MPOA) of rats. The right side of the MPOA represents the female, and the left side represents the male, phenotype. (A) The sexually dimorphic nucleus of the MPOA (SDN-POA), also referred to as the central part of the medial preoptic nucleus (MPNc), is larger and contains more neurons in males than in females. Other abbreviations: AC = anterior commissure; OC = optic chiasm. (B) Outside the sexually dimorphic nucleus of the MPOA, there are more dendritic spines protruding from the dendrites of neurons in the MPOA of males than in females.

This sexual dimorphism within the MPOA has been shown to develop under the influence of perinatal estradiol action on the brain as a result of the conversion of testicular testosterone to estradiol by aromatase [242]. For example, if male rats are castrated on day 1 of life, in adulthood they have a small SDN-POA. However, if males are castrated on day 1 of life and also injected with either estradiol or testosterone at this time, in adulthood they have a large SDN-POA. Further, when female rats are injected with either testosterone or estradiol during the first postnatal week, in adulthood they have a large male-like SDN-POA.

How might estradiol action within the MPOA during the perinatal critical period result in the formation of an SDN-POA that contains more neurons? Although many processes influence how the brain develops, a process referred to as cell death or apoptosis plays a crucial role. Many more neurons are born (neurogenesis) during early brain development than actually end up surviving, and if there is a sex difference in apoptosis within a particular neural region, then one sex will end up having more neurons in this region than the other sex [313]. Therefore, one important perspective on development of sexual brain dimorphisms, where a male has more neurons in a particular region than a female, as is the case for the SDN-POA, is that perinatal testicular steroids bind to either androgen receptors or estrogen receptors to ultimately affect gene transcription mechanisms so that proteins are expressed that suppress apoptosis. Two proteins are importantly involved in regulating cell death in the developing brain: Bcl-2, which is antiapoptotic; and Bax, which is proapoptotic [313]. Importantly, the Bcl-2 gene contains an estrogen response element in its regulatory region, which could be a mechanism that would allow estradiol to antagonize cell death, leading to a greater survival of those neurons in a particular brain region that contain estrogen receptors [977].

That differential cell death contributes to the sexual dimorphism within SDN-POA is supported by evidence [242,977]: during the first postnatal week, Bcl-2 levels are higher and Bax levels are lower in the SDN-POA of male rats when compared to the levels expressed by female rats; during the early postnatal period, the number of apoptotic neurons within the SDN-POA is greater in females than in males (apoptosis can be histologically verified through the detection of DNA fragmentation within the nucleus of a neuron).

Although estradiol converted from testosterone might depress apoptosis in SDN-POA during brain development through the direct activation of Bcl-2 transcription, other cellular mechanisms may also be involved. Important research suggests that during the perinatal period, estradiol also increases the expression of NMDA glutamate receptors within the SDN-POA and that subsequent glutamate activation of this receptor during the perinatal period stimulates an intracellular chemical cascade that, in turn, influences genetic mechanisms that increase the expression of antiapoptotic proteins [428,429]. In this case, estradiol would indirectly, rather than directly, stimulate Bcl-2 gene expression.

While differential cell death via the Blc-2/Bax system appears to be the mechanism involved in the development of the sexual dimorphism within SDN-POA of rats, such a process may not be involved in the sexually differentiated development of a presumably homologous brain region in mice, which indicates that there can be important species differences in the manner by which perinatal steroids influence brain development [334].

These studies show that perinatal testicular secretions act to increase the size of the SDN-POA. Given the important role of the MPOA in both the appetitive and consummatory aspects of male sexual behavior, it is tempting to conclude that the larger size and neuron number within the SDN-POA of male rats compared to female rats is related to the masculinization of male sexual behavior and that the SDN-POA is part of the MPOA neural circuitry that regulates male sexual behavior. The research reviewed so far, however, is correlational in nature and does not prove that SDN-POA neurons contribute to male sexual behavior. Indeed, when highly selective lesions are produced in adult male rats, lesions that destroy the SDN-POA, while sparing most of the remaining parts of the MPOA (see Figure 4.9(A)), male copulatory responses remain intact [39,233]. These results suggest that the SDN-POA does not contribute in significant ways to the consummatory aspects of male sexual behavior.

Although the exact function of the SDN-POA remains to be determined, recall that perinatal testicular secretions not only cause the masculinization of sexual behavior but also cause the defeminization of sexual behavior, which results in the suppression of lordosis in males. Perhaps perinatal testicular steroids, by preventing cell death in SDN-POA neurons, result in the development of neural circuits that depress the ability of males to show lordosis in response to estradiol plus progesterone. If this were the case, a larger SDN-POA would be part of the defeminization process [243,405,962]. The fact that the SDN-POA projects strongly to the VMNvl [903] and that most of the neurons in the SDN-POA are GABAergic [325] suggest a route and mechanism through which SDN-POA output could inhibit lordosis. The fact that the SDN-POA is larger in males than in females would suggest that the VMNvl would be under greater inhibition in males than in females, accounting for the fact that females are more likely to show lordosis than are males when each sex is treated with the same doses of estradiol and progesterone. There is some evidence to support this hypothesis. First, recall that MPOA lesions facilitate lordosis when *female* rats are treated with suboptimal doses of steroid hormones. Perhaps the small size of SDN-POA in females contributes to a small-to-moderate inhibition of lordosis, which would allow females to switch away from lordosis and consummatory responses in order to effectively pace their sexual interactions with the male. More to the point, however, large electrical or excitotoxic amino acid lesions of the MPOA, performed on adult *male* rats, have been found to facilitate the display of lordosis when such males are castrated and injected with estradiol and progesterone [405,725]. While the LQ shown by control males is very low (about 0.20), the LQ shown by males with large MPOA lesions is virtually indistinguishable from that shown by control females (about 0.80). Note, however, that these MPOA lesions were large and damaged most of the MPOA. No one, to my knowledge, has investigated whether selective lesions restricted to the SDN-POA would facilitate the display of lordosis in male rats. Interestingly, the MPOA lesions in these studies were so large that, in addition to enhancing lordosis, they should have also disrupted male sexual behavior, but this was not tested. As will be suggested below, it is possible that MPOA neurons outside the SDN-POA regulate male sexual behavior, while SDN-POA neurons inhibit lordosis.

The regulation of cell death by perinatal testicular steroids is one mechanism through which sexually dimorphic brain circuits can be constructed, but other processes have been shown to contribute to sexual differences in brain development [617]. One such mechanism involves the regulation of a neuron's phenotype [242]: a particular brain region may contain the same number of neurons in adult males and females, but the phenotype of these neurons might differ by sex. As examples, such phenotypic differences could include differences in neurotransmitter content or differences in the number of excitatory or inhibitory neurotransmitter receptor sites. In this context, perinatal steroid-induced epigenetic processes that affect the degree of DNA methylation and/or histone acetylation may influence whether particular genes are active in adult neurons, in this way causing sex differences in the phenotype of such neurons [439,693,879]. For example, the action of perinatal steroids may cause sex differences in the amount of DNA methylation at CpG sites within the regulatory region of a particular neuropeptide gene, which would then influence the degree to which this neuropeptide could be expressed within the affected neurons in the mature organism [879]. Although such research is relatively undeveloped, epigenetic processes are likely to be one of many ways through which perinatal steroids produce sex differences in the phenotype of developing neurons.

In a series of very interesting studies, McCarthy's group [22,23,541,874,962] has described an important phenotypic sex difference in the morphology of MPOA neurons that is produced by perinatal steroids and which may contribute to the masculinization, but not the defeminization, process in rats. This phenotypic sex difference does not appear to be regulated by epigenetic processes. There are three basic kinds of synapses in the brain: axosomatic, axodendritic shaft, and axodendritic spine synapses. The latter synapse, where an axon terminal synapses on a protrusion (spine) from a dendritic shaft, is a major site for excitatory neural transmission. Therefore, it is assumed that if a neuron contains dendrites with many spines, it receives strong excitatory inputs. Importantly, there are many MPOA neurons outside the comparatively smaller SDN-POA (see Figure 4.9(A)), and these MPOA neurons exhibit a higher density of dendritic spines in males than in females. This difference is schematically shown in Figure 4.9(B). This sex difference in dendritic spine morphology develops under the influence of perinatal testicular steroids: in neonatal male rats, testosterone is converted to estradiol, and estradiol action in the developing MPOA not only results in more neurons surviving in SDN-POA but also in the formation of more dendritic spines on the dendrites of MPOA neurons outside the SDN-POA. Importantly, the evidence indicates that the mechanism through which estradiol leads to more dendritic spines on MPOA neurons is distinct from the mechanism through which cell death is prevented within SDN-POA.

McCarthy's group has shown that perinatal estradiol increases the expression of cyclooxygenase-2 (COX-2) mRNA and protein within the developing MPOA. COX-2 is an enzyme that catalyzes the synthesis of the lipid prostaglandin E2 (PGE2). In an elegant series of studies (see [541,874] for reviews), it has been shown that estradiol-induced PGE2 synthesis is the first step in a complex series of subsequent biochemical signaling pathways downstream from PGE2 that ultimately results in the formation of increased dendritic spines on the dendrites of MPOA neurons. Importantly, this perinatal estradiol-induced development of dendritic spines is correlated with behavioral masculinization but not with behavioral defeminization. Table 4.3 provides an integrative summary of the results from McCarthy's group by showing the effects of various perinatal treatments on the development of MPOA morphology and sexual behavior. One of the perinatal treatments indicated in the table is the treatment of neonatal male rats with indomethacin, which is a chemical that inhibits COX-2 and therefore would depress PGE2 synthesis. In reviewing this table, some findings are obviously based on the research already reviewed: normal females have low MPOA dendritic spines, a small SDN-POA, and show high levels of lordosis but low levels of male copulatory behavior in adulthood; while normal males and females treated with estradiol on the first few days of life develop a high number of MPOA dendritic spines, a larger SDN-POA, and show high levels of male copulatory responses when injected with testosterone, but almost no lordosis behavior when treated with ovarian hormones, in adulthood. The novel results are observed when only dendritic spine density is affected by particular neonatal treatments. When neonatal females are treated with PGE2 they develop an increased MPOA dendritic spine density, but the size of their SDN-POA is still small. In adulthood, these animals

TABLE 4.3 The Effects of Various Perinatal Treatments on MPOA Morphology and Adult Sexual Behaviors

			Adult Sexual Behavior	
	MPOA Morphology		T-induced Male	E + P-induced
Neonatal Treatment	**Dendritic Spines**	**Size of SDN-POA**	**Sexual Behavior**	**Lordosis**
Normal male	High	Large	High	Low
Normal female	Low	Small	Low	High
Female + E	High	Large	High	Low
Female + PGE2	High	Small	High	High
Male + indomethacin	Low	Large	Low	Low

Abbreviations: E = estradiol; MPOA = medial preoptic area; P = progesterone; PGE2 = prostaglandin E2; T = testosterone.

show high levels of both male sexual behavior and lordosis. In contrast, when neonatal male rats are treated with indomethacin, the density of MPOA dendritic spines is low but the size of their SDN-POA is large. These males are relatively asexual, showing low levels of both male sexual behavior and lordosis. These results clearly indicate that the mechanism of masculinization is distinct from that of defeminization and that the density of dendritic spines in the MPOA is positively correlated with the development of masculine sexual behavior, while the size of the SDN-POA is positively correlated with the suppression of lordosis. One hypothesis to explain these results is that perinatal estradiol action on the developing MPOA has two important effects. It increases the expression of Bcl-2 and COX-2. The former effect decreases cell death in the SDN-POA, resulting in a larger SDN-POA in adulthood, which may, in turn, result in a greater inhibition of VMNvl and a suppression of lordosis (defeminization). The latter COX-2 effect triggers the development of increases in dendritic spine density on MPOA neurons, allowing such neurons to receive increased excitatory inputs, and this effect may underpin increases in male copulatory responses in adulthood (masculinization), perhaps by increasing MPOA functional inputs to the PAG and VTA in response to stimuli from estrous female rats.

Although these results are very exciting, they are still correlational in nature, and more experimental research will be needed to prove that these estradiol-induced morphological sex differences in the MPOA are indeed related to the masculinization and defeminization of sexual behavior. For example, if one were able to experimentally depress the expression of dendritic spines in the MPOA of normal *adult* male rats so that the expression level matched that of females, would this result in a decrease in male copulatory responses? And, if one were to selectively lesion the male SDN-POA in adulthood, would this increase the male's ability to show lordosis?

Taking a broader view, these overall results go far in providing a framework to understand how perinatal testicular steroids can influence brain development, just like they affect the development of the peripheral reproductive system, so that the sexual preferences and behaviors of the typical male and female mammal develop along different trajectories.

4.6.2.4 Brain Mechanisms: Additional Defeminization Processes

Research indicates that there are multiple brain mechanisms that suppress the ability of normal male rats (males that have been exposed to testicular steroids during the perinatal period) to exhibit lordosis in response to ovarian steroid treatment in adulthood. Specifically, in addition to MPOA lesions, lesions of either the dorsal raphe nucleus (DR) or the lateral septum (LS) also facilitate lordosis in male rats [469,978]. Therefore, the defeminization process appears to involve multiple neural pathways that act to suppress lordosis. Tsukahara and Yamanouchi [978] have provided evidence for the proposal that it is LS efferents to the PAG that act to suppress lordosis in males. Recall that the PAG is essential for the display of lordosis in females. Since the major output neurons of LS are GABAergic [953], LS synapses in PAG may act to suppress those parts of PAG that mediate lordosis. With respect to the DR, it is likely that serotonin neurons originating from the DR project to the VMNvl to suppress lordosis in males. First, DR lesions that facilitate lordosis in males also decrease 5-HT input to VMN [468]. Second, direct injection of 5,7-dihyroxytryptamine, a neurotoxin that destroys serotonin neurons, into the VMN region facilitates lordosis in adult male rats treated with ovarian steroids [663].

The fact that LS-GABA projections to PAG and DR-5-HT projections to VMN suppress lordosis in adult male rats suggests that these influences may act independently of one another, and evidence suggests that this is the case [469]. Normal female rats show an LQ of about 0.90

after treatment with ovarian steroids, while normal male rats are virtually unresponsive (LQ of about 0.10). Adult males with knife cuts that sever the ventral connections of the LS, or adult males with lesions of the DR, exhibit an LQ of about 0.50. However, males that receive combined lesions of both LS and DR exhibit an LQ of about 0.80, suggesting that the suppressive effects on lordosis of the relevant neural connections are additive in nature.

It is interesting to recall that in Chapter 3 evidence was presented that 5-HT input to the HAA was involved in suppressing aggressive behavior in male rats and that the VMNvl contains neurons that are part of the HAA. It is also likely that the VMNvl neurons that facilitate lordosis are distinct from those that promote aggression, and that the action of 5-HT on one group of these neurons suppresses lordosis, while action on a separate population suppresses aggression.

What might be the mechanisms that allow DR-5-HT to inhibit lordosis in male rats but not in female rats? The 5-HT input to the VMN is sexually dimorphic, with adult male rats having a higher density of 5-HT axon terminals in this region than do females. Importantly, treatment of neonatal female rats with estradiol results in increases in the 5-HT axon terminal density in the VMN to the levels observed in males [741]. A conclusion from these data is that an aspect of the perinatal testicular steroid-induced defeminization process is the development of greater 5-HT inhibition of VMNvl neurons involved in lordosis (cf. [517,741,1072]).

Similarly, how might perinatal testicular steroids result in a greater LS inhibition of lordosis in male rats than in female rats? One possibility is that vasopressin (AVP) input to the LS is involved in this effect. As indicated in Chapter 3, the MeA and BST contain AVP neurons, and some of these neurons project to the LS. Importantly, this projection is sexually dimorphic and under the control of perinatal testicular secretions. Normal male rats and female rats treated with testosterone during the perinatal critical period have more AVP cell bodies in MeA and BST, and a greater density of AVP axon projections to LS, in adulthood than do normal female rats or male rats that have been castrated during the perinatal critical period [242,1015]. Recall that in Chapter 3, I suggested that AVP input to LS might promote aggression (see Figure 3.5). Since the MeA processes olfactory input, perhaps when a normal adult male is exposed to pheromones from another adult male, olfactory input has two effects via AVP input to LS. On the one hand, aggression would be facilitated, particularly if the intruder male was unfamiliar, and on the other, the probability of lordosis to mounts by the male would be inhibited. Since AVP exerts excitatory effects on neurons, the synaptic circuitry mechanisms within the LS whereby AVP might promote aggression and inhibit lordosis would have to be different.

Although the hypothesis that the greater AVP input to LS in male rats might stimulate LS GABAergic efferents to PAG in order to depress lordosis in males to a much greater extent than in females is attractive, some evidence suggests that it may not be the case. Sodersten, Boer, De Vries, Buijs, and Melin [916] castrated male rats in adulthood and treated them with estradiol and progesterone in order to induce lordosis. One group of males received an ICV injection of an AVP receptor antagonist 6 h after the progesterone injection and 20 min before being exposed to a sexually active male rat. The control group of males received ICV injections of saline. The results indicated that the AVP receptor antagonist did not facilitate lordosis (both groups showed low LQ scores), casting doubt on the hypothesis that the strong AVP input to LS is involved in suppressing lordosis in males. However, a modification in the time of testing for lordosis after the ICV injection may have revealed a significant effect. Pedersen and Boccia [746] examined the potential inhibitory effects of AVP on lordosis in *female* rats. Adult females were ovariectomized and injected with suboptimal doses of estradiol that induced very low levels of receptivity (LQ=0.10). When such females received ICV injections of a V1a vasopressin receptor antagonist and were tested with a sexually active male rat 4–6 h later, their LQ scores increased to 0.60. Perhaps if Sodersten et al. [916] had tested their males several hours, rather than 20 min, after AVP receptor antagonist administration, they would have been able to facilitate lordosis in their male subjects.

4.6.2.5 Brain Mechanisms: Is Feminization Associated with Demasculinization?

The results just reviewed for male rats indicate that the lordosis circuitry is present in the male brain, but as a result of perinatal steroid-induced defeminization, this lordosis circuitry is under much stronger inhibition in males than in females. These findings give rise to the question of whether the neural circuitry for male copulatory responses is present in the brain of both sexes, but that it is under greater neural inhibition in female than in male rodents. More specifically, when the female brain develops in the absence of being exposed to perinatal testosterone/estradiol during the first week of life, in addition to feminization, perhaps a demasculinization process causes the development of neural circuits that suppress the reactivity of male sexual circuits to stimuli from receptive females. Although evidence for this proposal is sparse, the findings from Kimchi, Xu, and Dulac [487] are supportive. As will be described in more detail in the next chapter, two major olfactory systems exist in the nasal mucosa and olfactory bulbs of rodents and other mammals: the main olfactory system and the vomeronasal system, both of which converge on the MeA, which is a major amygdala nucleus involved in the processing of olfactory inputs (see Chapter 1). Kimchi et al. [487] found that when the vomeronasal organ in the nasal cavities was selectively destroyed in adult female mice, while leaving the main olfactory system intact, the incidence of male copulatory responses directed toward receptive females was dramatically increased. These results suggest that in adult female mice the neural circuitry for male copulatory responses (mounting an estrous female with pelvic thrusts) is present in the female brain but that it is under the tonic inhibition by neural circuits that are activated by vomeronasal organ neurons that detect pheromones emitted by estrous females. Just which circuits are affected by the vomeronasal system so that male copulatory responses are suppressed remains to be determined, as well as the mechanism through which the absence of perinatal testicular steroids results in this enhanced vomeronasal inhibition over male copulation. It would certainly be interesting to determine, based on the findings from McCarthy's research group, whether vomeronasal ablation in female rodents ultimately results in increases in the density of dentritic spines within the MPOA of the affected females.

4.6.2.6 Conclusions

The results reviewed in this section clearly show that exposure of the developing rodent brain to testicular steroids results in a masculinization and a defeminization process and that these two processes are separate and distinct. However, unlike the differentiation of the Mullerian and Wolffian systems, where masculinization causes the Wolffian system to develop, while defeminization prevents the development of the Mullerian system, it appears that the neural circuitry for male and female copulatory responses are present in the brains of both sexes, but that in the male the lordosis circuitry is under strong inhibition, while in the female the masculine copulatory circuitry is inhibited. Sex-specific olfactory processing may contribute to these effects: the manner in which sex-specific sensory stimuli are processed may influence whether the neural circuits related to male or female sexual responses are either stimulated or inhibited (cf. [596,877]). The results reviewed also show that sex differences in sexual behavior and preferences are the result of sexual differences at multiple sites in the neural circuitry that regulates sexual behavior.

Although the neural circuits for both male and female sexual behavior appear to exist in the brain of both sexes, it is obvious that perinatal neurodevelopmental processes result is an extreme sexual differentiation so that heterosexual activity, rather than homosexual activity, is the highly dominant outcome. It is also possible that the rewarding consequences of postpubertal heterosexual, but not homosexual, interactions cement in place this important sex difference in sexual behavior and partner preference, a difference that provides the behavioral basis for sexual reproduction.

4.6.3 Sexual Differentiation of Behavior in Primates

4.6.3.1 Introduction

In this section I will present the evidence that perinatal testicular secretions have organizational effects on the development of sexual behavior in primates. The social life of primates is more complex than that of rodents, and experiential factors play important roles in behavioral development. It is not my intention to promote a simplified and unitary explanation of sexual development in primates, where the organizational effects of perinatal steroids are the only important factor. My intention, however, is to show that the presence or absence of perinatal steroids does contribute, in important ways, to the development of sexual behavior and motivation in primates.

4.6.3.2 Nonhuman Primates

Most of the research on the sexual differentiation of sexual behavior in nonhuman primates has been done on rhesus monkeys; reviews on this subject have been written by Wallen [1006] and by Wallen and Hassett [1008]. Rhesus monkeys have a 168-day gestation period, and gonad differentiation occurs on about embryonic day 40 (E40). From E40 onwards, the testes secrete androgens while the fetal ovaries are quiescent. Therefore, fetal male rhesus monkeys are exposed to significantly higher levels of testosterone than are females. This fetal androgen secretion contributes to the masculinization and defeminization of the reproductive tract, and also influences the development of sexual behavior. Importantly, evidence indicates that the aromatization process that converts testosterone (T) to estradiol does not play a significant role in the sexual differentiation of primates. Prenatal treatment of rhesus females with dihydrotestosterone (DHT), which cannot be aromatized to estradiol, is just as effective as T treatment in masculinizing and defeminizing the reproductive system and sexual behavior.

As described in Wallen ([1006]; also see [780]), adult male rhesus monkeys, but not adult females, will mount a sexually receptive female at a high rate. Castration decreases such mounting in males. When adult castrated males and ovariectomized females are treated with testosterone, it is only the males that mount receptive females. With respect to female sexual behavior, when a tethered sexually active male is used as a stimulus, adult periovulatory females solicit the male's sexual attention: they approach the male, exhibit several solicitation behaviors, sit close to the male, and present their perineum to the male. Ovariectomy decreases such sexual solicitations, and this decrease is reversed by estradiol treatment. Importantly, adult males that are castrated and treated with estradiol do not engage in such solicitation behaviors in the presence of a sexually active male stimulus [781,960]. Therefore, clear sex differences in the adult sexual behavior of rhesus monkeys occur: in response to testosterone, males mount receptive females while females do not; and in response to estradiol, females solicit the sexual attention of a sexually active male, while males to not.

Importantly, when pregnant mothers are injected with either T or DHT between E40 and E100, the affected fetal rhesus females are masculinized and defeminized: such XX females are born with a penis and scrotum, and when they are treated with T in adulthood they show higher levels of mounting behavior than do control females [1006]; but when they are treated with estradiol, they show lower levels of solicitations directed toward a sexually active male than do control females. The extent to which masculinization of the external genitalia influenced some of these results remains to be determined, although other evidence suggests that such virilization of the external genitalia probably did not contribute significantly to the observed masculinization and defeminization of adult sexual behavior [1008]. It should be noted that female rhesus monkeys that have been exposed to androgens prenatally do not express the degree of male sexual behavior shown by normal males. For example, although they mount receptive females at a higher rate than control females, they still mount significantly less than control males. One might propose that such prenatally treated females have been partially masculinized and defeminized by androgens. The exact timing of androgen exposure and the doses employed may have resulted in these partial effects. Alternatively, developmental factors in addition to perinatal steroid exposure may contribute to the expression of adult sex differences in sexual behavior.

4.6.3.3 *Perinatal Steroids and Human Psychosexual Differentiation*

Psychosexual phenotypes in humans are complex and involve sexual orientation and gender identity. Sexual orientation refers to the sex of the individual that one is sexually attracted to, and gender identity refers to an individual's characterization of oneself as either masculine or feminine. Since gender identity is a subjective evaluation, it cannot be determined in nonhuman animals. In my review of the nonhuman animal literature, however, it was clear that perinatal steroids exert organizational effects on the development of sexual partner preferences. In order to present an integrated analysis of the human and nonhuman animal literature, this section will therefore be primarily concerned with an examination of the contribution of perinatal steroids to the development of sexual orientation (sexual partner preferences) in humans.

Several important reviews have been written on the potential organizing effects of perinatal steroids on the development of sexual orientation in humans, and the evidence clearly indicates that perinatal steroids do contribute to such developmental processes [17,56,72,412]. About 90–95% of humans are heterosexual and are sexually attracted to the opposite sex, while the remaining 5–10% are homosexual. In humans, the fetal gonads differentiate at about 6–8 weeks of gestation [186]. Similar to nonhuman animals, the testes, but not the ovaries, subsequently secrete hormones, which results in males being exposed to more testosterone throughout fetal development and during the first three postnatal months than females. Based on the nonhuman animal literature, one can propose that perinatal testosterone masculinizes and defeminizes the human brain so that the typical adult male is sexually attracted to women but not to men. In the absence of the organizing influences of testosterone, the brain is feminized so that the typical female is sexually attracted to men. This analysis suggests that one route to homosexuality is that homosexual men might be less affected by perinatal testosterone than heterosexual men and that homosexual women might be affected by developmental processes that mimic the effects of perinatal testosterone.

As in nonhuman animals, the development of the internal reproductive tract and the external genitalia is guided by whether or not fetal testes are functionally active [186,1037]. An illustrative example involves the effects of the androgen insensitivity syndrome (AIS) on the development of the human peripheral reproductive system. The androgen receptor (AR) gene is located in the long arm of the X chromosome [186], and various mutations within the coding region of this gene can result in a nonfunctional steroid-binding domain within the AR protein. In other words, such mutations would disrupt the ability of testosterone to bind to its receptor and therefore would interfere with the transcriptional cellular effects that are typically induced by testosterone (see Chapter 2). Therefore, even though testosterone is secreted, it is functionally inactive. An XY human fetus with complete AIS develops testes because SRY is active on the Y chromosome. The fetal testes secrete MIS and testosterone. MIS inhibits the development of the Mullerian system, but testosterone and DHT cannot exert their typical masculinizing effects on the Wolffian system and the external genitalia because the AR is not active and does not bind with T and DHT. With respect to the phenotypic characteristics of XY individuals with the AIS (refer to Figure 4.7), internally they possess neither a developed female nor male system (for example, neither a uterus nor seminal vesicles are present), but their external genitalia are feminine (vaginal labia and clitoris). These results indicate that just like in nonhuman animals, it is fetal testicular secretions that exert masculinizing and defeminizing effects on the development of the peripheral reproductive system. Therefore, even when testosterone is secreted normally, if the AR is not functioning properly, then atypical development will result.

What about the psychosexual development of XY humans with the AIS? Because these individuals have female external genitalia, they are raised as girls by their parents and treated as girls by their peers. Further, at puberty, the conversion of some testosterone to estradiol by aromatase induces breast development (normal estrogen receptors are produced in these individuals). Therefore, externally they look like normal women. However, they never menstruate because a uterus and ovaries are not present. In terms of sexual interests and attraction, their sexual orientation is directed toward men [56,412]. One interpretation of these results is that because of the inability of the AR to bind testosterone, the fetal and early postnatal brain is feminized, resulting in a female sexual orientation with erotic interests directed toward men. Such a proposal would also indicate that, as in other primates, it is testosterone, rather than estradiol, that masculinizes and defeminizes the human brain, and other evidence indicates that this is the case [261,1037]. These results also suggest that it is the action of perinatal testosterone on the brain, rather than the presence of genes on the Y chromosome, that determines sexual orientation toward women and prevents sexual orientation toward men.

A problem with interpreting the causes of the sexual orientation of XY humans with AIS is that not only is the brain of such people not functionally exposed to the effects of perinatal testosterone, but such individuals are also raised and socialized as females. Perhaps socialization processes rather than the absence of perinatal testosterone effects have the primary influence over the development of sexual orientation. To accept this view, one would have to argue that in humans the brain develops under unique processes that are distinct from those that influence the development of the peripheral reproductive system, while in all other mammals similar organizational processes influence peripheral reproductive tract development and brain development. It is not likely that this is the case.

An examination of the sexual orientation of XX women with congenital adrenal hyperplasia (CAH) provides an example where perinatal steroid exposure and rearing/socialization experiences oppose, rather than coincide with, one another, and these results support the involvement of perinatal steroids in the development of sexual orientation in humans. CAH is caused by an autosomal recessive mutation in a gene that regulates the synthesis of an enzyme involved in the conversion of steroid precursors to cortisol within the adrenal cortex [1037]. As a result of this enzymatic defect, toward the end of pregnancy the fetal adrenal gland of the affected individuals secretes testosterone. Such

XX girls are born with slightly masculinized external genitalia (clitoral hypertrophy and partial fusion of vaginal labia). Shortly after birth, these XX girls are surgically corrected to feminize their external genitalia and are also treated with cortisol, which lowers the abnormally high testosterone secretion from the adrenal cortex. Most importantly, these individuals are raised as girls by their parents. Therefore, although the brains of the affected XX girls were exposed to elevated testosterone during a portion of the presumed perinatal critical period, their rearing/socialization experience approximates that of a typical girl. As reviewed by Hines [412], CAH women are less likely than control women to be exclusively heterosexual. In one of the studies she reviewed, up to 44% of CAH women over the age of 21 had desired a long-term sexual relationship with another woman. These results certainly support the view that in spite of rearing conditions, exposure of the fetal brain to elevated testosterone increases the tendency for a human's sexual partner preference to be directed toward other women.

The fact that the majority of CAH women remain heterosexual can be interpreted in a number of ways. One possibility is that a more extended period of perinatal and postnatal testosterone exposure is needed to fully masculinize and defeminize the brain. In addition, socialization processes may co-act with perinatal steroid exposure to influence sexual orientation. If perinatal T exposure is mild and socialization as a female is intense, a heterosexual orientation may develop. Therefore, factors other than perinatal steroid action may contribute to the development of sexual orientation.

Although the examples of AIS and CAH support a role for perinatal steroid action in the development of sexual orientation, it should be understood that these are not the usual routes for the development of homosexuality. AIS and CAH are clinical syndromes, and these disorders are not the typical causes of the homosexuality that occurs in 5–10% of the human population. Any biological theory of homosexuality needs to explain how an XY male with testes, normal adult testosterone levels, and a male peripheral reproductive system, or an XX female with ovaries, a normal menstrual cycle, and a female peripheral reproductive system develops a homosexual orientation. We are a long way from answering these questions. Since most research on homosexuality has been done on males, I will try to organize the existing evidence to propose a rudimentary idea about how homosexuality in males might develop.

As indicated by Hines [412], there may be multiple routes through which homosexuality develops in humans. With the understanding that the peripheral reproductive system differentiates earlier in development than does the brain, any process that elevates T levels in the female brain, or lowers T levels in the male brain, toward the end of gestation and into the early postnatal period, has the potential to produce a sexual orientation that does not match the gender of one's peripheral reproductive system. With respect to the development of homosexuality in males, it may be that T levels are normal throughout the perinatal period but that there is a modification in the AR that temporally and spatially affects the action of testosterone so that during the critical period of brain differentiation, optimal levels of T action within the brain do not occur.

Genetic studies show that the heritability for homosexuality is high: the concordance rate for homosexuality in monozygotic twins is about 65% while for dizygotic twins it is only 15% [56,261,412]. Additional evidence points to the contribution of genes on the X chromosome for the development of sexual orientation in men. Sexual orientation appears to be transmitted through the mother's side of the family: if a gay man has gay uncles, they tend to be the mother's brothers rather than the father's [56]. Importantly, genetic linkage studies have indicated that genetic polymorphisms within the long arm of the X chromosome are related to variations in the sexual orientation of men and that particular variants increase the probability of homosexuality [105,430]. Although the sharing of particular alleles on the X chromosome among men increases the probability of homosexuality in these men, it should be noted that such alleles are neither necessary nor sufficient for the development of a homosexual orientation in the selected sample population [430], which supports the view that homosexuality can develop through multiple routes.

Although particular alleles on the X chromosome may influence male sexual orientation, the particular genes involved and their gene products have not been determined. Since the AR gene is located on the long arm of the X chromosome, it is attractive to speculate that some genetic process that influences the expression of the AR in the brain, but not in the periphery, during the perinatal period may contribute to homosexual development in men. Relevantly, Macke et al. [585] examined the coding region of the AR gene and did not find particular allele variants that were associated with male homosexuality. Therefore, if some modification in the AR gene is related to homosexuality in men, it may be that it is a variation within the regulatory region of the gene, a variation that influences the expression of the gene in the brain during the perinatal period.

An interesting study by Bocklandt, Horvath, Vilain, and Hamer [104] suggests that epigenetic events involving a skewed X-inactivation process in women may contribute to the development of homosexuality in their sons. X-inactivation refers to the process by which one of a female's two X chromosomes is inactivated in each diploid cell. This inactivation process is meant to equalize X chromosome gene dose effects in males and females, since the male has only one X chromosome in each cell. X-inactivation is typically considered to be a random process, so that within each cell, the maternal and paternal X chromosome in an XX cell each has an equal chance of being inactivated. A skewed inactivation pattern refers to a nonrandom pattern where one of the two X chromosomes is inactivated preferentially. Bocklandt et al. [104] related an extreme skewing of X chromosome inactivation (from white blood cell samples) in mothers, where the probability of inactivating a particular X chromosome was 90% rather than 50%, to the development of homosexuality in their sons. They found that 23% of mothers who had more than one gay son showed such extreme skewing of X-inactivation in comparison to only 4% of mothers without gay sons.

Several processes contribute to X-inactivation, and one of these involves increased DNA methylation of the selected X chromosome, which interferes with the transcription of genes on the inactivated X chromosome [15]. Importantly, during meiosis X-inactivation is reversed so that all haploid oocytes contain an active X chromosome [15]. However, I would like to develop an interesting hypothesis based on the results of Bocklandt et al. [104]. If X-inactivation is skewed in some mothers who have more than one homosexual son, then perhaps there is an anomaly in the epigenetic process that results in X-inactivation. One anomaly, of course, is the preferential selectivity of the inactivation process. A speculation on my part also includes the possibility that such X-inactivation may be particularly strong and hard to *completely* reverse during gametogenesis. If this were the case, and if a son inherited the X chromosome that had been inactivated, perhaps it would remain partially inactivated during the perinatal period, and such partial inactivation may, in turn, have lowered the expression of the AR in the fetal and early postnatal brain so that full masculinization and defeminization was prevented from occurring. One way to test this idea would be through a long-term longitudinal study where fetal cells from males could be obtained through amniocentesis in order to determine the degree of DNA methylation marks on the X chromosome, which could then be correlated with the subsequent development of homosexuality in the tested population. If these marks were shown to be on the AR gene, the results would be particularly dramatic. Perhaps there is a particular problem with the demethylation of the AR gene that prevails during fetal development and contributes to the development of homosexuality.

If such a speculative hypothesis were indeed accurate, why should the peripheral reproductive system, but not those brain areas concerned with sexual partner preferences, become masculinized and defeminized in homosexual men? First, it is clear that the expression of the AR can be decreased as a result of methylation at CpG sites within the regulatory region of the AR gene [439]. Second, using the estrogen receptor (ER) gene as a model, evidence indicates that multiple response elements, activated by different transcription factors, regulate the expression of the ER in different cell types [253]. Figure 4.10 presents a hypothetical model that could explain how hypermethylation at certain regulatory sites within the AR gene might decrease AR expression in the brain, but not in the periphery, a process that could lead to male homosexuality. The figure shows three response elements within the regulatory region of the gene, A, B, and C, and three transcription factors (TF): TFA, TFB, and TFC. TFA and TFB activate AR expression in the periphery, while TFB and TFC activate such expression in the brain. The model proposes that response element C of the AR gene on the X chromosome is hypermethylated, which would decrease expression of the AR in the brain but not in the periphery.

If this hypothesis were correct, and if such decreased expression of AR in certain parts of the brain persisted into adulthood, then perhaps differences in the expression of ARs between the brains of heterosexual and homosexual men could be measured with PET scans through the use of a radiolabeled ligand capable of binding to the AR. I am not aware of any studies that have examined this interesting idea. However, if such hypermethylation within the regulatory region of the AR gene were only maintained for the duration of

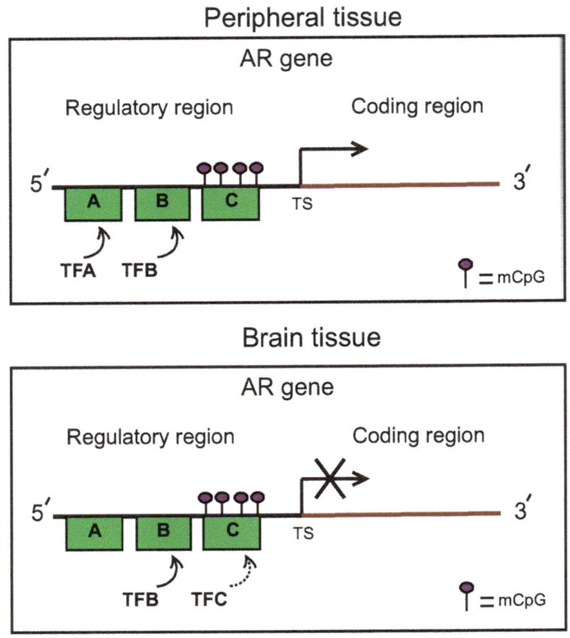

FIGURE 4.10 A hypothetical model that might explain how hypermethylation (an increase in methylated cytosine bases at CpG sites; see Chapter 2) at certain regulatory sites within the androgen receptor (AR) gene might decrease AR expression in the brain, but not in the periphery, of the developing male fetus. Such a process could lead to typical forms of male homosexuality, where the reproductive tract is masculine, while the brain areas regulating sexual orientation are feminine. Transcription factors (TF) A and B are proposed to act at response elements A and B of the AR gene to promote AR gene transcription in peripheral tissues, such as the Wolffian system and primordial external genitalia. Such AR gene transcription would not be affected by a proposed hypermethylation at CpG sites (mCpG) within response element C. It is also proposed that TFB and TFC act at response elements B and C to promote AR gene transcription in those brain regions that control sexual motivation and behavior. Since DNA methylation typically opposes transcription (see Chapter 2), hypermethylation within response element C would result in a decrease in the transcription of the AR gene in the brain because only TFB would be operative. Such a process could lead to a feminization, and a lack of masculinization, of those brain mechanisms that influence an individual's sexual orientation. The dashed arrow between TFC and response element C represents decreased binding of the transcription factor to the response element. TS = transcription start site. An X across the arrow that originates from TS indicates depressed transcription.

the perinatal period, then adult PET scans would probably not detect differences in AR expression between homosexuals and heterosexuals.

If a decrease in expression of the AR in the brain during the perinatal period were involved in the development of male homosexuality, is there any evidence that the brain is altered by such a process in a way that would be consistent with nonhuman animal research? Using postmortem brain tissue, Levay [546] reported on differences in the volume of a hypothalamic nucleus, the third interstitial nucleus of the anterior hypothalamus (INAH3), between heterosexual men and women, with the volume of this nucleus being larger in men. According to Byne et al. [142], INAH3 is likely to be homologous to parts of the MPOA of rats. Therefore, a sexual dimorphism, in terms of nuclear volume, is evident in the MPOA-AH of rats and humans. Importantly, Levay [546] reported that the size of INAH3 of homosexual men was smaller than that of heterosexual men and approximated the volume observed in heterosexual women. These findings on the volume of INAH3 have basically been confirmed by Byne et al. [142]. Interestingly, Byne et al. not only measured the volume of the INAH3 but also counted the number of cells in this region. Heterosexual men had significantly more neurons in this region than did heterosexual women, but the number of neurons in INAH3 of homosexual men did not differ from heterosexual men. Based on the rodent research from McCarthy's group, this last finding is quite interesting: the volume of INAH3, but not neuron number, is smaller in homosexual men when compared to heterosexual men, suggesting that the density of neurons in the region is higher in homosexuals. Byne et al. [142] conclude that there is a reduction in the neuropil within INAH3 of homosexual men. Such a reduction could be due to decreases in synaptic inputs to this region that might be correlated with decreases in INAH3 dendritic branching and dendritic spines.

Although the findings just presented offer a plausible neurobiological account of some processes that may contribute to the development of male homosexuality, a large part of the proposal is speculative, and it should be clear that much more research is required to understand psychosexual development in men. In addition, the proposals I have presented do not offer an explanation for the development of homosexuality in women.

4.7 CONCLUSIONS

Sexual behavior among mammalian conspecifics is the most basic form of prosocial interactions. Indeed, for solitary species, except for maternal behavior, sexual behavior is the only other form of prosociality, and such behavior obviously precedes parenting. In comparing the previous chapter on aggression with the current chapter, it should be clear that the occurrence of either antisocial or prosocial responses to conspecifics is dependent upon the neural pathways over which conspecific social stimuli are processed, and that the presence of particular hormones and neuropeptides influences such processing mechanisms. For example, in male rodents, testosterone action on MPOA acts to promote male sexual behavior toward estrous females, while testosterone within the VMNvl and other parts of the hypothalamic aggression area acts to promote aggressive responses toward unfamiliar male intruders (cf. [854]).

The neural circuitry that underpins the consummatory aspects of sexual behavior differs by sex in most mammals, and this can be related to the fact that male copulatory responses are different from female copulatory responses, necessitating the ultimate activation of different sets of spinal and cranial motor neurons. Interestingly, however, evidence from rodents implicates MPOA neurons, and their interactions with the mesolimbic DA system, as being involved in the appetitive aspects of both male and female sexual behavior. Whether the same or different MPOA neurons participate in the appetitive aspects of male and female sexual behavior remains to be determined. I will discuss this issue more extensively in the next chapter.

The amygdala, hypothalamus, midbrain, and NA-VP circuits that have been shown to influence sexual motivation in rodents also operate in primates, including humans. In other words, a basic core neural circuitry appears to regulate appetitive sexual motivation in all mammals, and this circuitry is under the influence of the activational effects of gonadal steroids. What distinguishes primates, particularly humans, from nonprimates, is the degree of control that prefrontal cortical mechanisms exert over this common core circuitry. In humans, prefrontal cortical mechanisms regulate and restrain when sexual behavior occurs and the types of stimuli that are erotically arousing. Dysfunctions in such top–down PFC regulatory mechanisms can lead to pathologies in sexual motivation, such as occurs in pedophiles.

In order for sexual reproduction to occur, males should be sexually attracted to females and females should be sexually attracted to males. This strong sexual differentiation of sexual behavior is significantly regulated by the organizational effects of perinatal steroids: fetal and early postnatal testicular secretions have been shown to have masculinizing and defeminizing effects on the development of brain circuits that regulate sexual behavior in all mammals, including humans. One effect of this developmental process is to influence the manner in which sex-specific sensory stimuli are processed in the brain of each sex. Although the neural circuits for both male and female sexual behavior are present in some form in the brain of both sexes, for the typical male mammal, the female circuits are under intense neural inhibition, while for the typical female mammal, the male circuits are under intense inhibition.

Homosexuality occurs in 5–10% of the human population. One factor that may contribute to the development of homosexuality in men is the selective interference with androgen action within particular brain regions, probably involving the hypothalamus, during the perinatal period.

CHAPTER

5

Parental Behavior

5.1 INTRODUCTION

Parental behavior is defined as any behavior of a member of a species toward an immature conspecific that increases the likelihood that the immature organism will survive to maturity. In mammals, depending on the species, parental behavior can include maternal, paternal, and alloparental behavior [703]. Maternal and paternal behaviors are those caretaking behaviors directed toward an infant by the biological mother or father, respectively, while alloparental behaviors are caretaking behaviors directed toward an infant by an individual other than a biological parent. In most mammals, maternal behavior is the dominant form of parental behavior, and this is related to the fact that it is the female that lactates. In fact, about 90–95% of mammalian species exhibit an exclusively uniparental maternal care system [498]; these species have a polygynous or promiscuous mating system, with males and females leaving each other after mating, and paternal behavior does not occur. In the remaining mammalian species, a biparental care system with both maternal and paternal behavior is evident, and this usually occurs in the context of a monogamous mating system. In some of these monogamous species, alloparental behavior may also be observed, where older subadult or adult siblings who have not emigrated from the family group help their mother and father care for more recently born young [427,625]. Such cooperative breeding species include prairie voles, marmosets, tamarins (New World monkeys), and early hominins.

Because maternal behavior is the dominant form of parenting in mammals, most of this chapter will be devoted toward exploring its underlying mechanisms. The nature of maternal behavior is influenced by both the degree of maturity of the young at birth and the type of social group within which the mother and her infant(s) are embedded [702,703]. Infants can be altricial (immature) and immobile at birth, precocial and mobile, or intermediate between these two extremes (semi-altricial). Rodents, sheep, and monkeys will serve as examples for each type, respectively. The typical infant rodent or pup is helpless and incapable of temperature regulation at birth. Such infants are usually born in a secluded nest site where the mother hovers or crouches over them in order to nurse and keep them warm. She also grooms the pups by licking them, which also facilitates urination and elimination. If the nest site is disturbed, or if pups become displaced from the nest, the mother shows retrieval or transport behavior, during which she carries individual pups, one at a time, in her mouth to a new nest site or back to the original nest. Most rodents do not form selective attachments to their own offspring, and one can cross-foster or exchange young between litters; a postpartum female will then care for young

Neurobiology of Social Behavior
http://dx.doi.org/10.1016/B978-0-12-416040-8.00005-5

that are not her own. The evolution of selective attachment mechanisms has not occurred in species with altricial young because under natural conditions the young are not capable of moving from one nest to another, and, therefore, mixing and confusion of own and alien young does not occur. In other words, species do not evolve mechanisms that would not play a role in promoting reproductive success under natural conditions. In contrast to rodents, sheep and other ungulates, such as goats, give birth to young that are relatively mature and mobile at birth, and the calves are born into herds composed of unrelated individuals. A rapid selective attachment mechanism between a mother and her offspring develops through learning processes within the first 12–24 h after birth, resulting in a mother that remains near her calf and that will only nurse her own calf while rejecting the advances of alien young. Such an attachment mechanism operates to prevent confusion between own and alien young under conditions where mixing of young from different mothers is very likely. Finally, the maternal responses of most monkey and ape species are adapted to young that are semialtricial at birth. Infants are typically born in a social group consisting of related and unrelated individuals. Initially, the mother is in continuous contact with her infant, whom she nurses and grooms while the infant clings to her for transport. As one example, this initial period of constant contact lasts for the first month of life in rhesus monkeys (see Chapter 3). Subsequently, as the infant develops, it will wander away from its mother, but she is always aware of its location and will initiate contact and reunion when necessary. Selective maternal attachments develop in most primates, although this development is not as temporally restricted as that which occurs in sheep, which is in line with the semialtricial characteristics of primate infants.

In comparing maternal behavior in species that form selective attachments to their own young with those that do not, it is inappropriate to assume that the mother–infant bond is *stronger* in those species with selective attachments. The mother–infant bond is probably just as strong in rats as it is in sheep or rhesus monkeys. What differs is the infant stimulus to which the mother is attached. In the rat, the mother is strongly attracted to a general pup stimulus, while in sheep and many primates the female becomes selectively attracted to a specific infant. In other words, there are two separate and distinct processes that influence how a mother responds to young, a recognition process and an attraction/attachment process, and the two should not be confused.

For most female mammals, where a uniparental maternal care system is evident, the endocrine events associated with pregnancy termination act on the brain to stimulate the onset of maternal responsiveness [703]. Therefore, the typical virgin estrous cycling nulliparous (a female who has not given birth) female will not show maternal behavior toward young of its species, but instead will avoid and reject infants and may even attack them. In contrast, newly parturient primiparous females are highly maternal and will care for their own or alien young. For those species that form selective attachments, learning mechanisms are subsequently involved. A postpartum ewe will care for *any* lamb that is presented to her at the time of birth, but once a bond is formed with that lamb, she will reject other lambs. The formation of such selective bonds is a result of olfactory learning mechanisms that allow the parturient ewe to recognize the specific odor (olfactory signature) of the particular lamb that she was exposed to near the time of parturition [702].

The hormonal stimulation of maternal behavior at parturition in female mammals is a prime example of how the same social stimulus (a young infant) can be processed over distinct neural pathways depending on the female's internal state. In the virgin female, whose brain has not been primed with pregnancy hormones, infant cues lead to antisocial rejection responses, while in parturient females infant stimuli lead to prosocial acceptance responses. A major goal of this chapter is to explain how hormones and other factors cause such a functional rewiring or rerouting of the neural circuits over which infant stimuli are processed.

Feral or wild mice behave like most other female mammals when exposed to young: virgin females avoid or attack young while parturient females care for them [703]. In contrast, as a result of selective breeding or inbreeding, many nulliparous virgin females of *laboratory strains* of mice show relatively prompt maternal behavior toward pups, indicating that experimental genetic selection can alter brain mechanisms so that infant stimuli are processed over prosocial neural pathways in the absence of priming the brain with the endocrine events associated with pregnancy termination [522,930]. What this example suggests is that under natural conditions, depending on the particular species and its ecology, natural selection mechanisms can have effects similar to experimental selection in laboratory mice, which could help explain the evolution of paternal and alloparental behavior, where parental responses are observed in individuals who have not been exposed to the hormonal events associated with pregnancy termination. As an example, in cases where long periods of infant dependency occur, and where sequential births result in multiple dependent offspring, a monogamous mating system with maternal, paternal, and alloparental behavior may have been necessary for infant survival. In fact, Hrdy [427] has made the convincing argument that alloparental behavior was essential for infant survival during early human (hominin) evolution in hunter–gatherer type societies. The evolution of brain mechanisms that allow for allomaternal behavior in nonpostpartum females may explain why maternal behavior may be relatively emancipated from hormonal control in women, allowing for adoption of infants and perfectly normal maternal behavior. It should be noted, however, that in all species in which allomaternal behavior has been observed in nulliparous females, which includes humans, empirical studies have shown that pregnancy hormones still operate to boost or enhance maternal motivation in postpartum females [703,851]. To take one example from mice, although naïve virgin female mice of many laboratory strains will promptly retrieve, hover over, and groom young pups that have been placed in their home cage, they are not likely to retrieve pups that have been placed in a novel environment. In contrast, postpartum mice will enter novel environments, which presumably arouse fearfulness, in order to retrieve young to their nest site. Interestingly, continued allomaternal experience with young in the absence of hormonal priming also appears to boost the maternal motivation of nulliparous allomothers, so that after several days of such experience, these mice will retrieve young from novel environments [930]. Extending this analysis to humans, nulliparous women may adopt young and show appropriate maternal behavior, but their maternal motivation and interest may still improve as a result of continued interactions with their adopted infants.

Although human maternal behavior and motivation may be relatively emancipated from hormonal control, that should not be taken to indicate, as some have argued [206], that the neural mechanisms and neural circuits regulating maternal behavior and maternal motivation in the typical hormone-dependent mammal are distinct from those underlying maternal behavior in women. As described in Numan and Insel [703], maternal behavior is a defining characteristic of mammals and is conceived as being regulated by evolutionary conserved limbic and hypothalamic neural circuits. Differences between the typical female mammal and human females in their reactions to infants may be based on the following. For most virgin female mammals infant stimuli access antisocial rejection circuits, while in parturient females infant stimuli access prosocial acceptance circuits. In contrast, for nulliparous and postpartum human females, infant stimuli have relatively easy access to prosocial acceptance circuits. Despite these differences, however, the neural circuits regulating the core aspects of prosocial maternal motivation are likely to be quite similar in all mammals, including rodents and humans. To make this point in an analogous way, I will compare laboratory rats with laboratory mice: while the maternal behavior of the former is hormone dependent and that of the latter is relatively emancipated from hormonal control, interfering with the proper functioning of the medial preoptic area (MPOA) in the hypothalamus disrupts the maternal behavior of both laboratory rats [714] and mice [804,979]. As will be discussed in detail later, there is evidence that the MPOA and its neural connections with other brain regions are also involved in human maternal behavior [656,947]. These comparisons do not exclude the fact, because of the tremendous expansion of the cerebral cortex, including the PFC, in primates, including humans, that cognitive and neocortical control of limbic and hypothalamic maternal circuits is more highly developed in primates than in rats [206]. Indeed, I made this point in my analysis of the brain control of sexual behavior in primates and nonprimates. But if something were to go wrong with the core limbic and hypothalamic circuits that underpin maternal motivation in mammals, then there would be a devastating disruption of maternal behavior in both humans and rats. For an analogous example, humans can go on hunger strikes while rats can't do that, presumably as a result of species differences in cognition, but interference with the hypothalamic processes that regulate hunger and food intake would disrupt the motivation to eat in both rats and humans.

The study of the neurobiology of maternal behavior is important for many reasons. The use of animal models to explore the neurobiology of maternal behavior will provide a framework for understanding how maternal circuits may become dysfunctional, which would lead to faulty maternal behavior. Such an understanding might be helpful for creating therapies or interventions to ameliorate or prevent the occurrence of child abuse or neglect by human mothers. Since the quality and amount of maternal behavior affects the social and emotional development of infants [466], some examples of which were described in Chapters 2 and 3, the prevention or reversal of faulty maternal behavior would have broad beneficial outcomes. Finally, most mammals exhibit a polygynous or promiscuous mating system and a uniparental maternal care system. Therefore, sexual interactions are short lasting, but the mother–infant bond is a long-term social bond that persists at least until the young are weaned. Such an analysis leads to the conclusion that the mother–infant bond is the most common *enduring* prosocial bond in mammals. In a small percentage of mammals, however, strong affiliative bonds do form between adults, such as occurs in monogamous mating systems. Further, the effective functioning of human societies is highly dependent upon a strong degree of cooperation and friendships between unrelated adults. Such an analysis leads to the possibility that when types of enduring social bonds other than the mother–infant bond are adaptive, the neural circuits regulating maternal motivation and the mother–infant bond may serve as a foundation upon which these other types of social bonds are built [70,238,698,743,767,838]. By understanding the neurobiology of parenting, one may be gaining insight into mechanisms that more broadly regulate prosocial behaviors. Maternal care in mammals is the primary example of one individual aiding another conspecific, and evolutionary processes may harness the associated underlying mechanisms to allow for the occurrence of broader forms of prosocial behaviors in mammalian social groups.

Finally, as one reads this chapter, it will become clear that my own research on maternal behavior has strongly influenced my conception of how the brain is organized with respect to both prosocial and antisocial behaviors.

5.2 NEURAL SYSTEMS REGULATING MATERNAL MOTIVATION IN RATS

5.2.1 Introduction

Most research on the neurobiology of maternal behavior, including my own research, has been conducted on rats. Rats exhibit a hormonally dependent uniparental maternal care system, and the neural systems regulating maternal behavior in this species can be taken as a model for understanding the involvement of evolutionarily conserved neural circuits in the control of parental behavior. My intent is to show how the brain controls maternal motivation, that is, goal-directed prosocial approach and acceptance responses toward young infants. Another goal is to show how pregnancy hormones and other factors can alter the processing and interpretation of infant stimuli so that infant rejection is switched to infant acceptance. Different species exhibit different types of specific maternal responses. For example, a mother rat retrieves or transports pups by carrying them in her mouth, while many primate mothers transport young that cling to them, and the mother may also carry the infant in her arms. By focusing on the neural mechanisms of motivation rather than on the control of specific behaviors, one will gain insight into those neural circuits that underpin maternal interest and attraction to young in a variety of mammalian species.

Figure 5.1 shows a simplified view of what may be happening in the brain of female rats during different reproductive states [697]. In naïve virgin females, pup stimuli activate defensive neural circuits that mediate withdrawal and rejection responses; although most virgins avoid young pups, some will attack them. In the parturient female, as a result of hormone action on the brain, pup stimuli activate neural systems that mediate maternal behavior. Interestingly, during the postpartum period and after the initiation of maternal behavior, pup stimuli continue to activate maternal neural circuits even though the hormonal events that triggered maternal behavior have waned [703]. This fact suggests that

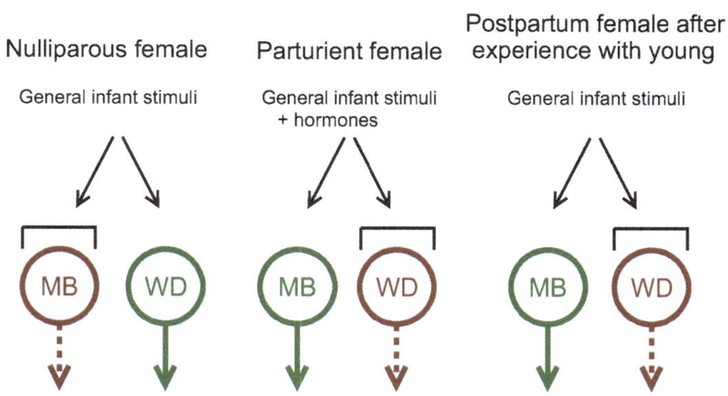

FIGURE 5.1 A simplified view of the brain mechanisms that mediate how a female rat responds to infants during different reproductive states. This general pattern appears to be applicable to most female mammals that raise altricial young and that do not form selective attachments to particular infants. Active brain systems are shown in green, and inactive systems are shown in red with a dashed arrow. A bracket over the cell body representing a neural system signifies that the system cannot be effectively stimulated by afferent sensory input from infants. For the naïve nulliparous female, a general infant stimulus activates a neural system that mediates withdrawal, rejection, and avoidance (WD). After the female's brain has been exposed to the endocrine events that occur at the end of pregnancy, general infant stimuli activate a maternal neural system (MB) and do not activate the WD system. This process continues into the postpartum period, after the hormonal stimulus that induced maternal behavior wanes. *Modified from Figure 1 in Numan [697] with permission from Elsevier.*

some persistent change occurs in the brain of the postpartum mother so that maternal responsiveness can be maintained in the absence of continued hormonal stimulation. Because rats give birth to altricial young that, under natural conditions, are reared in a secluded nest site, Figure 5.1 shows that the maternal rat responds to a general infant stimulus and does not form selective attachments to particular pups. But the mother–infant bond is an enduring one that lasts until the young are weaned at between 21 and 28 days of age.

5.2.2 The Sensitization of Maternal Behavior in Virgin Females

Although virgin female rats do not show maternal behavior upon initial exposure to pups, if one continuously exposes the virgin to freshly nourished pups on a daily basis by cohabitating them in a single cage (the pups to which the virgin is exposed are supplied by "donor" lactating females that were bred in advance), then several sequential behavioral changes occur [306,829]. Initially, the virgin female avoids the pups and will move to those parts of the cage where no pups have been placed. After about 3–4 days of continuous pup exposure, the female begins to tolerate the proximity of the pups and does not move away from them. During this tolerance stage, the female is not showing maternal behavior, but she is being exposed to proximal pup stimuli. Significantly, beginning about 7 days from the initiation of pup exposure, the virgin female starts to care for the young pups. She retrieves them to a single location, builds a nest around them, grooms them, and hovers over them in a nursing posture even though she cannot lactate. Such pup-induced maternal behavior has been called sensitized maternal behavior, and since it occurs in hypophysectomized virgin females, it is presumed to have a nonhormonal basis [829]. The actual number of days of pup exposure needed to induce maternal behavior in a particular female is referred to as the female's sensitization latency, and this value can vary among females. When comparing the average sensitization latency of virgins with that of primiparous parturient females, while the parturient female is immediately maternal on her first exposure to pups (sensitization latency = 0 days), the sensitization latency of virgins averages about 7 days.

Pup-induced or sensitized maternal behavior in virgins would never occur under natural conditions because virgins would not be in continuous association with another female's pups. It seems appropriate to conceive of the responsiveness of the nulliparous female to pups as being under the primary control of rejection and withdrawal neural mechanisms, which would prevent such females from caring for pups that are not their own if such pups were to be *briefly* encountered during exploration and foraging.

One mechanistic interpretation of sensitized maternal behavior in virgins is that novel pup stimuli initially arouse fearful defensive and withdrawal responses, but after a period of continuous exposure the female habituates to these fear-arousing aspects of the pups (since they are no longer novel), which then allows for proximal contact to occur. Proximal pup exposure, including tactile inputs, may then act to stimulate maternal responsiveness over a period of days. Rosenblatt and Mayer [833] proposed that maternal behavior occurs in rats when the tendency to approach and interact with pups outweighs the tendency to avoid or withdraw from pups. Given this analysis, the sensitization process in virgins is actually a dual sequential process of habituation followed by sensitization [703].

Under natural conditions, male rats do not show paternal behavior. But, as in virgin females, males can be sensitized to show parental behavior by continuously exposing them to pups, although their sensitization latencies tend to be longer than those of females [703]. These results suggest that a basic and intact neural circuitry for parental responsiveness may be present in all mammals of both sexes, but that this circuitry exhibits varying degrees of functional activity, from inactive to active, based on sex, genetic, developmental, hormonal, and experiential factors [257,697,703]. In the typical mammal, pregnancy hormones are needed to activate immediate parental responsiveness to infants, but in species where alloparental behavior is prevalent in virgin females, the parental circuitry may be functioning relatively normally without hormonal stimulation, while rejection/defensive circuits are inactive. Further, when paternal behavior occurs under natural conditions, mechanisms have evolved that also allow the behavior to be induced without the necessity of the hormonal events associated with pregnancy and pregnancy termination.

5.2.3 Hormonal Stimulation of Maternal Behavior and the Role of Oxytocin

Primiparous rats respond immediately to pups as soon as they are born. If the primiparous mother reacted like the virgin female, requiring several days of pup exposure before beginning to care for pups, her offspring would die of neglect. It is the hormonal events associated with pregnancy termination that stimulate immediate maternal care in most mammalian parturient females. The particular pattern of hormone secretion, as measured in blood plasma, that occurs over the 22 days of pregnancy in rats can be described as follows [703]. Progesterone is the dominant hormone of pregnancy, and it declines at the end of pregnancy. This progesterone decline is superimposed on high levels of estradiol and lactogens (peptide hormones that stimulate lactation: placental lactogens and pituitary prolactin). Importantly, each of these hormones can enter the brain via the blood supply, and receptors for each of these hormones are located in critical brain regions essential for maternal behavior [703]. Proving the importance of these hormonal events for the stimulation of maternal behavior in rats, early studies showed that if one systemically treats nulliparous female rats with a hormone regimen that includes increased levels of lactogens and estradiol superimposed on a background of progesterone withdrawal, the sensitization latencies of such females are reduced from 7 days to about 1–2 days [123,659,896].

Oxytocin (OT) is a neuropeptide hormone and neurotransmitter importantly involved in maternal physiology and behavior, respectively [702]. Factors that activate OT-containing neurons in the hypothalamic paraventricular nucleus (PVN) and supraoptic nucleus (SON) include the vaginocervical stimulation that occurs during the birth process and the suckling stimulus [714]. With respect to those stimuli, hormonal OT release from the neural lobe of the pituitary into the general blood supply acts on OT receptors (OTRs) on uterine cells to stimulate labor contractions and on OTRs in the mammary ducts to stimulate milk-ejection in response to suckling (see Chapter 1). OT as a hormone has poor penetrance across the blood–brain barrier [695], and therefore is unlikely to influence maternal behavior via this route. However, as a result of PVN activation, OT acting as a neurotransmitter is also released at synapses within the brain at the time of parturition. OT action on OTRs in critical brain regions then co-acts with the hormonal profile associated with pregnancy termination to stimulate the immediate onset of maternal behavior at parturition [703]. In support, intracerebroventricular (ICV) administration (to bypass the blood–brain barrier) of OT promotes maternal behavior in rats that have been systemically treated with a suboptimal hormone schedule (estradiol alone), which by itself does not shorten sensitization latencies [744], and ICV administration of an OTR antagonist, which would block endogenous OT action in the brain, disrupts the onset of maternal behavior in parturient rats [279,987].

In certain brain regions, the neuronal expression of OTRs is dependent upon estradiol action on the estrogen receptor, and estradiol superimposed on a background of progesterone withdrawal also stimulates the synthesis of OT by

PVN neurons [27,511,767]. Therefore, one of the ways in which declining peripartum progesterone levels superimposed on rising estradiol levels contributes to the stimulation of maternal behavior is by activating both OT and OTR expression in critical brain regions.

Figure 5.1 indicates that while the immediate onset of maternal behavior in rats at parturition requires hormonal stimulation, the maintenance of the behavior during the remainder of the postpartum period occurs in response to pup stimulation and is no longer tied to hormonal mediation. In support, postpartum female rats that have been hypophysectomized after their maternal behavior has become established continue to show normal maternal behavior even though they can no longer lactate [703]. Since the steroid hormones that stimulate the onset of maternal behavior also activate the brain's OT system, it is not surprising that evidence indicates that the brain's OT system is most important for the onset of maternal behavior in rats and is less important for its maintenance. For example, PVN neurons are the main source of OT released into diverse brain sites as a neurotransmitter, and PVN lesions performed during pregnancy disrupt the onset of maternal behavior at parturition [441], but similar lesions performed during the postpartum period after maternal behavior has become established do not have such disruptive effects [441,700]. Similarly, ICV administration of an OTR antagonist disrupts the onset of maternal behavior but does not prevent its maintenance [279].

Although the OT neural system of rats is not essential for the continuance of maternal behavior after a mother–infant bond has become established, research does show that OT modulates the quality and quantity of maternal behavior during the postpartum period. For example, ICV administration of OTR antagonists to postpartum rats decreases the amount of licking and grooming of the pups by the mother and alters the nature of the mother's nursing behavior; although total time spent nursing does not decrease, certain types of nursing postures, which are dependent upon suckling stimulation, are decreased [172,745A]. Perhaps suckling stimulation and other sensory stimuli from infants cause the release of OT into the brain, which then modulates maternal behavior. These results indicate that OT effects on maternal behavior need not be all or none. During the postpartum period, although maternal behavior can continue without OT, variations in the function of OT neural systems can result in significant variations in the quantity and quality of maternal behaviors and maternal attentiveness. Another way of looking at the role of OT during the postpartum maintenance period is that the central effects of OT act to increase the mother's responsiveness to the needs of her infants.

5.2.4 The Medial Preoptic Area and Maternal Behavior

The medial preoptic area (MPOA), located in the most rostral part of the hypothalamus (see Figure 4.1 in Chapter 4), is essential for all aspects of maternal behavior in rats [696,703,714,929]. Depression of MPOA neural activity, which includes neuron-specific excitotoxic amino acid lesions that spare axons of passage, disrupts retrieval behavior, nursing behavior, and nest building. Depending on whether the lesions are performed during pregnancy or during the postpartum period, such lesions will prevent or disrupt either the immediate onset of the behavior at parturition or the subsequent maintenance of the behavior after it has become established. MPOA lesions also disrupt sensitized maternal behavior in virgin rats. Therefore, unlike the PVN, MPOA activity is *essential* for both the onset and maintenance of maternal behavior.

As described in Chapter 1, motivated behaviors can be separated into appetitive and consummatory phases, with the appetitive phase consisting of goal-directed approach responses toward a desired stimulus. Goal-directed approach behaviors that help a female gain access to her pups can therefore be viewed as appetitive components of maternal behavior. Such appetitive behaviors include retrieval behavior, during which a female initiates contact with a displaced pup in order to transport it back to the nest site [716]. Retrieving can also be considered goal-directed because it functions to return pups to a safe and secure location. Nursing behavior, in contrast, is the main consummatory component of maternal behavior; it is elicited by proximal stimuli from nuzzling and suckling pups and is relatively reflexive in nature [925]. Since MPOA inactivation interferes with both retrieval and nursing behavior, it is involved in both components of maternal motivation. While postpartum rats with MPOA damage no longer retrieve pups, they will hoard or carry pieces of candy that approximate the size and weight of pups, which shows that the retrieval deficit is not the result of a general oral-motor dysfunction but instead is due to an interference with a goal-directed maternal response [703].

In further support for the involvement of the MPOA in appetitive maternal motivation, postpartum rats will learn to perform an operant response in order to obtain pups as a reward, but if such rats have MPOA lesions, they will not learn and perform the response; such MPOA-lesioned rats, however, will learn to perform the same operant response for a food reward [535].

In another paradigm that examines the appetitive or goal-directed aspects of maternal behavior, it has been shown that postpartum rats will learn a conditioned place preference when pups are used as a rewarding stimulus [305,615]. In this paradigm, over a series of training days, mothers are placed in a two-compartment cage, with pups present in one distinct compartment, while the other compartment is empty. Subsequently, during the test phase, postpartum females are placed in the two-compartment cage without pups, and it is observed that they spend more time in the compartment that had previously contained pups, which suggests that they are searching for pups that had served as a rewarding and attractive stimulus. Importantly, during the test phase, depression of MPOA neural activity blocks the expression of this conditioned place preference [754], and the affected rats do not show a preference for the compartment that was paired with pups.

Concerning the involvement of the MPOA in nursing behavior, research results have been variable [703]. Some studies have reported only a modest reduction in nursing after MPOA lesions, suggesting that the primary role of the MPOA relates to appetitive maternal motivation, since retrieval behavior is completely eliminated in these rats. Other studies, however, report an almost total elimination of both nursing and retrieval behaviors.

The absence of nursing after MPOA damage is not that easy to interpret. If a female is not interested in pups and is not attracted to them for more than a few seconds, which represents a lack of appetitive motivation, then it is difficult for her to be exposed to the suckling and nuzzling stimuli that reflexively elicit nursing. Taking this perspective, the absence of nursing may be the indirect result of the primary effect of a lack of appetitive motivation. It is also possible, however, that different subregions of the MPOA, with distinct neural projections, are involved in either the appetitive or consummatory aspects of maternal behavior. A similar proposal was presented in Chapter 4 when I discussed the role of the MPOA in male sexual behavior. Large MPOA lesions may damage both subregions, while smaller lesions may have more selective effects. The fact that some females with small MPOA lesions will nurse but will not retrieve their young, while large MPOA lesions disrupt both behaviors, is in accord with the possibility that different subregions of the MPOA regulate either the appetitive or consummatory aspects of maternal behavior.

While depression of MPOA activity decreases maternal responsiveness, hormonal activation of this region stimulates the onset of maternal behavior. In such studies, naïve females that have never been exposed to pups are systemically treated with suboptimal hormone regimens that do not induce immediate maternal behavior. Against such a suboptimal hormone background, additional hormones are applied directly to the MPOA in an attempt to stimulate full maternal behavior. These studies have shown that the MPOA is a site where estradiol, prolactin, and placental lactogens act to stimulate the onset of immediate maternal behavior in rats [125,281,703,710]. Importantly, all components of maternal behavior are activated, which supports a role for the MPOA in both the appetitive and consummatory aspects of maternal behavior.

The most convincing study supporting the involvement of estradiol action on MPOA in the stimulation of maternal behavior [710] employed the pregnancy-termination model developed by Rosenblatt's group [834,896]. In this preparation, it is found that if primigravid (first-time pregnant) rats are hysterectomized (H) and ovariectomized (O) on day 15 of pregnancy and then continuously exposed to pups beginning 48h after the HO procedure, they exhibit sensitization latencies of about 3days, which is substantially shorter than the virgin latency of 7days, but is not immediate maternal behavior. However, if such 15-day HO females (15HO) receive a systemic (subcutaneous) injection of estradiol (E) immediately after the HO surgery (15HO+E), then when pups are presented to them 48h later, they show immediate maternal behavior. On day 15 of pregnancy, progesterone is at high blood levels, and estradiol has begun to rise [123]. The HO procedure would cause both progesterone and estradiol to decline, but the action of estradiol and lactogens up to that point, coupled with the subsequent withdrawal of progesterone by the HO, may have been enough hormonal stimulation to reduce sensitization latencies from the virgin level of 7days to the 3day latency shown by 15HO females. The addition of estradiol in the 15HO+E females, so that estradiol remains high while progesterone declines, may have more closely simulated the events that typically occur at parturition, resulting in such females showing an immediate onset of maternal behavior.

Using this preparation, Numan et al. [710] injected estradiol locally into the hypothalamus of 15HO females. When estradiol was injected into the MPOA, females showed immediate maternal behavior, but when estradiol was injected into other hypothalamic regions, such as the posterior hypothalamus, or when cholesterol was injected into MPOA, sensitization latencies were 2–3days. These results indicate that estradiol action in the MPOA region is one factor that is involved in the immediate expression of complete maternal behavior when females are exposed to pup stimuli for the first time. These results should not be interpreted to mean that the MPOA is the only site where estradiol acts to stimulate maternal behavior. By day 15 of pregnancy, endogenous plasma estradiol has risen to moderately high levels [123] and estradiol action at more that one neural site may have primed the brain for maternal responsiveness, which, when coupled with continued estradiol action on the MPOA, resulted in full maternal responsiveness.

What might estradiol be doing to MPOA neurons to stimulate maternal behavior? Recall from Chapter 2 that when estradiol binds to its intracellular estrogen receptor (ER), the E-ER complex serves as a transcription factor and acts to affect the transcription of genes that contain an estrogen response element (ERE). With this understanding, one can view estradiol as contributing, along with other factors, to processes that alter the phenotype of MPOA neurons involved in maternal behavior. When estradiol increases the transcription of certain genes within MPOA neurons, with a resultant increase in the expression of certain proteins that may include neurotransmitter receptors and neurotransmitters, MPOA neurons may become more receptive to sensory inputs from pups, and the subsequent activation of MPOA neurons by such sensory inputs may then be associated with the release of a newly synthesized neurotransmitter into brain sites to which the MPOA projects. To add some specificity to these generalities, there is evidence that increased expression of OTRs within the MPOA at the time of parturition is involved in stimulating the immediate onset of maternal behavior. There are two types of ER, alpha and beta, and E binding the ER alpha stimulates the transcription of the OTR gene, with a resultant increase in OTR expression within neurons of the MPOA and other brain regions [172,1066]. Importantly, OTR expression increases within the MPOA near the time of parturition in rats [627,747], and the microinjection of an OTR antagonist directly into the MPOA of primiparous parturient rats disrupts the onset of both retrieval and nursing behavior [747]. Therefore, to the extent that the birth process and infant stimuli activate PVN neurons that project to and release OT into the MPOA, one role of estradiol is to prepare the MPOA to be responsive to OT. Since OT has excitatory effects on neurons [431,502,932,958], it can then act to stimulate MPOA neurons, some of which are likely to be output neurons with projections to other brain regions. Such a mechanism is schematically presented in Figure 5.2.

In my view, the increased expression of OTRs in the MPOA near the time of parturition is probably only one of many ways in which estradiol and the other hormonal events of pregnancy termination prepare the MPOA for its role in maternal behavior. An additional example involving the induction of Fos expression in MPOA will be presented below, and future research is sure to delineate other mechanisms as well. Further, the critical peripartum hormones that stimulate maternal responsiveness are likely to prepare other brain regions in addition to MPOA.

Fos expression increases in the MPOA, and other brain regions, of postpartum rats when these females respond maternally toward pups [703]. In mice, a knockout mutation of the Fos gene has been associated with severe maternal behavior deficits [133,521], suggesting that Fos expression in the MPOA may play an essential role in maternal behavior. In addition, the specific hormonal events associated with pregnancy termination in rats (increasing estradiol on a background of progesterone withdrawal) have been shown to activate high levels of Fos expression in MPOA in the absence of pup presentation and the occurrence of maternal behavior [890]. As described in Box 3.1, the *fos* gene is an immediate early gene that contains an ERE and a Ca/CRE in its regulatory region. The Fos protein, in turn, acts as a transcription factor to influence the transcription of so-called late-responding genes. Just which late-responding genes are affected by Fos in MPOA neurons to influence maternal behavior is not known, and this is an important area for future research. Therefore, in addition to affecting OTR expression, it can be suggested that the hormonal events of late pregnancy activate the ERE within the *fos* gene via the E-ER complex, and that the subsequent transcriptional effects of the Fos protein may then serve as a factor that further alters the MPOA phenotype to allow for the onset of maternal behavior. During the maintenance

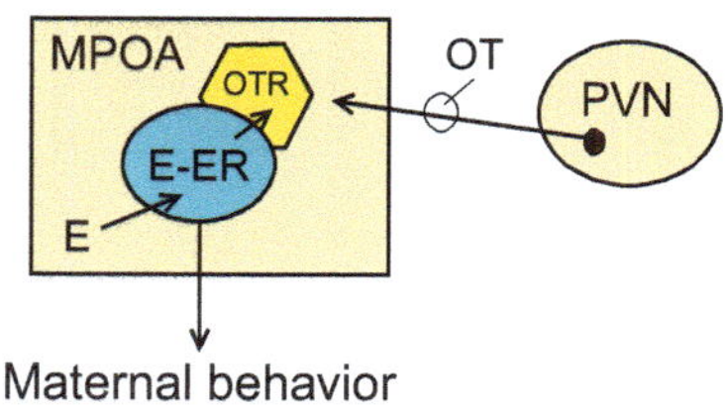

FIGURE 5.2 When estradiol (E) binds to estrogen receptors (ER) within medial preoptic area (MPOA) neurons, the E-ER complex becomes a transcription factor and acts within the regulatory region of the oxytocin receptor (OTR) gene (at an estrogen response element) to promote transcription of the gene, which ultimately increases the expression of OTRs within the cell membrane of MPOA neurons. This estradiol-induced change in the phenotype of MPOA neurons results in these neurons becoming sensitive to oxytocin (OT) input from the paraventricular nucleus (PVN) of the hypothalamus. OT stimulation of MPOA neurons facilitates the onset of maternal behavior. *Modified from Figure 4 in Numan and Stolzenberg [714] with permission from Elsevier.*

phase of maternal behavior, after the effects of peripartum hormones have waned, pup stimulation, perhaps via activation of Ca/CRE within the promoter region of the *fos* gene, maintains the continued expression of Fos, which in turn may contribute toward maintaining the phenotype of MPOA neurons that allows for the continued expression of maternal behavior [922].

Given that the MPOA is essential for maternal behavior, and that it is one site where estradiol, prolactin, and OT act to stimulate the onset of maternal behavior, Figure 5.3, which is an elaboration of Figure 5.1, presents a basic neurobehavioral model to explain the initiation of maternal behavior in rats [714]. The hormonal events of late pregnancy alter the phenotype of MPOA neurons so that they can respond to OT and to pup stimuli. Pup stimuli not only affect the MPOA by activating the PVN with a subsequent release of OT into MPOA but can also affect the MPOA through projections that include those from the amygdala and PFC: olfactory stimuli from pups can reach the MPOA from the medial amygdala, and processed multimodal sensory inputs can arrive at the MPOA from the infralimbic and prelimbic parts of the medial PFC [55,759,902,994]. The activation of MPOA efferents by pup stimuli is proposed to have two separate effects that result in the immediate onset of maternal behavior at parturition. One group of efferents is proposed to depress the activity of the default defensive neural circuit that is typically activated by pup stimuli in virgin females. By depressing this antisocial defensive circuit, the parturient female does not avoid, reject, or attack pups. Another group of efferents arising from the MPOA is conceived as stimulating neural circuits that allow pup stimuli to activate approach and acceptance/interaction circuits. Activation of these latter circuits underlies both the appetitive and consummatory aspects of maternal behavior. Two aspects of this model are worth noting. First, pup stimuli enter the system at many levels: MPOA, the defensive circuit, and the approach and interaction circuit. Second, the activation of MPOA efferents that results from a change in the female's internal state is a mechanism that can explain how the neural processing of the same social stimulus, in this case young pups, can be shifted away from pathways that result is social rejection and toward neural pathways that result in acceptance and affiliative responses. The model proposes that the MPOA may serve to block the processing of pup stimuli by the defensive-rejection circuit while it promotes the processing of such stimuli by the acceptance-interaction circuit.

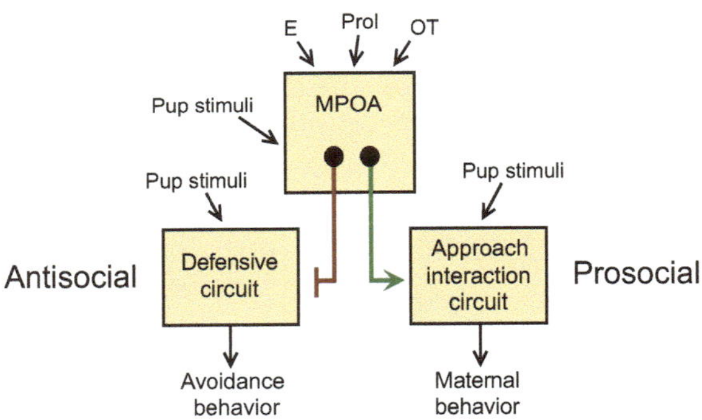

FIGURE 5.3 A basic neurobehavioral model of medial preoptic area (MPOA) function with respect to the onset of maternal behavior in rats. Estradiol (E), prolactin (Prol), and oxytocin (OT) act on MPOA neurons so that the responsiveness of these neurons to pup stimuli is increased. Activated MPOA efferents form two major functional routes, one of which inhibits an antisocial defensive neural circuit and another that excites a prosocial approach and acceptance neural circuit. The co-activation of both of these pathways at parturition allows for the immediate occurrence of maternal behavior toward pups as they emerge from the mother's birth canal. MPOA axons ending in a bar and shown in red are inhibitory, and those ending in an arrow and shown in green are excitatory. *Modified from Figure 5 in Numan and Stolzenberg [714] with permission from Elsevier.*

5.2.5 The Defensive-Rejection Circuit That is Active in Virgin Females

For the virgin female, olfactory stimuli from pups appear to be a major activator of the pup avoidance-rejection circuit: anosmia dramatically decreases sensitization latencies [307,703].

Virgin females whose ability to smell pups has been eliminated or decreased exhibit sensitization latencies of about 2 days instead of the standard 7-day latency shown by normal virgins. These data might lead one to conclude that the endocrine events associated with the end of pregnancy act in some way on the olfactory system to render the parturient female anosmic. Such is not the case however, as late-pregnant primigravid rats and parturient primiparae are actually attracted to pup odors [304,489]. Therefore, the hormonal events that stimulate maternal behavior act to change the valence of pup odors and presumably other pup stimuli from negative to positive. Since anosmic rats show maternal behavior toward pups, being attracted to pup odors is not essential for maternal behavior in rats as long as other attractive pup stimuli can be detected by the mother [694,703]. In other words, ongoing maternal behavior in rats is under multisensory control. However, the shift in the valence of pup odors from negative to positive should be viewed as one of the mechanisms that prevent pup stimuli from activating the rejection-avoidance circuit.

In those mammalian species that use olfaction during both social and nonsocial behaviors, which include rodents, there are two olfactory systems within the nasal mucosa that are affected by olfactory stimuli: the vomeronasal system and the main olfactory system [169,258]. The sensory neurons of the main olfactory nerve project to the main olfactory bulb (MOB) in the telencephalon, while vomeronasal sensory neurons project to the accessory olfactory bulb (AOB), which lies dorsal to the MOB (see Figure 4.1(B) for the locations of MOB and AOB). Both of these sources of olfactory input are involved in the inhibition of maternal behavior in virgin female rats, as interference with both systems shortens sensitization latencies to a greater extent than disruption of either system alone [309]. Importantly, the outputs of both the MOB and AOB, which are excitatory, converge in the medial amygdala (MeA), as depicted in Figure 5.4(A) [861,952]. Because of the typical association of the amygdala with fear-related processes, Fleming, Vaccarino, and Luebke [308] proposed that in virgin females olfactory inputs from pups activate amygdala circuits that give rise to defensive and avoidance responses. In support of this view, it has been found that both electrical [308] and excitotoxic amino acid lesions of MeA [706] facilitate maternal behavior in virgins. Such females do not avoid pups and tolerate their proximity, which results in sensitization latencies of about 2 days.

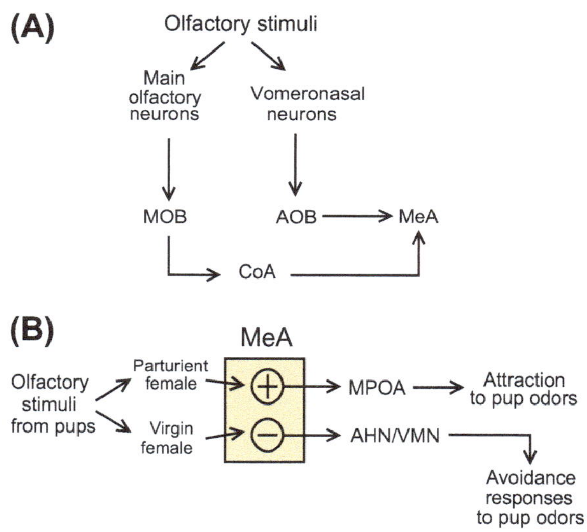

FIGURE 5.4 (A) The neural pathways that allow olfactory stimuli to converge on the medial amygdala (MeA). Other abbreviations: AOB = accessory olfactory bulb; CoA = cortical amygdala; MOB = main olfactory bulb. (B) A neural model that depicts how olfactory stimuli from pups might be differentially processed by virgin and parturient female rats. In virgin females, olfactory stimuli are proposed to activate medial amygdala neurons (MeA) with a negative valence that project to and activate defensive neural circuits that arise from the anterior hypothalamic nucleus (AHN) and ventromedial nucleus of the hypothalamus (VMN). In parturient females, due to the hormonal and neural changes that occur near the end of pregnancy, the model proposes that olfactory stimuli from pups activate positively valent MeA neurons that project to the medial preoptic area (MPOA). This shift in the neuronal circuits activated by pup odors causes the parturient female to approach, rather than to avoid, such odors.

In line with Figure 3.2, Figure 5.4(B) outlines two possible alternate circuits through which pup stimuli might be processed, which would give rise to either defensive/rejection/avoidance responses or attraction/acceptance responses. As previously reviewed, the MeA projects to diverse hypothalamic nuclei, which include MPOA, AHN, and VMN [154]. Since the MPOA is essential for maternal behavior while, as reviewed in Chapter 3, the AHN/VMN region contains intertwined, but separate, neural systems involved in either offensive aggression or defensive responses [153,363,556], the following can be proposed. In virgins, novel pup stimuli activate the MeA-to-AHN/VMN pathway that leads to either avoidance responses or attack (infanticide). Laboratory rats, because they have been bred to be tame, are more likely to show avoidance and escape responses, although infanticide does sometimes occur. In contrast to virgins, in parturient female rats, olfactory stimuli from pups are proposed to result in attraction and acceptance as a result of such stimuli being routed through the MPOA rather than through AHN/VMN. Subsequently, I will offer some ideas to explain how this switching of olfactory processing might occur. For the present, I want to reiterate that although the proposed switch from the MeA-to-AHN/VMN connection to the MeA-to-MPOA connection would serve to prevent olfactory stimuli from activating the rejection circuit, the MeA-to-MPOA connection is not essential for maternal behavior in laboratory rats; rats with MeA lesions show maternal behavior, presumably because other pup stimuli can gain access to MPOA via other pathways. Indeed, as already reviewed, processed sensory inputs can reach MPOA via the medial PFC. The following experiment, however, would be very interesting to perform (cf. [598]). As background, on day 22 of pregnancy and close to the time of parturition, when primigravid rats are placed in a two-compartment cage, with clean bedding on one side and bedding soiled by pups (presumably containing critical pups' odors) on the other side, such females, unlike virgins, are attracted to the pups' odors and spend more time in the part of the cage with the soiled bedding [489]. It would be interesting to determine whether MeA excitotoxic lesions would block this preference for bedding soiled by pups while not blocking attraction to a live pup. What this would suggest is that attraction to pup odors in parturient females is mediated via MeA neurons, as proposed in Figure 5.4(B), while other, nonolfactory pup stimuli, are attractive to such females via alternate neural routes.

There is some Fos expression evidence that supports the basic model shown in Figure 5.4(B) [888]. Hormone-treated nulliparous females and nulliparous control females were exposed to pups for 2h. While the hormone-treated females showed maternal behavior, the nonhormone-treated controls did not, but instead withdrew from the pups after first approaching and briefly sniffing them. When the brains of these females were analyzed for Fos expression, the maternal females had numerous Fos-expressing cells in the dorsal part of the MPOA and few Fos-expressing cells in AHN/VMN. The nonmaternal females, in contrast, displayed the reverse pattern: Fos expression was high in AHN/VMN and low in the dorsal MPOA. Since these females were exposed to live pups, both olfactory and nonolfactory pup stimuli were presumably involved in these effects.

In further support of a MeA-to-AHN/VMN circuit that activates defensive responses to pups in virgins, when nonmaternal nulliparous females with unilateral MeA lesions are exposed to pups, a subsequent Fos analysis of the AHN/VMN region shows decreased Fos expression on the same side of the brain as the MeA lesion [891]. Since MeA projections to the hypothalamus are primarily ipsilateral [154], these results suggest that the MeA lesion blocked the conduction of some pup stimuli, presumably olfactory, to AHN/VMN, resulting in decreased neural activation of this region. The most convincing evidence that activation of AHN/VMN by pup stimuli is associated with an inhibition of maternal behavior and the occurrence of defensive responses comes from research that has examined the effects of neuron-specific excitotoxic amino acid lesions on the maternal responsiveness of nulliparous females. In a manner similar to the effects of anosmia induction or MeA lesions, lesions of the AHN/VMN region reduce sensitization latencies in naïve nulliparae to about 2 days [124,891].

With respect to a MeA projection to AHN/VMN that excites the latter region to cause defensive responses, some might be perplexed by this proposition, since the classic view is that MeA projection neurons are GABAergic [952] and therefore should inhibit, rather than excite, AHN/VMN neurons. However, this is where an understanding of microcircuitry is so important. If the relevant projection of MeA to AHN/VMN were to contain gamma-aminobutyric acid (GABA), the question becomes: onto which AHN/VMN neurons do these GABA neurons synapse? If they synapse on AHN/VMN output neurons, that would be similar to causing an AHN/VMN lesion, but if they synapse on and inhibit local inhibitory interneurons that act to restrain AHN/VMN output, then activation of the MeA-to-AHN/VMN circuit should increase AHN/VMN output. If the critical MeA output neurons are indeed GABAergic, then the results that we have reviewed support the latter synaptic microcircuitry.

Moving beyond the classic view of MeA neurotransmitter content, recent evidence indicates that a substantial proportion of MeA neurons that project to the hypothalamus actually contain glutamate, an excitatory neurotransmitter, rather than GABA [95]. If this were the case for the critical MeA neurons that are operating in virgin females exposed to pups, then I would predict that they are directly exciting AHN/VMN projection neurons whose output in turn results in defensive responses toward pups. Finally, and in relation to the research on aggression reviewed in Chapter 3, there is also evidence that some MeA neurons that project to AHN/VMN to depress maternal responsiveness may contain substance P [711].

If an MOB/AOB-to-MeA-to-AHN/VMN circuit causes the virgin female to avoid pups and perhaps attack them, where might the AHN/VMN project to exert these effects? As outlined in Chapter 3, the HAA, which includes neurons within the AHN/VMN region, sends glutamatergic projections to the midbrain PAG to activate affective aggression. Further, Gross and Canteras [363] and Canteras [153] have defined medial hypothalamic defensive circuits that include AHN and VMN projections that directly or indirectly affect PAG neurons. Therefore, it appears likely that an MOB/AOB-to-MeA-to-AHN/VMN-to-PAG circuit contributes to the activation of either rejection/avoidance/defense responses or affective aggression responses in naïve virgins when they are exposed to pups. In support, activation of the dorsal PAG has been shown to depress the maternal behavior of female rats, while lesions to this region promote maternal responsiveness [941]. In addition to projecting to lower brainstem motor systems (medullary reticular formation), the PAG also has reciprocal projections with the MPOA [814]. Therefore, it is possible that downstream PAG circuits activate defensive escape and avoidance responses, while ascending PAG circuits act to inhibit MPOA neurons that might promote maternal behavior.

The PAG has been divided into dorsomedial, dorsolateral, lateral, ventrolateral, and ventral parts, and it should not be surprising that different PAG neurons with different projections are involved in different functions. As I have already emphasized in Chapters 1 and 3, the PAG is involved in affective aggressive responses, defensive immobility, and escape-avoidance responses. However, as described in Chapter 4, the PAG is also involved in regulating consummatory sexual responses that include the lordosis response in females [763] and penile erections in males [432]. Finally, as I will review later, the PAG also plays a role in the immobile arched-back nursing posture referred to as kyphosis [567]. One very general idea is that the output of the dorsomedial and dorsolateral PAG may be involved in regulating active aggressive responses and active defensive responses such as escape and avoidance (running away), while the ventrolateral PAG may be involved in promoting behavioral immobility [363]. Relevantly, a variety of stimuli may cause immobility, and not all of these are necessarily threatening. In addition to a perceived threat, suckling stimulation may cause the mother to remain immobile allowing for effective nursing, and male mounting behavior may activate immobility in sexually receptive females, which in turn allows for the occurrence of lordosis.

5.2.6 Inhibition of the Defensive/Rejection Circuit in Parturient Females

Figures 5.3 and 5.4(B) suggest that when the primiparous female's brain is exposed to the endocrine events associated with pregnancy termination, the processing of pup stimuli through the defensive circuit is inhibited while the processing of pup stimuli through the attraction/interaction circuit is facilitated. The purpose of this section is to describe the potential mechanisms mediating the former effect.

According to the proposed model in Figure 5.3, when pregnancy hormones modify the characteristics of MPOA neurons, one of the functions of certain MPOA efferents is to inhibit the defensive circuit. Although direct neurobehavioral evidence supporting this view does not exist, there is anatomical evidence supporting the possibility. Numan and Numan [704] injected the anterograde tracer, PHAL, into the dorsolateral MPOA of postpartum rats in order to trace the efferent projections of this MPOA region, a region that contains many neurons that express Fos during maternal behavior. Some of the results are shown in Figure 5.5. The following defined projection sites of MPOA are of general importance with respect to maternal behavior control, and those listed in italics are relevant to potential mechanisms whereby MPOA output might depress the defensive circuit: *AHN*, *PVN*, *VMNvl*, ventral tegmental area (VTA), and *PAG*. Since the MPOA projects to many of the brain regions that are part of the defensive circuit, it makes sense to propose that MPOA output may be inhibiting the activation of defensive responses to novel pup odors in parturient females. In support, Lonstein and De Vries [565] have reported that a significant proportion of MPOA neurons that express Fos during maternal behavior also contain the inhibitory neurotransmitter GABA, making it functionally possible for some "maternal" MPOA neurons to inhibit those AHN, VMN, and PAG neurons that mediate pup avoidance or infanticide. With respect to PAG, note that the dorsolateral part of the MPOA projects to the dorsal PAG at rostral PAG levels and to the ventrolateral PAG at more caudal PAG levels. When PHAL is injected more broadly into MPOA, and is not restricted to just the dorsolateral part of the MPOA, then MPOA projections to PAG are much more widespread than that shown in Figure 5.5 [814], demonstrating that MPOA output can strongly influence all parts of PAG.

In addition to the likelihood that MPOA output might directly inhibit the defensive circuit, other mechanisms may operate to have the same effect. In fact, multiple mechanisms probably co-act at the time of parturition to ensure that the helpless newborn pups do not evoke any kind of hostile or neglectful behavior on the part of the mother. Since the MPOA projects to PVN [704,903], perhaps activation of PVN-OT circuits by MPOA, by the birth process, and by suckling stimulation, results in an OT effect that inhibits the defensive circuit. I have already reviewed the evidence that the injection of an OTR antagonist into MPOA disrupts the onset of maternal behavior. Clearly, OT action on MPOA is necessary for the onset of maternal behavior, but no one has shown that it is sufficient. Since PVN-OT neurons

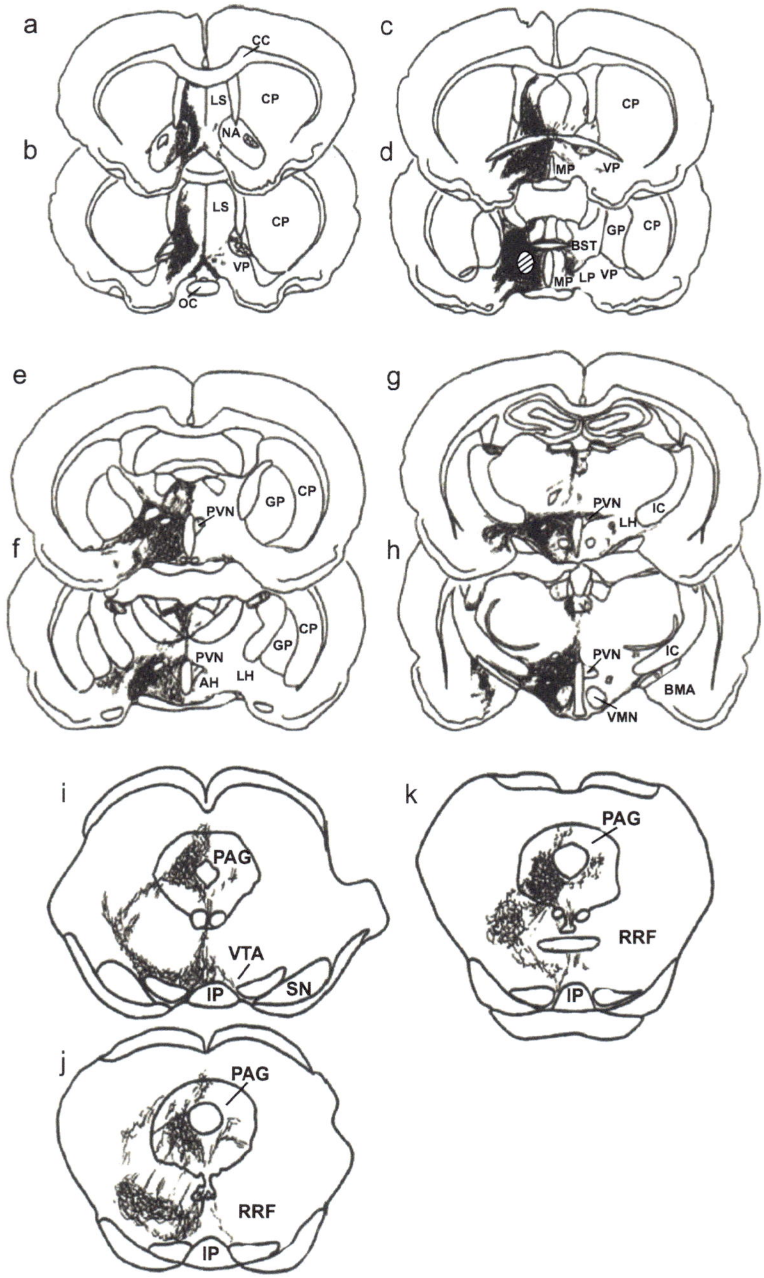

FIGURE 5.5 A series of frontal sections from rostral to caudal (a through k) through the rat brain showing the distribution of PHAL-labeled fibers after a unilateral iontophoretic injection of PHAL (an anterograde tracer) into the dorsolateral part of the medial preoptic area and adjoining ventral part of the bed nucleus of the stria terminalis (the injection site is indicated by striped lines in section d). Most of the projections are ipsilateral (on the same side of the brain) to the injection site. Noteworthy descending projections terminate in the paraventricular nucleus (PVN), anterior hypothalamus (AH), ventrolateral part of the ventromedial nucleus (VMN), ventral tegmental area (VTA), and periaqueductal gray (PAG). Other abbreviations: BMA = basomedial amygdala; BST = bed nucleus of the stria terminalis; CC = corpus callosum; CP = caudate/putamen; GP = globus pallidus; IC = internal capsule; IP = interpeduncular nucleus; LH = lateral hypothalamus; LP = lateral preoptic area; LS = lateral septum; MP = medial preoptic area; NA = nucleus accumbens; OC = optic chiasm; RRF = retrorubral field; SN = substantia nigra; VP = ventral pallidum. *Modified from Figure 6 in Numan and Numan [704] with permission © John Wiley & Sons.*

have widespread projections throughout the brain [502], which include projections to MeA, CeA, and basomedial/basolateral amygdala (BLA/BMA) in the amygdala, it may promote the onset of maternal behavior by acting at more than one site, and some of these actions may serve to depress avoidance and rejection responses toward pups [207].

To begin this analysis, I would like to first describe the important research that shows how OT action on OTRs in the lateral part of the central nucleus of the amygdala (CeAl) inhibits the conditioned fear response (CFR; see Chapter 1) in rats [431,502,999]. The basic conclusions from this research are shown in Figure 5.6(A). PVN efferents to CeAl release OT onto OTRs that excite CeAl inhibitory neurons that project to CeAm, blunting CeAm output to the PAG and decreasing the CFR to a conditioned aversive stimulus (CAS; a previously neutral stimulus that is subsequently paired with shock so that it acquires aversive properties). These results fit nicely with other research showing that OT action in the brain exerts anxiolytic effects [26,811,812,1058]. For example, rodents that are treated systemically with an OTR agonist that can cross the blood–brain barrier spend more time in the open arms of an elevated plus maze, suggesting that they are less fearful of novel external stimuli.

Within this framework, perhaps similar mechanisms are operative in MeA. Importantly, although the OTRs in CeAl are not estrogen dependent, those in MeA are [51,511]. Given that ER alpha is located within MeA neurons [894] and that the expression of OTRs increases in MeA of parturient rats [627], perhaps the late pregnancy rise in estradiol acts in MeA to increase OTR expression, resulting in increased MeA responsiveness to OT input from PVN. Since the MeA contains glutamatergic projection neurons and local GABAergic inhibitory neurons [95], Figure 5.6(B) outlines a potential mechanism to show how OT and estradiol may co-act on MeA neurons to cause a shift in the parturient

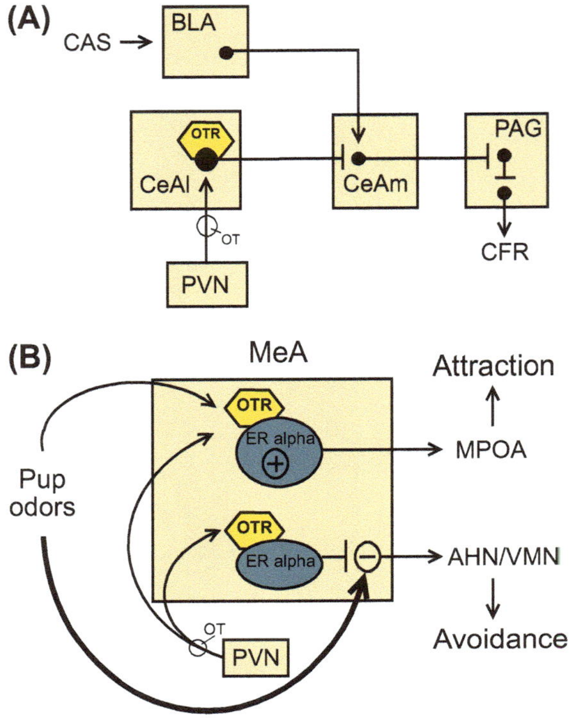

FIGURE 5.6 (A) Based on the research findings of Huber et al. [431], Knobloch et al. [502], and Viviani et al. [999], this figure shows how oxytocin (OT) input to neurons in the lateral part of the central nucleus of the amygdala (CeAl) can act to suppress the conditioned fear response (CFR) to a conditioned aversive stimulus (CAS). Axons ending in a bar are inhibitory, and those ending in an arrow are excitatory. Other abbreviations: BLA = basolateral amygdala; CeAm = medial part of the central nucleus of the amygdala; OTR = oxytocin receptor; PAG = periaqueductal gray; PVN = paraventricular nucleus of the hypothalamus. (B) Based on the mechanisms described in (A), a neural model is presented to explain how estradiol (E) and oxytocin (OT) action at the level of the medial amygdala (MeA) might alter the processing of pup odors in parturient females so that such odors activate positively valent MeA neurons that project to the medial preoptic area (MPOA) instead of activating negatively valent MeA neurons (as would occur in virgins) with projections to brain regions mediating defensive behaviors (AHN = anterior hypothalamic nucleus; VMN = ventromedial nucleus). Estrogen receptor (ER) alpha is shown as being present in positively valent MeA projection neurons and within inhibitory interneurons that suppress the output of negatively valent MeA projection neurons. Near the time of parturition, E action on ER alpha activates the transcription of the oxytocin receptor (OTR) gene and the subsequent expression of OTRs

female's responsiveness to pup odors, moving her away from avoidance and toward attraction. The figure proposes that ER alpha and OTRs are located on MeA projection neurons to MPOA and on local GABAergic interneurons that act to inhibit MeA projection neurons to the AHN/VMN. Given this circuitry, the excitatory effects of OT release into MeA would act to favor attraction toward pup odors and depress avoidance of such stimuli.

Because research has not been performed on the exact cellular location of OTRs within MeA, and because researchers have not examined whether estradiol and OT action at the level of MeA is important for the onset of maternal behavior, the neural model proposed in Figure 5.6(B) is simply an hypothesis. I have spent some time on this, however, primarily for heuristic reasons. The models shown in Figure 5.6 emphasize that knowledge of the neurochemical microcircuitry within a neural region, along with an understanding of the afferents and efferents of a brain region, will ultimately allow us to gain significant insight into the neural control of social as well as other behaviors, and can serve to explain how neuropeptides and other factors can act to modify the way an organism responds to certain stimuli. Figure 5.6(B) provides an excellent example of how steroid hormones and neuropeptides can alter the way social stimuli are processed by the brain and therefore alter the way an organism responds to a social stimulus.

In conclusion, although neurobehavioral evidence has clearly outlined an MOB/AOB-to-MeA-to-AHN-to-PAG circuit that opposes maternal behavior by stimulating avoidance and rejection of pup stimuli by nulliparous females, the exact mechanisms by which this system is depressed in parturient females has not been fully worked out, and most hypotheses are based on chemical neuroanatomical evidence rather than on neurobehavioral evidence. Even so, Figure 5.7 presents a summary model of the data reviewed to this point, filling in some of the gaps in Figure 5.3.

(see Figure 5.2), with the result that oxytocin release into MeA of parturient females stimulates positive MeA neurons and inhibits negative MeA neurons. The model also assumes that in the absence of OT action at OTRs, the dominant effect of pup odors is on negatively valent MeA neurons. Axons ending in a bar are inhibitory, and those ending in an arrow are excitatory. Other abbreviations: PVN = paraventricular nucleus of the hypothalamus.

5.2.7 MPOA Interaction with the Mesolimbic DA System Controls the Appetitive Aspects of Maternal Behavior

5.2.7.1 A Neural Model

In Figure 5.3, it was suggested that the hormonally primed MPOA not only inhibits the defensive circuit but also facilitates the processing of pup stimuli by an approach/interaction circuit that gives rise to appetitive and consummatory maternal responses. This section will review the very strong neurobehavioral evidence that indicates that MPOA interactions with the mesolimbic DA system control the *appetitive* aspects of maternal behavior in rats during both the onset of maternal behavior at parturition and during the continued maintenance of the behavior during the postpartum period [713,714]. I will start by presenting a neural model derived from this evidence, which is shown in Figure 5.8, and then will provide an overview of the supportive evidence. To fully appreciate this model, a review of Chapter 1 will be helpful. The figure shows that the hormonally primed MPOA of a newly parturient female becomes responsive to a variety of stimulatory inputs, which include OT from the PVN, glutamate inputs from MeA that are activated by pup odors that are now attractive, and multimodal pup stimuli that reach the MPOA via glutamatergic projections from the medial PFC (infralimbic and perhaps prelimbic cortex). An analysis of OT and MeA inputs to MPOA has already been described in previous sections of this chapter. As a result of these stimulatory inputs, it is proposed that MPOA glutamatergic projections

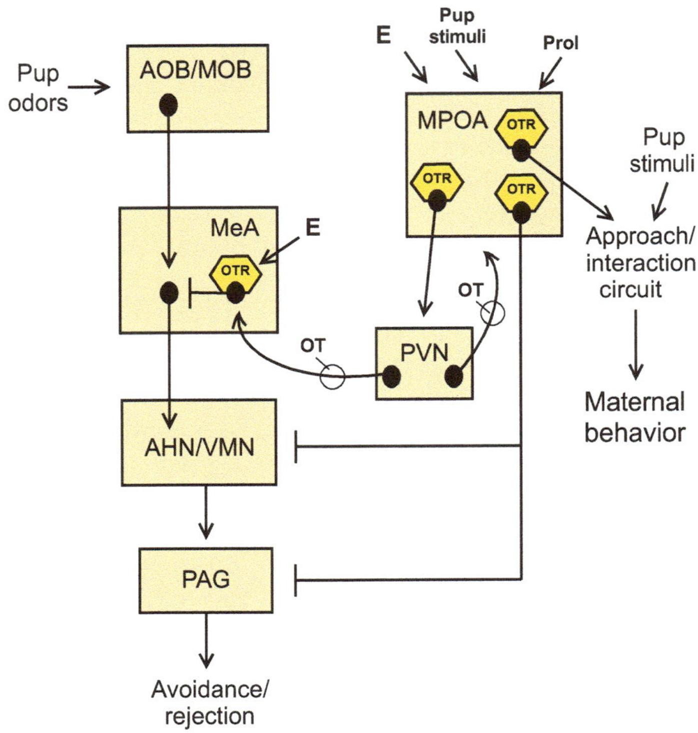

FIGURE 5.7 Based on a partial elaboration of Figure 5.3, a model of how separate outputs from the medial preoptic area (MPOA) can act to suppress antisocial and promote prosocial responses of the parturient female to pup stimuli. The suppression of antisocial responses is emphasized in this figure. See text for operational details. In comparing this figure with Figure 5.6(B), for simplicity, oxytocin (OT) action on the oxytocin receptor (OTR) at the level of MeA in the current figure only displays the potential inhibition of avoidance responses to pup odors. In addition to MeA, estradiol (E)-induced activation of OTR expression also occurs in MPOA (see Figure 5.2), and these OTRs are proposed to be located on MPOA neurons that promote maternal behavior and on those that inhibit avoidance and rejection responses to pups. Axons ending in a bar are inhibitory, and those ending in an arrow are excitatory. Other abbreviations: AHN = anterior hypothalamic nucleus; AOB = accessory olfactory bulb; MOB = main olfactory bulb; PAG = periaqueductal gray; Prol = prolactin; PVN = paraventricular nucleus; VMN = ventromedial nucleus.

to VTA-DA neurons are activated. The model also shows that the PVN sends a stimulatory OT input to VTA, which is proposed to co-act with MPOA glutamatergic inputs to stimulate VTA-DA neurons, which triggers a burst of DA release into the shell region of NA (NAs). The activated VTA-DA neurons, via axoaxonic synapses within NAs, are proposed to presynaptically inhibit glutamate release from BMA/BLA neurons that terminate on certain medium spiny GABAergic output neurons of NAs, with such inhibition being mediated by an action of DA on presynaptic D1 DA receptors. As described in Chapter 1, BMA/BLA can relay a variety of sensory inputs from pups, which would include olfactory, gustatory, somatic sensory, auditory, and visual stimuli, to *both* the NAs and ventral pallidum (VP). The presynaptic inhibitory action of DA on D1 receptors in NAs acts to depress a BMA/BLA excitatory drive on NAs medium spiny neurons that project to and inhibit the VP. The resultant disinhibition of VP increases the responsiveness of critical VP output neurons to pup stimuli that also arrive from BLA/BMA,

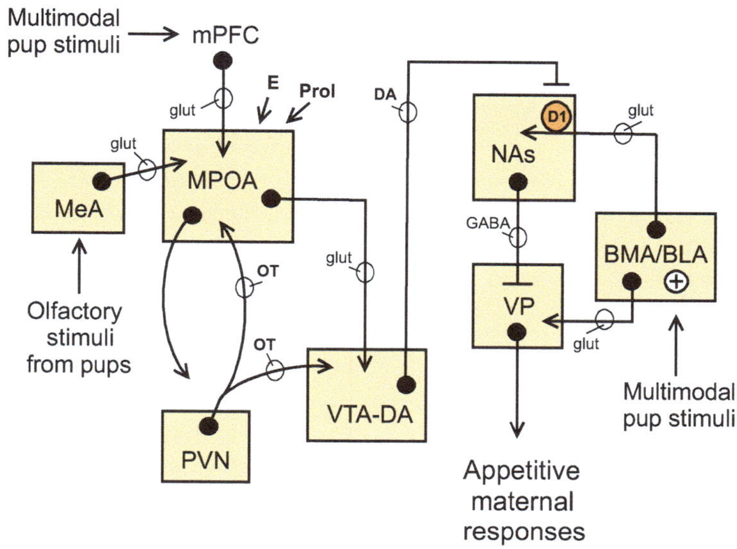

FIGURE 5.8 A neural model depicting how medial preoptic area (MPOA) interactions with the mesolimbic dopamine (DA) system promote the goal-directed appetitive aspects of maternal behavior in rats. See text for the operational details of this model along with the strong research evidence that supports it. The positive sign in the basomedial/basolateral amygdala (BMA/BLA) area indicates neurons that relay sensory inputs from pups to ventral pallidum (VP) neurons that mediate appetitive responses. Axons ending in a bar are inhibitory, and those ending in an arrow are excitatory. Other abbreviations: D1 = D1 dopamine receptor; E = estradiol; glut = glutamate; MeA = medial amygdala; mPFC = medial prefrontal cortex; NAs = nucleus accumbens shell; OT = oxytocin; Prol = prolactin; PVN = paraventricular nucleus; VTA = ventral tegmental area.

and the stimulation of VP projection neurons, which can affect both brainstem and telencephalic systems, is proposed to be necessary for the appetitive aspects of maternal behavior.

In this model, pup stimuli enter the neural circuits at multiple points, in this case at both the MPOA and within the NA-VP circuit. It is a modification in how pup stimuli are processed that results in changes in behavior. If MPOA is not responsive to pup stimuli, then DA is not released into NAs, which results in the VP being less likely to be activated by pup stimuli because it is under strong NAs inhibition. When the MPOA is primed by hormones and OT, which allows it to respond to pup stimuli, then DA is released into NAs, which then causes the VP to become responsive to pup stimuli through the mechanisms proposed above.

Since the mesolimbic DA system is a nonspecific motivational system, motivational specificity is arrived at by the types of stimuli that become capable of activating DA release into NA [696,716]. Figure 5.8 can be conceived as representing an interaction between a specific appetitive maternal motivational system (MPOA efferents to VTA that are activated by pup stimuli) and a nonspecific appetitive motivational system (VTA dopaminergic input to NA). According to the model, if pups are present in a female's environment, appetitive maternal responses, such as searching for and retrieving pups, will only occur if the specific MPOA system activates the nonspecific VTA-DA system. For the primiparous parturient female, it is her internal physiological state that allows pup stimuli to activate this critical MPOA-to-VTA connection.

Although the model shown in Figure 5.8 emphasizes the primary involvement of the mesolimbic DA system, the nigrostriatal DA system may also contribute to the regulation of postpartum maternal behavior [406].

5.2.7.2 Evidence Supporting the Neural Model

In describing the support for the model, it would first be important to show that MPOA neurons do indeed project to the VTA. As shown in Figure 5.5, anterograde tract-tracing studies show that this is the case (also see [329,903]). Further, a significant proportion of MPOA projections to VTA are glutamatergic [328], although it remains to be anatomically shown whether such excitatory MPOA input to VTA actually terminates on dopamine neurons. In further support of the proposal that a proportion of MPOA efferents to VTA are relevant to maternal behavior, estrogen-binding MPOA neurons project to the VTA [280], as do MPOA neurons that express Fos during maternal behavior [705].

Strong evidence supports the involvement of the mesolimbic DA system in the regulation of both the onset and maintenance of maternal behavior in rats (for reviews, see [714,716]). Microdialysis and in vivo voltammetry studies show that DA is released into NAs during maternal behavior. For example, Champagne et al. [171] found that the frequency with which postpartum dams licked and groomed their pups was positively correlated with DA release into NAs and that the increase in DA release preceded and predicted a bout of licking and grooming behavior. These results can be interpreted as suggesting that DA release into NAs increases a mother's interest and attention toward her pups [816].

Evidence shows that inactivation of VTA-DA neurons disrupts retrieval of pups, but not nursing behavior, during the postpartum maintenance phase of maternal behavior. Numan, Stolzenberg, Dellevigne, Correnti, and Numan [715] injected baclofen into the VTA of postpartum rats. Baclofen is a GABA-B receptor agonist that increases K^+ conductance and causes neural inhibition. Following the injection, females were administered a 20-min retrieval test, where pups were placed outside the nest and the number of pups retrieved back to the nest was recorded. After the retrieval test, any pups that were not retrieved were placed back into the female's nest, and 10 min later a 15-min nursing observation occurred. Control females that received either saline into the VTA or baclofen injections into a region dorsal to the VTA retrieved all of their eight pups back to their nests within about 5 min. For the females that were injected with baclofen into VTA, although they approached and investigated (sniffed) their pups, they were unable to complete the retrieval of most of their pups, although some females retrieved a few pups. In sharp contrast to retrieval behavior, there were no differences in the duration of nursing behavior between the three groups. These results indicate that inactivation of the VTA disrupts the appetitive, but not the consummatory, aspects of maternal behavior in postpartum rats. Although baclofen does not specifically target dopamine neurons, other research has shown that injection of 6-hydroxydopamine into the VTA, which destroys DA neurons, exerts similar effects [386]. In further support of a role for VTA neurons in appetitive maternal responses, in a study employing the conditioned place preference paradigm, Seip and Morrell [878] found that depression of VTA activity blocked the expression of a preference for a cage compartment that had previously been paired with pups and maternal behavior.

Depression of VTA-DA neurons would block DA release into the multiple neural sites to which these neurons project [949], and would not be selective for DA input to NA. However, experiments have shown that DA receptor blockade within NAs disrupts retrieving behavior in postpartum rats without interfering with nursing behavior [473,707]. Most importantly with respect to the neural model shown in Figure 5.8, Numan et al. [707] reported that blockade of D1 receptors in NAs, but not D2 receptors, disrupted retrieval behavior in postpartum rats. Whether such an effect is mediated via an action on presynaptic D1 receptors remains to be determined.

Given that blockade of VTA-DA input to D1 receptors in NAs disrupts the maintenance of appetitive maternal responses in postpartum rats, is there any evidence that *activation* of DA-D1 receptors in NAs can stimulate the onset of maternal behavior in rats? To test this possibility, Stolzenberg, McKenna, Keough, Hancock, Numan, and Numan [928] made use of the 15HO model described previously. Recall that when primigravid female rats are hysterectomized and ovariectomized on day 15 of pregnancy and then exposed to pups beginning 48 h later, they exhibit sensitization latencies of 2–3 days, which is significantly shorter than that shown by virgins, but is not equal to the immediate onset of maternal behavior shown by 15HO + E females. In other words, 15HO females display a suboptimal hormonal stimulation of maternal behavior, and it was asked whether D1 receptor agonist administration to NAs would be able to substitute for full estradiol stimulation of the MPOA and push 15HO females over the threshold for full maternal behavior. When either a DA-D1 or a DA-D2 receptor agonist was injected into the NAs of 15HO females over the first three days of pup exposure, the D1 agonist–injected females, but not the D2 agonist–injected females, exhibited significantly shorter sensitization latencies than control females that were injected with the water vehicle solution into NAs. While females injected with the D1 agonist were typically fully maternal on their first day of pup exposure, females injected with the D2 agonist or with water into NAs showed the 2-day sensitization latency that is typical of 15HO females.

Note that DA-D1 stimulation of NAs in 15HO females stimulated the onset of full maternal behavior, including retrieval and nursing behavior. The activation of both the appetitive and consummatory aspects of maternal behavior can be explained by the fact that the females were already partially primed to show full maternal behavior. Application of the D1 receptor agonist to the NAs in these females activated relatively immediate appetitive motivation and the pups were retrieved. The suckling stimulation that subsequently occurred in the nest area would then be able to elicit nursing behavior. Figure 5.9 presents a model to explain how D1 agonist administration to NAs might activate the immediate onset of maternal behavior in 15HO females. The figure suggests that only low levels of MPOA excitation of VTA occur when 15HO females are exposed to pups, resulting in a moderate facilitation of maternal behavior that is associated with a moderate release of DA into NAs. When a DA-D1 agonist is added to the NAs in such females, the amount of D1 stimulation of NAs approximates that received by 15HO + E females, and near immediate maternal behavior occurs.

The model shown in Figure 5.8 proposes that when DA acts on D1 receptors in NAs, it acts to depress the excitability of NAs GABA efferents to VP, which results in increased VP output, with the latter being essential for appetitive maternal responses. In other words, although DA acts on D1 receptors in NAs, it is VP output that is proposed to be essential for appetitive maternal responses. According to the model, NAs output actually opposes maternal behavior and must be depressed by DA action. What is the evidence for this proposal? First, bilateral inhibition of NAs neural activity does not depress the maternal behavior of postpartum primiparous female rats, indicating that NAs output is not essential for maternal behavior [708,755]. While D1 receptor antagonist injection into NAs depresses postpartum retrieval behavior [707], depression of NAs neural activity does not, and these results are most consistent with the view that DA acts to inhibit, rather than excite, NAs output. Finally, and importantly, while depression of NAs activity does not disrupt maternal behavior, bilateral neuron specific inactivation of VP does disrupt retrieval behavior in postpartum rats [701,708], indicating that it is the output of VP that is critical for maternal behavior. The mechanism and

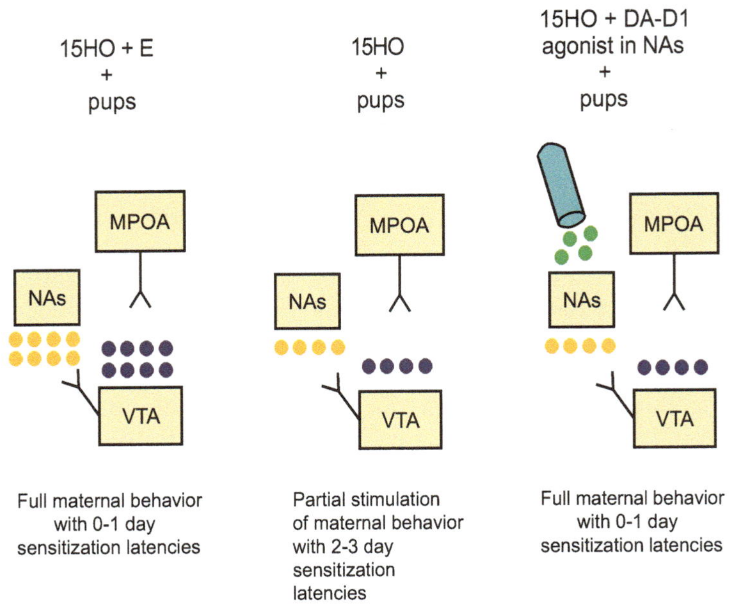

FIGURE 5.9 A diagram showing how microinjection of an exogenous dopamine (DA) D1 receptor agonist into the nucleus accumbens shell (NAs) might facilitate maternal behavior in female rats that have been exposed to suboptimal levels of "maternal" hormones. The diagram on the left shows that when the medial preoptic area (MPOA) is fully primed with the relevant hormones in females that have been hysterectomized and ovariectomized on day 15 of pregnancy and treated with estradiol (15HO + E), MPOA efferents to the ventral tegmental area (VTA) release large amounts of neurotransmitter (shown in blue; presumed to be glutamate) on to VTA-DA neurons, resulting in the release of large amounts of DA (shown in yellow) into NAs. This mechanism promotes full maternal behavior in these females. The middle panel shows the case where the female is suboptimally primed with hormones (15HO): pup stimuli are less effective in activating MPOA efferents to VTA, less DA is released into NAs, and maternal responsiveness is incomplete. As shown on the right, when 15HO females are also microinjected with a DA-D1 receptor agonist (shown in green) into NAs, full maternal behavior occurs. *Modified from Figure 10 in Numan and Stolzenberg [714] with permission from Elsevier.*

circuits through which VP output acts to promote the appetitive aspects of maternal behavior have not been determined, and this is an important area for future research.

Although neuroanatomical research shows us that MPOA efferents can affect the mesolimbic DA system, what is the neurobehavioral evidence that supports the proposal that MPOA interaction with the mesolimbic DA system, which is proposed to facilitate VP output, is actually essential for appetitive maternal responses? My laboratory has used the asymmetrical neural inactivation design to provide evidence for this part of the neural model [696,708,712]. The following information is necessary for understanding this design: (1) the neural circuits shown in Figure 5.8 are bilateral, meaning they occur on both sides of the brain; (2) the circuits shown in Figure 5.8 are primarily ipsilateral, which means that they are not crossed, but instead relay neural information within a single side of the brain. For example, the right MPOA projects primarily to the right VTA (see Figure 5.5), and the left MPOA projects to the left VTA. The same relationship holds for VTA projections to NAs, and for NAs projections to VP; (3) lesion studies indicate that in order to severely disrupt many basic motivated behaviors, bilateral inactivation of critical neural control regions within limbic, hypothalamic, or striatal circuits is necessary, while unilateral lesions have negligible and transient effects. For example, unilateral MPOA lesions do not severely disrupt maternal behavior, while bilateral lesions have devastating effects [921]. As long as the essential neural circuits on one side of the brain are functional, relatively normal behavior can occur. Based on these three facts, the asymmetrical neural inactivation design either bilaterally or unilaterally interferes with a proposed neural circuit at two different points in the circuit, with the two neural interventions located either contralateral (on opposite sides of the brain) or ipsilateral (on the same side of the brain) to each other. It is predicted that contralateral lesions, which would disrupt the critical circuit on both sides of the brain, should result in more severe behavioral deficits than ipsilateral lesions, which would leave the proposed critical circuit operative on one side of the brain.

Figure 5.10 shows how the asymmetrical lesion design was used to support the proposed maternal circuit for appetitive maternal responses. Numan and Smith [712] found that when postpartum rats receive unilateral lesions of the MPOA and VTA, severe deficits in retrieval behavior, but not nursing behavior, occurred when these lesions were located contralateral to one another, but not when they were ipsilateral to one another. Similar results occurred when a unilateral MPOA lesion was paired with a unilateral VP lesion: severe retrieval deficits occurred only when the lesions were on opposite sides of the brain, causing the proposed circuit to be damaged bilaterally [708].

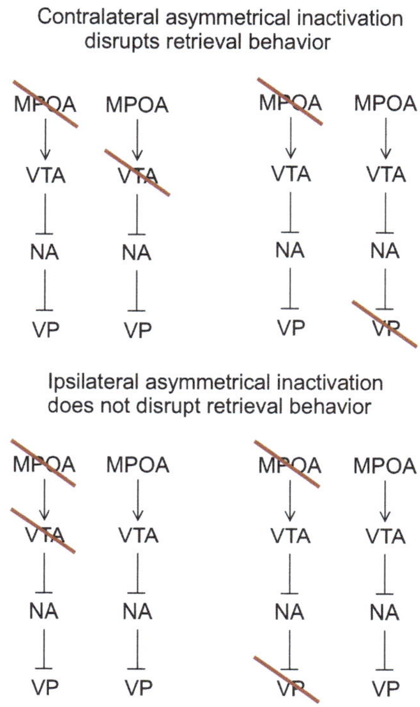

FIGURE 5.10 A medial preoptic area (MPOA)-to-ventral tegmental area (VTA)-to-nucleus accumbens (NA)-to-ventral pallidum (VP) circuit functions to regulate the appetitive aspects of maternal behavior in rats. The MPOA is proposed to stimulate VTA, which then depresses NA inhibitory projections to VP. This mechanism enhances the output of VP, which promotes appetitive maternal responses. The asymmetrical lesion design has been used in experiments to support this model. When unilateral MPOA and VTA lesions or unilateral MPOA and VP lesions are contralateral to one another (top panel), appetitive maternal responses (retrieval of pups) are disrupted. If such lesions are ipsilateral to one another (bottom panel), maternal behavior remains intact because the critical neural circuit is intact on one side of the brain. Axons ending in a bar are inhibitory, and those ending in an arrow are excitatory. *Modified from Figure 9 in Numan and Stolzenberg [714] with permission from Elsevier.*

In this research that has presented evidence for an MPOA-to-VTA-to-NA-to-VP circuit for appetitive maternal responses, retrieval behavior was used as the measure of appetitive maternal responses. Future research should be aimed at exploring whether MPOA interactions with the mesolimbic DA system are critical for MPOA-based operant responding when pups are used as the rewarding stimulus and for the MPOA-based expression of a conditioned place preference for a location that had been previously paired with pups.

As shown in Figure 5.8, OT projections from the PVN are proposed to play a pivotal role in modulating MPOA activation of the mesolimbic

DA system during the onset of maternal behavior. Not only is OT action on MPOA necessary for the onset of maternal behavior, as previously described, but microinjection of an OTR antagonist into the VTA also disrupts the onset of maternal behavior in parturient primiparous rats [747]. Therefore, OT action on the MPOA and VTA is essential for the onset of maternal behavior. Importantly, PVN OT neurons project to the VTA, where they act to stimulate DA release into NAs of postpartum rats [633,884]. As I have already reviewed, the PVN and OT neural systems are essential for the onset, but not the maintenance, of maternal behavior in rats, while the MPOA is essential for both phases of maternal behavior, as is the mesolimbic DA system. However, OT neural systems do play a modulatory role during the maintenance phase, increasing the mother's attention toward her pups. An interesting possibility is that MPOA glutamate stimulation of the VTA must co-act with PVN OT stimulation of VTA to cause enough DA release into NAs to stimulate the onset of maternal behavior. During the maintenance phase, MPOA glutamate stimulation of VTA-DA input to NAs can regulate a basic level of maternal responsiveness without the involvement of OT, but when OT is also active during this phase of maternal behavior, it boosts maternal responsiveness above this baseline level (see [884]).

There is also evidence that OT acts at the level of NAs to stimulate maternal behavior in prairie voles [396,472,720]. At present, there is no research available that has examined whether OT acts at the level of the NAs to stimulate the onset of maternal behavior in parturient rats. However, if the work on voles is applicable to rats, then OT would be acting at three critical nodes in the MPOA-to-VTA-to-NAs circuit to stimulate maternal behavior. In a larger context, although an essential role for OT in downregulating activity within the defensive circuit was basically conjecture, the research showing that OT acts at various nodes to activate the neural circuit regulating appetitive maternal responses is very strong.

An important aspect of the neural model shown in Figure 5.8 is that the BMA/BLA region of the amygdala is shown as relaying pup-related sensory inputs to both NAs and VP. Neuroanatomy indicates that all forms of sensory inputs can reach BMA/BLA and that BMA/BLA inputs to NA and VP are primarily ipsilateral [757,759,760,952]. Importantly, it has been reported that both electrical lesions and neuron-specific inactivation of the BMA/BLA region disrupts maternal behavior in postpartum rats [535,699]. Numan et al. [699] used muscimol to cause a neuron-specific inactivation of BMA/BLA, which would spare fibers of passage. Muscimol is a GABA-A receptor agonist that increases Cl^- conductance, which causes neural inhibition (hyperpolarization) at the level of dendrites and cell bodies. Bilateral muscimol injections into BMA/BLA caused a severe disruption of retrieval behavior, while leaving the consummatory aspect of maternal behavior, nursing behavior, relatively unaffected.

In order to test the proposal that BMA/BLA neurons interact with VP neurons to control appetitive maternal responses (Numan, unpublished findings), primigravid female rats were implanted with bilateral injection cannulas, one of which was located in the VP on one side of the brain, while the other was located in the BMA/BLA on the other side of the brain. The locations of the cannulas are shown in Figure 5.11. During the postpartum period, females received one of the following: control contralateral injections of saline into BMA/BLA on one side of the brain and into the VP on the opposite side; unilateral injections of muscimol into either the VP or BMA/BLA; or contralateral injections of muscimol into BMA/BLA on one side of the brain and into the VP on the opposite side. Only those females with inactivation of both the BMA/BLA and VP, on opposite sides of the brain, showed a severe retrieval deficit. Nursing behavior was relatively normal in these females.

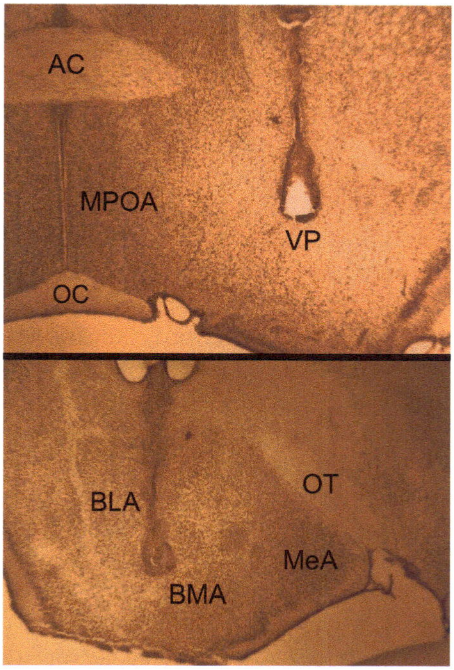

FIGURE 5.11 Photomicrographs of Nissl-stained frontal sections through the rat brain at the level of the ventral pallidum (VP; top panel) and the basolateral and basomedial amygdala (BLA, BMA; bottom panel). In support of the proposal that BLA and BMA neurons relay pup-related sensory inputs to VP, the output of which is essential for appetitive maternal responses, unpublished findings from the author's laboratory have shown that when muscimol (a GABA-A receptor agonist that inhibits neurons) is applied to the VP on one side of the brain (top panel) and to the BLA/BMA region on the contralateral side (bottom panel), retrieval behavior in postpartum rats is disrupted. Other abbreviations: AC = anterior commissure; MeA = medial amygdala; MPOA = medial preoptic area; OC = optic chiasm; OT = optic tract.

Such an inactivation would disrupt communication between BMA/BLA and VP on both sides of the brain, which distinguishes this contralateral muscimol group from the remaining three groups.

Since OTRs are located in the BMA/BLA region [197,992], it is interesting to speculate on an additional role for OT: its action at this site may play a role in regulating the functional routing of pup stimuli from BMA/BLA to VP in order to influence the onset and modulate the maintenance of maternal behavior (see [699]). As proposed in Figure 5.6(B) for MeA, perhaps OT action on OTRs in BMA allows pup stimuli to activate BMA neurons with a positive valence, while preventing pup stimuli from activating BMA neurons with a negative valence. The only major contrast between OTRs in BMA and those in MeA is that the expression of the latter receptors is estrogen dependent, while the expression of OTRs in BMA is not [511].

The last link in the chain to support the model shown in Figure 5.8 involves the proposal that medial PFC input to the MPOA may provide multimodal sensory inputs from pups that drive MPOA efferent neural activity to VTA. In support, Febo, Felix-Ortiz, and Johnson [286] have reported that inactivation of the medial prefrontal cortex through an intracerebral injection of a mixture of muscimol and baclofen disrupted retrieval behavior in postpartum rats (see also [10,755]). The injections were aimed at a region that would likely inactivate both the infralimbic and prelimbic medial PFC. To the extent that the muscimol/baclofen mixture depressed the activity of IL/PL projections to MPOA, the results would support the model in Figure 5.9. Note, however, that the IL and PL parts of the medial PFC have widespread projections, which also include projections to NA, VP, and BMA/BLA [994]. Therefore, the exact mechanism(s) through which medial PFC inactivation disrupts pup retrieval remains to be determined.

In Chapter 3, I presented evidence that IL cortex projections to the HAA may contribute to aggressive behavior. The HAA lies caudal or posterior to the MPOA. Therefore, while sensory inputs from IL cortex to MPOA may promote affiliative responses (maternal retrieval of pups), sensory inputs from IL cortex to HAA may promote antisocial aggressive responses. This example may again represent a case where differential sensory processing over distinct neural circuits can lead to either prosocial or antisocial behaviors (cf. [128]).

Although the involvement of the medial PFC in the maternal responsiveness of rats has been explored, the potential role of the orbital PFC has not been carefully investigated (cf. [294]). It is certainly possible that orbital PFC interactions with the BMA/BLA region of the amygdala modulate maternal responsiveness by affecting BMA/BLA input to the NA-VP circuit. The main point is that, as in aggression, PFC interactions with the hypothalamus and amygdala may exert an important modulatory control over maternal behavior.

5.2.8 The Mesolimbic DA System and Active Avoidance of Pups by Virgin Females

In Chapter 1, I presented research showing that the mesolimbic DA system is not only involved in appetitive reward-seeking behaviors but is also involved in goal-directed avoidance behaviors (see Figure 1.12). A labeled line point of view was presented where certain outputs from the NA-VP system would promote appetitive responses, while other distinct outputs would promote avoidance (or attack; see Chapter 3), and that VTA-DA input to NA could promote either response depending on the particular NA-VP circuits that are active in a particular situation. I have just reviewed the evidence that DA input to NAs, activated by MPOA efferents to VTA, promotes appetitive goal-directed maternal responses. The question for this section is whether evidence exists that the mesolimbic DA system is also engaged when naïve virgin females are actively avoiding pups.

For heuristic purposes, Figure 5.12 proposes circuits that may be active in either maternal rats (Figure 5.12(A)) or naïve virgins (Figure 5.12(B)) in response to pup stimuli. In Figure 5.12(A), the only additional anatomical information needed to understand the circuits is that MeA projects within the amygdala to influence activity in the BMA/BLA region [154]. This part of the figure proposes that in maternal rats, pup stimuli activate amygdala neurons (both within MeA and BMA/BLA) with a positive valence and that these amygdala neurons project to regions of NA and VP that are involved in appetitive reward-seeking behaviors. DA release into NAs, directed by the MPOA, facilitates the activation of VP neurons with a positive valence so that goal-directed maternal responses, such as pup retrieval, occur.

For Figure 5.12(B), the only additional anatomical information needed to understand the circuits is that PAG glutamatergic neurons project to VTA [328], which would be one route through which aversive stimuli could activate DA release into NA. The figure proposes that in naïve virgins, pup stimuli activate MeA and BMA/BLA neurons with a negative valence. One route activates consummatory escape and behavioral immobility responses (MeA-AHN/VMN-PAG-brainstem and spinal cord). Another route would activate goal-directed avoidance responses and would involve amygdala projections to NA-VP circuits that respond to aversive stimuli. DA release into NA, directed by PAG, would facilitate such proactive avoidance responses.

In examining both parts of Figure 5.12, note that the neural models give rise to either prosocial or antisocial responses because pup stimuli are routed over different circuits, depending on the physiological state of the female. OT and E actions at the level of the amygdala may be involved in shifting pup stimuli away from activating avoidance circuits and toward the activation of appetitive circuits (see Figure 5.6).

With respect to these proposals, evidence has already been provided that DA is released into NAs during maternal behavior. Is there evidence that DA is released into the NA in naïve virgin females that are exposed to pups? Afonso, King, Chatterjee, and Fleming [8] performed a relevant study. Postpartum maternal or naïve virgin female rats were placed in a chamber with a microdialysis probe implanted into the

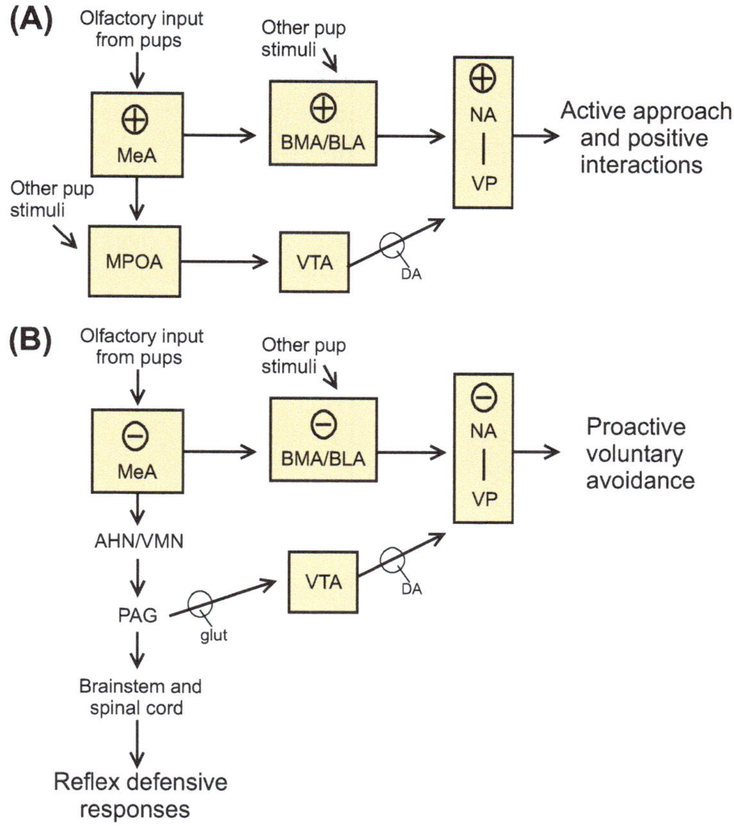

FIGURE 5.12 Interactions between the amygdala and the nucleus accumbens-ventral pallidum circuit (NA-VP) may contribute to appetitive maternal responses in postpartum rats (A) and to active avoidance responses to pup stimuli in virgin females (B), depending upon the particular neural circuits that are active. In postpartum rats, medial preoptic area (MPOA) activation of dopamine (DA) release into NA promotes the processing of pup stimuli across positively valent amygdala-NA-VP circuits. In virgin females, pup stimuli are processed over negatively valent amygdala circuits to result in reflexive defensive responses mediated by periaqueductal gray (PAG) projections to the lower brain stem and spinal cord. In addition, since PAG may also stimulate VTA-DA input to NA, it may promote proactive avoidance responses by favoring the processing of pup stimuli over negatively valent amygdala-NA-VP circuits. Other abbreviations: AHN = anterior hypothalamic nucleus; BLA = basolateral amygdala; BMA = basomedial amygdala; glut = glutamate; MeA = medial amygdala; VMN = ventromedial nucleus.

NAs. Baseline levels of DA release into the NAs were measured, and then the females were presented with pups and DA levels were measured over a 24-min period. The diameter of the chamber or cage where microdialysis was performed was small (about 10 inches). During the pup presentation period, postpartum mothers retrieved and nursed the pups within the chamber. In contrast, virgin females briefly sniffed and then remained away from the pups. The exact behavioral responses exhibited by the virgins during pup exposure were not provided. It would have been interesting to know if such females were active or immobile while in the presence of pups. The extracellular DA results indicated that basal DA levels within NA were significantly higher in the virgin females. However, once the pups were presented, DA levels in the postpartum females increased, while the DA level did not change in the virgins. Therefore, DA was released into NA in response to pup stimulation in the maternal females, but not in the virgins. Interestingly, however, in terms of the absolute amount to extracellular DA, the increased level of DA in the maternal females approximated the baseline level that was present in virgins.

These findings are difficult to interpret. Perhaps the high baseline level of DA in NAs acted on those circuits that promoted avoidance behavior in the virgins once they were exposed to pups, while a similar level of DA release in response to pups promoted maternal behavior in the postpartum females through an action on appetitive NA circuits. Several other interpretations are also possible: the increased signal-to-noise ratio with respect to DA levels in postpartum females (pup/baseline levels) may have been important for the occurrence of maternal behavior; DA may not have increased in NAs in response to pups in the virgins because in the small diameter chamber it was impossible for the virgins to actively avoid pups or show goal-directed avoidance responses; and, since research has shown that VTA-DA neurons are inhibited when aversive stimuli cause behavioral immobility or freezing [456], it is possible that for the virgin females in the study by Afonso et al. [8], the MeA-AHN/VMN-PAG-brainstem and spinal cord pathway might have been much more active than the MeA-AHN/VMN-PAG-VTA pathway in the presence of pups, resulting in behavioral immobility and an absence of a pup-induced increase in NA DA levels.

5.2.9 Changes in the Function of MPOA across the Postpartum Period

The maternal behavior of rats varies over the postpartum period [832]. Early in the postpartum period, up to day 10, pups that have been displaced from the nest are quickly retrieved and then nursed. During the mid-postpartum period (days 12–21), displaced pups are less likely to be retrieved, although nursing behavior remains high. Subsequently, nursing behavior also declines and weaning occurs by day 28 postpartum. A functional explanation of the decline in retrieval behavior after day 10 postpartum is that older pups are mobile enough to crawl to the nest area and do not require maternal transport. Therefore, young pups appear to activate retrieval while older pups do not. It is also interesting to note that if female rats are kept with young pups across the postpartum period (once a given litter reaches 10 days of age, it is replaced with a younger litter), they continue to show high levels of both retrieving and nursing for long periods of time [695].

Earlier in this chapter, I indicated that the MPOA is involved in both the appetitive and consummatory aspects of maternal behavior, since MPOA lesions depress both retrieving and nursing. Based on the course of maternal behavior over the postpartum period, as pups advance in age, one might conclude that young pups are able to activate both appetitive and consummatory MPOA mechanisms, older pups activate only consummatory processes, and even older pups activate neither mechanism, resulting in weaning.

In an interesting study, Pereira and Morrell [753] examined the effects of inactivation of the MPOA on the maternal behavior of female rats on either day 7 or day 14 postpartum. In order to inactivate the MPOA, they used local bupivacaine injections. Bupivacaine is similar to lidocaine and is a drug that blocks voltage-gated Na^+ channels and therefore blocks the production of action potentials. This fact is relevant because it indicates that bupivacaine would not only block MPOA neuron activity but would also block the activity of axons passing through the MPOA (axons of passage), but having their origins (cell bodies) elsewhere. Pereira and Morrell reported that bupivacaine injections into the MPOA disrupted retrieval behavior in rats when the injections were performed on day 7 postpartum, but that the injections actually stimulated retrieval behavior when they were performed on day 14 postpartum, at a time when such females typically do not retrieve their mobile pups. Pereira and Morrell [753] proposed that the functional role of the MPOA with respect to maternal behavior changes over the course of the postpartum period (cf. [756]).

These results are difficult to interpret. It is possible that bupivacaine primarily inhibits those neurons that are most active at the time of its injection. If fibers of passage through the MPOA, with origins and terminations distinct from MPOA neurons, were most active in day 14 postpartum females, and if such activity served to inhibit retrieval behavior, then retrieval might have been facilitated in the late postpartum rats not because of the depression of MPOA neurons but because of depression of an inhibitory system whose axons pass through the MPOA. Therefore, the possibility that the direct inactivation of MPOA neurons in the mid-postpartum period increases appetitive maternal responses will need to be confirmed through the use of methods that are neuron (cell body) specific and do not disrupt axons of passage, such as lesion production with excitotoxic amino acids or neuropharmacological inactivation with agents such as muscimol and/or baclofen. That the results obtained with bupivacaine are questionable relates to the finding by Pereira and Morrell [753] that bupivicaine injections aimed at the lateral preoptic area and ventral pallidum did not interfere with maternal behavior during the early postpartum period, while it is well known that excitotoxic amino acid lesions and muscimol inactivation of these regions do disrupt maternal behavior in such females [701,708].

A related explanation of the Pereira and Morrell [753] results arises from a recent finding by Smith, Holschbach, Olsewicz, and Lonstein [913] that medullary norepinephrine (NE) input to the dorsal MPOA region serves to depress retrieval behavior in postpartum rats, presumably by inhibiting MPOA neurons. Perhaps such NE input to MPOA is normally active in day 14 postpartum females and operates to depress the retrieval of older pups. Further, bupivacaine may have acted to selectively depress such NE input, so that older pups could be retrieved. Therefore, it is premature to conclude that bupivacaine injections into MPOA stimulate a resumption of retrieval behavior in day 14 postpartum females as result of depressing the activity of MPOA neurons. The observed effects could have resulted from depressing the activity of axons of passage through the MPOA or from depressing the activity of particular types of afferent inputs to the MPOA region.

5.2.10 Neural Regulation of the Consummatory Aspects of Maternal Behavior in Postpartum Rats

The PAG is involved in diverse functions that regulate a variety of reflexive-like behaviors via its projections to the lower brainstem, and such diversity is likely to be mediated by different PAG circuits (see Figure 5.13(A)). Research by Lonstein and Stern [566,567] has shown that the ventrolateral part of the PAG (vlPAG) is involved in the immobile arched-back crouch nursing posture referred to as kyphosis, which is a consummatory component of maternal behavior. Fos expression is activated in the vlPAG as a result of suckling stimulation, and vlPAG lesions abolish kyphosis, while leaving other aspects of maternal behavior intact. vlPAG-lesioned females will retrieve pups, group them in a nest area, and hover over them while licking them, but they will not exhibit the highly quiescent and immobile arched-back crouching response in response to pup suckling, which optimizes milk transfer [925]. Since the MPOA projects to the vlPAG (see Figure 5.5), this may be the route over which the MPOA influences some of the consummatory aspects of maternal behavior [921]. Therefore, while MPOA projections to VTA-DA neurons stimulate proactive appetitive

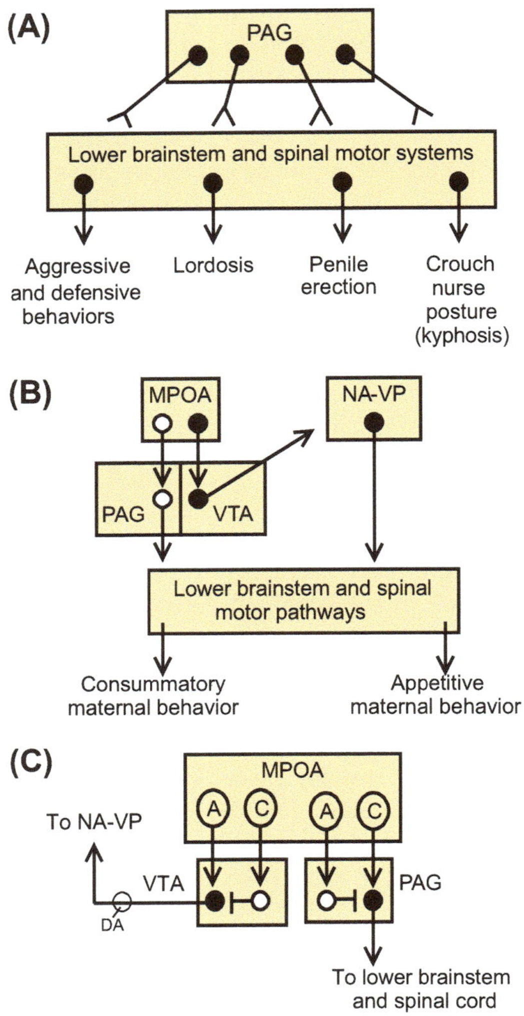

FIGURE 5.13 (A) The periaqueductal gray (PAG) controls a variety of reflex-like responses through its descending projections to lower parts of the brainstem, which include important terminations in the medullary reticular formation (MRF). MRF projection neurons, in turn, regulate the output of cranial and spinal motor neurons. Different PAG circuits to different parts of the MRF regulate different responses, some of which are active, while others are immobile responses. (B) A diagram showing how different populations of medial preoptic area (MPOA) neurons may regulate either consummatory or appetitive maternal responses. MPOA projections to the periaqueductal gray (PAG) are shown as influencing consummatory responses (cell bodies with open circles), while MPOA projections to the mesolimbic dopamine system, involving ventral tegmental area (VTA) projections to the nucleus accumbens-ventral pallidum circuit (NA-VP), influence appetitive responses (cell bodies with closed circles). Note the similarities between these circuits and those that were proposed to regulate the consummatory and appetitive aspects of male sexual behavior as indicated in Figure 4.2 in the previous chapter. (C) Consummatory (C) and appetitive (A) medial preoptic area (MPOA) neurons may serve dual functions. Consummatory MPOA neurons may stimulate ventrolateral periaqueductal gray (PAG) projection neurons while also inhibiting the output of ventral tegmental area (VTA) dopamine (DA) neurons to the nucleus accumbens-ventral pallidum circuit (NA-VP). In contrast, appetitive MPOA neurons may stimulate the output of VTA-DA neurons to NA-VP while inhibiting the output of the PAG. Axons ending in a bar are inhibitory, and those ending in an arrow are excitatory. VTA-DA axons are shown with an arrow to indicate that the output of VP is stimulated.

maternal responses, MPOA projections to vlPAG may stimulate kyphosis. It is even conceivable that a population of MPOA efferents that activate the vlPAG may also inhibit VTA-DA neurons in order to promote immobile quiescent nursing (cf. [961]), while another population of MPOA neurons may inhibit the vlPAG and activate VTA-DA neurons, which terminates quiescent nursing and leads to more attentive pup-directed appetitive responses. Some of these ideas are portrayed in Figure 5.13(B) and (C).

5.2.11 The Specificity of MPOA Neurons for Maternal Behavior

In comparing the research on male and female sexual behaviors presented in Chapter 4 with the maternal behavior research presented in this chapter, Figure 5.14 presents an integrative summary. For both male sexual behavior and female maternal behavior, it appears that the MPOA contains separate populations of neurons that regulate either the appetitive or consummatory aspects of each behavior (see Figure 4.2). In contrast, for female sexual behavior, the MPOA regulates appetitive responses while the VMN controls the consummatory lordosis response (see Figure 4.5). Existing evidence also indicates that the PAG is an important hypothalamic

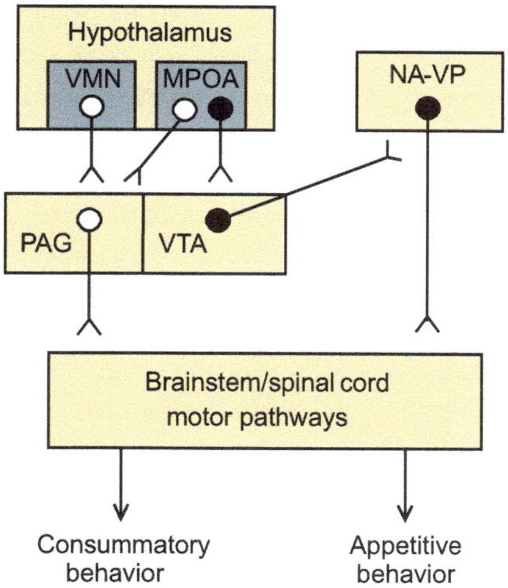

FIGURE 5.14 A summary diagram that integrates research findings on the neural regulation of maternal behavior (Figure 5.13(B)), male sexual behavior (Figure 4.2), and female sexual behavior (Figure 4.5) in rodents. Circuits that regulate the appetitive aspects of each of these reproductive behaviors are indicated with black neuronal cell bodies. Neurons contributing to consummatory circuits are shown with open (white) cell bodies. Populations of medial preoptic neurons (MPOA) that interact with the mesolimbic dopamine system appear to be involved in the appetitive aspects of maternal behavior, and male and female sexual behaviors. For maternal behavior and male sexual behavior, additional groups of MPOA neurons with projections to the periaqueductal gray (PAG) regulate the consummatory responses associated with these two behaviors. For the consummatory aspects of female sexual behavior, neurons in the ventromedial nucleus of the hypothalamus (VMN) with projections to the PAG are involved. Other abbreviations: NA-VP = nucleus accumbens-ventral pallidum circuit; VTA = ventral tegmental area. *Modified from Figure 1 in Stolzenberg and Numan [929] with permission from Elsevier.*

target involved in the regulation of the consummatory aspects of male and female mating behaviors and female maternal behavior. With respect to the appetitive (reward seeking) aspects of maternal behavior and male and female sexual behaviors, depression of MPOA activity disrupts each of these. A case can be made that the MPOA regulates the appetitive aspects of all three reproductive behaviors via interactions with the mesolimbic DA system, although the strongest evidence on this point is derived from research on maternal behavior [929]. The most important conclusion from these data is that the MPOA is critically involved in a broad range of reward-seeking affiliative social responses related to reproduction.

In this context, the issue of the specificity of the MPOA, and its connections with the mesolimbic DA system, for appetitive maternal motivation can be questioned [694]. Newman [683] has proposed that a general social behavior neural network, consisting of an integrated group of hypothalamic and limbic nuclei, regulates a variety of social behaviors, such as maternal behavior and sexual behaviors. Although a similar group of brain nuclei may be involved in a variety of reproductive behaviors, an important issue is whether the same population of neurons within this network regulates all of the basic social behaviors, or whether there are distinct populations of neurons intermixed within the overall network, each of which is selectively involved in a particular behavior. Although current research, particularly with respect to the MPOA, has not settled this issue, my bias is toward a labeled-line point of view, where *distinct* MPOA neurons regulate the appetitive aspects of maternal behavior, male sexual behavior, or female sexual behavior (see [719]).

An important behavioral study by Agrati, Fernandez-Guasti, and Ferreira [12] fits with the proposal that separate and distinct MPOA neurons regulate the appetitive aspects of female sexual motivation and maternal motivation in rats. As background, estrous cycling female rats that are induced to show maternal behavior as a result of pup-induced sensitization will also engage in sexual behavior during behavioral estrus. In addition, although lactating female rats generally do not engage in sexual behavior, a postpartum estrus does occur on day 1 postpartum. Therefore, on the first day of lactation females are capable of showing both maternal behavior and female sexual behavior. During periods when both female sexual behavior and maternal behavior were able to occur, Agrati et al. [12] examined the preference of the female for pups or for a male rat using a three-point star-shaped maze, with each chamber containing either pups, a sexually active male, or no stimulus, respectively. This would be a test of reward-seeking or appetitive motivation. Sensitized maternal females in behavioral estrus preferred males to pups, suggesting a stronger attraction to males than to pups. In contrast, parturient females (who had given birth before) preferred pups to males during their postpartum estrus. These behavioral findings seem to be more consistent with separate neural populations regulating maternal motivation and female sexual motivation, although conclusions with respect to the MPOA cannot be made.

To get at the issue of MPOA specificity, note that Fos is expressed in the MPOA during both maternal behavior [615,921] and female sexual behavior [203]. Anatomical methods are available to determine whether such Fos expression occurs in the same or different MPOA neurons. Immediately after engaging in a particular behavior, Fos mRNA is expressed primarily in the nucleus of cells, while 35 min later it is expressed mainly in the cytoplasm, where translation of the message into protein occurs (see [556]). The following experiment could take advantage of this temporal change. During the postpartum estrus, first allow females to actively seek out a sexually active male for 5 min. Thirty minutes later, allow the females to actively seek out and retrieve pups for 5 min, and then process their brains to localize Fos mRNA. If separate populations of MPOA neurons are involved in each behavior, then some MPOA neurons should show nuclear Fos mRNA ("maternal neurons"), while a separate population should exhibit a cytoplasmic localization of the Fos mRNA ("sexual neurons"). If the same population of MPOA neurons regulates both appetitive behaviors, then most MPOA neurons should contain both nuclear and cytoplasmic Fos mRNA. Such an experiment will help us understand how the brain controls motivational specificity and will allow us to determine whether specific MPOA neurons are involved in specific motivational states or whether the MPOA plays a more general and nonspecific role in reproductive motivation, with the specificity of reproductive behavior being determined by mechanisms that do not require distinct MPOA neurons. In either event, the MPOA appears to be importantly involved in the regulation of social attraction between individuals, and it is possible that this process extends beyond the context of reproduction.

5.2.12 The Maintenance of Maternal Behavior, Maternal Memory, and Plasticity within Maternal Circuits

Although estradiol, prolactin, and OT are essential for the immediate onset of maternal behavior at parturition in primiparous rats, they are no longer necessary for the maintenance of the mother's attraction to her young during the remainder of the postpartum period. Therefore, hormonal and neuropeptide action on the brain, *in conjunction with an initial maternal experience with pups*, causes a persistent modification of brain function that results in an enduring mother–infant bond that is maintained in the absence of an essential hormonal and oxytocinergic contribution [714].

The continuance of maternal behavior during the postpartum period without the need for hormonal and oxytocinergic mediation is analogous to a process referred to as maternal memory, which has been defined in the classic studies of Bridges [120,121,122] and Orpen and Fleming [732]. Maternal memory refers to the fact that although first-time primiparous parturient rats require hormonal and OT stimulation to initiate prompt maternal behavior, once a critical duration of mother–infant interaction (maternal experience) occurs, subsequent episodes of maternal behavior, even after long periods of mother–infant separation, become relatively emancipated from hormonal control. In the typical maternal memory experiment, one group of primiparous females is allowed to interact with pups for 1h immediately after birth and then the pups are removed, while a second group has its pups removed immediately after birth and is not allowed maternal experience. A retention test is conducted 10 days later, at which time all females are exposed to healthy young pups (obtained from "donor" mothers) on a daily basis until sensitized maternal behavior occurs. The parturient females that experienced the 1-hr maternal interaction typically exhibit sensitization latencies of 1–2 days, while the parturient females that were not allowed this postpartum maternal experience typically exhibit sensitization latencies of about 7 days, which does not distinguish them from naïve virgin females.

I view both the maintenance of maternal behavior and the process of maternal memory as representing the establishment of an enduring mother–infant bond that attracts a primiparous female rat toward a general pup stimulus. Therefore, the neural mechanisms and circuits underlying maternal behavior in rats are not fixed and inflexible, but can be modified by hormones and maternal experience in first-time mothers, resulting in a more efficient maternal circuit that can be activated by pup stimuli alone, resulting in a strong attraction of the mother toward the infant. But how might the brain be modified by a maternal experience so that a long-term attraction of a female toward infants develops? An important guiding principle is that the sites of neural plasticity should be located at points within the neural circuits that regulate maternal behavior: certain circuits that are active during the hormonally stimulated onset of maternal behavior might become strengthened by the maternal experience (experience-dependent plasticity), underpinning an enduring mother–infant bond. Although several potential mechanisms may be involved [58,703], I want to present some evidence in support of the hypothesis that a postpartum maternal experience modifies the brain in a way that allows the MPOA, in the absence of hormonal stimulation, to subsequently activate ventral pallidum output necessary for appetitive maternal responses. Figure 5.15 presents a potential model, which is based on the neural circuitry underlying appetitive maternal responses. In first-time mothers, hormones and OT act on the MPOA to facilitate the stimulation of MPOA efferents by pup stimuli. As a result of an initial experience, it is hypothesized that the synapse between BMA/BLA and VP becomes strengthened, allowing pup stimuli to more easily activate the VP. Such synaptic strengthening now allows a presumed weaker MPOA activation of the mesolimbic DA system (because of a lack of hormonal stimulation of MPOA) to disinhibit the VP sufficiently to allow for appetitive maternal responses. Such experienced females will quickly approach, interact with, and retrieve pups, and the nuzzling pups will then stimulate nursing-like postures.

To provide some support for this model, recent work has shown that DA and OT action at the level of NAs is necessary for maternal memory formation in primiparous rats [222,735]. In both of these studies, primiparous female rats were allowed 1h of postpartum maternal experience and then they were separated from their pups. Immediately after the 1-hr maternal experience, various neuropharmacological agents were microinjected into NAs. Ten days after the maternal experience, females were exposed to young pups on a daily basis and sensitization latencies were measured. Figure 5.16(A) shows an experiment that examined the effects of DA receptor antagonist injection into NAs on maternal memory formation [735], where it was found that blockade of *both* D1 and D2 receptors was the most effective treatment necessary to prevent maternal memory formation (also see [145]). Similarly, Figure 5.16(B) shows that microinjection of an OTR antagonist into NAs immediately after a maternal experience also blocked maternal memory formation [222]. Therefore, it appears that DA action on D1 and D2 receptors, co-acting with OT action on OTRs within NAs, contributes to maternal memory formation. Several points are worth noting. Recall that DA action on D1 receptors in NAs is sufficient

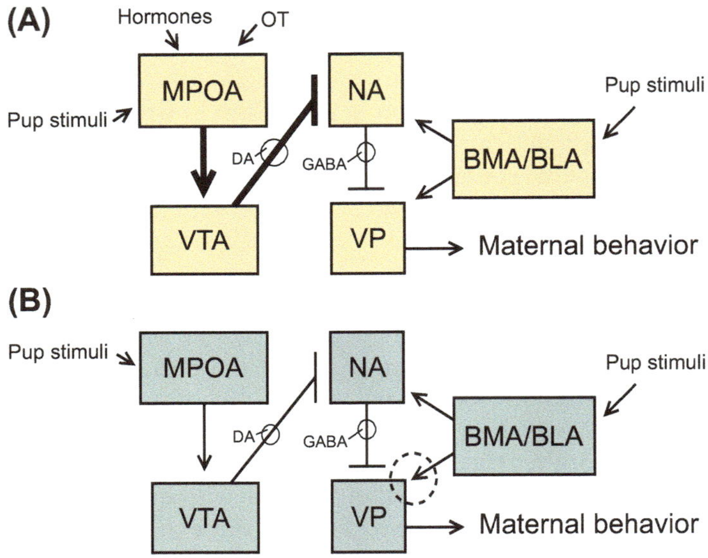

FIGURE 5.15 Proposed synaptic plasticity that may occur in postpartum rats to allow for the continuance of maternal behavior after its hormonal onset without the necessity of continued hormonal stimulation. Such plasticity may also represent one of the neural modifications that mediates maternal memory. (A) During an initial maternal experience at parturition in primiparous females, hormonal and oxytocin (OT) action in the medial preoptic area (MPOA) allows pup stimuli to strongly activate MPOA projections to ventral tegmental area (VTA) dopamine (DA) neurons that, in turn, connect to the nucleus accumbens (NA)–ventral pallidum (VP) circuit. Such strong activation is indicated by thick axons along the MPOA-to-VTA-to-NA circuit. Activation of these circuits allows pup stimuli to effectively stimulate basolateral (BLA) and basomedial (BMA) amygdala efferents to VP, leading to maternal behavior. (B) As a result of the initial neural activity across the circuits shown in (A), it is proposed that the synapse between BLA/BMA and VP is strengthened (dashed circle), which allows maternal behavior to continue in the absence of hormonal and OT action in MPOA, and in the absence of intense activity across the MPOA-VTA-NA circuit (a certain amount of activity across this circuit is still necessary for postpartum maternal behavior, but it does not need to be intense). For (A) and (B), axons ending in a bar are inhibitory, and those ending in an arrow are excitatory.

to stimulate the onset of maternal behavior in 15HO females, while DA action on D2 receptors is not [928]. Similarly, DA action on D1 receptors in NAs is necessary for ongoing maternal behavior over days 5–7 postpartum, while DA action on D2 receptors is not [707]. However, it does appear that DA action on either D1 or D2 receptors in NAs is necessary for maternal memory. These results indicate that DA mechanisms involved in the *synaptic plasticity* underlying maternal memory need not be exactly the same as those that regulate the onset and maintenance of maternal behavior. With respect to OT, OTRs are located in NAs [992] and PVN OT neurons project to NAs [502,836]. Given that OT action on NAs facilitates maternal memory, it would be interesting to know whether OT action at this site is also involved in stimulating the onset of maternal behavior at parturition.

A very basic model to explain the potential mechanisms of maternal memory is shown in Figure 5.17 (also see [696,714]). This figure is based on the research reviewed in Chapter 1 that the neuromodulatory effects of DA action on D2 receptors on NA medium spiny output neurons (MSNs) are primarily inhibitory, while the existence of D1 receptors on incoming glutamate axons from the amygdala to the NA would allow DA to exert a presynaptic inhibitory effect of those axon terminals. Finally, although the

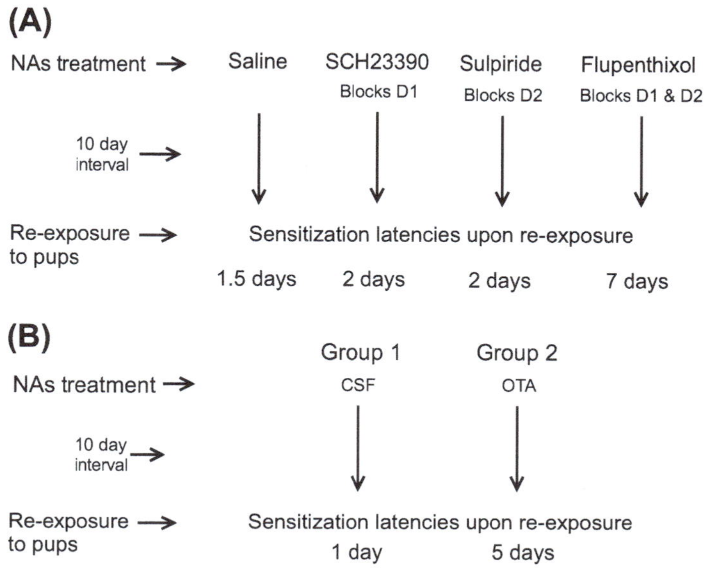

FIGURE 5.16 Dopamine action on D1 and D2 receptors (A), and oxytocin action on oxytocin receptors (B), in the nucleus accumbens shell (NAs) are each necessary for the establishment of maternal memory in rats. Primiparous rats are allowed 1 h of postpartum experience with their young, after which the pups are removed. Immediately thereafter, various neurochemicals are microinjected into NAs in different groups of rats. Ten days later, the rats are reexposed to pups, and sensitization latencies are measured. As shown in (A), blockade of both D1 and D2 receptors in NAs with flupenthixol disrupts maternal memory formation. As shown in (B), injection of an oxytocin receptor antagonist (OTA) into NAs also blocks maternal memory formation. Other abbreviations: CSF = artificial cerebrospinal fluid.

exact cellular locations of OTRs within NAs need to be clarified, a recent study has presented evidence that within the NA, D2 receptors and OTRs can form heteromers and that OT binding to such OTRs increases the affinity of DA to the associated D2 receptor, in this way potentiating the inhibitory effect of DA at this site [822]. The figure shows that during an initial maternal experience at parturition, DA action on D1 and D2 receptors and OT action on OTRs in NAs all act to depress MSN inhibitory output to VP. Such depression would allow for a superexcitation of BMA/BLA input to VP, which then strengthens that particular synapse via a Hebbian-like activity-dependent facilitation. Subsequently, when hormones are no longer acting at the level of MPOA, a weaker activation of pup-stimulated MPOA output to the mesolimbic DA system, in conjunction with DA action on D1 receptors in NAs, is enough to stimulate full maternal behavior because of the strengthened ability of pup stimuli to also activate the VP.

In this model, I emphasize the possibility of plasticity at the BMA/BLA-to-VP synapse. However, as reviewed in Numan and Insel [703], it is likely that maternal memory is mediated by synaptic plasticity and strengthening at multiple neural sites (also see [58]). A good example could be changes that occur in the MPOA as a result of maternal experience that would allow certain MPOA neurons to be more easily activated by pup stimuli in the absence of hormonal priming [931].

There are remarkable similarities between the empirical findings with respect to the neural basis of maternal memory in rats and the empirical findings on the neural basis of pair bond formation between mating partners in monogamous vole species. In Chapter 6, I will elaborate on these comparisons in order to make the case that the neural circuits underlying

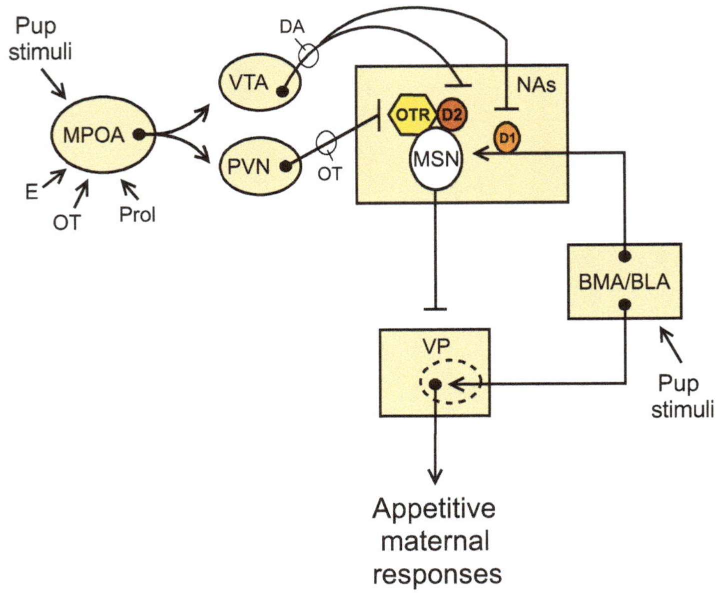

FIGURE 5.17 A neural model proposing how oxytocin (OT) and dopamine (DA) action on the nucleus accumbens shell (NAs), driven by medial preoptic area (MPOA) activity during an initial maternal experience, might promote synaptic strengthening (dashed circle) across synapses between the basomedial/basolateral amygdala (BMA/BLA) and the ventral pallidum (VP). This synaptic strengthening is proposed to be important for maternal memory formation (compare with Figure 5.15). An important underlying assumption of the model is that strong depression of NAs medium spiny neuron (MSN) inhibition of VP allows the VP to be superexcited by BMA/BLA inputs, which then strengthens the BMA/BLA-to-VP synapse. The MPOA is shown as activating DA and OT release into NAs. DA action on presynaptic D1 receptors depresses BMA/BLA excitation of MSNs. DA action of D2 receptors directly depresses MSNs that project to VP. D2 dopamine receptors may form heteromers with oxytocin receptors (OTR) on MSNs, and OT action at such OTRs appears to act as a neuromodulator, enhancing the inhibitory effects of DA at the associated D2 receptor. These combined effects should result in a profound depression of neural activity between NAs and VP. Axons ending in a bar have inhibitory effects, and those ending in an arrow are excitatory. Other abbreviations: E = estradiol; Prol = prolactin; PVN = paraventricular nucleus of the hypothalamus; VTA = ventral tegmental area.

maternal behavior, and the plasticity within this circuit that mediates the formation of an enduring mother–infant bond, may have provided the neural foundation from which other types of social bonds are constructed.

5.2.13 Serotonin and Maternal Behavior

In Chapter 3 on aggression, it was suggested that 5-HT neural systems, by depressing aggression and antisocial responsiveness, might indirectly promote the occurrence of prosocial behaviors. Serotonin neural systems may act at certain neural sites to shift social responsiveness away from aversion and rejection and toward acceptance and affiliation. From this analysis, it could be concluded that an interference with 5-HT action at particular neural sites might disrupt maternal behavior. In comparison to research on the involvement of DA neural systems in maternal behavior, not much research has been done on the role of 5-HT, but the evidence that does exist supports a positive role. In an early study on rats, Barofsky, Taylor, Tizabi, Kumar, and Jones-Quartey [62] microinjected a specific neurotoxin for serotonin neurons (5,7-dihydroxytryptamine) into either the dorsal or median raphe (DR, MR, respectively), which resulted in a partial destruction of 5-HT neurons in the injected area. While DR injections did not interfere with postpartum maternal behavior, MR lesions did. Females in the latter group exhibited decreases in nursing and retrieval behavior and increases in infanticide. More recent research that has *systemically* injected neuropharmacological agents that block central 5-HT receptors has also reported a disruption of postpartum maternal behavior in rats [1074].

Recall from Chapter 3 that resident males of mouse lines with a null mutation of either the *Pet-1* gene or the *Tph-2* gene are hyperaggressive in resident-intruder tests. Both of these mutations would result in a disruption of serotonin neural transmission, the former by disrupting the normal development of 5-HT neurons and the latter by preventing the synthesis of 5-HT within serotonin neural systems. Importantly, *Pet-1* −/− and *Tph-2* −/− dams show severe deficits in postpartum maternal behavior [16,542]: they are inattentive to their young and do not retrieve scattered young. In addition, the *Tph-2* −/− mothers showed an increase in infanticidal attacks on their pups. In the Alenina et al. [16] study, it was determined that the *Tph-2* −/− females were capable of lactating but exhibited poor maternal behavior toward their own and toward healthy foster young. Further, the offspring of *Tph-2* −/− dams were capable of being reared normally by foster mothers. These studies indicate an inverse relationship between offensive aggression, as measured in males, and maternal behavior, as measured in females, in each of these mutant strains.

All of these studies point to the involvement of 5-HT neural systems in maternal behavior in rodents. The genetic studies are difficult to interpret because 5-HT plays such an important role in the neural development of diverse brain systems [544]. Further, none of these studies offers any insight into 5-HT's neural site and mechanism of action. One possibility, as suggested above, is that 5-HT might be an additional factor that decreases aversive and rejection responses toward novel pups, which then permits pup stimuli to activate neural systems that regulate prosocial acceptance responses. In this context, Yoshida et al. [1058] have presented evidence that MR 5-HT axon terminals in MeA and BMA may contain presynaptic OTRs and that OT may act in MeA and BMA to promote 5-HT release. Perhaps OT and 5-HT interactions in MeA and BMA play a role in allowing novel pup stimuli to gain access to positively valent neural systems, while depressing the access of pup stimuli to negatively valent neural systems.

5.3 NEURAL SYSTEMS AND MATERNAL BEHAVIOR IN SHEEP: MATERNAL RESPONSIVENESS AND THE FORMATION OF SELECTIVE ATTACHMENTS

5.3.1 Introduction

In addition to rats, the other species on which much research on the neural basis of maternal behavior has been conducted is sheep [548,703,778]. Ewes give birth to precocial young in large herds composed of unrelated individuals. Due to seasonal breeding, many females give birth around the same time, which increases the likelihood that mobile lambs will seek to suckle from postpartum females other than their own mothers. Since it is not advantageous, in terms of individual reproductive success, for a mother to care for unrelated young, evolutionary forces have resulted in neural mechanisms that result in the formation of a selective attachment between a ewe and the lamb that she gives birth to: the mother learns the specific olfactory characteristics of her infant as a result of interactions near the time of parturition. As background, estrous cycling ewes reject all lambs with head butts and do not allow lambs near them. In contrast, the recently parturient ewe will accept *any* newborn lamb, she will approach it, lick it, and allow it to suckle, and while she is interacting with the lamb, she learns its olfactory characteristics. Subsequently, she will only accept and care for the lamb that she was exposed to at parturition, and she will reject the advances of all other alien young. Therefore, an enduring *and exclusive* mother–infant bond is formed through experience. Please note that under natural conditions, the lamb that a ewe is exposed to at parturition is her biological offspring, but experimental research shows that the ewe is initially responsive to any lamb. Figure 5.18 depicts the basic neural mechanisms of maternal behavior in sheep (compare with Figure 5.1 for rats).

Various testing paradigms have been employed to study maternal behavior in ewes. Allowing a lamb to nurse is a measure of consummatory maternal behavior. In order to measure the appetitive aspects of maternal behavior, the following separation–reunion paradigm has been used [549]: postpartum ewes are separated from their lambs, with the lambs being placed in an adjoining pen; subsequently, a gate is opened and the time it takes a mother to reunite with her lamb and the amount of time she spends with the lamb in the adjoining pen is recorded.

The following procedure is used to test for the establishment of a selective attachment between a mother and her lamb [474]: after a postpartum ewe has been with her lamb for several hours, the mother is separated from her lamb and then is presented in succession with her own lamb or an alien lamb (the order of lamb presentation is counterbalanced across the different postpartum ewes that are being tested). A selective bond or attachment is demonstrated when a ewe allows her own lamb to suckle while she head butts, rejects, and avoids the alien lamb.

With respect to the onset of maternal behavior at parturition, a time during which a ewe will accept any lamb, many of the same hormonal and neuropeptide events that trigger maternal behavior in rats are operative in sheep: declining progesterone levels, rising estradiol levels, and the central release of OT from PVN neurons [695,703].

Learning mechanisms within the main olfactory system, but not within the vomeronasal/accessory olfactory system, are essential for the formation of a selective maternal bond [703].

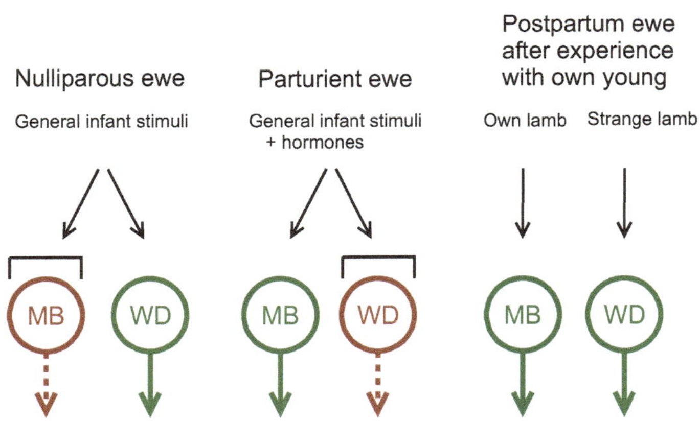

FIGURE 5.18 A simplified view of the brain mechanisms that mediate how ewes respond to lambs during different reproductive states. This model may be applicable to most female mammals that raise precocial young and that form selective bonds to the infants that they are exposed to at birth. Active brain systems are shown in green, and inactive systems are in red with a dashed axon. A bracket over the cell body of a neural system indicates that the system cannot be stimulated by afferent sensory inputs from lambs. For nulliparous ewes, a general lamb stimulus representing any lamb activates a withdrawal and rejection neural system (WD). Therefore, all lambs are rejected. In newly parturient ewes that are exposed to lambs for the first time, a general lamb stimulus representing any lamb activates the maternal behavior system (MB), and the WD system is not operative. Therefore, any lamb will be accepted. After the ewe interacts with a particular lamb for several hours after birth, an olfactory learning process occurs so that only this specific lamb activates the ewe's MB system, while unfamiliar lambs activate the WD system. *Modified from Figure 1 in Numan [697] with permission from Elsevier.*

In parturient ewes, interference with neural activity within the main olfactory bulb (MOB) does not disrupt the onset of maternal behavior but does disrupt the establishment of selective maternal bonds. Such anosmic ewes continue to care for any young and do not learn to discriminate between their own and alien young.

5.3.2 Brain Mechanisms Regulating Maternal Responsiveness and the Formation of a Selective Maternal Attachment

Early studies on the neural mechanisms underlying the onset of maternal behavior in parturient ewes had a primary focus on OT. ICV administration of OT was shown to stimulate maternal behavior in steroid-primed ovariectomized ewes [478]. Subsequent research indicated that under natural conditions, the vaginocervical stimulation that occurs during parturition triggers the release of OT from the PVN into multiple brain regions, including the MPOA, where OT acts to contribute to the onset of maternal responsiveness toward lambs [207,479,695]. As in the rat, although PVN OT neural systems are necessary for the onset of maternal behavior at parturition in sheep, they do not appear to be essential for the maintenance of established maternal behavior [207,692].

Recent research has focused on the role of the MPOA, and evidence indicates, as in rats, that the MPOA is crucial for both the onset and maintenance of maternal behavior in sheep. Perrin et al. [756] injected lidocaine, a drug that inhibits neural activity by blocking voltage-gated Na^+ channels, into the MPOA of parturient and postpartum ewes and reported that such ewes completely lost interest in their lambs. The MPOA lidocaine treatment disrupted both the appetitive and consummatory aspects of maternal behavior. The affected ewes did not nurse their young, and during the separation–reunion test, they did not approach and spend time with their lamb.

In comparison to rats, research into the neural mechanisms that mediate the rejection of all lambs by estrous cycling females that have not been primed by the hormones of pregnancy and by oxytocin has not been conducted. The neural circuits involved in lamb rejection might be similar to the avoidance/rejection pathways described in rats. Such circuits are also likely to mediate the rejection of alien lambs by postpartum ewes.

What about the neural mechanisms underlying the formation of a selective maternal attachment to the particular lamb that a mother is exposed to at parturition? Recall that the MOB is necessary for the formation of such selective attachments. In reference to Figure 5.4(A), note that the MOB has efferent connections to CoA and MeA. Keller, Perrin, Meurisse, Ferreira, and Levy [474] continuously infused lidocaine or a control solution (CSF) into the CoA and MeA of parturient and postpartum dams. Although lidocaine injections into these regions did not interfere with maternal responsiveness, they did interfere with the formation of a selective attachment. Such lidocaine-treated females showed maternal behavior toward both their own and alien lambs, while control females only cared for their own lamb.

In an attempt to integrate the rat research with the research on sheep, a hypothetical neural mechanism that might contribute to the formation of selective maternal attachments in sheep is presented in Figure 5.19. The figure indicates that in parturient ewes olfactory stimuli from any lamb activate MOB and MeA neurons that have a positive valence (neurons that respond to pleasant stimuli) and that the MeA activates MPOA maternal mechanisms. In addition, nonolfactory lamb stimuli of a general nature, through other neural circuits that might involve the PFC (but see [128]), also activate MPOA. Since anosmic ewes are maternal, as are females with lidocaine injections to the MeA, the sensory inputs to the MPOA from nonolfactory sources are presumed to be the main stimulus for maternal responsiveness in those cases. Once a normal ewe is allowed to interact with a specific

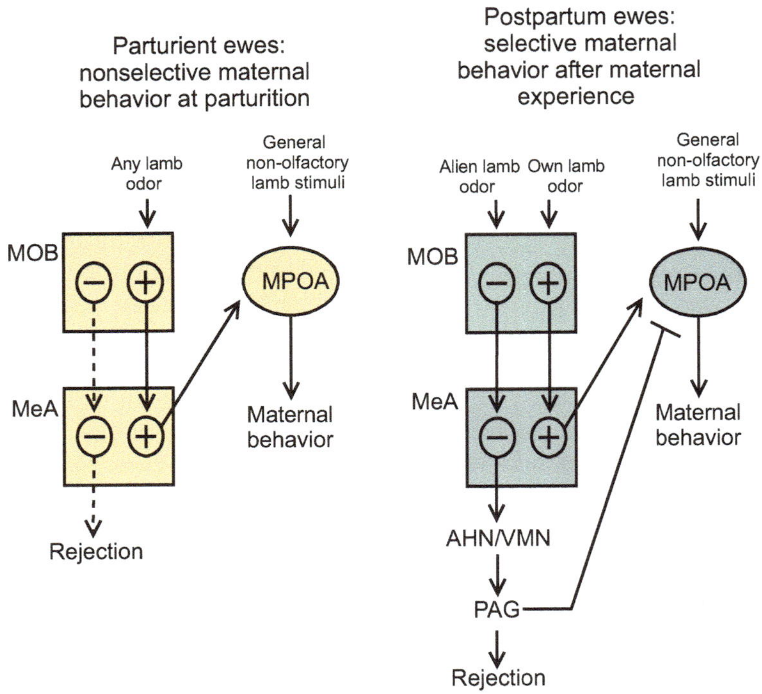

FIGURE 5.19 For maternal sheep, potential mechanisms that might mediate the differential neural processing of olfactory stimuli from lambs prior to (left panel) and after (right panel) the formation of a selective mother–infant bond. Circuits shown with a dashed line are not operating. Main olfactory bulb (MOB) and medial amygdala (MeA) neurons that have the potential to respond to aversive lamb odors are shown with a negative sign, while those that respond to pleasant lamb odors are shown with a positive sign. At parturition, a ewe will accept any lamb, and odors from any lamb activate positively valent MOB and MeA neurons that stimulate the medial preoptic area (MPOA) and maternal behavior. After a postpartum experience with a specific lamb, odors from this lamb (own lamb odor) activate positively valent MOB and MeA neurons that stimulate MPOA. Strange (alien) lamb odors now activate negatively valent MOB and MeA neurons that cause rejection via projections to the anterior hypothalamic nucleus (AHN), ventromedial nucleus (VMN), and periaqueductal gray (PAG). PAG is also shown as inhibiting MPOA. Axons ending in a bar are inhibitory, and those ending in an arrow are excitatory. See text for additional details.

lamb, however, learning mechanisms occur in the MOB (see [703]) and perhaps in CoA and MeA that narrow the range of lamb stimuli that are capable of activating prosocial circuits. The resultant neural plasticity results in only "own" lamb odors retaining the capability of activating MOB-to-MeA-to-MPOA neural circuits with a positive valence. In contrast, "alien" lambs now activate MOB-to-MeA neural circuits with a negative valence, and such negative MeA neurons are proposed to activate the AHN/VMN-to-PAG circuit previously described for rats. The activation of PAG by alien lamb odors results in rejection and withdrawal and is also proposed to inhibit the MPOA's ability to respond to general nonolfactory stimuli from young. In the case of lidocaine injections to the MeA, alien young would not be able to activate rejection and the PAG would not inhibit the MPOA, allowing the ewe to respond to general infant characteristics so that she would care for both her own and alien young. Although not indicated in the figure, this model would also propose that for estrous cycling ewes, lamb stimuli would only activate rejection neural circuits. Further, it is assumed that the MPOA influences appetitive maternal responses in sheep through its interactions with the mesolimbic DA, although this has yet to be investigated.

Assuming that the mechanisms that determine the enduring attachment of a ewe to its lamb are similar to the mechanisms mediating maternal memory in rats (see Figure 5.17), it is also possible that maternal selectivity in sheep results from the types of lamb stimuli that are relayed to the VP. Perhaps BMA/BLA neurons relay general lamb stimuli to the VP, while MeA relays own lamb olfactory stimuli to VP. With this understanding, interference with MOB and/or MeA connections would disrupt selectivity by preventing specific olfactory lamb stimuli from reaching MPOA and VP, while general lamb stimuli could still access these areas, which would result in a nonselective yet enduring attachment to general lamb stimuli. In other words, interfering with MOB-MeA learning mechanisms results in sheep behaving like rats in that they form an enduring attachment to a generic stimulus.

5.3.3 Discussion

It has often been argued that research on rats is not relevant to the neurobiological mechanisms that mediate the establishment of enduring mother–infant bonds or attachments [692]. However, such an argument is based on the assumption that the establishment of an enduring bond between a mother and her young requires selectivity. But I have already shown that the rat mother does form an enduring bond to general infant characteristics. It is simply not necessary, in terms of reproductive fitness, for a rat mother to learn the specific characteristics of her offspring. But such selectivity does have adaptive significance for sheep. Therefore, the rat forms a strong and enduring attachment to general pup stimuli, while sheep acquire a strong and enduring attachment to a specific lamb stimulus. The difference between the species relates to the types of stimuli that gain access to and activate maternal attraction circuits, and not to the nature of the attraction circuits themselves. Olfactory learning mechanisms may occur in sheep but not in rats. However, MPOA activation of the mesolimbic DA system, and modifications within the NA-VP circuit, are likely to be involved in the strong and enduring attraction of both sheep and rat mothers to their young. Therefore, if one were interested in the neural mechanisms that underpin strong and enduring social attractions, and not necessarily the selectivity of such attractions, then the rat model would be as good as the sheep model. An interference with, or dysfunction in, such maternal attraction mechanisms would disrupt maternal behavior in both rats and sheep. Based on Harlow's ideas [392], the mother–infant bond is sometimes referred to as the maternal affectional (love) system. One should not confuse the neural mechanisms that underlie love with those that regulate the nature of the stimuli that have access to the affectional system. The neural mechanisms underlying individual recognition interact with, but are distinct from, those that underlie an attraction toward a social stimulus.

5.4 MATERNAL BEHAVIOR IN NONHUMAN PRIMATES

There has not been much research on the neurobiology of maternal behavior in nonhuman primates, and most of this research, particularly with respect to neurochemistry, has been correlational in nature [851]. With respect to OT, Maestripieri, Hoffman, Anderson, Carter, and Higley [589] measured blood OT levels in a free-ranging social group of rhesus monkeys. They observed a strong positive correlation between the amount of time a mother nursed and groomed her infant (their measure of maternal warmth) and blood OT levels. Although this finding can be used to support the proposal that

OT action within the brain promotes maternal attentiveness and motivation in rhesus monkeys, as it does in rats, the following caveats put constraints on such an interpretation. Since suckling stimulation activates OT release from the posterior pituitary, the detected correlation may reflect causality in the opposite direction: increases in maternal behavior may have caused increases in plasma OT. This study also measured blood, rather than CSF levels of OT. Since the peripheral release of OT is not always correlated with its central release [467,529,695], one cannot determine if OT release and action within the brain was positively correlated with "maternal warmth".

A few studies have correlated central 5-HT activity with maternal behavior in lactating rhesus monkeys. There is interindividual variability in the maternal behavior of rhesus monkeys, and a significant proportion of mothers (about 5–10%) abuse their infants [587,590]. Infant abuse occurs in short bouts that are embedded within long periods of normal maternal behavior. This abuse, such as infant dragging, throwing, or crushing, usually occurs during the first 3–5 months of the infant's life, and is typically triggered by social stressors such as intragroup aggression. In addition to abusive behavior, which can severely harm the infant, there are also other variations in maternal behavior that reflect the level of maternal motivation. One of these is a measure of maternal rejection, defined as preventing an infant from making contact by pushing the infant away and by preventing or terminating infant suckling. Some studies have recorded a negative correlation between the level of infant abuse and/or rejection and the level of 5-HIAA in CSF [588,591], and this correlation has been interpreted as indicating that low central release of 5-HT correlates with abusive and rejecting maternal responses. As in rats, normal 5-HT levels might indirectly promote prosocial responses by depressing affective/impulsive aggression. Decreases in central 5-HT release may promote an overreactivity to social stressors, leading to infant abuse and rejection. I am not aware of any research that has attempted to reverse abusive parenting by injecting rhesus mothers with drugs that would enhance 5-HT neural transmission.

In an important study, McCormack, Newman, Higley, Maestripieri, and Sanchez [619] determined the 5-HTTLPR genotypes of abusive and nonabusive rhesus monkey mothers. They reported that the proportion of mothers carrying the short allele of the 5-HTT gene was significantly higher in abusive mothers than in controls. Recall from Chapter 3 that the short allele was proposed to result in a dysregulation in the development of 5-HT neural systems, particularly under conditions where the carrier of the short allele is also subjected to early adverse social experiences. In this regard, McCormack et al. [619] reported that although most nonabusing mothers did not carry the short allele, some of them did, suggesting the importance of a gene–environment interaction in the development of an abusive maternal genotype.

Two lesion studies support the involvement of the amygdala and PFC in the maternal behavior of rhesus monkeys, which is in accord with the results reviewed for rats. Toscano, Bauman, Mason, and Amaral [966] produced neuron-specific lesions of the amygdala or control lesions in neonatal female infant monkeys. The amygdala lesions included the basolateral/basomedial region. At approximately 3.5 years of age, these young, nulliparous, nonpregnant females were exposed to a mother–infant dyad. The control nulliparous females showed an interest in the newborn babies; they attempted to approach and touch the infants while producing affiliative vocalizations. In contrast, the nulliparous females with amygdala lesions were not interested in the infants, did not attempt to make contact with them, and showed a significantly lower level of affiliative vocalizations. This study suggests that the amygdala is involved in the attraction of rhesus females to infants that are not their own. The fact that the lesions were performed while the subjects were neonates, which might have produced effects different than if the lesions had been performed in adulthood, and that the effects of the lesions on natural postpartum maternal behavior were not examined, make an overall interpretation of these findings difficult.

Franzen and Myers [316] produced aspiration lesions of the medial and orbital PFC in postpartum rhesus monkeys, and such lesions caused drastic and long-term deficits in maternal behavior. The lesions were produced when the mothers' infants were 2 months old and were beginning to show some independence from complete maternal contact. In comparison to females with control lesions, the PFC-lesioned females exhibited a loss of protective retrieval of infants during threatening situations, increases in maternal withdrawal from infants, increases in infant punishment, and decreases in active infant cuddling. The lesions produced by Franzen and Myers were large and nonspecific and undoubtedly interfered with maternal behavior by disrupting many different functions (see Chapter 1). Based on the literature reviewed in this chapter and in the chapter on aggression, PFC lesions may have decreased maternal behavior by disrupting the proper functioning of the amygdala and by disrupting important inputs to the hypothalamus. That is, the effective lesions may have blocked the ability of multimodal infant stimuli from reaching the MPOA and may have caused a hyperreactivity of the mother to stressful stimuli.

Although research on the neurobiology of maternal behavior in nonhuman primates is quite limited, the research that does exist can be interpreted in ways that are consistent with the larger and more concrete findings obtained from rodents. In rodents and primates, OT, 5-HT, the BLA/BMA, and PFC appear to be positively involved in maternal responsiveness. I am not aware of any studies that have directly examined the involvement of the MPOA in nonhuman primate maternal behavior.

5.5 PATERNAL AND ALLOPARENTAL BEHAVIOR

5.5.1 Introduction

Although the hormonal induction of the onset of maternal behavior at parturition is the primary parental behavior mechanism in mammals, paternal behavior also occurs in about 5–10% of mammalian species, and alloparental behavior occurs in cooperatively breeding rodents (e.g., prairie voles) and primates (e.g., marmosets, tamarins). Since paternal behavior and alloparental behavior are not induced by the endocrine events associated with pregnancy termination, there must be alternate routes to allow infant stimuli to gain access to caregiving neural circuits. Although very little research has been done on the neural and neurochemical bases of paternal and alloparental behavior, I will review evidence that indicates that the neural mechanisms that underlie these caregiving parental responses are similar to those that mediate maternal behavior.

5.5.2 The Female Laboratory Mouse and Allomaternal Behavior

Recall that in most laboratory strains of mice (derived from *Mus musculus*, the house mouse), naïve nulliparous virgin females show prompt maternal behavior toward pups, which is sometimes referred to as spontaneous maternal behavior. Because of this fact, such females can serve as an important model to explore the alternate routes through which the proposed common parental neural circuitry can be activated without the involvement of pregnancy hormones. First, Fos is activated in the MPOA of virgin female mice that are showing maternal behavior [133,147,979]. Next, MPOA lesions disrupt maternal behavior in postpartum and virgin mice [979]. The maternal behavior shown by virgin female lab mice toward pups is not dependent upon estradiol, as ovariectomized mice and aromatase knockout mice show prompt maternal behavior toward pups [930,931]. A null mutation of the aromatase gene would block the synthesis of estradiol within the brain and other tissues. Interestingly, however, in a recent study, Ribeiro et al. [804] microinjected a viral vector containing an interfering RNA into the MPOA in order to block the synthesis of ER alpha at that site, and such treatment drastically suppressed maternal behavior in mice (also see [718]). These results indicate that the MPOA is necessary for the prompt maternal behavior shown by virgin female laboratory mice and that ER alpha may also be involved, although it may be activated by a ligand other than estradiol itself (see [714]).

Recall that prolactin action on MPOA at the end of pregnancy is one of the critical hormonal events essential for the onset of maternal behavior at parturition in rats. Since virgin lab mice show prompt maternal behavior outside the hormonal profile of pregnancy, it is not surprising that virgin female mice with a knockout mutation of the prolactin gene show spontaneous maternal behavior when presented with pups [424]. In contrast, however, Lucas, Ormandy, Binart, Bridges, and Kelly [577] have reported that nulliparous females of a mouse line with a null mutation of the prolactin *receptor* gene show significant deficits in maternal behavior. Analogous to the work with ER alpha, this study suggests that perhaps there is an endogenous central nervous system ligand, in addition to prolactin, that is capable of binding to central prolactin receptors in MPOA to stimulate maternal behavior in nulliparous mice.

OT action on the OTR at several brain sites is important for the onset of maternal behavior in parturient rats and sheep. Research on OT and OTR knockout (KO) mice has yielded results somewhat similar to those found with respect to mouse lines with a null mutation of either prolactin or its receptor. Virgin and postpartum OTKO mice show normal maternal behavior when tested in standard laboratory cages [687]. However, when tested under novel or stressful conditions, OTKO virgin mice show deficits in maternal behavior [749,793]. With respect to the OTR, when tested in standard laboratory cages, both virgin and parturient female OTRKO mice show major deficits in maternal behavior [957]. One interpretation of these results is that postsynaptic OTR activity is essential for the initiation of maternal behavior in both virgin and postpartum mice. Under nonstressful conditions, another ligand, for example, vasopressin, may effectively activate the OTR to allow for normal maternal behavior (see [687]). However, under conditions of stress and anxiety, it is likely that full activation of OTR by its natural ligand, OT, is necessary for the normal display of maternal behavior.

The purpose of this section was to provide examples of how parental responses might be regulated under conditions where the behavior occurs without the hormonal stimulus associated with pregnancy termination. Although feral virgin female mice are infanticidal rather than maternal [449], virgin laboratory mice are spontaneously maternal, and the examples presented above offer some insights into how selective breeding or inbreeding might have altered the brain to allow such parental responses to occur without concurrent pregnancy and lactation. For species that show alloparental behavior under natural conditions, natural selection may have operated in similar ways. Although the research reviewed here is sparse, it does suggest that a similar neural circuitry and neurochemistry underlies the natural onset of maternal behavior in rats and the spontaneous allomaternal behavior shown by virgin female laboratory mice. The MPOA, ER alpha, prolactin receptor, and the OTR appear to be involved in both cases. One mechanism through which evolutionary forces may have promoted the occurrence of natural allomothering in cooperatively breeding species, including early hominins, is by allowing a variety of alternative ligands to activate the critical neural receptor systems involved in maternal behavior

5.5.3 Paternal Behavior

Paternal behavior is rare in mammals, but existing evidence indicates that the same neural circuits that mediate maternal behavior also regulate paternal behavior. When natural paternal behavior occurs in a monogamous biparental species, where the male aids the female in raising young (the father can do everything that the mother does, except lactate), processes other than the endocrine events of pregnancy termination must prime the parental circuits. For both sexes in most mammals, it is likely that dual neural circuits control how an adult responds to infants. One circuit regulates defensive and withdrawal responses, and the other regulates parental responses. For the typical female mammal, the hormonal events associated with pregnancy termination inhibit the defensive circuit and activate the parental circuit. For the typical male mammal, the defensive circuit is always dominant, inhibiting the display of paternal behavior. In those species where paternal behavior occurs, some event or events other than pregnancy-related factors must activate the parental circuit and depress the defensive circuit so that paternal behavior occurs.

It is interesting to note that while virgin female laboratory mice are spontaneously maternal, male laboratory mice are highly infanticidal [703,866]. The defensive/rejection circuit is dominant in these males, and the activation of this circuit by pup stimuli is mediated by vomeronasal organ input to the MeA-AHN-PAG pathway [955]. Therefore, selective breeding and inbreeding appear to have suppressed this defensive mechanism in a sex-specific fashion so that only virgin female mice of laboratory strains show spontaneous parental behavior.

Under natural conditions, rats show a promiscuous or polygynous mating system and paternal behavior does not occur; it is only the mother that is involved in infant caretaking. Does this mean that the neural circuit for parental behavior does not exist in males? Research has conclusively shown that neural circuits for parental responses to infants exist in male rats, and the critical neural regions are the same as those that regulate maternal behavior: (1) male rats, just like female rats, show sensitized parental behavior [829]. Continuous exposure to pups induces parental responses in males after about 7–10 days. Such males will retrieve pups and hover over them; (2) if one treats male rats with a hormone regimen that simulates the endocrine events associated with pregnancy termination in females (rising estradiol and prolactin superimposed on progesterone withdrawal), sensitization latencies in such males are reduced to about 3 days [576,831,940]; (3) MPOA lesions disrupt sensitized and hormone-stimulated parental behavior in male rats [831,940]; (4) estradiol acts on the MPOA to stimulate parental responses in male rats [830].

What these results indicate is that the neural circuits that mediate parental responses to infants are present in male rats and are likely to be the same circuits as those that regulate maternal behavior. Paternal behavior does not occur under natural conditions in male rats (or in most other male mammals) because they are never exposed to the experimental conditions, such as continuous exposure to young or exposure to pregnancy hormones, that have been shown to activate their parental responses in the laboratory. However, the important point is that latent parental circuits probably exist in the brains of all male mammals, which would then allow natural selection to operate on these circuits to promote the occurrence of paternal behavior if such behavior had adaptive significance. In those species where ecological conditions necessitate a biparental care system for effective infant survival, natural selection would create a brain where factors other than the hormonal events associated with pregnancy termination, or continuous exposure to young over long periods of time, could activate the parental circuit in fathers.

With respect to biparental species where paternal behavior occurs naturally, important research has been done on the California mouse (*Peromyscus californicus*). Unmated males of this species are infanticidal, but after mating with a female, the male becomes paternal upon the birth of his young [232,972,973]. It is likely that both copulatory stimuli and pheromones from the pregnant and parturient female in some way prime the neural processes that promote paternal behavior [703]. Importantly, increased Fos expression occurs in the MPOA of paternal males of this species [232]. Further, many of the neural circuits involved in maternal behavior are involved in paternal behavior. Lee and Brown [531,532] have shown that MPOA lesions and BLA lesions disrupt both maternal and paternal behavior in California mice. Significantly, NA lesions do not disrupt parental behavior in this species, which maps on to the effects observed in rats, where NA is conceived as playing an inhibitory role in parental behavior by depressing the output of VP [708]. Unfortunately, the involvement of VP in the parental responses of California mice has not been examined.

If copulatory stimuli, cohabitating with a mating partner, and parturient female pheromones prime parental circuits so that paternal behavior occurs in California mice, what mechanisms are involved? Only partial answers to this question are available. One mechanism might involve a reduction in vomeronasal inhibition of paternal behavior (see [955]). Furthermore, Trainor, Bird, Alday, Schlinger, and Marler [971] found that aromatase activity is increased in the MPOA when paternal males are compared to unmated males. Since additional research has shown that testosterone stimulates paternal behavior in California mouse males via its conversion to estradiol [972,973], the following mechanism is possible: olfactory and other stimuli from parturient females may promote aromatase activity in MPOA of males. Increased aromatase function would then allow circulating testosterone to be converted to estradiol locally in MPOA, and estradiol action may then alter the phenotype and function of MPOA neurons so that they become responsive to pup stimuli, allowing for the occurrence of the immediate onset of paternal behavior once the young are born.

Trainor et al. [971] also provided evidence that progesterone, most likely secreted from the adrenal cortex, was higher in nonpaternal than paternal males, suggesting that progesterone may inhibit paternal behavior and that it must decline in order to allow for estradiol-stimulated paternal behavior. These results are interesting because progesterone has a clear inhibitory effect on maternal behavior in rats [709], and because males of a laboratory mouse line with a KO mutation of the progesterone receptor showed decreased infanticide and increased parental responses toward young pups [866].

Although estradiol may be important for paternal behavior in some species, evidence indicates that this is not a universal mechanism for paternal behavior [433,1046]. Although the priming mechanisms that activate parental circuits, allowing them to become responsive to infant cues, may differ by sex and by species, the underlying neural circuitry that regulates paternal behavior in mammals appears to be the same as that which regulates maternal behavior.

In my review of maternal behavior, I stressed the importance of OT action at several neural sites in the regulation of the onset of maternal behavior in rats and sheep. In the chapter on aggression, I reviewed the role of vasopressin neural pathways in the stimulation of male offensive aggression. However, there is also a large body of research implicating vasopressin neural pathways in the regulation of prosocial responses. Further, research has sometimes supported the view that OT is a neuropeptide that is primarily involved in female prosociality, while AVP is the dominant neuropeptide involved in male prosociality [1065,1067]. With respect to the prosocial and antisocial effects of vasopressin in male mammals, one would expect that distinct neural pathways would be involved in these opposing effects (see Chapter 3). As one example, since MeA and BST contain groups of AVP neurons, perhaps AVP projections to MPOA influence prosocial responses, such as paternal behavior, while AVP projections to AHN regulate male aggressive responses.

There is some support for the view that vasopressin neural pathways within the brain potentiate paternal behavior in monogamous species, such as prairie voles and California mice, where biparental care of young occurs under natural conditions, although the exact vasopressin neural pathways involved have not been clearly delineated [92,736,1017]. However, there is additional evidence that *both* AVP and OT neural pathways promote paternal behavior in male prairie voles [53]. Further, although most research on maternal behavior in rats has emphasized OT, research shows that AVP also acts to stimulate maternal behavior in rats [113]. It has sometimes been difficult to differentiate the effects of OT from AVP on social behavior because many of the agonists and antagonists that affect OTRs also affect the AVP V1aR [595]. Given the current state of knowledge, it is probably best to conclude, with respect to prosocial behaviors such as maternal and paternal behaviors, that both OT and AVP are involved in each sex, although their sites of action may be different.

5.5.4 Conclusions

The results reviewed above show that the neural circuits that regulate paternal behavior and alloparental behavior are likely to be the same as those that regulate maternal behavior. Significantly, parental caregiving circuits might also share similarities with the neural circuits that mediate other types of affiliative prosocial responses. I would like to provide some *behavioral* evidence to suggest that this might be the case.

In two important studies, the prosocial behaviors shown by two different species of primate have been compared. The term *unsolicited prosociality* has been used to refer to the likelihood that an individual will perform a response that allows a conspecific to obtain food even if the individual performing the response does not get any food [141]. When one compares chimpanzees [898] with marmosets [141], marmosets are much more likely to engage in unsolicited prosocial food sharing. The idea here is that marmosets, but not chimps, aid other members of their species even under conditions where the donor of the aid does not personally receive a benefit. In comparing the social organization of chimpanzees with marmosets, although chimpanzees live in troops composed of males and females, an exclusive uniparental maternal care system exists, and alloparental and paternal behaviors do not occur [427]. In contrast, marmosets are cooperative breeders with a monogamous mating system, and maternal behavior, paternal behavior, and alloparental behavior are all observed [317,427,703]. Based on the idea that marmosets exhibit a greater tendency toward mutual interdependence than chimpanzees, particularly with respect to raising offspring, perhaps this factor contributes to their greater unsolicited prosociality [141,898]. One very interesting possibility is that the parental care neural system has become a relatively open system in cooperative breeders, such that conspecific social stimuli from either infants or adults can gain access to the system so that caregiving behavior is evoked in the recipient of those stimuli. If this were indeed the case, then parental care neural circuits may have provided the neural foundation for other types of affiliative responses, such as the unsolicited food sharing described above. Hrdy [427] has argued that early hominins were likely to be cooperative breeders, which might account for the high level of prosociality in humans.

The above analysis is not meant to suggest that chimpanzees do not show certain types of prosocial behaviors. Chimps are highly cooperative socially, but such cooperation is typically based on reciprocity [141].

The following point is very important: the brain and cognitive abilities of the chimpanzee are vastly more developed in comparison to the small-brained marmoset. What this suggests is that prosociality and the tendency to provide care to members of one's species, irrespective of whether such care also aids the donor, may be less related to cognitive development and more related to whether evolutionary forces have allowed conspecific social stimuli to gain access to a caregiving neural system that may be importantly derived from core parental neural circuits [141,698].

5.6 DEVELOPMENTAL INFLUENCES ON THE MATERNAL BEHAVIOR OF MAMMALS

5.6.1 Introduction

Numan and Insel [703] have reviewed the literature on the intergenerational continuity of abnormal maternal behavior in humans (also see [87,1033]). This research indicates that children who have been abused (physically, emotionally, or sexually) or neglected (not cared for and protected) by their parents tend to become abusive or neglectful parents themselves. As indicated by Maestripieri [587], approximately 30% of abused children become abusive parents. Two aspects of these findings on humans are worth considering. First, these studies do not resolve whether the intergenerational continuity of abnormal maternal behavior is due to genetic inheritance, the early adverse effects of poor parenting on the child's brain development, or both. Second, being abused does not destine one to become an abusive parent, which suggests that gene–environment interactions are involved in the development of faulty maternal behavior.

In order to determine whether the quantity and quality of maternal care influences the development of maternal behavior in the offspring, I will first review experimental behavioral studies on nonhuman animals that clearly show that the way a mother cares for her offspring influences the nature of the maternal behavior shown by her female offspring in adulthood. This behavioral review will then be followed by an exploration of the underlying mediating mechanism for the described developmental effects.

5.6.2 Behavioral Studies on the Development of Maternal Behavior

5.6.2.1 The Seminal Work of Harlow

In a series of studies, Harlow and his collaborators [845] compared the adult maternal behavior of rhesus monkey females who had been reared in one of three ways: mother-reared (MR), peer-reared (PR), and social isolates (SIs). The basic rearing conditions of MR and PR females were described in Chapter 3, and each of these two groups had some type of social interactions with other monkeys during the first 6 months of their lives. In contrast, the SIs were reared in total social isolation (except for being fed by humans during the first month of life) until 6 months of age. In adulthood, the females from these groups mated with males or were artificially inseminated, and their subsequent maternal behavior toward their own offspring was examined. Using the MR females as the control baseline, the socially isolated females showed a severe and near-complete disruption in their maternal behavior, with 75% of such females either ignoring or abusing their offspring. Although the maternal behavior of PR females was better than that observed in the social isolates, it was still inferior to that of the MR females, with approximately 25% of PR females physically abusing their infants. Abusive females would not only prevent the infant from nursing but would also violently reject infant contact.

These results show that complete social deprivation has devastating effects on the development of maternal behavior in the affected offspring, and that selective maternal deprivation also has a significant impact. The latter finding indicates that there is a positive developmental effect of being reared by a normal mother that cannot be substituted for by peer–peer interactions, and it is this selective effect that I want to emphasize. If the absence of maternal care in PR females leads to an increased likelihood of abusive parenting when these females have their own offspring, what mechanisms might be involved? Several possibilities exist [703], and two will be presented here. First, being raised without a mother, or being raised by an abusive or neglectful mother, might interfere with the development of maternal motivation and behavior in the affected offspring by directly affecting the development of those neural systems that regulate maternal behavior, such as MPOA projections to VTA. Second, being raised by a neglectful/abusive parent might modify the neural circuits over which social stimuli are processed, resulting in an increase in the aversive consequences of social interactions and a decrease in the reward value of such interactions. That is, a result of poor parenting on an offspring's brain development might be that social stimuli become more likely to be processed over aversion-rejection neural pathways than over attraction-acceptance neural pathways. In support, recall from Chapter 3 that maternal deprivation or separation heightens affective aggression in primate and rodent offspring. In addition, in Chapter 2, data were reviewed that showed that lower levels of maternal care are associated with the development of enhanced levels of anxiety and stress reactivity. In fact, PR rhesus monkeys are more aggressive than MR monkeys, and perhaps such heightened aggression contributes to an increased tendency toward infant abuse in postpartum females. A recent study favors the view that social stimuli might be differentially processed in MR and PR rhesus monkeys [291]. The MR monkeys were raised in a social group with their mothers while the PR monkeys were raised without mothers but with a peer through 7 months of age. At 7 months, the PR monkeys were placed in a larger social group. After 3 years of living in this larger social group, the PR monkeys showed significant behavioral differences when compared to the MR monkeys. PR monkeys were much less likely to initiate an act of sitting next to another monkey, suggesting an aversion of social affiliation. Such monkeys also showed increased levels of stereotyped behaviors such as rocking, pacing, digit sucking, and self-grasping. These stereotyped behaviors are usually considered to be measures of anxiety and were virtually absent in MR monkeys. One conclusion is that PR monkeys found social living to be an aversive experience, resulting in avoidance of contact with other monkeys even after 3 years of living in a large social group.

PR monkeys are a model of maternal neglect, separation, or deprivation. But what impact does maternal abuse have on the development of maternal behavior in the offspring? In an important study, Maestripieri [587] asked the following questions: if an infant female rhesus monkey is abused by its mother, will she be more likely to abuse her own offspring and, if so, is this effect due to genetic or experiential transmission? Remember, within normal social groups of rhesus monkeys, about 10% of mothers physically abuse their infants, with such abuse occurring in short bouts that alternate with relatively normal maternal behavior. Abusive mothers tend to remain abusive across successive births, indicating that it is a stable maternal characteristic. In a cross-fostering study, multiparous females were utilized, which allowed the maternal abuse status of each mother to be known ahead of time: Maestripieri [587] raised infant rhesus females with a known abusive foster mother or with a known nonabusive control foster mother. The biological mothers of about half of the infants examined in this study were known abusers, while the remaining biological mothers were known to be nonabusive mothers. The cross-fostered infants subsequently grew into adulthood, and their maternal behavior toward their own offspring was examined. Infants that were born to abusive mothers but were cross-fostered at birth to control mothers did not abuse their own offspring. In contrast, 50% of the infants that were born to control mothers but cross-fostered to abusive mothers were observed to abuse their own offspring. These results support an experiential, rather than a genetic, mode of intergenerational transmission of faulty maternal behavior. However, as emphasized above, being abused by one's mother does not destine one to become an abusive parent since 50% of the infants raised by abusive mothers did not abuse their own offspring. Therefore, it is possible that a gene–environment interaction is involved: infants with certain genotypes might be more susceptible to the negative impact of abuse on the development of their maternal behavior.

5.6.2.2 Research on Rodents

In a series of studies by Fleming and collaborators, the effects of maternal separation on the development of maternal behavior in female rat offspring have been studied. In one group of studies [346], baby pups were separated from their mothers on day 4 of life and raised in complete social isolation until 21 days of age. These artificially reared (AR) females, who were fed through a gastric tube, were compared to control females that were reared normally by their mothers until weaning at 21 days of age (MR females). At 21 days of age, all females were caged with a same-sex conspecific until adulthood (70–100 days of age), at which time they were mated with males and their subsequent maternal behavior toward their own pups was measured. A second paradigm employed a partial maternal separation (MS) procedure [573]: pups were separated from their mothers over postnatal days 1–17 for 5 h/day, and their subsequent maternal behavior toward their own pups was compared to that shown by control females that remained with their mothers over days 1–17. Both studies found significant disruptions in the subsequent maternal behavior of the offspring that had been exposed to experimental models of maternal deprivation or neglect: AR and MS postpartum females licked, groomed, and nursed their young at lower levels than did control females. These results in rats indicate that maternal deprivation does not eliminate maternal behavior in the affected offspring, nor does it result in abusive maternal behavior. Rather, in rats, maternal deprivation seems to decrease the subsequent maternal attentiveness of the affected females toward their own young.

In a mouse model of partial maternal deprivation, Kikusui, Isaka, and Mori [481] subjected mice to early weaning. While control mice were weaned at the standard age of 21 days, the experimental females were weaned at 14 days of age. In adulthood, the early weaned mice licked/groomed their pups less than control females, which conforms to the results from Fleming's laboratory. Kikusui et al. also reported that early weaned mice were more fearful, as measured in the elevated plus maze.

In rats, Roth, Lubin, Funk, and Sweatt [840] used a model of maternal maltreatment of offspring to study its effects on the development of the offspring's maternal behavior. For 30 min each day during the first postnatal week (days 1–7), pups were either cross-fostered to a normal mother or cross-fostered to a mother (maltreatment condition) who had been placed in a novel cage that contained limited nesting materials. Pups within the maltreatment condition received significant amounts of "abusive" maternal behaviors: the maltreated pups were roughly handled, actively rejected, dropped during retrieval, and stepped on by their foster mothers. The maltreated pups also received significantly less licking/grooming and nursing from their mothers. The pups from each of these two treatment conditions were mated with males in adulthood, and then their maternal behavior was studied under normal home cage conditions with proper amounts of nesting materials. In comparison to controls, mothers that had been maltreated as pups exhibited significantly greater amounts of abusive behaviors and less licking and nursing of their own pups.

Using a similar maternal maltreatment paradigm, Raineki, Cortes, Belnoue, and Sullivan [797] explored additional treatment effects. Forty-five-day-old rats were placed in a two-chambered cage that contained a same-sex conspecific in one of the chambers. In a 10-min test, the amount of time spent in the chamber with the social stimulus was measured, and it was found that rats exposed to early maltreatment spent significantly less time in the chamber with the conspecific. Furthermore, when exposed to a physical stressor (immersion in a tank of water), the rats with a history of maltreatment showed greater activation of the basolateral amygdala and central nucleus of the amygdala, as measured by the number of Fos-expressing cells, than did control females.

Taken together, these results in rodents conform to important aspects of the primate work. Maternal deprivation and maternal abuse in rodents and primates depresses the maternal behavior of the affected offspring and may also give rise to an abusive phenotype. Such maternal deficits are also correlated with an increase in stress-reactivity and fear-related responses to social and nonsocial stimuli. The decrease in prosocial behaviors may be due to both an increased aversion of social interactions and a direct dysregulation of prosocial neural pathways.

Kaffman and Meaney [466] have made the important point that lower levels of parental bonding to offspring (deceased maternal attentiveness) that are not considered abusive or neglectful, and are within the normal range of maternal responsiveness, may still have significant influences on the social and emotional development of offspring. It is this more nuanced view of the influence of the mother on the development of her offspring that has given rise to an examination of the effects of normal variations in maternal behavior on offspring development. Research from the laboratories of Meaney and Champagne, aspects of which were introduced in Chapter 2, has demonstrated that rat mothers show individual differences in the duration of particular types of maternal behaviors [173]. Mothers that are referred to as high licking and grooming (HLG) females spend more time licking and grooming their pups during the first postpartum week when compared to low licking and grooming (LLG) mothers. Since all of these mothers raise their young to weaning, these phenotypes do not represent dysfunctional parenting. However, HLG mothers might be considered as being more attentive to their young than LLG dams. Importantly, there is an intergenerational transmission of these phenotypic differences, since the daughters of HLG mothers grow up to become HLG mothers when they have their own offspring, and the daughters of LLG mothers become LLG mothers [173]. Somewhat similar to the Maestripieri [587] findings on rhesus monkeys, this intergenerational continuity of maternal style is experientially transmitted: female rat pups born to HLG dams but cross-fostered to LLG dams at birth develop the LLG phenotype, while females that are born to LLG mothers but cross-fostered to HLG dams become HLG mothers [173]. Clearly, maternal treatment effects influence the development of neural circuits that directly or indirectly affect the offspring's subsequent maternal behavior.

Champagne, Diorio, Sharma, and Meaney [172] have reported an important relationship between virgin sensitization latencies and postpartum licking and grooming in rats. Virgin females that were raised by HLG moms showed shorter sensitization latencies (mean = 4 days) when they were continuously exposed to pups in adulthood than did virgin females that were raised by LLG mothers (mean = 8 days), and when these virgins were mated and had their own pups, they demonstrated higher levels of licking and grooming that did their counterparts that exhibited longer sensitization latencies as virgins.

How might one interpret the positive relationship between sensitization latencies in virgin rats and the maternal attentiveness of postpartum mothers? One possibility is that an increase in maternal motivation underlies both processes. Perhaps enhanced activation of MPOA circuits to the mesolimbic DA system by pup stimuli mediates both shorter sensitization latencies in virgins and increased levels of LG in postpartum females. In partial support, Champagne et al. [171] have reported that HLG dams exhibit a greater release of DA into NAs than do LLG dams, and that the DA release precedes and predicts a bout of licking and grooming.

An additional possibility is that females that exhibit short sensitization latencies and high levels of LG are less fearful and stress reactive than their long-sensitization latency-LLG counterparts. As reviewed in Chapter 2, pups that experience high levels of LG during their first 10 days of life develop a decreased hypothalamic–pituitary–adrenal response to stressful stimuli, and the epigenetic mechanisms underlying that effect were described. Additional research shows that offspring that receive high levels of LG as neonates exhibit decreased fear-related responses in novel environments when compared to offspring of LLG mothers [149,150,750]. Perhaps such decreased fearfulness and stress reactivity shortens the avoidance stage of the sensitization process in virgins and also allows postpartum dams to efficiently attend to their young rather than become distracted by innocuous external environmental stimuli [703]. In this context, it is worth pointing out that postpartum rats exhibit a general reduction in anxiety and fearfulness when compared to their nonpostpartum (estrous cycling) counterparts, and that OT neural systems play a role in such anxiety reduction [564,716]. Even so, the anxiety or fearfulness level of LLG moms may still be higher than that of HLG mothers.

What the above analysis suggests is that there may be normal variations in the extent to which social and nonsocial stimuli are processed over aversive neural pathways and in the extent to which social stimuli can gain access to prosocial neural pathways, including those mediating maternal motivation. These variations appear to develop under the influence of the type of maternal care an individual receives from its parent(s).

5.6.3 The Developmental Neurobiology of Maternal Behavior

5.6.3.1 Neurobiology of the Intergenerational Continuity of Normal Variations in the Maternal Behavior of Rodents

Research by Champagne and Meaney has shown that the experience-based intergenerational transmission of normal variations in the maternal behavior of rats is related to experience-induced interindividual differences in OTR expression in MPOA, which, in turn, appears to modulate the strength of MPOA activation of the mesolimbic DA system [170]. Recall that adult estrous cycling virgin female rats that received high levels of LG from their mothers display shorter sensitization latencies than adult virgins that were raised by LLG mothers. In an additional experiment, virgin females that were raised by either HLG or LLG mothers were ovariectomized in adulthood and treated with estradiol. OTR expression in MPOA is stimulated by estradiol, and Champagne et al. [172] reported that estradiol stimulated a greater expression of OTRs in MPOA of the virgins that were raised by HLG mothers. Therefore, the shorter sensitization in these females may have been due to an increased expression of OTRs, which may have allowed endogenous OT to shorten sensitization latencies by stimulating maternally relevant MPOA output circuits.

As described earlier in this chapter, the estradiol peak at the end of pregnancy is involved in stimulating both OTR expression in MPOA and the onset of maternal behavior at parturition. These processes appear to be within the normal range for both HLG and LLG dams, since the onset of maternal behavior at parturition is not delayed in LLG mothers. For the remainder of the postpartum period, during the maintenance phase of maternal behavior, endogenous estradiol levels are low, and therefore estradiol cannot operate to stimulate OTR expression. However, OTR expression in MPOA is higher in those postpartum dams that display HLG of their pups in comparison to the level of MPOA OTR expression in those postpartum dams that show LLG of their pups [172]. These results suggest that a mechanism other than estradiol is involved in maintaining the high postpartum OTR expression in HLG dams. That such elevated expression of OTRs in the MPOA is also involved in stimulating HLG receives support from the finding that ICV administration of an OTR antagonist (OTA) depresses the licking and grooming levels of postpartum HLG dams to that shown by LLG mothers. Since DA release into NAs precedes a bout of HLG [171], these results suggest that high levels of OTR in MPOA may enhance the functional connectivity between MPOA and VTA as a result of OT stimulation of MPOA efferents. Please note that since adequate maternal behavior still occurs in LLG dams, OT's role in this proposed upregulation of the MPOA-to-VTA pathway in postpartum dams is modulatory in nature and it not essential for ongoing maternal behavior. That is, variations in OTR expression are related to normal variations in maternal behavior. The only caveat I have with respect to OT's site of action is that since it was ICV OTA injections that decreased the LG level of HLG postpartum mothers, a site of action on the MPOA is not conclusive. It would be important to show that direct injection of OTA into MPOA could decrease LG behavior in postpartum mothers (cf. [884]).

What is the mechanism that results in greater estradiol-induced OTR expression in MPOA of females that received HLG as neonates? Epigenetic mechanisms have been shown to be involved [170,173,174]. Neonatal female rat pups that receive high levels of LG from either their biological or foster mothers show decreased DNA methylation at CpG sites within the regulatory region of the ER alpha gene of MPOA neurons. In contrast, neonates that received LLG show enhanced methylation at these sites. This epigenetic process involving hypomethylation at CpG sites, which was described in detail in Chapter 2, results in a higher expression of ER alpha protein in MPOA of female rats that: (1) received HLG as neonates, and (2) develop an HLG phenotype.

Putting all these findings together, the likely sequence of events underlying the experience-based intergenerational continuity of normal variations in the maternal behavior of laboratory rats can be described as follows: female neonates that receive more maternal attention (licking and grooming) from either biological or foster mothers have low levels of cytosine methylation at CpG sites within the regulatory region of the ER alpha gene in MPOA neurons, resulting in a greater expression of ER alpha in MPOA in comparison to females that receive LLG from their mothers. Such an enhanced expression of ER alpha allows for greater estradiol-induced OTR expression in MPOA, and this higher level of MPOA OTR expression is subsequently maintained during the postpartum period in the absence of continued estradiol stimulation. It is certainly possible that ligand-independent activation of ER alpha maintains OTR expression during the postpartum period [170,767]. This greater expression of OTRs in MPOA results in the development of an HLG phenotype, which is presumably due to increased OT activation of MPOA projections to VTA.

Although differences in OT-induced excitation of MPOA efferents to VTA may explain some of the interindividual variability in the maternal behavior of rats, other mechanisms may also be involved, particularly since neonates that have received HLG from their mothers not only develop an HLG phenotype but also show a decrease in fear-related responses in adulthood. Since OT not only stimulates the appetitive aspects of maternal behavior in rats but also has anxiolytic effects, perhaps the enhanced maternal behavior of rats that received HLG as neonates is based in part on OT stimulation of MPOA efferents that operate to both decrease fearfulness and increase maternal behavior. Both of these factors may contribute to the enhancement of maternal attentiveness in postpartum females, the reduction of sensitization latencies in virgin females, and to the reduced fearfulness exhibited by postpartum females. A potential neural circuitry for this proposal is described in Figure 5.20, which depicts MPOA efferents terminating in VTA and in the AHN-PAG circuit. Inhibitory projections to the latter circuit would depress avoidance or fear responses to certain stimuli (also see Figure 5.7), while excitatory projections to the VTA-DA system would enhance appetitive maternal responses and maternal attentiveness. Increases in the number of ERs and OTRs within MPOA neural circuits, resulting from an experience-based epigenetic process, would cause increased functional activity in each of these circuits. Further, although OT stimulation of MPOA connections to AHN and PAG may be involved in certain aspects of fear reduction, particularly to social stimuli, the anxiolytic effects of OT, particularly in the face of environmental novelty, may also be exerted at other sites (see Figure 5.6(A)). In this regard, neonatal females that received high LG from their mothers not only explore the center area of a novel open field in adulthood to a greater extent than do neonates that received

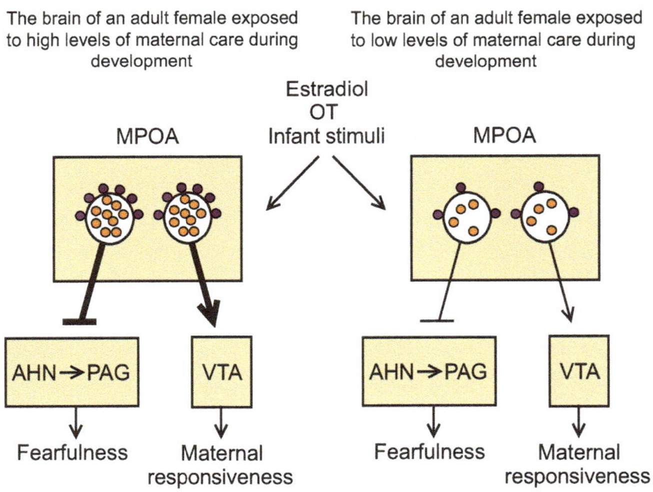

FIGURE 5.20 Experiential influences on the development of medial preoptic area (MPOA) neurons and on the development of maternal behavior. Female rat offspring that are exposed to high levels of maternal care, when compared to those that receive lower levels of maternal care, grow up to show higher levels of maternal care to their own offspring, and such females are also less fearful. Further, female young that receive high levels of maternal care also develop more estrogen receptors (ERs = orange circles) and oxytocin receptors (OTRs = purple circles) in MPOA. It is proposed that such enhanced expression of ERs and OTRs allows MPOA neurons to more effectively inhibit anterior hypothalamic nucleus (AHN)-to-periaqueductal gray (PAG) defensive circuits and to more effectively stimulate ventral tegmental area (VTA) dopamine neurons. Axons ending in a bar are inhibitory and those ending in an arrow are excitatory. Axon thickness represents the strength of a particular connection. Other abbreviations: OT = oxytocin. *Reproduced from Figure 5 in Numan [697] with permission from Elsevier.*

low LG [750] but also exhibit greater OTR expression in the central nucleus of the amygdala [172], where OT has been shown to exert anxiolytic effects [502]. Finally, the decreased fearfulness of the offspring of HLG dams may also be related to the epigenetic processes described in Chapter 2 that decrease HPA and CRH responsive to stressful stimuli. Therefore, increased activity within brain OT systems and decreased activity within brain CRH systems may both exert important anxiolytic effects.

Jia, Tai, An, and Zhang [457] explored the development of maternal behavior in Mandarin voles that were subjected to *paternal* deprivation as neonates. Mandarin voles are monogamous and both maternal and paternal care of young are observed. Jia et al. allowed one group of neonates to be normally reared with biparental care (BP). In the paternal deprivation (PD) group, the father was removed as soon as the young were born, resulting in the young being raised only by the mother. Importantly, the mother's behavior did not compensate for the absence of the father, so the total amount of LG received by the PD neonates was less than that received by the BP neonates. In adulthood, the female voles that were raised with only a mother showed lower levels of licking, grooming, and nursing of their own pups than did the females who had been reared by both a mother and a father. This difference in maternal behavior was correlated with differences in the expression of ER alpha within MPOA neurons: the females that received biparental care had greater expression of the ER than did their PD counterparts. One would assume that OTR expression was also higher in the MPOA of the adult females that had received biparental care, but this factor was not examined in this study.

5.6.3.2 The Developmental Neurobiological Effects of Maternal Deprivation and Maternal Abuse

An implication of the data on the underlying mechanisms that mediate the experience-based intergenerational continuity of normal variations in maternal behavior is that similar, but more extreme, processes might contribute to the development of pathological forms of maternal behavior that result from severe early life stress, such as parental abuse, neglect, or deprivation. For example, total maternal deprivation in some species, or chronic and intense maternal abuse, may result in a more extreme disorganization of MPOA output circuits, particularly in individuals with certain genotypes, with the result that the adult maternal responsiveness of the affected offspring might be more severely disrupted.

There is some evidence that total maternal deprivation may disrupt MPOA interactions with the mesolimbic DA system. Afonso, King, Novakov, Burton, and Fleming [9] examined mesolimbic DA function in postpartum rats that had either been reared with their mothers (MR) or artificially reared (AR) without their mothers. As previously described, the AR rats exhibited deficits in their maternal behavior in adulthood, and this study also reported that AR postpartum dams displayed reduced pup-stimulated dopamine elevations in the nucleus accumbens when compared to MR rats. These findings could be interpreted to support the view that maternal deprivation early in life (an animal model of parental neglect), interferes with the development of normal MPOA connectivity with the mesolimbic DA system, resulting in maternal behavior deficits in adulthood. Since research also indicates that maternal deprivation has other neural effects, with dysfunctions developing in multiple neural systems [205], these additional effects may also influence the development of maternal behavior. As one example, maternal deprivation appears to directly disrupt the general functions of the mesolimbic DA system, resulting in a decrease in the responsiveness of the affected offspring to a broad range of rewarding stimuli [448,614]. Further, in a manner similar to the effects of being reared by an LLG dam, maternal separation/deprivation also results in increased fearfulness, stress reactivity, and aggression in the affected offspring (see [703] and Chapters 2 and 3). Therefore early adverse experiences such as child abuse and neglect are likely to interfere with the development of multiple neural systems, which then impact the development of maternal behavior. The experience of abuse or neglect may result in one or more of the following: a direct disruptive effect on the development of maternal neural systems; a general dysfunction in the organism's responsiveness to rewarding or appetitive stimuli; an increase in stress reactivity, fearfulness, and affective aggression. Not surprisingly, research indicates that aspects of the latter process involve the faulty development of serotonin neural systems.

In Chapter 3, I reviewed the evidence that early maternal deprivation increases affective aggression by causing a dysfunction in serotonergic neural systems, and that CRH hyperactivity may contribute to this effect. In this chapter, I reviewed the evidence that a disruption in 5-HT neural transmission disrupts maternal behavior. It is likely, therefore, that early life stress, which would include maternal abuse and/or neglect, might disrupt the development of maternal behavior in the affected offspring by causing a dysregulation within 5-HT systems. Such an effect might disrupt maternal behavior directly, although it is more likely that the increases in affective aggression and stress reactivity that result from a disruption of 5-HT systems indirectly interfere with maternal behavior.

In a series of studies, Maestripieri has presented evidence that the intergenerational transmission of maternal rejection and abuse from mothers to daughters in rhesus monkeys is likely to involve an experience-based disruption of 5-HT neural function: biological or foster mothers that show high levels of abuse and rejection of their infants have lower CSF levels of 5-HIAA than do mothers that show normal maternal behavior, and the female infants reared by these abusive/rejecting mothers also exhibit lower CSF 5-HIAA levels later in life when compared to control infants that were not subjected to abuse or neglect [588,591]. Since McCormack et al. [619] found that the proportion of mothers carrying the short allele of the 5-HTTLPR was higher in abusive than in nonabusive rhesus monkey mothers, and since not all rhesus monkey females that experience abuse as infants grow up to become abusive mothers [587], a gene–environment interaction may be involved. Remember from Chapter 3 that PR monkeys are more likely to become hyperaggressive if they carry the short allele of the 5-HTTLPR. Perhaps the same relationship holds with respect to the development of an abusive/rejecting maternal phenotype in infants who have been maltreated by their mothers. An interesting study on humans may be relevant to this idea [74]. This study found that there was increased cytosine methylation at CpG sites within the regulatory region of the 5-HTT gene in adults who retrospectively reported that they were abused by their parents when they were children. These authors note that the amount of variance in 5-HTT mRNA produced by the different 5-HTTLPR alleles is modest, which suggested to them that other regulatory factors might influence 5-HTT expression. Although these authors did not differentiate between 5-HTTLPR variants, it is very intriguing to speculate that early abuse and neglect might result in a hypermethylation of the short allele, resulting in an even greater decrease in the expression of the 5-HTT. As described in Chapter 3, the resultant high levels of synaptic release of 5-HT during brain development, because of low 5-HT reuptake, would give rise to a hyperaggressive phenotype that might contribute to the cycle of violence where the children of abusive parents have an increased likelihood of becoming abusive parents.

Given the rodent research on the effects of early life experiences on the development of OT neural systems, is there any evidence in nonhuman primates that maternal deprivation influences OT systems? I am aware of only one relevant study, and this research examined the effects of MR versus PR on *male* rhesus monkeys through 3 years of age [1039]. In comparison to MR males, PR males showed increased aggression, decreased affiliation (grooming; sitting in contact with a conspecific), and increased solitary behaviors. Baseline CSF levels of OT were lower in the PR monkeys when compared to their MR counterparts, and for both groups, CSF levels of OT were positively correlated with the occurrence of affiliative social behaviors. Also noteworthy was that plasma OT levels did not correlate with CSF OT levels, and there was not a significant difference in plasma OT based on rearing conditions, indicating that in this experiment the central and peripheral release of OT were independent of each other. These results suggest that maternal deprivation impairs the normal development of neural systems that regulate the central release of OT in male rhesus monkeys and that this effect is correlated with decreases in prosocial and increases in antisocial behaviors. To the extent that similar effects occur in PR female rhesus monkeys, it is likely that a central OT dysfunction contributes to the maternal deficits that develop in such females. In the next section on human maternal behavior, additional evidence in support of this view will be presented [399].

Additional factors have been related to the development of experience-based intergenerational transmission of faulty maternal behavior. In the previously described study by Roth et al. [840], where neonatal rats were exposed to an abusive foster mother who had limited access to nesting material, not only did these neonatal rats grow up to show faulty maternal behavior toward their own offspring but they also displayed decreased expression of brain-derived neurotropic factor (BDNF) in their PFC. The decreased expression of BDNF in PFC was related to a hypermethylation of cytosine at CpG sites within the regulatory region of the BDNF gene. Since BDNF is involved in the development of neural circuits, an abnormal development of PFC neural circuits may have been caused by maternal maltreatment, which then influenced the development of faulty maternal behavior in the affected offspring. Since the PFC is involved in both maternal behavior and in emotional regulation, including the regulation of affective aggression, the development of faulty maternal behavior is likely to be caused by multiple processes. A recent study by Provencal et al. [788] has also reported increased methyl-DNA in several genes within the PFC of PR rhesus monkeys in comparison to their MR controls. Therefore, PFC dysregulation appears to be another likely mechanism that mediates the experience-based intergenerational transmission of abnormal maternal behaviors.

5.6.3.3 Conclusions

The way a mother treats her offspring can influence the development of maternal behavior in the offspring, and several mechanisms are likely to be involved, including effects on the development of maternal and oxytocinergic neural systems, fearfulness and affective aggression neural systems, general appetitive motivational neural systems, and PFC regulatory mechanisms. Gene–environment interactions are likely to be involved, as are epigenetic processes. Early adverse experiences that decrease OT and 5-HT neural transmission may decrease prosociality and increase antisociality, respectively. The antisocial effects of 5-HT dysfunctions may also be coupled with hyperactivity within the brain's CRH system (see Chapters 2 and 3).

A summary proposal with respect to the intergenerational continuity of abnormal maternal behavior can be offered. Maternal treatment effects that depress the normal development of those neural circuits that mediate maternal motivation in the affected offspring may contribute to the intergenerational continuity of maternal neglect; maternal treatment effects that result in the development of heightened stress reactivity, anxiety, and impulsive aggression in the affected offspring may contribute to the intergenerational continuity of maternal abuse.

Importantly, the fact that the quality and quantity of maternal care influences the development of such a diverse set of neural processes indicates that its impact extends far beyond influences on how maternal behavior develops in offspring and includes a critical effect on broad aspects of socioemotional development [466]. Given the intergenerational continuity of maternal care, particular socioemotional characteristics may be transferred across generations through the vehicle of the type of maternal care that is received.

5.7 THE NEUROBIOLOGY OF MATERNAL BEHAVIOR IN HUMANS

5.7.1 Introduction

Research on the neural systems related to maternal responsiveness in humans has increased dramatically over the past several years [933,946]. This research, primarily using fMRI technology, has explored the brain regions activated when mothers (and non-mothers) are exposed to various types of infant stimuli. This research has also focused on the involvement of DA and OT in the human maternal brain and on the maternal effects of genetic polymorphisms that modulate the functioning of DA and OT neural systems. Within these boundaries, research has also examined whether dysfunctions in particular neural systems might be related to less than adequate maternal behavior in humans, and whether the kind of maternal care received by a young female affects the development of critical neural systems that influence her subsequent maternal behavior in adulthood.

Based on the presumed importance of alloparental behavior in human evolution and the inherent attractiveness of infant facial features [427], some research has contrasted the brain regions activated in nulliparous women when viewing either infant faces or the faces of adults. In an fMRI study conducted on nulliparous women, Glocker et al. [342] found that infant faces activated the nucleus accumbens (NA) to a greater extent than did the faces of older children, and Kringelbach et al. [512] reported that unfamiliar infant faces, in comparison to unfamiliar adult faces, selectively activated parts of the orbital PFC. Glocker et al. argue that the inherent attractiveness of infantile visual features may be the result of the ability of such stimuli to activate the mesolimbic DA system. In reference to Box 1.2, which discusses how one should interpret an enhanced BOLD response in a particular neural region, one possibility is that the increased BOLD response in NA was the result of increased action potential frequency in VTA-DA axon terminals within the NA [872]. I do want to mention another important issue in the context of these findings. In many cases, the spatial resolution of current fMRI technology is not sensitive enough to differentiate nearby regions from each other. Therefore, cases that demonstrate an increased BOLD response in NA may actually also be measuring an increased BOLD response in the nearby ventral pallidum (VP). In fact, in some of the brain images shown in the Glocker et al. study, the increased BOLD seems to be in the VP rather than in NA (see Figure 2 in Glocker et al. [342]).

Recall from Chapter 1 that Barbas et al. [61] proposed that the neural circuit from the basal amygdala nuclei to the mediodorsal thalamus to the orbital PFC might be the route through which emotions are subjectively experienced. Given that proposal, the activation of the orbital PFC observed in the Kringelbach et al. [512] study might be related to positive affective experiences triggered by infantile facial features.

Although the mesolimbic DA system and the orbital PFC exhibit greater activity when nulliparous women view infant faces in comparison to adult faces, other research indicates that the activation of these two systems is even greater when postpartum women view infant faces, particularly if those faces are of their own infants [70,688,689], suggesting that these two systems may be upregulated by the physiological, experiential, and neural changes associated with motherhood. Importantly, the greater activation of the orbital PFC that occurs when mothers view their own versus unfamiliar infants is positively related to the mother's pleasant mood ratings, supporting the contention that orbital PFC activity may be the basis of positive maternal feeling states [689].

Not surprisingly, given the involvement of the insular cortex in a variety of emotional experiences and empathy (see Chapter 4), several studies have reported increased BOLD responses in this region when mothers view images of their own infants [70,529,689A]. Indeed, the posterior orbital PFC and adjoining anterior insular cortex may operate together in the mediation of positive affective states. The broader significance of this process will become evident in Chapter 7.

It should be clear from Chapters 1 and 4 that the mesolimbic DA system, PFC, and anterior insula are parts of general motivational and/or emotional systems that would be activated by a variety of highly attractive and pleasant social and nonsocial stimuli. Evolutionary forces may have caused infant facial features to easily activate general reward systems within the brain, even in nulliparous women, but it is unlikely that such activation in NA-VP, anterior insula, and orbital PFC is outlining areas specifically related to maternal motivation (cf. [739]). In other words, one needs to ask about specific maternal motivational systems that allow infant stimuli to access the nonspecific motivational and emotional systems. In relation to the nonhuman animal research, it would have been interesting to determine whether hypothalamic regions that included the MPOA were strongly activated in *nulliparous* women by attractive infantile facial features. If the MPOA had been shown to be highly active in response to infant stimuli, that effect might represent a stimulation of specific maternal circuits that could underpin alloparental motivation in humans. Even better, it would be important to determine whether a positive correlation might exist between MPOA BOLD and NA-VP BOLD responses.

A recent study by Moll et al. [656], although not specifically focusing on maternal behavior, is very relevant to the issues being discussed. They asked the following question: are there brain regions that are active during affiliative emotions (feelings of warmth and tenderness toward kin) that can be separated from those that are active during other hedonic emotions (pleasant emotions with a positive valence)? To this end, while in an fMRI scanner, women and men read short descriptions of various scenarios that included the following three types. (1) A positive valence scenario that was associated with kin and that evoked strong feelings of warmth and tenderness (Positive-affiliative: you read a book to your child, who fell asleep in your lap). (2) A positive valence scenario not associated with a relative and which did not evoke strong feelings of warmth and tenderness (Positive-nonaffiliative: your boss was impressed by your performance). (3) Neutral scenarios that did not give rise to strong pleasant feelings and that did not evoke feelings of warmth and tenderness (Neutral: you went to lunch with coworkers). The results indicated that, in comparison to neutral scenarios, all scenarios with a positive valence (affiliative and nonaffiliative) activated the NA and orbital PFC. Importantly, positive-affiliative scenarios selectively activated the septal area and preoptic hypothalamic region while positive-nonaffiliative scenarios did not.

The results of Moll et al. [656] suggest that neural activity within the basal forebrain region that includes the septum and preoptic area is specifically related to strong affiliative motivations and emotions, and they suggest that the activation of this region along with the concurrent activation of the ventral striatum (and orbital PFC) by positive-affiliative scenarios represents an interaction between a specific affiliative social motivational system and a more general reward-related motivational system. This analysis is similar to that suggested by Numan [696] with respect to the interaction between the MPOA and the mesolimbic DA system in the control of the appetitive aspects of maternal behavior in rats. Moll et al. [656] emphasize that the spatial resolution of their fMRI scanning procedure was not sensitive enough to separate out the individual contributions of the septal area and the preoptic region to the overall BOLD response. They go on to suggest that more refined MRI techniques might be able to define partially distinct subcortical responses that are related to specific types of social affiliations. Note how this proposal relates to my discussion of the specificity with which the MPOA influences the appetitive aspects of maternal behavior and male and female sexual behaviors.

5.7.2 Brain Systems Associated with Maternal Behavior in Postpartum Women

5.7.2.1 Introduction

In the typical fMRI scanning study meant to associate neural activations (BOLD increases) with maternal responsiveness, postpartum women are scanned while they are exposed to different types of stimuli. In one common procedure, women are exposed to either faces, cries, or laughter from either their own infants or from unrelated infants [945]. Since humans form relatively selective attachments to their own offspring, the assumption is that those brain regions that respond most strongly to one's own infant are likely to contribute to the neural circuits regulating maternal motivation and the strong mother–infant bond. Realize, however, that such studies are primarily correlational in nature, and therefore cause–effect relationships cannot be proven. As one example, if a mother shows a greater BOLD response in a particular brain region when viewing her own infant's face in comparison to an unfamiliar infant, the active brain region may not be specifically related to maternal motivation but may instead be related to another process, such as the activation of a specific episodic memory by the infant's face. Because of the correlational nature of most of these studies, the current analysis will focus on the overlap between those brain regions that are active in postpartum women when viewing infant stimuli and those brain systems that have been experimentally shown to be important in the regulation of maternal behavior in nonhuman animals. This kind of convergent analysis is very likely to uncover the core brain regions that are involved in the regulation of maternal motivation in postpartum women.

In reference to some of the neural systems and mechanisms that regulate goal-directed maternal behavior in nonhuman animals, as shown in Figures 5.6–5.8, the attempt will be made to map the evidence from human studies on the neural basis of maternal behavior onto the circuits shown in these figures. MPOA output circuits are shown as activating the mesolimbic DA system to regulate goal-directed maternal responses toward infant stimuli. The MPOA also activates PVN-OT systems, and OT feeds back onto MPOA to potentiate its output, while other OT circuits also stimulate the VTA and influence how the amygdala processes infant stimuli. Such OT actions on MPOA, VTA, and amygdala would promote appetitive maternal responses and maternal attentiveness. Recall that during the postpartum period, OT neural systems also have anxiolytic effects. Therefore, in the human case, postpartum OT neural systems may act to increase the level of maternal attentiveness, the mother's sensitivity to her infant's needs, and the ability of the mother to cope with stressful situations.

Probably the best single study in terms of its fit with a substantial portion of the model shown in Figure 5.8 is that by Strathearn, Fonagy, Amico, and Montague [934]. Initially, primigravid women were interviewed during pregnancy, and their attachment relationships with their own mothers were assessed. As a result, females were classified as either having a secure or an insecure attachment to their own parents. This part of the study is related to the potential experience-based intergenerational transmission of maternal behavior and is based on the possibility that if a particular female had a less-than-adequate attachment relationship with her own mother, then such an experience might influence the development of her maternal neural system and her corresponding maternal behavior. In the next part of the study, which was conducted at 7 months postpartum, plasma OT levels were measured after a 5-min mother–infant free play interaction. In the last part of the study, which was conducted at 11 months postpartum, fMRI brain images were obtained while mothers viewed the faces of their own or unfamiliar infants. The most important findings can be summarized as follows. (1) Plasma OT levels were higher in the secure mothers than in the insecure mothers during the free play period with their infants. (2) When compared to viewing an unknown infant face, the hypothalamic BOLD response increased when viewing one's own infant and this increase was significantly greater for secure than for insecure mothers. (3) The BOLD nucleus accumbens (NA) activation that occurred when mothers viewed their own infant was also greater in secure than insecure mothers. This increased BOLD response in the secure mothers might have been due to increased neural activity within VTA-DA axons terminating in the ventral striatum [872]. (4) When all females were considered, there was a significant positive correlation between plasma OT levels measured during free play at 7 months postpartum and the NA BOLD response measured at 11 months postpartum while mothers viewed the face of their own infant.

Note that in this study, plasma blood levels of OT were measured as a proxy for central brain OT release and activity. As indicated previously, such an assumption is controversial, with some studies showing independent regulation of pituitary and central release of OT, while other studies suggest that the peripheral and central release of OT may be coordinated and positively correlated [527,836]. To the extent that the plasma OT levels in the Strathearn et al. [934] study were positively correlated with central OT release, the following integration of their findings with the nonhuman animal research is possible. The experience of a poor attachment relationship with their own mothers may have impacted the development of the MPOA and PVN-OT neural systems in the affected offspring. In adulthood, as a result, infant stimuli may have been less effective in activating hypothalamic mechanisms that stimulate the mesolimbic DA system in the insecure mothers.

A criticism of the Strathearn et al. [934] study is that the nature of the mother–infant interaction at 7 months postpartum was not explored. The results of this study would have been more powerful if it had been shown that the insecure mothers showed less-than-adequate maternal responses to their own infants and that such deficits could be correlated with OT levels and hypothalamic and nucleus accumbens BOLD responses. Other studies, however, have reported a positive correlation between plasma OT levels in postpartum women and the quality of their maternal behavior, with higher OT levels being associated with better maternal behavior [290,350,935]. Unfortunately, these findings do not inform us as to whether OT increases maternal attentiveness or whether increases in mother–infant interactions cause enhanced OT release.

Using this introduction as a foundation, in the remaining parts of this section I will present additional data that supports the model shown in Figure 5.8.

5.7.2.2 The MPOA-Hypothalamus and Maternal Behavior in Women

In addition to the findings of Strathearn et al. [934], other studies have reported increases in the hypothalamic BOLD response when mothers are exposed to stimuli from their own infants [947,948]. The fMRI techniques in these studies, however, were not sensitive enough to discriminate between the individual contributions of particular hypothalamic nuclei to the total BOLD response.

Kim et al. [485] applied a voxel-based morphometry (VBM) analysis to MRI images of the brains of postpartum females. VBM is used to measure changes in gray matter volume within brain regions, and increased gray matter volume may be representative of increased functional activity (cf. [480]). In this study, all of the mothers were breast-feeding and MRI images were obtained at two time points: 2–4 weeks and 3–4 months postpartum. As the postpartum period progressed over these time points, the gray matter volume within the hypothalamus, which included MPOA, increased. Increases in gray matter volume also occurred in other regions, including the VTA-substantia nigra region, amygdala, and the globus pallidus/ventral pallidum. Importantly, a mother's positive perception of her infant at 2–4 weeks significantly predicted the increases in gray matter volume within these regions at 3–4 months postpartum. Since these findings are correlational, they are open to multiple interpretations. Based on the animal research, however, the following mechanistic possibility should be considered. A more positive attitude toward one's infant may have resulted in better maternal behavior across the postpartum period with a resultant increase in neural activity across the maternal circuit, which would have included PVN input to MPOA and MPOA activation of VTA, which, in turn, would promote a stronger amygdala-to-ventral pallidum connection.

In a preliminary study, Lahey et al. [525] examined the interactions of mothers with their 4–6-year-old children during a free play period, and maternal behaviors were characterized as representing either positive parenting (praise, positive affect, affectionate touch) or negative parenting (criticism, negative affect, harsh touch). Each mother was also genotyped for the ER alpha gene (ESR1 gene). The ESR1 gene contains several single nucleotide polymorphisms (SNPs), giving rise to a variety of different alleles. This study found that certain SNPs were associated with negative parenting, such as the A (adenine) allele at rs1884051. This SNP of ESR1 has also been associated with premenstrual dysphoria in menstrual cycling women [435]. Although this SNP is located in an intron of the gene, it is possible (but not confirmed) that such genetic variation might influence the transcription and subsequent expression of ER alpha. It would be very exciting if it could be shown that the level of expression of ER alpha in the hypothalamus of postpartum women influences their maternal behavior (cf. [767]).

5.7.2.3 *Amygdala and NA-VP Activation in Postpartum Women*

Several studies have recorded increases in VTA and NA-VP BOLD responses when mothers view images or videos of their own infants in comparison to other infants [44,70,948]. With respect to the amygdala, Barrett et al. [68] showed mothers, at 3 months postpartum, photographs of their own or unrelated infants while their brains were being scanned. Each mother also rated how she felt while viewing the photographs. Mothers reported feeling more positive when viewing their own infants, and the basolateral/basomedial amygdala BOLD response increased as the intensity of positive maternal affect increased. These combined results can certainly be interpreted as supporting the view that increased DA input to NA promotes the functional connectivity between the BLA/BMA and VP, which influences positive maternal motivation and emotion. BLA input to the orbital PFC may also be involved.

The postpartum period is associated with an increased vulnerability to mood disorders, and 10–20% of women develop major depression, referred to as postpartum depression, during the first 6 weeks after parturition [703]. It should not come as a surprise that postpartum depression is associated with negative maternal feeling states and deficits in maternal behavior [703]. Laurent and Ablow [528] have examined the brain changes that might contribute to the faulty maternal behavior of depressed women (also see [529]). Postpartum women were exposed to three different stimuli while in an fMRI scanner: own infant cry; other infant cry; control sound. Nondepressed mothers showed greater BOLD signals in the NA and VTA/SN region in response to their own infants when compared to the control sound. In contrast, depressed mothers did not show an enhanced BOLD response in these brain regions above that of the control sound. In addition, when listening to their own infant's cry, the BOLD response in these critical motivational regions was significantly greater in the nondepressed than in the depressed mothers. The authors conclude that the failure of depressed mothers to engage the mesolimbic DA system in response to their infants may contribute to social bonding difficulties and deficits in maternal behavior.

The human dopamine transporter gene (DAT1 gene) contains a 40-bp variable number tandem repeat polymorphism in the 3′-UTR of exon 15, and variations in this aspect of the DAT1 genotype have been associated with variations in maternal behavior in women [534]. Genotypes at this site include one or two copies of a 9-repeat or a 10-repeat allele, and the most common genotypes are 9/9, 9/10, and 10/10. The parental behavior of mothers with their 4–7-year-old children was observed in the laboratory during a free play period and then during a period where the mother instructed the child to complete certain tasks, such as returning toys to their proper locations. Behaviors shown by the mothers were characterized as demonstrating positive parenting or negative parenting. Mothers with the 9/9 genotype showed the most positive parental responses, while the 9/10 and 10/10 genotypes exhibited higher levels of negative parenting. Although not all studies agree, Mill, Asherson, Browes, D'Souza, and Craig [647] have reported that DAT1 mRNA levels increase with increases in the number of 10-repeat alleles. To the extent that this latter finding is accurate, 9/10 and 10/10 genotypes should produce more DAT protein than 9/9 genotypes. Since DAT is involved in the reuptake of DA after its synaptic release, mothers with the 9/9 genotype should exhibit enhanced synaptic DA activity in several brain regions. In accord with the rodent research, these findings can be used to support the proposal that enhanced release and action of DA in the NA contributes to more positive maternal responses. Other findings have similarly related variations in dopamine system genes to variations in human maternal behavior [646,986].

5.7.2.4 OT Systems and Maternal Behavior in Women

Several additional studies have supported the view that OT neural systems are positively associated with maternal responsiveness in women. In the context of the animal research, OTRs in the human brain have been detected in the anterior cingulate cortex, septal-preoptic region, ventral pallidum, central, cortical, and basolateral amygdala, and the substantia nigra pars compacta, including the medial part that adjoins the VTA [103,571,572].

Kim et al. [483] compared postpartum women who either breast-fed their infants or bottle-fed their infants. This study was based on the assumption that breast-feeding would be associated with greater central OT neural system activation than would bottle-feeding, which might lead to differences in brain activity and maternal behavior. Brain scans were conducted at 1 month postpartum while the mothers listened to their own baby cry or to a control baby cry. At 3–4 months postpartum, mother–infant interactions were observed during a 5-min observation period in the mother's home, and the level of maternal sensitivity was measured. Maternal sensitivity included such behaviors as: positive affect, affectionate touch, gazing directly at the infant, and vocal clarity. Mothers who breast-fed their infants and those that did not showed enhanced BOLD responses in several brain regions when listening to their own infants, but the magnitude of the BOLD increase over that which occurred during the control baby cry was greater for the breast-feeding mothers. Among the activated neural regions, noteworthy BOLD increases were detected in the amygdala, globus pallidus-ventral pallidal regions, and the insular cortex. Further, the mothers who breast-fed their infants showed higher maternal sensitivity scores than the mothers who did not breast-feed, but this difference only approached, but did not reach, statistical significance. Importantly, when both groups were considered together, a positive correlation ($r = 0.53$) occurred between the degree of maternal sensitivity and the BOLD response detected in the amygdala and globus pallidus. One interpretation of these results, based on the experimental animal findings, is that breast-feeding, in comparison to bottle-feeding, more potently activated brain OT systems throughout the postpartum period and that such enhanced OT activity strengthened the functional connectivity between the amygdala and the pallidum, leading to enhanced maternal motivation and an enhanced mother–infant bond. Please note, however, that all mothers were showing normal maternal behavior and that any OT effect would have been modulatory in nature. An alternative interpretation, however, is that mothers who chose to breast-feed began the postpartum period with a higher level of maternal sensitivity, and therefore such increased sensitivity would have occurred prior to breast-feeding, and the presumed breast-feeding-induced increases in central OT levels.

In an interesting longitudinal study, Strathearn, Mamun, Najman, and O'Callaghan [936] compared the incidence of child maltreatment (neglect and/or abuse) in over 6000 mother–infant pairs in an attempt to relate the likelihood of maternal maltreatment of their offspring to the duration of breast-feeding. About 10% of this cohort was suspected of abusing or neglecting their children (based on government child protection agency records) and the authors hypothesized that the absence of breast-feeding during the infant's first 6 months of life would predict maternal maltreatment. In support, they found that the probability of maltreatment increased as the duration of breast-feeding decreased and that the probability of maternal maltreatment for non-breast-fed infants was about five times greater than that for breast-fed children. The authors suggested that breast-feeding might enhance maternal responsiveness and mother–infant bonding by stimulating central OT release. However, as in the study by Kim et al. [483], it is just as likely that the mothers who decided to breast-feed were on average better mothers to begin with.

Given the potential involvement of oxytocin in the maternal behavior of women, variations in maternal behavior may not only be related to variations in central OT release but may also be related to variations in the OTR gene (OXTR gene), which in turn may control the sensitivity of the OTR protein to endogenous levels of OT. Several SNPs of unknown functionality are present within the nucleotide sequence of the OXTR gene. One of these SNPs, at site rs53576, will be my primary concern. At this site, which is located within the third intron of the gene, either guanine (G) or adenine (A) can be present, and the resultant genotypes containing one of these two different alleles on each chromosome would be GG, AA, AG [967]. The A allele has been considered a risk allele with respect to social behavior because when healthy men and women are administered a personality questionnaire to measure prosocial temperament, homozygous GG individuals exhibit the highest prosociality scores and AA homozygotes have the lowest scores [967]. Interestingly, voxel-based morphometry measures of MRI scans show that hypothalamic gray matter volume is decreased in individuals with the AA genotype when compared to the GG genotype [967].

With this background, research has examined whether this particular genetic polymorphism is associated with variations in maternal behavior. In a laboratory setting, Bakermans-Kranenburg and van Ijzendoorn [48] observed mothers interacting with their 2-year-old child while the child was attempting to solve a difficult problem during a 15-min test session. The mother's maternal sensitivity was rated on a 7-point scale based on the mother's support, intrusiveness, and clarity of instruction. Mothers with the AA or AG genotypes of the OXTR gene were significantly less sensitive than GG mothers. The observed effect size, however, was small, and variation in OXTR genotype only explained 3% of the variance in maternal sensitivity. This latter result should not be surprising, as maternal responsiveness is undoubtedly affected by multiple factors. Also, any interpretation is constrained by the fact that the effect of this particular SNP on OTR function is not known, although there is some indirect evidence that the A allele at rs53576 might lower the sensitivity of the OTR protein to OT [604]. In addition, other evidence has shown that transcription factor binding to nucleotide sequences within intron 3 of the OTR gene may decrease transcription of the gene [652].

In addition to the longitudinal breast-feeding study by Strathearn et al. [936], is there any other evidence that might support the view that a decrease in central OT neural transmission might be related to faulty maternal behavior? In relation to the fact that women with postpartum depression (PPD) show deficiencies in their maternal behavior [703], there is evidence that has linked PPD to OT systems. Skrundz, Bolten, Nast, Hellhammer, and Meinischmidt [908] measured plasma OT levels in women during the third trimester of pregnancy. At 2 weeks postpartum all women were administered a depression scale, and a score of 10 or more was indicative of a risk for PPD. The results showed a negative relationship between plasma OT and depression scores: women at risk for PPD had significantly lower plasma OT levels than those not at risk. Assuming that plasma OT reflected central OT release, and that a causal relationship exists between central OT and depression, the authors suggested that enhancing OT release during pregnancy might be a potential therapy to prevent the development of PPD.

Based on the research described in Chapter 3, where intranasal application of vasopressin was found to increase CSF levels of the neuropeptide, several studies have examined the effects of intranasal application of OT on a variety of social behaviors. With respect to the findings of Skrundz et al. [908], one might ask whether intranasal OT could ameliorate PPD symptoms. In a study that employed a within-subjects design with women diagnosed with PPD, at 3–12 months postpartum women received intranasal OT or placebo, 1 week apart and in random order [592]. OT administration did not improve the mood of the women, but it did improve the mothers' perception of the quality of their relationship with their infants. Since PPD is caused by many factors, which are likely to include dysfunctions in 5-HT systems, it is possible that OT administration cannot reverse these effects. However, OT administration during pregnancy and over a longer time period might have been more effective. Finally, although this study is very preliminary, one might conclude that OT is more important for maternal responsiveness per se than for the mother's mood state (cf. [604]).

In my review of the animal literature, a case was made that the experience-based intergenerational continuity of maternal behavior was due, in part, to maternal treatment effects on the development of OT neural systems in the offspring. In the context of the findings that children who have been abused or neglected by their parents have an increased probability of maltreating their own children, is there any evidence that a disruption of OT neural activity might mediate aspects of this effect? In a preliminary study, Heim et al. [399] measured CSF levels of OT in healthy women who were either exposed or not exposed to early childhood abuse and/or neglect (based on retrospective self-reports). The women who reported being exposed to maltreatment by their parents had significantly lower levels of CSF OT than did the women who were not maltreated. The maternal behavior of the participants was not a subject of this study, and although the women ranged in age from 18 to 45 years, no data was given as to whether any of the participants had children of their own.

These studies highlight the possibility that OT administration might have therapeutic effects for certain types of maternal behavior dysfunctions in women. However, the appropriate experimental examination of such a proposal awaits the development of selective and long-acting nonpeptide OTR agonists that can be administered systemically (orally) and that can cross the blood–brain barrier.

5.7.2.5 *Conclusions*

In this selective review of the neurobiology of human maternal behavior, I emphasized those studies that could be related to the animal literature. While the human studies are primarily correlational, the overlap of those findings with the experimental animal research is impressive. Although much more research needs to be done, a case can be made that hypothalamic interactions with the mesolimbic DA system underpin maternal motivation and behavior in women and that such interactions are influenced and modulated by oxytocinergic mechanisms. Further, variations in the function of these systems, which may be influenced by the maternal care that women have received from their own mothers, may cause variations in maternal behavior, and a major disruption in the operation of these systems may cause serious disturbances in appropriate maternal responsiveness and mother–infant bonding.

At this point I would like to briefly discuss the possible involvement of vasopressin in human maternal behavior, since there is evidence that vasopressin is involved in both rodent maternal behavior [113,114] and in aggression. Recall from Chapter 3 on Aggression that there are polymorphic repetitive sequences in the regulatory region of the AVPR1A gene and certain alleles within the RS3 site are associated with aggression and with heightened amygdala reactivity to aversive social stimuli. More specifically, the longer RS3 alleles, including allele 334, have been associated with heightened emotionality and with increased transcription of the AVP1a receptor. In relation to these issues, a recent study found that postpartum women who were homozygous carriers of long RS3 alleles showed lower levels of maternal sensitivity when interacting with their children during free play and teaching sessions [98]. Further, this effect was only significant for those mothers who were exposed to early adversity (child abuse) when they were young. Although variations in the AVPR1A gene may directly influence maternal behavior, it seems more likely that such variations indirectly influence maternal responsiveness by affecting stress reactivity, emotionality, and aggressiveness.

These results emphasize that a complex phenotype such as maternal behavior is likely to be influenced by many genes and environmental factors. Although the effects of variations in any one factor may be mild, the combined effects of multiple factors may produce more dramatic results.

5.8 GENERAL CONCLUSIONS

By integrating research on the neurobiology of maternal behavior in mammals, including humans, a fairly comprehensive picture of the core neural and neurochemical mechanisms that regulate parental motivation and behavior has emerged. Although most of the detailed experimental work has been done on rats and sheep, the comparative analysis presented here indicates that there is a strong overlap in these basic mechanisms among nonhuman animals and humans. Further, the neural circuits regulating maternal, paternal, and alloparental behavior appear to be similar, although the specific manner in which such circuits are primed and activated, so that they can be accessed by infant-related stimuli, differs among these three types of parental behavior. Dysfunctions at critical nodes in this core circuitry would be expected to cause disruptions in parental behavior in all mammals, including humans.

A uniparental maternal care system is the most common form of parenting in mammals, and most of this chapter has been devoted to its neurobiology. In the typical mammal, the hormonal events accompanying the termination of pregnancy, co-acting with central oxytocinergic systems, prime and stimulate the brain so that the parturient female is immediately responsive to her young. Such immediate maternal responsiveness contrasts sharply with the response of virgin females to young, who reject and avoid infants. The analysis that I have presented clearly shows how physiological events, such as those associated with pregnancy termination, can reroute infant stimuli within the brain so that instead of being processed by "antisocial" rejection and avoidance neural pathways, they are received and processed by "prosocial" approach and acceptance neural circuits.

Although hormonal and oxytocinergic mechanisms stimulate the onset of maternal behavior in the typical parturient mammal, these physiological events are less important for the maintenance of maternal behavior throughout the remainder of the postpartum period. These facts suggest that some kind of neural plasticity underlies the enduring nature of the mother–infant bond, allowing infant stimuli to continue to gain access to the neural circuits regulating maternal behavior in the absence of hormonal and oxytocinergic priming and stimulation. The point was also made, however, that although OT is not essential during the maintenance phase of postpartum maternal behavior, variations in the activity of OT neural systems do modulate various types of maternal attentiveness, with increases in OT activity causing increases in the attentiveness of mothers to their infants.

The MPOA plays a dominant role in parental motivation, and I have presented evidence showing that one can achieve a concrete and mechanistic understanding of how MPOA interactions with other neural systems promote maternal responsiveness. In this analysis, I have also proposed models emphasizing that an understanding of the detailed neurochemical microcircuitry within neural regions, such as the amygdala, can help us understand how social stimuli might be differentially processed under different physiological conditions.

The interaction of the MPOA with the mesolimbic DA system regulates the appetitive aspects of maternal behavior, and I have described this interaction as one between a specific maternal motivational system (MPOA) and a nonspecific motivational system (the mesolimbic DA system). However, I also emphasized the possibility that MPOA neurons might be more generally involved in affiliative social motivation, although different MPOA neurons may contribute to different types of social affiliations, such as those between mates, and those between mothers and infants. Since the mother–infant bond is the most common enduring social bond in mammals, its neural circuitry may have provided a foundation for other types of strong social bonds, such as those between monogamous mating partners, and MPOA circuits may be an important element in this neural foundation. I will explore these ideas in more detail in the next chapter.

Because most experimental research on the neurobiology of maternal behavior has been conducted on animals, the core maternal circuits that I have emphasized have been related to goal-directed maternal behaviors that attract mothers to their infants. In humans, however, strong positive maternal *feeling* states are also an important element of what it means to be a mother, and a breakdown of such positive feelings is likely to contribute to maternal neglect. Since activity in the orbital PFC and the anterior insular cortex may be the basis of subjective feeling states, including maternal feelings, it is relevant to emphasize that the basal amygdala can directly influence neural activity in each of these regions (see Chapters 1, 4, and 7). Therefore, subdivisions of the same neural systems that promote goal-directed maternal responses (amygdala connections to NA-VP circuit) may also influence subjective maternal feelings in humans (amygdala connections to OFC and anterior insula). Further, orbital PFC and anterior insula influences on the MPOA, mediated by orbital PFC projections to the medial PFC, may be a route over which maternal feeling states influence goal-directed maternal responses (see Chapter 7).

The kind of care a young mammal receives from its parent(s) has dramatic impacts on the social and emotional development of the infant. For example, maternal neglect or abuse, through an experientially based process, can result in the development of increased aggressiveness and decreased maternal responsiveness in the affected young. Such effects on social and emotional development occur through multiple mechanisms. On the one hand, the specific neural circuits regulating maternal behavior may be affected by maternal abuse, neglect, or more subtle variations in the maternal care received by young. Further, maternal abuse or neglect also appears to more generally impact social behavior by increasing the likelihood that social stimuli will be processed over antisocial rather than prosocial neural pathways, and some of these effects are likely to be mediated by a down-regulation of oxytocinergic and serotonergic neural systems. Because of the experience-based intergenerational continuity of maternal behavior, there is also an intergenerational continuity of more general aspects of social and emotional development. Therefore, by understanding the neural basis of maternal behavior, potential interventionist strategies might be formulated that could prevent faulty maternal behavior, and this would then have broad ameliorative effects on the social and emotional development of offspring across multiple generations.

Note Added to Proof:

As the *Neurobiology of Social Behavior* was going to press, a research article was published that is relevant to several important issues discussed in Chapter 5: Wu Z, Autry AE, Bergan JF, Watabe-Uchida M, Dulac CG. Galanin neurons in the medial preoptic area govern parental behaviour. Nature 2014;509:325–330. Based on temporal changes in the cellular localization of Fos mRNA, as described in Section 5.2.11 of this book (also see [719]), Wu et al. present evidence that sexual behavior and parental behavior activate distinct MPOA neuron populations in the mouse, supporting a labeled-line point of view with respect to MPOA involvement in sexual and parental motivation. Furthermore, adding to the initial correlational findings of Tsuneoka et al. [979], Wu et al. provide experimental evidence that galanin neurons within the MPOA promote both maternal and paternal behavior in mice.

CHAPTER

6

Monogamy and the Formation of Enduring Social Attachments between Mating Partners

6.1 INTRODUCTION

When comparing mating systems in animals, one discovers that monogamy is rare in mammals but common in birds [131]. In a monogamous mating system, one male engages in sexual activity with one female, and then the male–female pair stays together and shares in the responsibility of raising offspring. Monogamy, therefore, represents the formation of an enduring and selective social attraction between two specific adult individuals who recognize each other and remain together after sexual activity has been consummated. The general view is that monogamy evolves under ecological conditions where both maternal and paternal behaviors are essential for infant survival and, therefore, the reproductive success of each of the mating partners [131].

About 95% of mammalian species are polygynous (or promiscuous), where one male typically mates with several females (or where both males and females mate with several partners), after which the mating partners leave each other without forming an enduring attachment [497]. In these cases, the impregnated female, upon giving birth, raises the offspring on her own. This system is the typical mammalian uniparental maternal care system, and the social bond that forms between the mother and her young, which may or may not be selective, represents the most common enduring social bond in mammals.

Approximately 5% of the remaining mammalian species exhibit a monogamous mating system, and many of these species are primates (gibbons, marmosets, tamarins, and titi monkeys, for example). Humans also represent a relatively monogamous species [225]. When describing monogamy, sometimes researchers differentiate social monogamy from genetic monogamy [225,770]. Although pair bonds are formed and maintained between mating partners, sexual "infidelity" may occur. In addition, species such as humans may also be described as engaging in serial social monogamy, where a long-term social attachment is formed between mating partners, but this bond is not necessarily permanent [225]. My concern in this chapter is with the neural mechanisms that underpin social monogamy or the formation of a stable and enduring (although not necessarily everlasting) social attraction between mating partners.

Neurobiology of Social Behavior
http://dx.doi.org/10.1016/B978-0-12-416040-8.00006-7

Although many examples of monogamy occur in primates, it is not practical to concentrate experimental neuroscience research solely on primates. In order to obtain a mechanistic understanding of the neurobiology of pair bonding and the formation of social attachments between adult mammals, experimentation with other species, more amenable to laboratory research, has been invaluable [625]. In particular, some rodent species within the *Microtus* genus (microtine vole species) are socially monogamous with mating partners forming enduring social bonds to each other, and research on such species has provided us with the largest body of experimental research on the underlying neural mechanisms involved [1060]. The first part of this chapter will review the vole research, and then I will present a comparative analysis in order to understand the relevance of the vole research to social pair bonding in other species, including humans.

Understanding the neurobiology of enduring social attachments across mammalian species will also provide us with insights into the neural basis of long-lasting love in humans [225]. The love that is expressed between two long-term partners (spouses) involves both prosocial feeling states or sentiments and prosocial behaviors. The nonhuman animal research presents evidence on the neural basis of the latter process—the formation of a long-term attraction between mating partners, which is one crucial element in long-lasting love. In Chapter 7, I will discuss the neural circuits that appear to allow prosocial sentiments, such as feelings of love toward one's partner, to activate prosocial behavioral attraction mechanisms.

Although the strongest social bonds in mammals are formed between a mother and her young, and between mating partners in monogamous species, humans also form rich, complex, and enduring social attachments to individuals other than their children and spouses. The extent to which an understanding of the neurobiology of monogamy is relevant to these broader aspects of human sociality will be considered in this chapter within a discussion of the neurobiology of autism. Chapter 7 will present a more extensive analysis of human sociality.

6.2 THE VOLE MODEL SYSTEM

6.2.1 Introduction

Species within the *Microtus* genus exhibit different forms of social organization ranging from promiscuous and maternal to monogamous and biparental [625,1067]. Table 6.1 compares prairie voles and pine voles on the one hand with montane voles and meadow voles on the other. Prairie and pine voles are highly social, and mating partners form lasting pair bonds after mating. The male also develops selective aggression or mate guarding after mating where he shows aggressive responses toward unfamiliar conspecifics that enter the pair's territory. Recent work suggests that the same may be true for females [1061]. Therefore, once a pair bond is formed, the mating partners develop a selective attraction to each other and an aversion toward unfamiliar conspecifics. Furthermore, the breeding pair nests together, and both parents care for the young. The offspring remain in the breeding pair's territory for several weeks after they are weaned and may engage in alloparental behavior toward subsequent litters. Therefore, both prairie and pine voles represent cooperatively breeding social units. In contrast, the montane and meadow voles are relatively asocial, breed promiscuously without forming pair bonds, the sexes separate after mating, paternal behavior does not occur, and the female shows maternal behavior until the young are weaned.

Importantly, these four vole species are closely related genetically and are very similar in physical appearance, but in nature they inhabit different ecologies (see Table 6.1). A monogamous mating system, biparental care, and delayed dispersal of offspring appears to

TABLE 6.1 Comparisons between the Social Organization of Four Different Vole Species

Characteristics	Meadow Vole and Montane Vole	Pine Vole and Prairie Vole
Mating system	Promiscuous	Monogamous
Parental care	Maternal	Biparental
Partner preference	No	Yes
Male selective aggression (mate guarding)	No	Yes
Alloparental behavior	No	Yes
Ecology	High density of resources	Low density of resources
Offspring dispersal from mother or family unit	Early	Delayed

be adapted to the low density of suitable territories and food resources in the prairie and pine vole environments, while plentiful resources in the montane and meadow vole environments permit a nonmonogamous system, uniparental maternal care, and dispersal of young after they are weaned so they can acquire their own territories [164,784]. Therefore, natural selection has presumably affected the social organization of these closely related vole species, and research has been aimed at discovering the nature of the genetic and neurobiological changes that underlie these species differences.

Several steps are obviously involved in the formation of pair bonds between mating partners in such species as the prairie vole. The first process involves the development of individual recognition: each mating partner must learn to recognize the specific stimulus characteristics of its mate so that the partner can be distinguished from unfamiliar conspecifics. The next step involves two processes: (1) the development of a persistent selective attraction to the mating partner, which persists in the absence of continued sexual activity, and (2) an aversion or rejection of other unfamiliar conspecifics. Since these are complex behavioral processes, they are likely to be represented by complex and diverse neurobiological mechanisms. For example, the neural mechanisms involved in individual recognition are probably at least partially distinct from those that underpin social attraction. Most of the detailed research on pair bonding in voles has been aimed at uncovering the mechanisms underlying the latter process, which leads to the formation of an enduring attraction between partners.

In Chapter 4, I discussed the neurobiology of sexual behavior, and evidence was presented that neurons in the medial preoptic area (MPOA) are involved in the appetitive or attraction aspects of both male and female sexual behavior, presumably through the interaction of particular MPOA neurons with the mesolimbic dopamine (DA) system. Yet in most mammals, although the two sexes are attracted to each other during mating, which is a rewarding experience, once mating is complete the sexes separate and a pair bond is not formed. Indeed, the typical polygynous male mammal, after mating with one female, will then search for additional receptive females to mate with [303]. What is it about the neurobiology of monogamous species that changes this basic mammalian pattern? What are the mechanisms that allow the mating partners to remain attracted to each other outside the context of mating behavior? How is the brain of the prairie vole, for example, different from that of the montane vole so that a stable pair bond is formed in the former but not the latter species? Can the mechanisms underlying monogamy in certain vole species be extended to explain monogamy in other mammals, or are there different routes to achieve monogamy in different mammals? The answers to these and related questions are important goals of this chapter.

In Chapter 5, I reviewed the neurobiology of maternal behavior. As previously indicated, several scholars have suggested that the neural circuitry and mechanisms utilized to form the mother–infant bond may have been co-opted to form a neural foundation upon which other strong social bonds, such as the pair bond in monogamous species, are built. If this is the case, although one should not expect complete commonality, there should be important similarities between the neural processes that contribute to the formation of mother–infant bonds and pair bonds. Aspects of this chapter will be aimed at uncovering such similarities. Although most research on the neurobiology of maternal behavior has been done on rats, the phenomenology of sheep maternal behavior fits most closely with that of pair bonding in monogamous species. While maternal rats form a long-term attraction to a general pup stimulus through a process that does not include individual recognition, the attachment of the maternal ewe to her lamb involves both individual recognition and attraction. Therefore, a neural model of mother–infant bond formation that represents a synthesis of the rodent and sheep research will be compared to a neural model that explains pair bond formation in voles. Such a comparison will allow us to judge the extent of overlap between the two systems.

For the most part, the research on the neurobiology of pair bond formation is more extensive than the research on the mother–infant bond and maternal memory, and in a chronological sense our understanding of the detailed neurobiological mechanisms of pair bond formation preceded that of the mother–infant bond. However, since the mother–infant bond is the most common strong social bond in mammals, it made heuristic sense to have presented the maternal research first in Chapter 5 in order to now compare it to the research presented in this chapter. To the extent that an important commonality exists, ongoing maternal behavior research may suggest neural possibilities for other types of strong prosocial bonds in mammals, and research on adult monogamy and pair bonding may not only be generally relevant to the neurobiology of sociality but may also suggest neural mechanisms that might contribute to the formation of mother–infant bonds.

6.2.2 The Laboratory Study of Pair Bonding and Partner Preference Formation in Voles

Mating in vole species usually involves 15–30 bouts of copulation during a 24h period. The formation of a partner preference has been used to study pair bond formation in the laboratory. As shown in Figure 6.1, a male and a female are allowed to mate over a 24h period, and then a partner preference test is conducted in a three-chambered apparatus [444,675,1067]. Let's say one wanted to test whether a male formed a preference or selective attachment to his mate. In this case, his partner (the female he mated with) would be tethered in one chamber of the apparatus and an unfamiliar female would be tethered in another chamber. The male would then be placed in a third (central) neutral chamber from which he could move around freely through all parts of the apparatus, and over a 3-h period the amount of time the male spends with each female would be recorded. A similar test could be conducted to measure bond formation in females, but in this case familiar and unfamiliar males would be tethered in different chambers and the female would be free to move around and choose which male to remain in contact with. When prairie voles are tested in this way, both males and females choose to

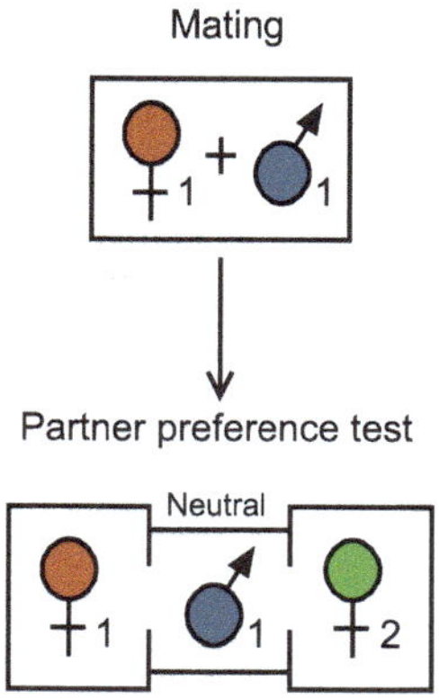

FIGURE 6.1 Laboratory tests for partner preference formation in males of vole species. A particular male (male 1) is allowed to mate with a particular female (female 1) over a 24-h period. Following this mating period, the male is placed in a three-compartment cage, with his mating partner tethered within one chamber, and a strange female (female 2) tethered in another chamber. The male is initially placed in a center neutral chamber and is then allowed to move around freely. The amount of time male 1 spends with each female, or in the neutral chamber, is recorded. A similar test could be conducted to measure pair bond formation in female voles, but in this case familiar and unfamiliar males would be tethered and the mated female would be free to move around and choose which male to remain in contact with.

spend most of the 3-h test period in contact with the individual with whom they mated. In contrast, nonmonogamous voles, such as montane or meadow voles, do not differentiate between familiar and unfamiliar conspecifics but instead spend approximately equal amounts of time with each social stimulus, and an even larger amount of time alone in the neutral chamber [442,554].

In order to examine the brain mechanisms underlying pair bond formation in monogamous voles, two basic procedures have been used. In a disruption procedure, a male and a female are allowed to mate for 24 h while the brain of one of the partners is manipulated in an attempt to interfere with the bonding process without disturbing sexual behavior and mating. In a facilitation procedure, a male and a female cohabitate for about 6 h without the occurrence of mating (for example, the female could be ovariectomized). Such a suboptimal cohabitation experience does not result in the formation of a partner preference. Using this suboptimal paradigm, the brain of one of the paired conspecifics would be manipulated in an attempt to promote the bonding process. Specific examples of these basic methods will be described below.

6.2.3 The General Importance of Oxytocin and Vasopressin for Pair Bond Formation in Prairie Voles

Early studies that used intracerebroventricular (ICV) injections of oxytocin (OT) or vasopressin (AVP) agonists or antagonists indicated that OT and vasopressin neural systems were importantly involved in pair bond formation in prairie voles, but that an important sex difference existed: OT action on the OTR was found to be more important for the development of a partner preference in females, while vasopressin action on the V1a receptor was more important for the development of male bonding to his female partner [442,1038]. These results are shown in Table 6.2. Briefly, ICV injection of an OT receptor antagonist (OTA) immediately prior to a 24-h mating episode blocked partner preference formation in female prairie voles but not in males. In contrast, ICV injection of a V1a receptor antagonist (VPA) immediately prior to mating blocked partner preference in males but not females. It is worth noting that each of these antagonists did not disrupt sexual behavior even though the antagonists remained bound to their respective receptors for most of the 24-h mating period. Similar conclusions were reached with the facilitation paradigm. Male prairie voles were cohabitated with ovariectomized females so that mating could not occur. Continuous ICV infusions of OT during the cohabitation period only facilitated pair bond formation in female prairie voles, while similar infusions of AVP only facilitated pair bond formation in males.

TABLE 6.2 The Effects of Manipulating OT and AVP Neural Systems on Pair Bond Formation in Prairie Voles

	Disruption Paradigm: 24 h of Mating + ICV Treatment					
	Males			**Females**		
Treatment I.	**CSF**	**OTA**	**VPA**	**CSF**	**OTA**	**VPA**
Partner preference formation (as assessed during a subsequent 3-h test)	Yes	Yes	No	Yes	No	Yes
	Facilitation Paradigm: 6 h Cohabitation without Mating + ICV Treatment					
	Males			**Females**		
Treatment II.	**CSF**	**OT**	**AVP**	**CSF**	**OT**	**AVP**
Partner preference formation (as assessed during a subsequent 3-h test)	No	No	Yes	No	Yes	No

Abbreviations: AVP = vasopressin; CSF = artificial cerebrospinal fluid; ICV = intracerebroventricular; OT = oxytocin; OTA = oxytocin receptor antagonist; VPA = vasopressin receptor antagonist.

This sex difference, with OT being more important in females and AVP being more important in males may, in part, be related to the finding that male prairie voles have more AVP-containing neurons than do female prairie voles in extrahypothalamic sites, such as medial amygdala (MeA) and bed nucleus of the stria terminalis (BST) [1018]. Subsequent studies, however, have cast some doubt on this strict sexual distinction between the functions of OT and AVP, and instead have suggested that OT and AVP may both be involved in the formation of partner preferences in male and female prairie voles [37,181,730,1016]. For example, using a suboptimal facilitation paradigm, Cho et al. [181] paired male and female voles together for only 1 h and found that ICV OT or AVP could promote pair bond formation in each sex and that concurrent administration of either OTA or VPA could block the facilitation effect of either OT or AVP.

The interpretation of these results is complicated by the facts the OT and AVP are very similar in their amino acid sequence, which allows each neuropeptide to bind to the receptor of the other peptide to some extent, and that many "selective" OTR antagonists and V1a receptor antagonists have some ability to block V1a and OT receptors, respectively [408,595].

Several aspects of these findings are worth discussing further. Since ICV injections were utilized, the neural site(s) where AVP and OT act to facilitate pair bonding cannot be determined. The results also imply that endogenous OT and AVP are released into the brain during mating behavior in order to act somewhere to facilitate bonding. In females, it is well known that the vaginocervical stimulation that accompanies sexual activity activates OT release into the brain, and microdialysis studies have confirmed such release into the female prairie vole brain during mating [836]. Definitive research has yet to be presented with respect to whether AVP is released into the male brain during sexual activity (cf. [555]). Interestingly, there is evidence for the central release of OT during male sexual behavior [111,199]. With respect to neural sites of action, it is certainly possible that OT and/or AVP act at one or more neural sites to promote individual recognition of one's mating partner and at different neural sites to promote the development of the strong attraction between mating partners that persists after the completion of sexual activity. Along these lines, I will make the case that OT may act at the level of the amygdala to facilitate individual recognition, while OT and AVP act within the nucleus accumbens–ventral pallidum (NA-VP) circuit to promote the development of the persistent attraction that prairie vole partners exhibit toward each other. The results reviewed so far indicate that OT and AVP action on the brain during an initial mating experience is necessary for the formation of the pair bond. An interesting question is whether the continued release of OT and/or AVP is necessary for the maintenance of the pair bond after its formation and after mating has been consummated. Recent evidence has indicated that continued release of AVP into critical brain sites is necessary for the maintenance of a male prairie vole's preference for the female partner that he mated with [254], but evidence as to whether continued OT release is necessary for the maintenance of an established pair bond is not yet available for prairie voles. Finally, note the contrast between the role of AVP in pair bond formation and its role in aggression. AVP action in the anterior hypothalamus stimulates aggression, an antisocial response, but its action within the NA-VP circuit will be shown to be important for pair bonding, a prosocial response. Obviously, the same neurotransmitter can produce different functional effects that are mediated by distinct neural circuits. Remember, however, that pair bonding is associated with mate guarding. Therefore, the bond formation effects of AVP acting at one neural site appear to be supported by the pro-aggressive effects of this neuropeptide at other sites, which affects the way a bonded male interacts with unfamiliar conspecifics. Therefore, the coordinated release of AVP at different neural sites may affect different behaviors (aggression toward unfamiliar conspecifics versus attraction to a familiar female), but may influence a singular adaptive outcome, the maintenance of monogamy.

6.2.4 The Mesolimbic DA System and Partner Preference Formation in Prairie Voles

From Chapter 4, recall that DA release into the nucleus accumbens (NA) regulates the appetitive, but not the consummatory, aspects of sexual behavior in male and female rats. Rats are a nonmonogamous species, and such DA release is involved in maintaining an attraction between the sexes during the mating period only, but a permanent bond is not formed. Interestingly, several studies have shown that activation of mesolimbic DA input to the NA shell (NAs) serves an additional function in monogamous rodents such as the prairie vole—it is also important for the formation of partner preferences and long-term pair bonds in *both* sexes [35,36,335,1019]. Some of these results are shown in Table 6.3. The basic findings of these studies showed that: (1) DA is released into NAs during mating in both male and female prairie voles; (2) when male and female prairie voles are allowed to mate over a 24-h period, microinjection of a D2 dopamine receptor antagonist, but not a D1 antagonist, into NAs during the mating period blocks the formation of a partner preference; (3) when male and female prairie voles cohabitate together for 6h without mating, microinjection of a D2 dopamine receptor agonist, but not a D1 agonist, into NAs during the cohabitation period facilitates partner preference formation. These results show that

TABLE 6.3 The Effects of Dopamine on Pair Bond Formation in Male Prairie Voles

	Disruption Paradigm: 24h of Mating + NA Microinjection			
Treatment I.	**Vehicle**	**D1 Antag**	**D2 Antag**	**D1 Agon**
Partner preference formation (as assessed during a subsequent 3-h test)	Yes	Yes	No	No
	Facilitation Paradigm: 6h Cohabitation without Mating + NA Microinjection			
Treatment II.	**Vehicle**	**D1 Agon**	**D2 Agon**	**D2 Agon + D1 Agon**
Partner preference formation (as assessed during a subsequent 3-h test)	No	No	Yes	No

Based on results from Aragona et al. [36].
Abbreviations: D1 Agon = D1 dopamine receptor agonist; D1 Antag = D1 dopamine receptor antagonist; D2 Agon = D2 dopamine receptor agonist; D2 Antag = D2 dopamine receptor antagonist; NA nucleus accumbens.

DA action on D2 receptors in NAs during mating is necessary for pair bond formation and the subsequent selective attraction of the partners to one another during the preference test. Note that although D2 antagonist injections into NAs block pair bond formation, such injections do not disrupt copulation, presumably because in a small cage situation sexual behavior primarily consists of consummatory elements.

Aragona et al. [36] have reported additional interesting results. Not only is DA action on D1 receptors in NAs not necessary for pair bond formation but such action actually impairs social bonding in male prairie voles. D1 agonist injection into the NAs of males during a 24-h mating period blocks pair bond formation and also blocks the facilitation of pair bond formation induced by a D2 agonist when a male cohabits with a female for 6h without mating.

A perplexing issue pertaining to these results is that when DA is released into NAs during mating, one would expect it to act on both D1 and D2 receptors, and if that were the case, DA action on D1 receptors should counteract the facilitation effects of D2 receptor activation on pair bond formation. Aragona and Wang [37] have suggested a way out of this dilemma. They propose that the amount of DA that is released into NAs during mating may be modest rather than high, which would allow DA to preferentially activate high affinity D2 receptors to a much greater extent than its activation of low affinity D1 receptors. Future research should be aimed at testing this hypothesis.

Given that DA is released into the NAs of both sexes during a mating episode, and that such release occurs in both monogamous and nonmonogamous species, why do pair bonds only form in monogamous species? One possibility, suggested by the DA data, is that nonmonogamous species, such as meadow and montane voles, may have a higher density of DA D1 receptors in NAs than do prairie voles, which allows DA action on the D1 receptor to counteract the pair bond promoting action of DA action on the D2 receptor. In fact, Aragona et al. [36] have found that meadow vole males have a higher density of D1 receptors in NAs than do prairie voles. However, when a D1 receptor antagonist was injected into NAs of male meadow voles during mating with a female, pair bond formation did not occur, and such males did not develop a selective attachment to their mates. These results show that a decrease in DA activation of D1 receptors is not sufficient to promote pair bond formation in male voles and that additional factors are needed. As I will show, DA action of D2 receptors in NAs interacts with OT and AVP neural systems to promote the formation of long-lasting pair bonds, and such an interaction is restricted to monogamous vole species.

6.2.5 Species Differences in OT and Vasopressin V1a Binding Sites in the Brain Correlate with Species Differences in Social Organization and Mating Systems of Voles

As previously reviewed for rats, the main source of OT neural projections that target brain regions is derived from neuronal cell bodies in the paraventricular nucleus (PVN), while AVP neural projections to diverse brain regions are derived from neuronal cell bodies that include the PVN, MeA, and BST. Research has indicated that monogamous and nonmonogamous vole species do not exhibit significant differences in OT and AVP immunoreactive cell bodies and axons within the brain [836,1020]. However, there are significant species differences in distribution and density of OTRs and vasopressin V1a receptors within the brain that distinguish vole species on the basis of mating systems. Some of these differences are catalogued in Table 6.4 [443,552,1065,1068]. In vitro radioligand receptor autoradiography for OTRs and vasopressin V1a receptors indicates that both male and female prairie voles have a greater density of OTRs in the NA and V1a receptors in the ventral pallidum (VP) than do male and female montane voles. What these results suggest is that even if OT and AVP were to be released into the brain during a mating episode in both monogamous and nonmonogamous species, such release would undoubtedly have differential effects on behavior due to differences in the degree to which particular target neural regions would be affected by each of the neuropeptides. As I will review, the selective action of OT within the NA, and AVP within the VP, of vole species that exhibit a monogamous mating system, in combination with DA action on D2 receptors in NA, may affect the NA-VP reward circuit it such a way that the attraction of the mating partners to each other persists long after mating behavior has ceased.

With respect to the source of OT input to the NA, evidence suggests that the PVN provides the major input, although some contribution may also be made by the supraoptic nucleus [502,836]. The source of the AVP input to the VP has not been conclusively determined, and there is evidence that MeA, BST, and PVN neurons might all be involved [555,825]. Since the MeA is a recipient of olfactory input, it makes sense to propose that as the sexes use olfaction

TABLE 6.4 OTR and V1aR Density in Selected Neural Regions of Montane Voles (Promiscuous) and Prairie Voles (Monogamous)

Brain Region and Receptor	Montane Vole	Prairie Vole
VP-V1aR	Low	High
LS-V1aR	High	Low
NAs-OTR	Low	High
LS-OTR	High	Low
BLA-OTR	Low	High

Abbreviations: BLA-OTR = density of oxytocin receptors in the basolateral amygdala; LS-OTR = density of oxytocin receptors in the lateral septum; LS-V1aR = density of V1a vasopressin receptors in the lateral septum; NAs-OTR = density of oxytocin receptors in the nucleus accumbens shell; VP-V1aR = density of V1a vasopressin receptors in the ventral pallidum.

to investigate one another during mating bouts, that olfactory input may activate AVP release into the VP, and in those species with dense V1a binding sites in VP, the action of vasopressin may play a role in creating a stable and enduring attraction between the sexes.

An interesting older study by Insel and Shapiro [443] provides some information relevant to a comparison of the mechanisms underlying pair bond formation with those underlying mother–infant attachment. As indicated in Table 6.4, not only do sexually mature prairie voles have more OTR binding sites in NA than do montane voles but they also have a greater density of such binding in the BLA. However, Insel and Shapiro found that OTR binding increases in the BLA of maternal postpartum montane voles, reaching a level equal to that observed in sexually mature non-postpartum female prairie voles. These findings have several implications. The fact that prior to mating the number of OTRs in BLA is greater in sexually mature female prairie voles than in their montane vole counterparts suggests that OT action in BLA may contribute to pair bond formation. The fact that OT binding in BLA increases in postpartum montane females suggests that OT action on BLA may promote the mother–infant bond. Experimental evidence to support these proposals is not available, but if OT did act on BLA to facilitate both types of bonds, it would be a good example of how the neural mechanisms regulating the mother–infant bond might overlap with those that regulate pair bond formation.

6.2.6 The Neural Sites of Action of OT and AVP in the Regulation of Pair Bond Formation

6.2.6.1 The Amygdala as a Potential Site Where OT May Act to Affect the Individual Recognition Process of Pair Bond Formation

In understanding the mechanisms of partner preference formation in monogamous voles, at least two neural processes are occurring. A mating partner must first learn to recognize its mate, and then a persistent attraction to that mate's characteristics must develop. Research on prairie voles has not been aimed at clearly differentiating the neural mechanisms that underlie experience-based individual recognition from those that underlie the creation of a persisting attraction to one's mating partner. However, I will try to make the case that OT action at the level of the amygdala during mating may allow the stimulus characteristics of one's mate to activate amygdala neurons with a positive valence. The stimulus characteristics of strange females would not have this property but, instead, would activate amygdala neurons with a negative valence, leading to withdrawal and rejection. Once this recognition mechanism is "stamped in", research suggests that conspecific mating partner stimuli are relayed to the NA-VP circuit where the action of OT on NA and AVP on VP, in conjunction with D2 receptor activation in NA, results in the development of a long-lasting attraction toward the stimulus characteristics of one's mate.

Research has shown that prairie voles, like other rodents, use olfactory cues to distinguish between conspecifics and that such cues are importantly involved in mate recognition [684,1035]. Since the amygdala is a major recipient of olfactory input, it would seem to be a prime candidate for regulating individual recognition. An interesting study by Kirkpatrick et al. [491] explored the role of the amygdala in the social attraction that develops between conspecific prairie voles. Male prairie voles were paired with an ovariectomized female for 2 days, during which time the male presumably became familiar with the female's stimulus characteristics. The males then received either large neuron-specific lesions of the amygdala that damaged MeA and BMA, small amygdala lesions restricted to MeA, or sham lesions. Subsequently and over a period of several days, each male was placed in one compartment of a *two-compartment* cage, while the familiar female was placed in the other compartment. Control males and males with MeA lesions, but not males with the large lesions, spent most of their time in the compartment with the familiar female, and this duration of physical contact with the familiar female was significantly greater than that exhibited by males with the combined MeA and basomedial (BMA) lesions. The test utilized by Kirkpatrick et al. [491] was not a social discrimination test. It was a social attraction test, since the male had to choose between interacting with a familiar female or remaining alone. The MeA/BMA appears important for the attraction to the female, but one cannot determine whether these regions are also important for individual recognition. That information would have been available if a three-compartment cage had been employed and the male had to choose between the familiar and a novel female. In relation to the chapter on parental behavior, perhaps BMA projections to the VP regulated the attraction process in the Kirkpatrick et al. study. A further contrast concerns the research on maternal selectivity in sheep, where lidocaine injection into MeA at parturition was shown to disrupt the ability of the maternal ewe to distinguish between her own and an alien lamb without disrupting maternal behavior. If the males in the Kirkpatrick et al. study had been given a preference test rather than a social attraction test, would selective MeA lesions have disrupted individual recognition and therefore partner preferences?

Research on nonmonogamous rodent species, such as rats and mice, has indicated that vasopressin action in the lateral septum contributes to the development of individual recognition in males, while oxytocin (OT) action in MeA contributes to social recognition in both males and females [96,185,321,580,853]. Given my proposal that OT action in the amygdala may contribute to the individual recognition process that participates in pair bond formation, I will restrict my analysis to OT action on MeA. In the typical social recognition test that is used in rats and mice, a test subject is exposed to a stimulus conspecific for about 5 min, and the amount of time that the test animal spends in olfactory investigation (anogenital sniffing) of the stimulus conspecific is recorded. The stimulus animal is then removed, and after a short interval (about 30 min) the test animal is exposed to either the familiar conspecific or a new stimulus animal. Individual recognition is demonstrated by the finding that the test animal shows a dramatic decrease in olfactory investigation of the familiar conspecific, but shows a high level of investigation of novel stimulus animals. Note that this test is not a measure of a persistent selective attraction between two individuals. It is just a simple measure of individual recognition. Indeed, the rats and mice used in these studies have social systems that are not characterized by the formation of persistent and selective social attractions between individuals.

Ferguson, Aldag, Insel, and Young [292] compared the social recognition process in wild-type (WT) and OT knockout mice (OT−/−). While WT male mice developed social recognition as described above, OT−/− mice continued to investigate both "familiar" and novel stimulus animals at the same high level. When OT was microinjected directly into MeA 10 min before the first 5-min encounter, it restored social recognition as measured by the fact that olfactory investigation of the original stimulus animal declined on the second encounter. In contrast, microinjection of an OTR antagonist into the MeA of WT mice disrupted social recognition in that these test male mice investigated the same stimulus animal to a high degree on both the first and second encounters (also see [853]). These results indicate that in male mice, OT action on MeA at the time of an initial encounter is necessary for the test animal to subsequently recognize the individual that he was exposed to. Results consistent with these effects in male mice also occur in female mice [185].

Given the role of olfaction and the MeA/BMA in the attraction of familiar prairie voles to one another, and the role of OT action on MeA in social recognition in male and female mice, it is surprising that research on the role of OT at the level of the amygdala in pair bond formation in prairie voles has not been conducted. Lim and Young [555] have ruled out a role for AVP action on V1a receptors in MeA in pair bond formation, but no one to my knowledge has explored the effects of microinjection of an OTR antagonist into MeA/BMA at the time of mating on pair bond formation in prairie voles. In making comparisons between pair bond formation in voles and maternal selectivity in sheep, it would also be interesting to know whether OT action on MeA is necessary for the maternal ewe to develop a discrimination between her own and an alien lamb.

6.2.6.2 OT Action on the NA and the Persistent Attraction Process of Pair Bond Formation in Female Prairie Voles

It has been hypothesized that pair bond formation in voles results from an association between a rewarding mating experience and the specific olfactory characteristics of the partner [838,1065]. Although all rodents may learn the olfactory characteristics of their mates, and although DA is released into NA and mating is rewarding for all rodents, only certain species form pair bonds because only certain species develop a persistent attraction to their mate's stimulus characteristics. Given the early research that emphasized the importance of OT for partner preference in females and AVP for partner preference in males, and given that OTRs in NA and vasopressin V1a receptors in VP are higher in both sexes of monogamous than nonmonogamous voles, research has focused on whether OT acts on NA to promote partner preference formation in female prairie voles, and whether AVP acts on V1a receptors to promote partner preference formation in male prairie voles.

In an important study, Liu and Wang [560] showed that OT acts on NA to promote pair bond formation in female prairie voles and that such action also involves concurrent activation of DA D2 receptors. Using a facilitation paradigm, female and male prairie voles were placed in a cage for 6h of cohabitation without mating. Microinjections of OT into NA of females during the 6-h cohabitation period facilitated partner preference formation as measured during the subsequent 3-h test with the familiar male and a novel male. Importantly this facilitation of female pair bonding was blocked if either an OTR antagonist or a D2 DA receptor antagonist was concurrently injected with OT. In a subsequent experiment, the facilitation of female pair bonding induced by injection of a D2 receptor agonist into NA was blocked by the concurrent administration of either a D2 receptor antagonist or an OTR antagonist.

These results indicate that the combination of OTR activation and DA-D2 receptor activation in NA is involved in the formation of a female prairie vole's persistent attraction to her male partner. An important question is the particular cellular mechanism through which OT and DA interact within NA. Although I will discuss this issue in more detail later, research does indicate that OTRs are located postsynaptically on NA neurons in prairie voles rather than being located presynaptically on DA axon terminals originating from the ventral tegmental area (VTA) [1063]. In terms of cellular architecture, it would be interesting to know whether OTRs are colocalized on NA medium spiny output neurons (MSNs) that also contain D2 receptors, allowing for some kind of direct interaction. Recall from Chapter 5 that D2 receptors and OTRs can form heteromers in the NA, which then allows OT to enhance the effects of DA on the associated D2 receptor. Another interesting possibility is that OTRs are located on inhibitory interneurons in NA that, in turn, inhibit MSNs that contain D1 dopamine receptors. Recall that OT action on OTRs typically exerts excitatory effects and that DA action on D1 receptors in NA appears to antagonize pair bond formation. Also, recall from Chapter 1 that DA action on

MSN D1 receptors is excitatory. Therefore, if OT could act to depress the output of MSNs with D1 receptors, it could promote the facilitation of pair bond formation that results from a presumed DA action on MSNs with D2 receptors.

Given the more recent research that suggests that OT may be involved in pair bonding in both male and female prairie voles, it would be interesting to explore the potential involvement of OT action on the NA in the development of mating-induced partner preferences in male prairie voles. Aragona and Wang [37] do cite unpublished research supporting the view that OT action on NA is important for partner preference formation in male prairie voles.

Recent research also indicates that epigenetic events may contribute to the facilitation of pair bonding in prairie vole females. Cohabitation with a male increased histone acetylation within the regulatory region of the OTR gene in NA, and this epigenetic process was associated with increased levels of OTR mRNA and protein in NA [1016]. Therefore, through epigenetic mechanisms, the already high levels of OTR expression in NA of unmated female prairie voles may be further enhanced by interactions with a mating partner.

The results of these studies also shed light on one of the reasons why partner preferences do not develop in nonmonogamous vole species, such as meadow and montane voles. Although DA may be released into the NA during mating in these species, the low density of NA OTRs would prevent the critical interaction between OT and DA action on D2 receptors in the NA that is necessary for the formation of a persistent attraction. Therefore, even if an individual recognition process were to develop, a persistent attraction to the recognized stimulus would not. In this regard, Ross et al. [837] asked whether an increase in OTR expression in the NA would be *sufficient* to promote partner preference formation in the naturally nonmonogamous female meadow vole. This study made use of recombinant adeno-associated viral (rAAV) vector gene transfer (see [674]). In this procedure, a particular gene of interest is inserted into a replication deficient virus, whose own structural genes have been removed. The rAAV can then be microinjected into a particular neural region of a host organism, and the host's RNA polymerase will then transcribe the inserted gene, which in turn, would increase the expression of its protein. Ross et al. [837] microinjected a rAAV that contained the coding region of the prairie vole OTR gene attached to a neuron-specific promoter into the nucleus accumbens of female meadow voles. Receptor autoradiography confirmed that the injection of the adenovirus with the OTR gene code did increase the density of OT binding sites in the NA of meadow vole females. However, this upregulation of OTR expression was not sufficient to promote partner preference formation after 24 h of cohabitation and mating with a male. Indeed, even 48 h of cohabitation with the male did not facilitate partner preference formation in the affected female meadow voles.

One interpretation of these results is that although OT action on the NA may be necessary for pair bond formation, it is not sufficient. Some other difference(s) between prairie voles and meadow voles are likely to contribute in important ways to differences in their mating systems. With respect to the finding that meadow voles have a higher density of D1 receptors in NA than do prairie voles [36], perhaps upregulation of OTRs in NA combined with the inhibition of DA action on D1 receptors in NA would have facilitated partner preference formation in meadow voles.

6.2.6.3 Vasopressin Action on V1a Receptors in Ventral Pallidum and the Persistent Attraction Process of Pair Bond Formation in Male Prairie Voles

Given that prairie voles exhibit high levels of vasopressin V1a binding sites in the VP and that vasopressin may play a more important role in the formation of partner preferences in males than in females, Lim and Young [555] examined the effects of injections of a selective V1a receptor antagonist into VP on partner preference formation in male prairie voles. Males were allowed to mate and cohabitate with females for about 24h. Immediately prior to mating, males received site-specific brain injections of either a long-acting V1a antagonist or cerebrospinal fluid (CSF) control injections. In comparison to CSF injections into VP or injections of the V1a antagonist into control sites, injection of the antagonist into VP disrupted pair bond formation (also see [67]). It would be interesting to determine whether AVP action on V1a receptors in VP would also be necessary for pair bond formation in female prairie voles.

In a remarkable study, Lim et al. [554] examined whether rAAV vector transfer of the V1a receptor gene into the VP of male meadow voles would act to facilitate partner preference formation in this nonmonogamous species. In contrast to the work by Ross et al. [837] on viral vector transfer of the OTR gene to the NA of female meadow voles, Lim et al. [554] found that viral vector transfer of the coding region of the V1a receptor gene to the VP of male meadow voles both increased the expression of the V1a receptor in this critical brain region and also allowed the male meadow vole to form a partner preference with the female that he mated with. Therefore, in meadow vole males, upregulation of the expression of a single gene, the V1a receptor gene, at a specific neural site, the VP, appears sufficient to promote the formation of a partner preference. It would be interesting to explore whether such a manipulation would also be effective in female meadow voles.

6.2.7 A Neural Model of Pair Bond Formation in Prairie Voles: Comparisons with Mother–Infant Bonding

6.2.7.1 Pair Bonding in Prairie Voles

Based on the research described above, I have devised a hypothetical neural model that may underlie pair bond formation in prairie voles (see Figure 6.2). This model combines the findings that OT may act on the amygdala to promote individual recognition and that OT and AVP may act within the NA-VP circuit to facilitate social attraction. The model also underscores the concept that the stimulus an organism is attracted to is dependent upon the particular stimulus that is actually relayed to the NA-VP circuit.

The model proposes that during mating behavior in both sexes, the MPOA activates the VTA and the PVN. Activation of the mesolimbic DA system during mating is common to all mammals, but the expression of high levels of OTRs in NA and V1a receptors in VP distinguishes monogamous from nonmonogamous vole species. Activation of the PVN by MPOA is proposed to activate the release of OT into both the amygdala and the NA. Olfactory investigation of one's mating partner is shown as providing neural inputs to both MeA and basomedial/basolateral amygdala (BMA/BLA), and some MeA neurons are shown to contain AVP. In the model, OT action in MeA and BMA/BLA acts to strengthen the connection between the olfactory characteristics of one's mate and positively valent amygdala neurons. This process would comprise the individual recognition step of partner preference formation, since one's mate would be able to strongly activate these amygdala neurons, while a stranger would not. BMA/BLA glutamatergic neurons with a positive valence are shown to project to appetitive NA-VP neural circuits, while positively valent MeA neurons that contain AVP are shown as projecting to VP, where they exert excitatory effects on V1a receptors. DA action on D2 receptors and OT action on OTRs in NA are proposed to depress the activity of NA GABAergic MSNs that project to VP, resulting in a disinhibition of VP. Decreased inhibition on VP allows the amygdala-derived stimulus input from the mating partner to strongly activate the VP, causing a strong attraction between the mating partners. In addition,

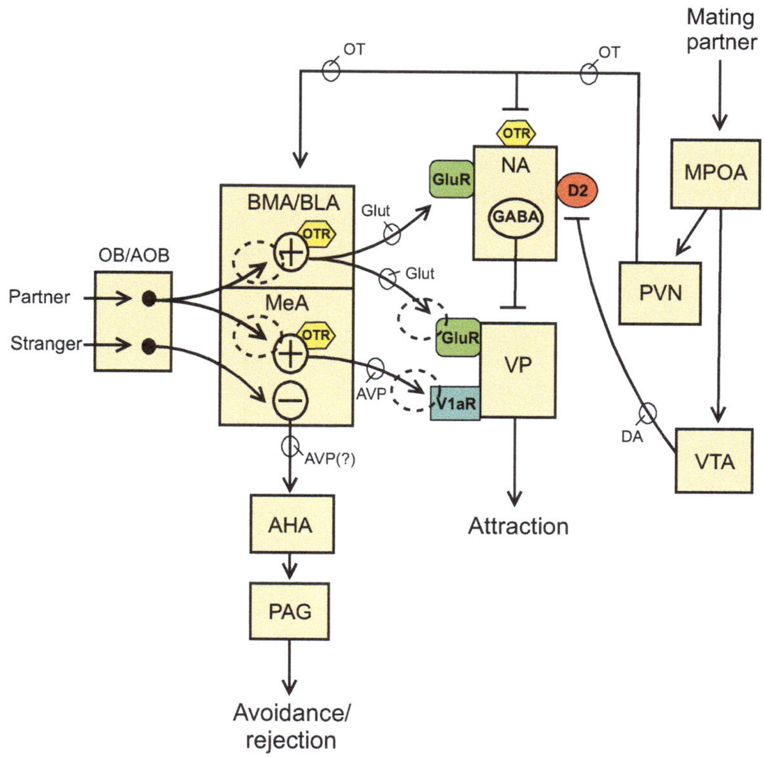

FIGURE 6.2 An integrative neural model representing the events that may underlie pair bond formation in prairie voles. During mating, medial preoptic area (MPOA) activity stimulates both paraventricular nucleus (PVN) oxytocin (OT) neurons and the ventral tegmental area (VTA) dopamine (DA) neurons, resulting in OT and DA release into nucleus accumbens (NA), and also OT release into the amygdala. Within the amygdala, OT action on oxytocin receptors (OTR) mediates an individual recognition process, whereby olfactory and other stimuli from the mating partner acquire the ability to activate positively valent basolateral (BLA), basomedial (BMA), and medial (MeA) amygdala neurons that project to the NA–ventral pallidum (VP) circuit. The synapses in the amygdala that are strengthened by mechanisms that include OT action are outlined with a dashed circle. Within the amygdala, the stimulus characteristics of a strange female, stimuli that have not been associated with OT release, are shown as activating negatively valent neurons that project to defensive circuits involving the anterior hypothalamic area (AHA) and periaqueductal gray (PAG) and that mediate aggressive and avoidance responses. The model also proposes that during a mating session, synapses within VP are persistently strengthened (dashed circle), and that such strengthening is necessary for the enduring attraction that occurs between mating partners after mating has been consummated. As described in the text, OT action on OTRs and DA action of D2 receptors in NA acts to suppress NA inhibition of VP. During a mating session, stimulatory inputs from one's partner activate MeA vasopressin (AVP) neurons that excite VP via an action on vasopressin 1a receptors (V1aR); these inputs also activate BLA/BMA glutamatergic (Glut) neurons that excite VP via an action of glutamate receptors (GluR). The resultant superexcitation of VP neurons is proposed to cause both the indicated synaptic strengthening and the development of a persistent attraction to one's mating partner. Axons ending in a bar are inhibitory, and those ending in an arrow are excitatory. Other abbreviations: AOB = accessory olfactory bulb; OB = main olfactory bulb.

the model proposes that the synapses between the BMA/BLA and VP and between the MeA and VP become persistently strengthened by these neurochemical interactions. Once mating is consummated, and a mate is given a partner preference test, the olfactory and other stimulus characteristics of the partner are now capable of strongly activating the VP over long periods of time, creating an enduring bond. In contrast, the stimulus characteristics of a strange or unfamiliar conspecific, which had not been associated with the neural events just described, are shown as activating negatively valent MeA neurons that project to the antisocial anterior hypothalamic nucleus–periaqueductal gray (AHN-PAG) circuit to cause avoidance and rejection behavior. As described in Chapter 3, AVP action within the AHA of male prairie voles promotes aggression toward novel females, although the source of such AVP input has not yet been determined. This model demonstrates how AVP can either promote attraction or rejection depending on the neural circuits affected and upon AVP's neural site of action.

Although this model may be applicable to both sexes, the action of AVP on V1a receptors in VP may be more important for pair bond formation in male prairie voles, while OT action on NA may be more important for bond formation in females.

In a further elaboration of the model, Figure 6.3 shows the microanatomy that may be operative within the NA-VP circuit. This model is based on the data presented in Chapter 1, where it was shown that DA action on postsynaptic D1 receptors in NA facilitates MSN GABA input to VP, while DA action on postsynaptic D2 receptors in NA depresses MSN GABA input to VP. Based on the assumption that VP output is necessary for social attraction, the model shows how DA action on D2-containing MSNs and OT action on inhibitory NA interneurons and on OTRs that form heteromers with D2 receptors act in concert to inhibit the inhibitory projection from NA to VP. This disinhibition allows for a strong co-action of glutamate and AVP on particular VP neurons, which strengthens the neural circuit that relays the stimulus characteristics of one's mating partner from the amygdala to the VP. Note how this model also explains how the activation of D1 receptors in NA could depress pair bond formation.

The proposed model is very elaborate, and although it is based on research evidence, it clearly contains hypothetical elements that are in need of future experimental support. The research on pair bond formation in prairie voles has been so focused on the roles of AVP and OT that certain basic research questions have not been investigated. For example, is VP output essential for pair bond formation and maintenance? Would neural inactivation of VP, with muscimol for example, disrupt the formation and maintenance of partner preferences? As another example, does the MPOA play a role in pair bond formation and maintenance? Given the importance of the MPOA for the appetitive aspects of male and female sexual behavior, and the fact that MPOA efferents to VTA may contribute to the release of DA into NA during sexual interactions [929], it would be very interesting to determine whether MPOA inactivation would disrupt the *maintenance* of pair bonds in prairie voles after sexual activity has been consummated (recall that MPOA inactivation disrupts the maintenance of the mother–infant bond). I emphasize investigating the role of the MPOA in pair bond *maintenance* because the study of the role of the MPOA in bond *formation* would be difficult, particularly in males, because MPOA inactivation disrupts male sexual behavior. However, the role of the MPOA in female prairie vole bond formation could be studied, since MPOA inactivation does not interfere with lordosis (see Chapter 4).

6.2.7.2 Comparisons between Pair Bonding and Mother–Infant Attachment

The general behavioral and physiological characteristics of the formation of pair bonds

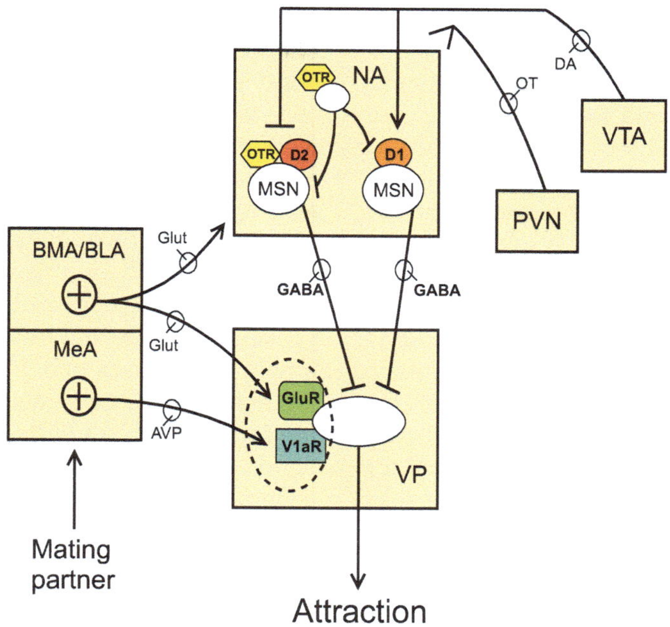

FIGURE 6.3 Based on Figure 6.2, a neural model of the possible microcircuitry within nucleus accumbens (NA) that would allow the co-action of oxytocin (OT) on oxytocin receptors (OTR) and dopamine (DA) on D2 receptors during a mating session to strongly depress the output of medium spiny neurons (MSNs) to the ventral pallidum (VP). OTRs are shown as located at two different sites in NA: (1) forming heteromers with D2 DA receptors on MSNs. A neuromodulatory action of OT at this site presumably enhances the inhibitory effect of DA on its D2 receptor; and (2) on inhibitory interneurons within NA. OT action at these receptors exerts OT's typical excitatory effects, which would then allow these interneurons to inhibit MSNs that contain either D1 or D2 receptors. See text for remaining details. Axons ending in a bar are inhibitory, and those ending in an arrow are excitatory. The axons of paraventricular nucleus (PVN) OT neurons are shown as ending in typical axon terminals (Y), since according to the model, OT can exert postsynaptic excitatory effects on interneurons and inhibitory neuromodulator effects on D2-containing MSNs. Strengthened synapses within VP (indicated by a dashed circle) mediate the attraction between mating partners that persists after mating has been consummated. Other abbreviations: AVP = vasopressin; BLA = basolateral amygdala; BMA = basomedial amygdala; GluR = glutamate receptor; Glut = glutamate; MeA = medial amygdala; V1aR = vasopressin 1a receptor; VTA = ventral tegmental area.

in prairie voles and the formation of mother–infant bonds in rats and sheep are quite similar. In each case, hormonal events activate the appetitive and consummatory aspects of sexual behaviors and maternal behaviors. After these hormonal stimuli subside, prairie vole sexual partners remain attracted to each other, and mothers remain attracted to their infant(s). Rats develop an enduring attraction to a generic pup stimulus, which suggests that the formation of this attraction occurs in the absence of a selective individual recognition process. In sheep maternal behavior, however, both an individual recognition process and the development of a long-term attraction to a specific lamb are the basis of the mother–infant bond, which closely approximates pair bonding in prairie voles. In comparing Figures 5.17 and 5.19 with Figures 6.2 and 6.3, one can see that there are many commonalities between the proposed neural mechanisms of mother–infant bonding and prairie vole pair bonding. Synaptic plasticity within the NA-VP circuit has been proposed as essential for the development of the persistent attraction that underpins both bonding mechanisms. A role for OT interaction with DA action on D2 receptors in NA appears important for maternal memory in rats and for pair bond formation, at least in female prairie voles. The BMA is important for maternal attachment in rats and for social affiliation in prairie voles, and processes occurring in MeA may be involved in the selective recognition aspects of both maternal attachments in sheep and partner preferences in prairie voles. Given the importance of AVP for pair bonding, there is even some evidence that vasopressin, along with OT, may be involved in maternal memory in rats [679]. Interestingly, although AVP action on VP is important for pair bond formation on male prairie voles, it does not seem to be essential for the alloparental behavior that males show toward unfamiliar infants [66]. AVP action on VP may be primarily involved in the formation and maintenance of a strong attraction to a familiar social stimulus rather than in an unconditioned attraction to a novel social stimulus. I suggest, however, that VP inactivation, with muscimol for example, would disrupt both pair bonding and alloparental behavior (see Chapter 5).

Another commonality between the neural mechanisms underlying pair bond formation in prairie voles and maternal behavior in rats is worth mentioning. The activation of μ opioid receptors in NAs is another process that is essential for pair bond formation in female prairie voles [802], and a similar process may be involved in the formation of maternal memory in rats [144,696]. These findings are interesting because NA MSNs with D2 receptors not only contain GABA, but also contain enkephalin (see [696]).

These commonalities support the proposal that the neural framework for mother–infant bonding may have provided the scaffold upon which pair bonding neural circuits were built. This commonality does not rule out the likelihood that differences between the two processes are sure to exist. For example, DA action on D1 receptors in NA is important for mother–infant bonding in rats but actually antagonizes pair bond formation in voles. Perhaps this difference is related to the specific location of DA-D1 receptors within the NA of rats and prairie voles (compare Figure 5.17 with Figure 6.3) Further, although the MPOA is necessary for the onset and maintenance of maternal behavior, no one has investigated its role in the formation and maintenance of pair bonds in prairie voles. One possibility is that MPOA activation of the mesolimbic DA system and PVN OT systems during mating in prairie voles may be essential for the formation, but not the maintenance, of pair bonds. However, MPOA neurons may be more generally involved in social affiliation and therefore may also be important for the maintenance of pair bonds, and this needs to be investigated.

With respect to maternal behavior in sheep, the model presented in Figure 5.19 emphasizes MeA projections to MPOA as playing a dominant role in determining maternal selectivity, while for pair bonding, the models shown in Figures 6.2 and 6.3 emphasize the particular social stimuli that are relayed to VP. If it is determined that the MPOA plays a role in pair bonding, then I would suggest that it is highly likely that both sets of circuits (MeA-to-MPOA and MeA/BMA-to-VP) are strengthened during the onset of maternal behavior in sheep and during mating in prairie voles, and that the selective activation of such circuits by specific social stimuli underlies maternal selectivity in sheep and partner preferences in prairie voles.

Given that moderate numbers of OTRs exist in the NA of rats [992], why do rats form enduring mother–infant bonds but not pair bonds with their mates? One possibility is that the release of OT into NA occurs during the onset of maternal behavior but not during sexual behavior. This explanation is unlikely since the vaginocervical stimulation that accompanies mating should cause the release of OT into the female rat's NA [836]. Another consideration is related to the fact that although OTRs exist in the NA of sexually mature virgin rats, their density is lower than that observed in sexually mature virgin prairie voles [721]. Perhaps OT-binding sites increase in NAs at parturition in rats, with this increase being necessary for OT's role in maternal memory formation (cf. [1016]). In an analogous manner, recall that OTRs are expressed at high levels in the BLA in sexually mature prairie voles and at lower levels in sexually mature montane voles, but that at parturition a significant increase in these receptors occurs in association with maternal behavior in montane voles [443].

6.2.8 The Genetics of Species Differences in Vasopressin V1a Receptor Distribution Patterns within the Brain and Their Role in Pair Bonding

6.2.8.1 Introduction

Prairie voles have high levels of expression of V1a receptors in the VP while montane and meadow voles do not. Experimental research has clearly shown that AVP action on V1a receptors in VP is necessary for pair bond formation in male prairie voles. In addition, rAAV vector transfer of the V1a receptor gene into the VP of male meadow voles, which increases the expression of the V1a receptor, facilitates pair bonding in male meadow voles. These findings have raised several interesting questions. How general is the relationship between the degree of V1a receptor expression in the VP and species differences in mating systems? Is AVP action on the VP necessary for male pair bonding in most or all monogamous mammalian species? What are the genetic mechanisms that account for species differences in V1a receptor expression that correlate with a species mating system?

In viewing Table 6.4, what accounts for tissue specific variation in V1a receptor expression so that the receptor is expressed at high levels in the VP and low levels in the lateral septum of the prairie vole, while the opposite expression pattern is present in the montane vole? When one compares the nucleotide sequence within the coding region of the AVPR1A gene, there is almost perfect homology between prairie and pine voles on the one hand and montane and meadow voles on the other [382]. This fact indicates that the V1a receptor protein that is actually translated from the mRNA code should be virtually identical for monogamous and promiscuous vole species. This analysis leads to the likely possibility that nucleotide sequence differences within the regulatory region of the AVPR1A gene between different vole species may determine differences in the distribution and density of V1a receptors in different brain regions. Hypothetically, let's say that transcription factor A is only present within VP neurons and that this transcription factor acts on Response Element A within the regulatory region of the AVPR1A gene to stimulate transcription of the gene. If the prairie vole contained the nucleotide sequence of Response Element A within the regulatory region of its AVPR1A gene, while the montane vole did not possess this response element, then one would expect greater expression of the V1a receptor in the VP of prairie voles than in montane voles.

Analysis of the AVPR1A gene regulatory region in four vole species (prairie, pine, montane, and meadow voles) has revealed an important genetic difference that might account for species differences in the expression pattern of the gene and for species differences in mating systems [382,552,1062]. The monogamous prairie and pine vole AVPR1A gene contains a complex microsatellite sequence of approximately 500 base pairs (bp) within the 5′ upstream regulatory region of the gene. This complex microsatellite is virtually absent, or is present in a highly truncated form, in the promiscuous montane and meadow vole AVPR1A gene. Complex microsatellites are composed of simple nucleotide sequences that form tandem repeats (such as GA repeats) and that are interspersed with nonrepetitive nucleotide sequences. Importantly, the presence of microsatellites within the regulatory region of a variety of genes has been shown to influence transcription, and both the length of the microsatellite as well as the particular nucleotide sequences that are repeated have been shown to be important [196].

It has been found that variation in the length of the microsatellite within the regulatory region of the AVPR1A gene does influence transcription of the gene in cell culture [382,383]. However, can such effects be shown to influence the mating tactics of vole species? I will approach this question in the following two ways. First, I will review studies that have asked whether *intraspecies* variation in the length of the microsatellite can influence mating strategies and pair bond formation within prairie voles. Then I will analyze those studies that have examined the contribution of variation in microsatellite length and V1a receptor expression in the VP to *interspecies* variation in mating systems.

6.2.8.2 Microsatellite Length Polymorphisms within the AVPR1A Regulatory Region and the Regulation of Pair Bonding in Male Prairie Voles

Hammock and Young [383] note that not only do interspecies differences occur in the structure of the AVPR1A gene but there is also intraspecies variation. Even within monogamous species, interindividual differences exist in the length of the microsatellite. Note, however, that differences in microsatellite length within prairie voles (a difference of about 20 base pairs) is much smaller than the length differences that occur between monogamous and promiscuous voles (hundreds of base pairs).

In laboratory tests, Hammock and Young [383] explored whether variation in the size of the microsatellite within a population of prairie vole males can affect V1a receptor expression and partner preference formation. The genotyping of prairie voles described a short and a long microsatellite allele. The two alleles differed in length by only 19 nucleotides within the microsatellite, primarily caused by an expanded GA repeat in the longer allele (note that even the short allele of the prairie vole would represent a large microsatellite when compared to that of the montane or meadow vole). Prairie voles were bred together based on their genotype in order to obtain offspring that were homozygous for either the short or the long microsatellite allele of the AVPR1A gene. Partner preference tests were subsequently conducted on male homozygous offspring when they were adults. Test males were paired with sexually receptive female prairie voles for a shortened 18h (instead of the typical 24h) cohabitation period in an effort to challenge the ability to form a partner preference. In the 3-h partner preference test that followed, the long allele males spent more time with their partner than with the stranger, while the short allele males did not. Therefore, with this abbreviated cohabitation period, long, but not short, allele males formed a partner preference. It is unfortunate that the short allele males were not tested after a 24-h cohabitation period in order to determine whether they could form a pair bond under optimal mating conditions. This issue will be relevant to the examination of mating strategies in prairie voles under naturalistic, rather than laboratory, conditions. In addition, it would have been interesting to examine partner preference formation in short and long allele females.

After the partner preference test, V1a receptor density was examined in a variety of brain regions, and these data were correlated with the genotypic differences. In some brain regions, the long allele was associated with a higher level of V1a receptors, while in other regions it was associated with lower levels. I just want to emphasize two aspects of these findings. The V1a receptor density in the VP did *not* distinguish the long allele males from short allele males, but the long allele males did have higher V1a receptor densities in the olfactory bulb. One interpretation of these data is that if the short allele males were allowed a full 24h cohabitation period, they would have formed a partner preference, and the fact that the long allele males formed a partner preference in spite of an abbreviated cohabitation period might have been the result of a greater AVP-induced olfactory investigation of their partner (see [383]). Since AVP may act at the level of the olfactory bulb to enhance individual recognition processes [1003], and since MeA receives input from the olfactory system, selective partner preference formation in the long allele males may have been promoted because of enhanced individual recognition and the ability to discriminate the partner from the stranger, rather than to an enhanced social attraction process. Indeed, Hammock and Young [383] reported that the long and short allele males spent equal amounts of time in total social contact with females during the partner preference test, which suggests that the short allele males were attracted to the females but presumably could not discriminate between them. Note how these findings for the short allele males coincide with the mother–infant bonds that develop in sheep when MeA activity in the mother has been disrupted—nonselective attachments to lambs develop.

These results, which are primarily correlational, demonstrate the difficulties in establishing cause–effect relationships between gene differences and the brain mechanisms regulating behavior. Multiple steps are involved in partner preference formation. The action of AVP at the level of the VP is an important step, but it is not the only important factor. Optimum levels of V1a receptor expression may exist in VP, but if a male has a faulty individual recognition process, or if the male is too anxious or aggressive—all processes that may be regulated by AVP action at sites outside the VP—then bond formation may still be impaired. Multiple factors in addition to high levels of V1a receptor expression in VP may influence pair bond formation in male prairie voles. These proposals also highlight the idea that AVP action at different neural sites can have different functional effects.

These ideas are supported by research that has examined the mating strategies of male prairie voles under seminatural conditions. As noted by Ophir, Wolff, and Phelps [731], under field conditions, about 70% of male prairie voles form pair bonds and cohabitate with a single female while defending a territory. In contrast, about 30% of males are referred to as wanderers. These wandering males have large home ranges that overlap with the smaller territories of pair-bonded voles. Wandering males do not form pair bonds but do engage in "sneak" copulations—they enter a pair-bonded male's territory in order to mate with his female partner, resulting in an extra-pair fertilization. It is likely that such wanderers are subordinate males who cannot compete for their own territories and mates. The fact that they are polygamous is likely related to the fact that they never cohabitate with a female for a long enough duration for the pair bonding process to operate. However, this idea needs to be experimentally verified. For example, if one were to capture wandering males and bring them into the laboratory, would they be capable of forming a partner preference if they were allowed to mate with a single female over a 24-h period without competition from other males? Another interesting question is whether, under natural conditions, wanderers ultimately become monogamous later in life. Perhaps younger males cannot compete effectively for territories and therefore engage in "sneak" copulations as an adaptive strategy. As males grow older and stronger, perhaps they can compete more effectively for good territories, which would favor a monogamous mating strategy. Field studies have indicated that male prairie voles do alternate between these two mating styles [917], suggesting that under the proper conditions a wanderer is indeed capable of forming a pair bond.

Ophir et al. [731] have reported that V1a receptor binding density did not differ within the VP of pair-bonded and wanderer male prairie voles. In a second study that examined microsatellite length within the promoter region of the prairie vole AVPR1A gene, Ophir, Campbell, Hanna, and Phelps [729] reported that allelic variation from 508 to 585 bp occurred in their population of voles. However, paired males were just as likely to exhibit a short as a long genotype and allele length did not differentiate pair-bonded resident males from wanderers (also see [599]).

It has been suggested that the results from these studies that have been conducted under natural field conditions, instead of in the laboratory, indicate that AVPR1A polymorphism is unlikely to have a major influence on the maintenance of social monogamy in feral prairie voles [599]. This view, of course, disagrees with the laboratory results of Hammock and Young [383], where microsatellite length did influence partner preference formation in male prairie voles when they were tested under a shortened cohabitation period. It is possible, however, that microsatellite length, per se, is not the critical variable [1062]. It may be that one needs to look more closely at the particular nucleotide sequence that is repetitive and that leads to a longer microsatellite. Perhaps only certain repetitive sequences enhance V1a receptor expression within certain brain regions, and therefore only some, but not all, large microsatellites would be associated with a facilitation of pair bonding.

Another consideration is that OTR binding in the brain should be evaluated alongside V1a receptor binding. To the extent that both OT and AVP play a role in pair bonding in male prairie voles, it may be that variation within both neuropeptide systems contributes to individual differences in pair bonding potential under natural conditions. Importantly, Ophir, Gessel, Zheng, and Phelps [730] have reported that OTR binding in the NA is significantly greater in the bonded monogamous male prairie voles than in nonbonded wanderers under seminatural field conditions. Although it is likely that both intraspecies and interspecies differences in OTR expression in NA are due to variations within the regulatory region of the OTR gene, results are not available to prove this point.

Finally, intraspecies variation in the AVPR1A microsatellite region within prairie voles is very small, and any effects that such variation may have on pair bonding are likely to be diluted by the many other factors that affect whether pair bonding will be successful under natural conditions, such as a male's maturity and strength, aggressiveness, and anxiety. However, the more controlled conditions of the laboratory do suggest that relatively small variations within the promoter region of the AVPR1A gene of prairie voles can modulate the intensity of pair bonding, although this effect appears to be mediated outside the VP. Indeed, most studies have shown that the small variation in microsatellite length within prairie voles does NOT influence the in vivo expression of V1a receptors within VP, which may account for the relatively small effect such variation has on the occurrence of pair bonding in males.

This research on intraspecies variability leads us to the more important question: does the large variation in microsatellite size that occurs between vole species play a role in determining the type of mating system that the species exhibits?

6.2.8.3 Interspecies Comparisons: Microsatellite Length Variation within the AVPR1A Regulatory Region, V1a Receptor Expression in Ventral Pallidum, and the Regulation of Pair Bonding

In a paper entitled "Mammalian monogamy is not controlled by a single gene," Fink, Excoffier, and Heckel [301] examined the structure of the AVPR1A gene in 21 microtine rodent species. As expected, the prairie and pine voles, two monogamous species, had an expanded microsatellite within the regulatory region of the gene, while the promiscuous montane and meadow voles did not. Surprisingly, however, the remaining 17 species, all of which were promiscuous, also exhibited an expanded complex microsatellite within the regulatory region of the AVPR1A gene. They argue that if the expanded microsatellite were important for monogamy in microtine rodents, and perhaps other mammals, it should only be present in about 5% of species. Instead, however, they find that this complex microsatellite is present in most microtine rodents. Other data indicate that the long microsatellite is actually the ancestral genotype and that the truncated form evolved more recently in montane and meadow voles.

In a related study, Turner et al. [980] argue that monogamy has evolved through multiple mechanisms and that one should not focus only on the expression of the AVPR1A gene within the VP. In comparing eight species of deer mice (*Peromyscus* genus), some of which are monogamous while others are promiscuous, they found no differences in V1a receptors in the VP, which were expressed at high levels in all species.

How can we integrate these findings with the experimental data that clearly shows that AVP action on VP is necessary for bonding in male prairie voles and that upregulation of the V1a receptor in the VP of meadow vole males permits this typically promiscuous male to form a pair bond? The distinction between necessary and sufficient needs to be discussed. Although high AVP binding to the V1a receptor in VP may be necessary for pair bonding and monogamy, it may not always be sufficient. As already discussed, many factors contribute to pair bond formation, and a species, or an individual within a species, may have many V1a receptors in VP but may not form a pair bond because other factors interfere with the pair bonding process. However, none of the studies that I have reviewed rules out the possibility that AVP binding to V1a receptors in VP is necessary or essential for the formation of a stable pair bond. If it could be shown that monogamy occurs in a mammalian species that has low-to-nonexistent V1a receptor expression in VP, then one could conclude that it is not always necessary for AVP to act within VP to contribute to social monogamy, and that other mechanisms are able to regulate monogamy in the absence of the AVP-VP connection. With respect to genetic mechanisms that influence V1a receptor expression in VP, although variations in the complex microsatellite may be involved, it is also possible that other variations within the regulatory region of the AVPR1A gene, which have yet to be uncovered, may play a role.

Interestingly, in an older study, Loup et al. [571] have reported OT-binding sites, but not AVP-binding sites, in the human VP. Perhaps OT action within the NA-VP circuit regulates the serial social monogamy that occurs in humans. However, once highly selective and safe nonpeptide radioligands, which can pass through the blood–brain barrier, for OT receptors and AVP receptors are developed, then the use of these ligands in conjunction with PET neuroimaging will give us a more up-to-date understanding of the distribution of OT and AVP receptors within the human brain [440]. More broadly, however, it is certainly possible that some species depend more strongly on OT action on OTRs for forming social attachments, while other species depend on AVP action on V1a receptors, and still other species utilize both neuropeptide mechanisms.

6.3 OT AND VASOPRESSIN NEURAL SYSTEMS AND THE REGULATION OF SOCIAL MONOGAMY IN BIRDS AND PRIMATES

6.3.1 Monogamy and Pair Bond Formation in Birds

Unlike in mammals, monogamy is the dominant mating system in birds, yet research on its neural mechanisms is almost nonexistent. In birds, the homologues of OT and vasopressin are mesotocin and vasotocin, respectively [349]. In two preliminary studies that have explored pair bond formation in zebra finches in an aviary setting, chronic antagonism of central OT-like receptors (mesotocin receptors) has been found to disrupt pair bond formation in both males and females. These studies administered an OTA either systemically via the intramuscular route [748] or centrally via ICV injections [495]. Therefore, the neural site of action remains unknown, and it is also not clear whether the systemically administered OTA was able to cross the blood–brain barrier to affect brain mechanisms directly. Importantly, vasotocin does not appear to be necessary for pair bond formation in male zebra finches [465], which further emphasizes the involvement of mesotocin for pair bonding in both male and female zebra finches.

Goodson et al. [349] note that the social organization of finch species is diverse. While all finches are monogamous and form pair bonds, some are gregarious while others are territorial. Zebra finches form pair bonds and also live in large colonial flocks with other conspecifics. In contrast, other finch species, such as the Melba finch, form pair bonds but are highly territorial and will not associate in close proximity with other conspecifics. Such a comparative analysis leads to the conclusion that *sociality* is a broad term and that different types of sociality may be regulated by different neural mechanisms [728]. The neurobiology that allows two relatively unfamiliar conspecifics to remain in close proximity without fighting should be expected to be different from the mechanisms that regulate a long-term attraction between two mating partners (see [81]). To state this in more anthropomorphic terms, gregariousness and sociability are different from love. Relevantly, Kelly et al. [475] have reported that projections from vasotocin neurons in BST to the lateral septum play a positive role in regulating flocking behavior in male zebra finches. One could conclude, therefore, from the data reviewed in this section, that mesotocin action in the finch brain is essential for pair bonding while vasotocin action is involved in flocking. Kelly et al. [475] suggest that vasotocin action in the lateral septum decreases anxiety-like behavior in male finches and that this action permits close physical association with other males, allowing for the formation of flocks. Note how this research contrasts with the research on aggression in rats that was reviewed in Chapter 3, where it was suggested that vasopressin action in the lateral septum might increase aggression (see Figure 3.5). However, Chapter 3 also suggested that AVP neurons in BST might depress aggression by activating inhibitory projections to the hypothalamic aggression area, which suggests some partial overlap between rats and zebra finches in the neural regulation of social proximity (see Figure 3.5).

Another take-home message is that vasopressin- and OT-like neuropeptides can have a variety of behavioral effects (pair bonding, flocking and gregariousness, aggression, anxiolysis, maternal responsiveness) depending on the neural circuits and neural sites of action that are involved (cf. [348]).

6.3.2 Pair Bonding in Primates

6.3.2.1 *Introduction*

When discussing the neurobiology of long-term social relationships in primates, and particularly in humans, one is dealing with a tremendously complex process [225]. One can still ask whether the detailed research on the neurobiology of pair bonding in prairie voles has relevance to the mechanisms regulating social monogamy in primates. It makes sense that the core neural mechanisms that regulate pair bonding within monogamous vole species have been at least partly conserved during the evolution of social monogamy across all mammals. The limited research that does exist for humans and other primates suggests that this is the case.

6.3.2.2 *Nonhuman Primates*

The marmoset and the titi monkey represent New World monkeys that are socially monogamous and form long-term pair bonds between mating partners. An autoradiographic study examining the location of OT and vasopressin V1a binding sites in the marmoset brain has described high OT binding sites in NA and high V1a binding sites in VP [871]. A PET scan study that measured glucose uptake in the brain of titi monkeys found that glucose uptake, presumably reflective of changes in neural activity, occurred in both the NA and VP of male titi monkeys during the formation and maintenance of pair bonds with their female partners [54]. Smith, Agmo, Birnie, and French [912] have examined the importance of OT for pair bonding in marmosets. Male and female marmosets were paired together over a 3-week period, and each sex received one of the following treatments: intranasal administration of OT, oral administration of a nonpeptide OT receptor antagonist, or control (placebo) treatments. The results showed that central OT systems appear to be involved in both the formation and maintenance of pair bonds in each sex—compared to the control conditions, marmosets initiated huddling and contact with their partners more often after intranasal OT, but reduced proximity and huddling after OT antagonist administration.

6.3.2.3 Humans

In an fMRI study, Acevedo, Aron, Fisher, and Brown [3] showed men and women who were married for over 20 years facial images of their partner or control images, such as of a highly familiar acquaintance. Images of the loved partner resulted in a significantly greater blood-oxygen-level dependent (BOLD) response in several regions, which included the hypothalamus, VTA, NA, globus pallidus (which probably also encompassed the VP), and amygdala. Note how these are just the brain regions, based on the vole work, that one would expect to be active during the formation of pair bonds. Since the tested subjects were in love with each other for years, these fMRI data support the view that interactions among these critical regions are also involved in the maintenance of a long-term bond mediating a persistent attraction. Also note that activity in each of these regions is also associated with the mother–infant bond, supporting the view of overlap in the neural circuitry regulating these two important social bonds.

Schneiderman, Zagoory-Sharon, Leckman, and Feldman [867] measured plasma OT levels in men and women who were either in a romantic relationship for 3 months or who were non-attached singles, and found significantly higher OT levels in both men and women who were in a relationship. Interestingly, 6 months later some individuals remained in their romantic relationship, while others separated from their partners, and it was found that OT levels at 3 months predicted whether a couple would remain together 6 months later. Individuals with higher OT levels were more likely to remain together. To the extent that plasma OT levels are reflective of the amount of OT released into critical brain sites, these results suggest that brain OT is likely to play a positive role in the formation and maintenance of a romantic relationship in both men and women.

Two studies have examined the effects of intranasal OT administration on pair bonds in men and women. In a laboratory setting, Ditzen et al. [250] examined the social interactions among heterosexual couples, that were married or living together for at least 1 year, while they discussed an area of disagreement in their relationship, such as financial issues. One group of couples received a single intranasal OT treatment, while a placebo was administered to the remaining couples. Each member of the pair self-administered the treatment. The results showed that OT significantly increased the duration of positive communication behaviors (verbal and nonverbal) in relation to negative behaviors during the couple conflict discussion. These results can certainly be used to support the contention that OT promotes the maintenance of a stable pair bond in men and/or women. But other interpretations are possible, since intranasal OT likely affected multiple neural sites and therefore more than one function. As an example, the well-known anxiolytic effects of OT may have calmed couples down, which may have promoted a less argumentative and more conciliatory social interaction.

In an incredibly interesting laboratory study, Scheele et al. [862] examined the effects of intranasal OT or placebo on the distance a man maintains between himself and an unfamiliar attractive woman during an initial encounter. One group consisted of men in a monogamous relationship (note, however, that the partners were not present during the experiment), while the other men were single. Men in a monogamous relationship, but not single men, maintained a greater distance between themselves and the attractive woman after OT treatment. OT did not affect the distance maintained between the subjects and a novel man. The authors suggest that these results imply that OT release within the brain promotes the maintenance of monogamy by decreasing the likelihood that a man will engage in romantic interactions with other women, in this way maintaining fidelity. Since monogamous men given OT maintained a larger distance between themselves and the novel attractive woman than did monogamous men given placebo, who did not differ from single men, it may be that the presence of one's partner would normally cause endogenous OT release into the male's brain, which may then function to depress flirting behavior in the presence of one's mate.

These studies have several implications. OT seems to be involved not only in the formation but also in the maintenance of pair bonds in humans. Second, OT appears to have this function in both males and females. Lastly, OT effects are not entirely prosocial. While OT may increase attraction between partners, it may act to decrease the social attraction between a bonded partner and a strange conspecific. This effect is reminiscent of the effects of AVP on pair bonding in male prairie voles, where AVP stimulates mate guarding in bonded males so that he rejects both novel males and females. A very interesting question is the extent to which intranasal OT affects V1a receptors in the brain.

Two studies by Walum and colleagues have examined the relationship between genetic polymorphisms in either the AVPR1A gene [1014] or the OXTR gene [1013] on the strength of the pair-bonded relationship in humans. These studies find that variations within the AVPR1A gene were associated with the strength of the male bond to his female partner, while variations in OXTR gene were associated with the strength of the bond in women. These results fit nicely with the prairie vole research that indicated the greater importance of AVP and OT for male and female bonding, respectively.

Although there is no sequence in the regulatory region of the human AVPR1A gene that is homologous to the complex microsatellite found in the corresponding prairie vole gene, humans do have repetitive sequences in this region that are polymorphic in terms of length and nucleotide composition. Recall from the chapters on Aggression and on Parental Behavior that certain alleles within the RS3 repetitive sequence are associated with heightened amygdala reactivity to aversive social stimuli and with lower levels of maternal sensitivity. More specifically, the longer RS3 alleles, including allele 334, have been associated with these characteristics. In the Walum et al. [1014] study, allele 334 was also associated with higher levels of "marital discord" in men, but not in women. This study examined over 1000 men and women that had been in a heterosexual relationship for at least 5 years, and each individual was administered a Partner Bonding Scale questionnaire that measured the strength of the relationship between the couples. This questionnaire was based on retrospective recollections. For men only, carriers of one or two copies of allele 334 performed more poorly on the Pair Bonding Scale. Further, males who carried allele 334 were more likely to be living with their partner than married to their partner, and had spouses who were more likely to perceive difficulties in their long-term relationship with their male partners. Significantly, males with two copies of 334 had lower scores on the Pair Bonding Scale than males carrying one copy of allele 334, who in turn had lower scores than males who did not carry allele 334. With respect to the actual scores, the effect size for homozygous 334 men versus men who did not carry allele 334 was relatively modest ($d = 0.40$), suggesting that polymorphic variation within RS3 of the AVPR1A gene is just one of many factors that might influence the strength of the social bond between a man and his partner. Walum et al. [1014] suggested that the observed effects might have been larger if men were considered who were not bonded to a partner. In other words, it was hypothesized that allele 334 might occur at a higher frequency in single unattached males than in males who were in a long-term relationship with a woman.

The question of how this polymorphism might influence bonding to a partner in men remains unknown. It is intriguing to propose that allele 334 in some way decreases the effectiveness of AVP action on the V1a receptor in the VP. Recall from Chapter 3, however, that in a postmortem study, allele 334 at RS3 was associated with increased expression of the V1a receptor [500]. Since men who carry the 334 allele exhibit heightened activation of the amygdala in response to aversive social stimuli [639], it appears likely that the effect of this allele on partner bonding in men might be indirectly mediated by the proaggressive effects of AVP. Men that are prone to be more impulsive may overreact to negative interactions with their partners, creating greater levels of marital discord.

In the study on the OXTR gene [1013], a single nucleotide polymorphism within the 3′ UTR of the gene, rs7632287, was associated with partner bonding strength in women, but not in men. Adenine (A) is the minor allele (present in a smaller proportion of the population) at this SNP, while guanine (G) is the major allele. Couples were administered the Partner Bonding Scale and also a scale that measured the participants' affection for their partner. Women with one or two copies of the A allele scored lower than women with the GG genotype on each scale, indicating greater marital discord and less affection for one's partner. In addition, men who were married to women carrying the A allele reported a lower marital quality level when describing their relationship.

Just how variation at rs7632287 might influence OTR expression and pair bonding strength remains unknown. The fact that it is located in the 3′ UTR suggests that allelic variation might be able to affect the expression of the OTR (see Chapter 2).

Interestingly, variation within rs53576, which is located within the third intron of the OXTR gene, was not associated with the strength of the partner relationship in either women or men [1013]. Recall from the chapter on Parental Behavior that variation within this SNP was associated with individual differences in maternal sensitivity [48]. Ditzen, Bradley, and Heim [249] present evidence, however, that the rs53576 variation might have been associated with variation in partner bonding if the early life experiences of the participants had been considered. They suggest that women who carry the A risk allele (see Chapter 5) at rs53576 and who were also exposed to early adverse life experiences, such as abuse or neglect, may have demonstrated poor partner bonding in adulthood. These kinds of considerations show the difficulties involved in teasing apart the role of genes in human behavior.

Based on the research previously reviewed, it certainly would be interesting to determine whether a particular OXTR genotype was capable of influencing partner bonding in both women and men. Also, future studies need to integrate genetic analyses with PET scans that measure OTR and V1aR expression in the human brain.

The research in this section focused on the role of OT and AVP in human pair bonding. It is worth adding here that there is also evidence that certain alleles of the D2 DA receptor gene, which result in decreased expression of the D2 receptor, have been associated with insecure attachments in humans [333]. This finding fits with the prairie vole research that has emphasized the importance of D2 receptors in NA for the formation of pair bonds.

Taking a wider view, this research on the relationships between the strength of a monogamous bond in humans and genetic polymorphisms within genes related to the OT, AVP, and DA systems may be important for understanding the sexual infidelity that can occur in socially monogamous humans. Infidelity within mating partners occurs in human and nonhuman species, and the occurrence of such sexual cheating may be related to the strength of the social bonds that exist between mating partners, which, in turn, may be related to genetic polymorphisms within critical genes [1059].

6.4 OT, VASOPRESSIN, AND PAIR BONDING: CONCLUSIONS

The detailed mechanistic research on the neurobiology of social monogamy and pair bonding has been done on prairie voles. The formation of pair bonds and enduring partner preferences involves several processes, including the development of individual recognition of one's partner and the formation of a long-term attraction to that partner. Based on the prairie vole literature, a hypothetical neural model was proposed where OT action on MeA was viewed as critical for the individual recognition process, while OT action in NA and AVP action in VP, interacting with DA action in NA, was viewed as being critical for cementing the connection between the social stimuli of one's partner and a neural attraction mechanism mediated by VP output, resulting in an enduring attraction to one's partner that persists outside the mating context. This neural model was compared to the neural model hypothesized to underlie the mother–infant bond, and many commonalities were noted, suggesting significant overlap in the two bonding processes.

In comparing the monogamous prairie and pine voles to the promiscuous montane and meadow voles, the monogamous species have more OTRs in NA and more V1a vasopressin receptors in the VP than do the nonmonogamous species. Some research suggested that variations within the regulatory regions of the AVPR1A and OXTR genes may result in variations in the distribution of V1a receptors and OTRs in the brain, and that such variations could lead to species differences in mating systems and other aspects of social behavior. One particular focus has been on a complex microsatellite within the regulatory region of the AVPR1A gene that is present in prairie and pine voles, but is absent in montane and meadow voles. However, subsequent research showed that 17 other promiscuous vole species also possessed the complex microsatellite. Further, in a comparison of different Peromyscus rodent species, V1a vasopressin receptors were expressed at high levels in the VP of both promiscuous and monogamous species. These interspecies comparisons have led some to conclude that multiple neural and genetic mechanisms can lead to monogamy. I made the point that one should distinguish necessary conditions for monogamy from sufficient conditions. High levels of expression of OTRs in NA and/or V1a receptors in the VP may be necessary, but not sufficient, for social monogamy. To the best of my knowledge, a monogamous mammalian species has not been found that exhibits low levels of OTRs and/or V1a receptors in the NA-VP circuit.

Research on pair bonding in prairie voles has emphasized the importance of OT for female bonding and AVP for male bonding, although more recent research has indicated that OT action on the NA may be important for the formation of the pair bond in both sexes. The scant research that has been done on monogamy and pair bonding in birds and primates also suggests the importance of OT neural systems in the bonding process. Furthermore, a cross-species comparison indicates that OT may be involved in both the formation and the maintenance of partner preferences and pair bonding between mates. With respect to comparisons between pair bonding mechanisms and the mechanisms regulating the mother–infant bond, recall that for maternal behavior, OT stimulates the onset of maternal behavior and is also implicated in the synaptic plasticity involved in maternal memory. Further, research indicated that while OT neural systems modulate the intensity of maternal behavior after its onset, they are not necessary for the maintenance or continuance of maternal behavior. Perhaps a similar process is involved in pair bonding, where OT action on the NA influences the synaptic plasticity necessary for pair bond formation, but, subsequently, for established bonds, OT may modulate the intensity of the bond without being absolutely necessary for a continued partner preference.

6.5 THE RELATIONSHIP OF OT NEURAL SYSTEMS TO AUTISM SPECTRUM DISORDER

6.5.1 Introduction

Autism or autism spectrum disorder (ASD) is a phenotypically complex human disorder that is typically detected early in an individual's life and, in many cases, is associated with lifelong disabilities. The broad characteristics of ASD include deficits in social behavior and social interaction, deficits in language and communication, and the occurrence of repetitive and stereotypical behaviors [248,680,1001]. The deficits in social behavior have always been considered a central characteristic. In comparison to age-matched controls, individuals with ASD show deficits in orientation toward social stimuli and are not attracted to such stimuli, they show a lack of attention toward faces and have difficulty recognizing emotions based on visual and vocal cues, and they are less likely to provide a prosocial response toward another individual who is in distress [680]. It is not difficult, therefore, to understand why individuals with ASD have varying levels of difficulty, depending on the severity of their symptoms, in forming stable social relationships with others.

ASD occurs in about 1% of the human population, it is more prevalent in males than in females by a ratio of 4:1, it is highly heritable, and many genes have been implicated in its etiology [63,263,264,540,553]. Although some cases of autism involve single gene effects, even in these cases, deletions of a different gene in different individuals can each result in the disorder. Monogenic forms of ASD account for only about 10% of the cases; in the large majority of affected individuals, a polygenic form of inheritance contributes to development of the disorder.

Because of the involvement of OT and AVP neural systems in the formation of pair bonds, and because social deficits are a core feature of ASD, much research has been devoted toward examining the contribution of dysfunctions within OT and AVP systems to the etiology of ASDs. Although some studies have examined the role of AVP systems [486,637,639], most studies have explored the involvement of OT, and therefore my focus in this section will be concerned with OT.

Throughout this book, I have emphasized the effects of OT on a variety of behavioral processes: anxiolysis, anti-aggressive effects, maternal responsiveness, social recognition, and pair bonding. Indeed, OT may act at different neural sites to have these diverse effects. In the case of maternal behavior, for example, OT acts on the MPOA to stimulate an initial attraction to infant-related stimuli at parturition, and acts on NA to affect maternal memory and the formation of a long-term mother–infant bond. Therefore, the involvement of OT in pair bond formation should not be conceived as the raison d'être for its involvement in ASD. Let me offer an instructive example from laboratory mice. Recall that mice are promiscuous and do not form pair bonds. Pobbe et al. [775] placed WT and OTR knockout (OTR−/−) mice in a visible burrow system and recorded their social behaviors. While WT mice investigated and interacted with one another, OTR−/− mice spent most of their time alone and did not interact at high levels with other mice. This study shows that OT neural systems are involved in attraction to social stimuli even in a species that does not form pair bonds (also see [579]).

Given that OT has broad effects on sociality, dysfunction within OT systems may affect the development of ASD through multiple mechanisms. However, a disruption of OT's effects on the development of strong and stable social bonds is likely to contribute to ASD. While this section emphasizes the relationship between OT and ASD, the final chapter in this book will discuss human sociality more broadly and will examine the role of OT in other forms of human social behavior, such as cooperation, altruism, and empathy.

Figure 6.4 shows a relatively simple and condensed neural network model that may regulate social behavior. In part (A) of the figure, brain areas are not labeled, and it is emphasized that OT acts at three neural nodes, B, C, and D, and that it acts at these nodes in conjunction with other factors, X, Y, and Z, respectively. Part (B) of the figure equates node B with BMA/BLA and MeA, node C with MPOA, node D with NA, and node E with the VP. OT may act in MeA and other parts of the amygdala to regulate the valence characteristics of social stimuli, social recognition processes, and anxiolysis; OT action on MPOA may influence attraction toward social stimuli via MPOA activation of mesolimbic DA input to the NA, which would then affect the output of the VP; finally, OT action on NA may influence the formation of enduring social bonds. Clearly, on the basis of this model, a disruption in OT function should have broad disruptive effects on social behavior.

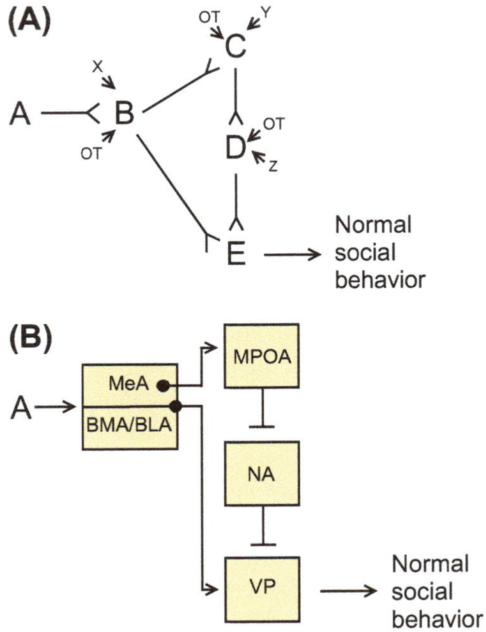

FIGURE 6.4 Neural networks within which oxytocin (OT), along with other factors (X, Y, Z), might act to influence social behavior in humans. In the top panel (A), brain areas are not labeled and OT is shown as acting at sites B, C, and D. The bottom panel (B) presents a simplified and condensed neural network, which equates area B with the basomedial (BMA), basolateral (BLA), and medial (MeA) amygdala, area C with the medial preoptic area (MPOA), area D with the nucleus accumbens (NA), and area E with the ventral pallidum (VP). OT action at the level of the amygdala may influence social behavior by affecting individual recognition processes and processes related to anxiolysis; OT action on MPOA may promote social attraction via activation of the mesolimbic dopamine system; OT action at the level of NA may promote the formation of enduring social bonds. Axons ending in a bar are inhibitory, and those ending in an arrow are excitatory. See text for the significance of these sites of OT action with respect to OT dysfunctions that may contribute to autism.

In examining Figure 6.4, what kinds of OT system dysfunctions might be related to ASD, and in those cases, could OT treatments serve as therapies for ASD if the model were accurate? If a neural disruption causing ASD were upstream from the proposed sites of OT action, site A for example, then OT treatments might be therapeutic, but if a neural deficit existed downstream from the proposed sites of OT action, at site E, for example, then OT treatment would probably not be therapeutic. What about dysfunctions within the OT system itself? If autism were influenced by deficits in OT synthesis or OT release into the brain, then OT treatments should be therapeutic. If dysfunctions at the level of the OTR contributed to ASD, then OT treatments might be effective in some cases but not others. If genetic alterations affected either the density of OTRs in the brain or the affinity of OT for the OTR, then OT treatment, by raising OT levels, might have therapeutic effects. If genetic alterations made the OTR completely nonfunctional, then OT treatment would probably be ineffective. In the sections that follow, I will explore the involvement of OT deficits in ASD and the potential therapeutic effects of OT administration. This research should be appreciated in the context of understanding that multiple causative factors contribute to ASD [264].

6.5.2 OT-Related Genes and ASD

Several studies have found that, in comparison to control subjects, certain alleles of various SNPs within OXTR gene are overtransmitted within ASD patients [152,263,1048,1049]. Overall, these results support the idea that genetic variants of the OXTR gene and changes in OTR function may contribute to the etiology of ASD. It is not my purpose to review this vast literature; I will restrict myself to just two SNPs within the OXTR gene that have been associated with autism, rs53576 and rs1042778, in order to make some general points.

In a Chinese sample, Wu et al. [1045] reported an association between the A allele at rs53576 and ASD, but other studies with Caucasian subjects have failed to replicate this effect [152]. In contrast, Campbell et al. [152] reported an association between the G allele at rs1042778 and ASD. Recall that rs53576A, which is located in the third intron of OXTR gene, has been associated with decreased maternal sensitivity (see Chapter 5) and that rs1042778G, located within the 3′ UTR of the gene, has been associated with heightened aggression in boys (see Chapter 3, and note that guanine is the complement of cytosine). Just how and whether these SNPs influence OTR expression is not known.

How can we account for the fact that certain studies have found an association between rs53576A and ASD while other studies have not (see [263])? Given that many genes typically contribute to ASD, each gene probably exerts a small effect, and any one allele may not necessarily be associated with ASD if other defective genes are not also inherited. To state this in another way, rs53576A may contribute to alterations in social behavior that would *not* be considered autistic per se, but when compounded with other social deficits, would contribute to ASD. Two examples follow that show that rs53576A influences social phenotypes outside the category of ASD but could be conceived as contributing to ASD if other social deficits were also present. In a population of normal healthy males (non-ASD), homozygous rs53576AA individuals scored lower on measures of positive affect and higher on measures of social loneliness than did their AG and GG counterparts [578]. In another very interesting study [503], judges were shown 20-s silent video clips of normal subjects interacting with one another. On the basis of nonverbal cues, the judges ranked men and women carriers of the rs53576A allele as less prosocial than rs53576GG subjects.

A more abstract analysis of why some studies have associated the rs53576A with ASD and others have not involves the model shown in Figure 6.4. At each node, OT is shown as co-acting with another factor, X, Y, or Z. Depending on the sample population, it is possible that these proposed cofactors might compensate for mild deficits in OTR function in some, but not all, samples. Perhaps the ethnicity of a sample population, by affecting the genetic background of the subjects, might be relevant to such a proposed process.

A final factor to consider is that although OT system dysfunction may contribute to individual cases of ASD, based on Figure 6.4, there should be many other ways in which autism could occur without alterations at rs53576. For example, within a particular sample population, individuals with the non-risk rs53576GG genotype may have some other deficit that influences the occurrence of autism, but the fact that such autistic individuals carry the GG genotype would decrease the strength of association between the A risk allele and ASD. These other deficits could include different genetic effects within the OT system or genetic variation in non-OT-related genes that influence the function of the hypothetical network shown in Figure 6.4.

A study by Gregory et al. [358] clearly demonstrates that there are significant additional ways to affect OXTR gene function and ASD that do not involve SNPs. In a comparison of 20 normal and 20 ASD subjects, patients with ASD had a significant increase in the methylation of CpG sites within the regulatory region of OXTR gene (detected from blood cells), and in a subsequent postmortem analysis on brain tissue from four patients with ASD and four normal controls, a similar increase in the methylation at CpG sites was detected in neurons from the temporal cortex, and this, in turn, was correlated with a decrease in OTR mRNA in temporal cortex of male subjects. As reviewed in Chapters 2 and 5, increased methylation at CpG sites in the regulatory region of genes is a mechanism that can depress gene transcription. The results of this study suggest that OXTR gene transcription was affected by this mechanism in ASD patients. The potential involvement of the temporal cortex, which contains both the superior temporal sulcus and the fusiform face area, is interesting, as the former has been proposed to relay social stimuli to the amygdala and ventral striatum, while the latter is involved in face perception [2,680].

Another gene related to the OT system has been associated with ASD: an SNP (rs3796863C) within the CD38 gene [263]. CD38 is a transmembrane protein that is expressed in many cells, including white blood cells and neurons. It is involved in regulating the release of Ca^{2+} from intracellular storage sites, and this mechanism within the brain contributes to the release of OT at synapses [409]. Importantly, ASD patients, when compared to controls, express lower levels of CD38 mRNA within blood lymphocytes [263]. If a similar decrease occurs in the brains of certain ASD patients, then a deficit in OT release may contribute to the disease.

6.5.3 Plasma OT and ASD

OT, in a manner similar to other neuropeptides, is initially translated from mRNA as a large precursor prohormone that is subsequently proteolytically cleaved to form the biologically active peptide that contains nine amino acids [322]. The extended precursor to OT has been referred to as OT-X, and several prohormone convertase enzymes are likely to be involved in the final production of OT [355]. Importantly, Green et al. [355] measured plasma levels of OT and OT-X in a sample of ASD male subjects and control males. OT levels were significantly lower, while OT-X levels were significantly higher, in the blood plasma of the ASD males. If the same events occurred within the brain, OT should have decreased effectiveness in affecting brain circuits. Therefore, it is possible that genetic polymorphisms that affect the function of prohormone convertase enzymes are involved in the etiology of some cases of ASD.

6.5.4 OT Treatment as a Therapy for ASD

The above studies suggest that a variety of alterations within the OT system may contribute to the occurrence of the social deficits of ASD. These alterations are likely to include modifications in OTR expression, OT synthesis, and OT release. All of these studies, however, are correlational, showing an association between the OT system and ASD. If dysfunctions in the OT system were causally related to ASD, then OT treatment might provide a therapy. Indeed, if deficits in OT synthesis or release contribute to autism, then OT treatment might provide a remedy. Even in cases where the OTR is involved, OT treatment might still be therapeutic; increases in OT levels resulting from the treatment might be able to compensate for changes in OTR structure and function.

A few experimental studies have examined the effects of intranasal OT treatment on the social behavior of ASD subjects. These studies show that OT treatment can have beneficial effects. Note, however, that the effects of intranasal OT are short lasting [110] and that such a treatment mode would not likely result in a practical clinical therapy. However, once long-acting nonpeptide OT agonists are developed that can pass through the blood–brain barrier, then systemic administration of such drugs may prove to be effective clinical treatments for some forms of ASD.

Andari et al. [28] examined the social behavior of subjects with Asperger syndrome, which is a high-functioning form of ASD. In a computer-simulated ball toss game, participants had to throw a ball to one of three fictitious partners. The partners were defined as good, bad, or neutral depending on their cooperation with the participant. A good partner exhibited a high probability of throwing the ball to the participant, while a bad partner excluded the participant. The dependent variable was the participant's ball toss choices. Healthy subjects, as one would expect, were significantly more likely to toss the ball to good partners. Asperger subjects that received intranasal placebo administration did not distinguish between the three partners and tossed the ball equally to good and bad partners. In striking contrast, Asperger subjects that received intranasal administration of OT preferred to toss the ball to the good partner. Although it is difficult to interpret the underlying mechanisms through which OT administration altered the social behavior of the Asperger patients, it appears that the treatment in some way improved their ability to discriminate the social cooperativeness of the different partners, leading to an appropriate level of social reciprocity. For example, perhaps OT increased the experience of social reward derived from interacting with a particular partner, leading to a proper reciprocated response [28].

Guastella et al. [369] administered a Reading the Mind in the Eyes Task (REMT) to male patients with ASD. This task is a test of emotion recognition where the subjects indicate the emotion associated with facial expressions from the eye region. On a computer screen, subjects are shown different images of a person's eye region, and under each image there are several different statements describing what the person might be thinking or feeling. The subject is requested to choose the most appropriate label. Guastella et al. [369] found that ASD patients who received intranasal OT showed more correct choices than did the placebo control patients, and this effect was particularly strong for relatively easy discriminations. Interestingly, in a similar study using normal healthy male subjects, intranasal OT treatment also improved emotion recognition in the REMT test, but in this case the effect was particularly robust for difficult discriminations [251]. One interpretation of these results is that OT treatment increases the salience of social stimuli and attention to facial/eye cues as a first step in evaluating the emotional state of another individual [28,324]. Understanding the emotional state of another person, of course, plays an important role in adaptive responsiveness to others. Relevantly, intranasal OT increases the amygdala BOLD response to facial stimuli in subjects with ASD [250A]. In comparing the Domes et al. [251] study with the Guastella et al. [369] study, the following is important. For healthy subjects in the former study, both placebo and OT-treated subjects made 85% correct responses when easy items were presented. For the ASD subjects in the Guastella study, for the easy items, patients treated with OT only made about 60% correct choices. Therefore, although OT treatment improves performance on this test, the level is still below normal.

To conclude, these few initial studies are promising. The Andari et al. [28] study, in particular, suggests that treatment with OT receptor agonists may be effective in enhancing adaptive social interactions of high-functioning ASD subjects in a group setting.

6.5.5 OT and Brain Mechanisms of ASD

In examining Figure 6.4, what is the evidence that a disruption in the function of the particular neural circuits emphasized in part (B) of the figure might be associated with ASD symptoms (cf. [248])? In a neuroimaging study that employed a voxel-based morphometry analysis of gray matter volume (GMV) in the hypothalamus, Kurth et al. [523] reported that GMV in children with ASD was lower than that of controls, and this decrease was located directly adjacent to the optic chiasm. Based on this location, the presumed dysfunction could include the PVN/SON, preoptic area, and perhaps the VP [608]. There is also evidence showing that abnormal amygdala morphology is associated with ASD [21].

Scott-Van Zeeland et al. [875] compared high-functioning ASD boys with age-matched controls in a simple learning task while fMRI scans were obtained. In one part of the study, the subjects had to categorize pictures into one of two categories, and if a correct response was made, a social reward was administered (a smiling face plus a text message indicating that the response was correct). The ASD subjects did not improve their performance on this task over trials while the healthy controls did, and the learning deficit in the ASD subjects was associated with a lower BOLD response in the ventral striatum. These findings support the view that a dysfunction in the mesolimbic DA system is correlated with ASD. These results, of course, could simply be related to the fact that social and nonsocial stimuli and rewards are simply less salient to ASD subjects under these testing conditions and that these attention deficits may cause performance deficits and decreases in the access of stimuli to the mesolimbic DA system and the NA-VP circuit (cf. [2]).

That OT dysfunction might be related to socially reinforced learning deficits in ASD subjects is supported by studies that have shown that intranasal OT administration facilitates socially reinforced learning (smiling faces for a correct response) in normal healthy subjects, while not affecting learning when a nonsocial reinforcer (green light for correct response) is used [436]. The authors suggest that OT action on the amygdala might be involved in facilitating performance in socially reinforced learning tasks, although an action on NA cannot be ruled out. Relevantly, Kohls et al. [505] have reported a decreased amygdala BOLD response in ASD subjects during a learning task with a social reward.

To conclude this section, the evidence reviewed, although open to multiple interpretations, can certainly be seen as consistent with the condensed neural model shown in part (B) of Figure 6.4. It is intriguing to view the social deficits associated with ASD as being, in part, the result of the fact that individuals with ASD do not find social stimuli attractive and rewarding and that this deficit might be influenced by disruptions in OT action at the level of the hypothalamus, amygdala, and the NA-VP circuit.

6.5.6 Sex Differences in the Occurrence of ASD

ASD is much more common in males than in females, although the cause of this difference remains largely unknown. Several hypotheses have been offered, and I will briefly review some of them [63,163,855]. Then I will offer a novel hypothesis of my own. As reviewed in Chapter 4, since males have one X chromosome while females have two, a genetic dysfunction on one X chromosome would be expected to have more drastic effects in males than in females. Although genetic polymorphisms and single gene deletions located on the X chromosome are involved in some cases of ASD, these genetic effects only result in about a 5% contribution to ASD risk [553]. Another possibility is that the higher fetal testosterone levels that the male brain is exposed to may interact with other factors to increase the male's susceptibility to developing ASD. Recall that the number of AVP-containing neurons in the MeA and BST is higher in male than in female mammals and that this increased expression of AVP in certain neuron groups is the result of perinatal androgen exposure [163]. Although I have not emphasized the involvement of AVP in ASD, because most of the research points to a greater role for OT involvement, research indicates that AVP is involved in both social bonding and aggression, particularly in males. Therefore, a modification in AVP function, perhaps involving variations in fetal testosterone secretion, may contribute to the higher incidence of ASD in male subjects.

Finally, genes on the Y chromosome may influence ASD. The SRY gene is not only involved in testes development but is also active in dopamine neurons in the VTA of adults, where the SRY protein has been found to enhance tyrosine hydroxylase expression and DA synthesis [246]. Therefore, males might exhibit greater DA release into NA than females. The prairie vole research indicates that DA action on DA receptors in NA regulates pair bonding. While moderate levels of DA release in NA may preferentially bind to D2 receptors, which play a positive role in bonding, higher levels of DA release also engage the lower affinity D1 receptor that has been shown to antagonize the formation of pair bonds. To the extent that the neural model for pair bonding in prairie voles shown in Figure 6.3 has relevance to social attachments in humans, one can see how hyperactivation of D1 receptors in NA would depress ventral pallidal activity, with a concomitant decrease in an individual's attraction toward social stimuli and the formation of social bonds. Such an effect, acting in conjunction with other factors, such as a disruption within the OT system, might contribute to the sex difference in ASD. Just how such a hypothesis is related to the finding that the NA BOLD response is decreased in ASD subjects during a social learning test remains to be explained, but one possibility, because of spatial resolution issues, is that the measured BOLD response area included the ventral pallidal and/or globus pallidus region.

CHAPTER

7

Human Sociality

7.1 INTRODUCTION

In contrast to other mammals, in addition to parental attachments and the occurrence of social monogamy, humans also form long-lasting social attachments and friendships to individuals other than their children and spouses, and many of these social bonds are with non-kin. Human prosociality includes the development of trust between individuals within social groups, the occurrence of cooperation, where individuals work together to achieve a common goal, and feelings of empathy for those in need of aid. Altruism also occurs in human social groups, where one individual may aid another, without any immediate direct benefit accruing to the donor of the aid. Therefore, unrelated individuals within human social groups can form long-lasting social bonds to one another, which then enhance the occurrence of empathy, cooperation, and altruism.

One basic and relatively unique aspect of human prosociality involves caregiving, or aid and helping behaviors, directed toward unrelated (non-kin) individuals within one's social group. Since parental behavior is the primordial caregiving system, I have argued that the neural systems underlying maternal behavior may have served as a foundation for other types of prosocial bonds, and a strong case for this idea was made when I compared the neural mechanisms of maternal care with those involved in pair bond formation between mating partners in monogamous species (Chapter 6). It is not surprising, therefore, that others have argued that the maternal care neural system probably provided the rudimentary neural foundation for the broader types of prosocial bonds that form within human social groups between unrelated individuals, giving rise to highly developed forms of cooperation and altruism [132,227,245,698,785]. This view does not deny that complex cognitive decision making plays an important role in human social interactions. However, cognitive and affective neural systems probably interact with a modified maternal neural system, which then allows thought processes and emotional experiences to be translated into caregiving prosocial responses directed toward others.

In this chapter, I will first describe the neural and neurochemical underpinnings of human prosociality. This analysis will then be followed by a discussion of the neurobiology of psychopathy, where a breakdown of prosocial neural systems gives rise to extreme antisocial behavior in humans.

7.2 AN EVOLUTIONARY PERSPECTIVE

Although certain forms of cooperation and prosocial behaviors toward non-kin are present within nonhuman primate social groups, such prosocial behaviors are much more highly

http://dx.doi.org/10.1016/B978-0-12-416040-8.00007-9

developed in humans [245,963]. It is clear that unique ecological and other environmental conditions, along with essential preadaptations, allowed for the evolution of high levels of cooperation, teamwork, and altruism during early hominin evolution [1036]. Such high levels of prosociality were adaptive and favored by natural selection. As an analogy, most mammals are polygamous, but about 5% are monogamous. As described in Chapter 6, the particular environmental conditions encountered by certain microtine rodents favored the development of a monogamous mating system and a cooperative breeding system. As a result of the selection pressures acting on these individuals, genetic changes occurred that contributed to a reorganization of brain mechanisms so as to favor the development of pair bonds, mate guarding, and paternal/alloparental behavior. In a similar fashion, one can envision that the particular ecological and social factors encountered by early humans, which were distinct from those of other primates, selected for genes that affected the development of brain mechanisms that favored high levels of empathy, cooperation, and altruism among individuals, whether kin or non-kin, within particular social groups (see [1036]).

The driving force behind evolution is natural selection and three kinds of natural selection have been recognized, depending upon the level at which the selection is operating: individual selection, kin selection, and group selection [131,1036]. Individual selection refers to the selection and evolution of characteristics that favor the survival and reproductive success of a particular individual in comparison to other individuals. Kin selection results in the evolution of traits that favor the success of groups of genetically related individuals (kin) over other competing kin-based groups, and group selection results in the evolution of characteristics that favor the success of one group, which may be composed of related and unrelated individuals, over a competing group.

Biological evolution causes changes in the frequencies of genes within a population of organisms, and these genetic changes, in turn, contribute to the phenotypes or characteristics of the individuals within the population. Table 7.1 shows evolutionary descriptions of selfishness, cooperation, and altruism [131]. The evolution of genes that influence the development of selfish behavioral characteristics is a common form of individual selection. In this case, the presumed reproductive success (number of offspring produced) of a donor individual who exhibits a particular behavioral trait increases during interactions with a recipient, while the recipient's reproductive success decreases. A good example would be two individuals competing for a territory, or aggressive encounters that establish dominance hierarchies within groups. The more aggressive individual will ultimately have priority access to resources in short supply, and his genes and aggressive traits should become more common in the population since his reproductive success increases at the expense of others. Therefore, selfishness, or exclusive self-interest, is the outcome of competition, and there are winners and losers. Cooperation is another example of individual selection. Two (or more) individuals act together to achieve a goal, and the outcome is mutually beneficial for the individual reproductive success of each participant. A good example of cooperation, which undoubtedly occurred during early hominin evolution [1036], is group hunting and the sharing of captured prey. Single individuals that do not cooperate are not

TABLE 7.1 Evolutionary Descriptions of Selfishness, Cooperation, and Altruism

	Selfish Behavior	Cooperation	Altruism
Donor	+	+	–
Recipient	–	+	+

+ = personal or individual gains, usually measured by increases in individual reproductive success.
– = personal or individual losses or costs, usually measured by decreases in individual reproductive success.

likely to capture large prey, while individuals that cooperate would be more successful, with such cooperative behavior increasing the individual reproductive success of all participants, assuming that the food is shared. Under such conditions, "cooperative" genes, in comparison to "selfish" genes, should increase in frequency within the population based on individual selection. Pair bonding and cooperative breeding would be another example of the evolution of cooperation by individual selection; if maternal and paternal care are essential for the survival and reproductive success of one's offspring under certain environmental conditions, then monogamous partners that cooperate in raising offspring will each be more reproductively successful than individuals that do not pair bond and where paternal behavior is absent.

The occurrence of altruistic behavior has been the most difficult social behavior for evolutionary biologists to explain. Altruism occurs when a donor individual gives some type of aid to a recipient who is not an offspring of the donor and there is also no direct and immediate benefit to the donor. The evolutionary conundrum revolves around the issue of how genes that favor altruism could increase in frequency in a population when the individual reproductive success of the aid giver is likely to decrease while that of the recipient increases. Three biological explanations for the evolution of altruism have been offered, and probably all three have played a role in its evolution: kin selection [381], reciprocal altruism [975], and group selection [116,1036].

Kin selection theory proposes that a donor who possesses altruistic genes can aid a recipient without any benefit to him/herself, and those altruistic genes may increase in frequency in the population if the recipient of the aid is a relative of the donor. Relatives share a certain proportion of their genes in common; for example, siblings share, on average, 50% of their genes in common, and cousins share 25% of their genes. Therefore, if a donor of aid incurs a small cost to him/herself while providing a large benefit to a sibling or cousin, then the giving of aid to kin who are not one's offspring, and the altruistic genes associated with such behavior, would be able to increase in frequency within the population. Alloparenting directed toward younger siblings is a good example of a behavior that has undoubtedly been influenced by kin selection. Therefore, kin selection can explain the evolution of altruism toward relatives (nepotism), but it is not a likely explanation for the evolution of aid behavior, caregiving, and compassion toward members of one's group who are nonrelatives.

Reciprocal altruism is a mechanism to explain the occurrence of altruistic aid to nonrelatives. Reciprocal altruism can be viewed as delayed cooperation, and it is therefore a form of individual selection. For reciprocal altruism to work, a donor gives some sort of aid to a recipient without any direct and immediate benefit to him/herself, but at some point in the future, the recipient returns aid to the donor. Therefore, over time, both participants benefit from the interaction. However, reciprocal altruism breaks down if the original recipient cheats and does not reciprocate. Because such cheating may occur, due to selfishness, the original donor is likely to provide only a small amount of aid to the recipient during their first interaction. If the recipient reciprocates, then the donor may give a higher amount of aid to the recipient in the future, with the expectation that such a level of aid will be reciprocated. Therefore, with reciprocation, trust develops between the interacting partners, and social memory obviously plays an important role in this process. I think this is a good approximation of how friendships develop between individuals—as trust develops between interacting individuals, so do strong social bonds between them. Such individuals are then more likely to cooperate and engage in reciprocal altruism in the future, helping one another out when a need is obvious.

Group selection is a controversial issue in evolutionary biology. Some have argued that kin selection and reciprocal altruism are the only evolutionary mechanisms needed to explain the occurrence of altruism in human social groups [245], while others have proposed that group selection played a dominant role in the evolution of human social behaviors [1036]. Wilson [1036] has been a strong proponent of the view that during hominin evolution multilevel selection has molded human characteristics. According to this perspective, during evolution, humans were exposed to both intense individual selection and intense group selection, which caused humans to possess a combination of selfish traits and cooperative/altruistic traits. Within social groups, individuals probably competed to some extent for limited resources and mates, giving rise to selfish tendencies. However, Wilson also argues that early *Homo sapiens* formed well-organized groups that competed with other human groups (tribes) for territory and scarce resources, and that this gave rise to intense between-group competition. He proposes that groups with altruistic/cooperative genes, and the corresponding traits that promoted cooperation, altruism, and teamwork, probably outcompeted groups composed primarily of selfish individuals, and that such a group selection mechanism could favor the evolution of altruistic genes even in those cases where individuals give their lives for their group during territorial disputes with other groups (also see [116]).

An important aspect of the group selection point of view is that it suggests that human prosociality and caregiving impulses are not necessarily indiscriminately aimed. The group selection view emphasizes the distinction between in-groups and out-groups, with prosocial behaviors being directed toward in-group members, while antisocial behaviors are directed toward out-group members. Reciprocal altruism and kin selection would also result in prosocial behaviors and altruism being primarily directed toward members of one's group.

To conclude, it seems clear that evolutionary forces have influenced the high level of prosociality, caregiving, cooperation, and altruism that is exhibited in human societies. The resultant genes, interacting with environmental inputs, create our social brains and our social behaviors. It is also obvious that individual differences exist in human social behavior, with some of us being more selfish than others and some of us being more altruistic than others. Genetic polymorphisms, differences in environmental inputs throughout development, and epigenetic factors are likely contributors to such phenotypic variability.

7.3 IMPORTANT POINTS FROM PREVIOUS CHAPTERS

It will be worthwhile to highlight some of the important principles and facts reviewed in the previous chapters, as these points will be relevant to the topics discussed in this chapter. There are core neural circuits that regulate social behavior in mammals, and these circuits control basic approach and avoidance responses with respect to social (and nonsocial) stimuli. I have tried to show in concrete ways how social stimuli can be processed over either approach/acceptance/prosocial neural pathways or over avoidance/rejection/antisocial neural pathways, and that such processing can be influenced by physiological state, genetic factors, and experience. In this context, and with respect to human sociality and the evolutionary views presented above, one can conceive that there should be a bias for social stimuli emanating from members of one's in-group to be processed primarily over prosocial circuits, while out-group social stimuli would not gain easy access to prosocial circuits, with a bias for such stimuli to be processed over antisocial circuits. If one refers to Figure 5.12(A and B) that outline the neural systems that are hypothesized to be activated by infant stimuli in postpartum and virgin female rats, respectively, one will get a sense of what I am proposing. Figure 5.12(A) might represent components of the neural circuitry activated by in-group members, while the circuits shown in Figure 5.12(B) might represent components of the circuits activated by out-group members.

The core neural circuits that underlie social behavior across mammals include interactions between the hypothalamus and amygdala-striatal-pallidal circuits, and I have emphasized the involvement of medial preoptic area (MPOA) interactions with the mesolimbic dopamine (DA) system for a variety of prosocial behaviors. These core circuits, in turn, interact with prefrontal cortical mechanisms as a result of prefrontal cortex (PFC) projections to the hypothalamus, ventral striatum, and amygdala. Importantly, PFC mechanisms are able to either upregulate or downregulate prosocial and antisocial neural circuits, with the particular functional effect depending on the detailed underlying neuroanatomy (see Chapters 1 and 3). Although the PFC exerts such modulatory effects in all mammals, this function is highly developed in primates and can serve as a route over which cognitive and emotional processes influence the operation of the basic circuitry that regulates social behavior. Indeed, in humans, such a route is likely to influence the processing of in-group social stimuli over prosocial neural circuits and out-group social stimuli over antisocial neural circuits, or to prevent the access of out-group stimuli to prosocial circuits. This route, for example, could allow religious, ethnic, or political ideologies, and their associated emotional states, to affect how we respond socially to those who are similar or dissimilar to ourselves [698]. That is, PFC mechanisms could operate to increase hostile behaviors toward out-group members, while decreasing hostility and increasing prosocial responding to in-group members.

Behavioral neuroscientists who study social behavior in nonhuman animals primarily investigate how the brain controls behavioral responsiveness to social stimuli, and an emphasis has been on the interaction between the hypothalamus and amygdala-striatal-pallidal circuits. In contrast, cognitive and affective neuroscientists who study the neural underpinnings of sociality in humans have primarily employed imaging methods to explore cortical brain activity in social situations, and the involvement of subcortical structures such as the hypothalamus has been given much less attention [139,909]. The emphasis on cortical mechanisms in human studies is probably related to a primary interest in the thought processes and feeling states (conscious experiences) associated with human sociality. However, the purpose of conscious cognitive and emotional processes is to affect the way we behave in social situations, and one goal of this chapter is to link up cortically mediated emotional/affective feeling states with hypothalamic-amygdala-striatal-pallidal behavioral circuits in order to understand the neurobiological mechanisms that allow prosocial feeling states to be converted into prosocial behaviors in humans.

The amygdala appears to be a significant, although not exclusive, entry point over which social stimuli gain access to the neural circuits that control social behavior, and depending on the particular amygdala neurons that are engaged, such social stimuli can be processed over either prosocial or antisocial circuits. Interestingly, in *adult* humans amygdala damage due to disease causes significant, but not severe and extreme, social deficits [211,398]. Since such lesions occurred in adulthood, and since the completeness of the damage is not clearly defined from MRI analysis, it is likely that compensatory mechanisms allowed for the occurrence of certain aspects of social behavior and its associated affective/cognitive states in such patients. In particular, a cognitive appreciation of how to behave in social situations, acquired prior to the lesion and represented in neural sites outside the amygdala, may regulate somewhat normal social responses during the postlesion period.

The MPOA has been shown to play a critical role in the appetitive aspects of male and female sexual behavior and maternal behavior. Therefore, it is centrally involved in reproductive prosociality. To the extent that maternal behavior is the primordial caregiving system, it is likely that the MPOA also participates in the broader aspects of human prosocial behavior, and I will review the limited evidence in favor of this point of view. Finally, throughout the previous chapters, the role of oxytocin (OT) in social behavior has been emphasized. OT, through its actions at multiple levels within hypothalamic-amygdala-striatal circuits, influences maternal behavior, pair bonding, and social recognition. OT also has antiaggressive and anxiolytic effects. Not surprisingly, in this chapter it will be shown that OT is importantly involved in the broader aspects of human prosociality, including empathy, trust, and friendship formation.

7.4 THE INSULAR CORTEX AND EMPATHY: A POTENTIAL STARTING POINT THROUGH WHICH FEELING STATES INFLUENCE PROSOCIAL BEHAVIORS

7.4.1 Introduction

Empathy, which may be a unique human experience, refers to the ability of one individual to vicariously experience the emotional state of another individual [89,227]. Empathy can involve both negative and positive emotional experiences. For example, when one observes another person in a painful or aversive situation, one may imagine and experience the pain that the other is suffering. Similarly, one may share in another person's happiness. In the discussion that follows, I will use the term *natural emotional experience* to refer to the emotions that one experiences when one is directly subjected to positive or negative emotional events, while empathy will refer to our ability to share in the natural emotional experiences of others.

If empathic responses are strong enough, they may pass a threshold to promote prosocial behaviors [227,244,347]. This perspective has usually been taken in the context of observing another individual in distress, which may first induce an empathic experience in the observer followed by an aid-giving response. Similarly, empathic emotional experiences may contribute to cooperative or mutual aid behaviors that promote the teamwork necessary to achieve a particular goal. It is interesting to speculate that some general moral rules that are common across cultures, such as the Golden Rule, derive from the ability of humans to experience empathy.

Imaging studies have shown that the insular cortex, and particularly its anterior agranular (allocortical) part (anterior insula; AI), is typically activated, as measured by an enhanced blood-oxygen-level dependent (BOLD) response, during a variety of natural and empathic feeling states, and that the neural BOLD response in AI is positively correlated with the intensity of one's emotional experiences [204,365,858]. For example, when one is directly experiencing pain, or is observing pain in others, the AI is active in both cases. Most studies have obtained similar results for positive emotional states ([283]; see Chapter 4 for sexual experiences), although not all studies agree on this point [664]. Of course, the AI is not active in isolation during natural and empathic emotional states but is part of a complex set of active brain regions, and I will describe this larger circuitry below.

In rodents, the insular cortex is located on the lateral surface of the cerebral cortex (see Figure 1.14), but in primates it invaginates and becomes buried within the depths of the lateral fissure due to the increased size of the frontal, temporal, and parietal lobes [685]. Therefore, in primates the insular cortex can only be observed if one separates the overlying parts of the frontal, parietal, and temporal lobes. The location of the human insular cortex at the level of the rostral hypothalamus is shown in Figure 7.1. In terms of anatomical connectivity [179,182,415,685], the insular cortex receives a variety of sensory inputs, which include viscerosensory (autonomic) inputs, directly from the thalamus and also from sensory regions of the cerebral cortex. The AI is considered to be part of the posterior orbital frontal cortex (OFC), and it also receives inputs from other parts of the OFC. Since the lateral PFC projects to the OFC, the AI can also get inputs from the more "cognitive" components of the PFC (see Figure 1.15). Importantly, the AI is also reciprocally connected with the amygdala, particularly basolateral amygdala/basomedial amygdala (BLA/BMA). Therefore, the AI is ideally situated, with respect to its inputs, to integrate autonomic and other sensory inputs with contextual/cognitive information and emotion-related amygdala inputs to create an emotional experience state. In terms of its outputs, in addition to projections to the amygdala, the AI cortex also projects to parts of the medial PFC, with a major projection to the anterior cingulate cortex (ACC = area 24). Since the ACC projects to the more ventral parts of the medial PFC ([423]; area 32 = prelimbic cortex and area 25 = infralimbic cortex), the AI can indirectly influence these areas as well. Indeed, imaging studies have found that during a variety of natural emotional and empathic feeling states in human participants, the AI and ACC are co-active [204,271]. Craig [204] has proposed that the AI is a sensory-related region and mediates emotional and empathic experiences, while the ACC serves as a motor-related region and drives motivated behaviors. Therefore, projections from the AI to the ACC might be one route over which empathic states

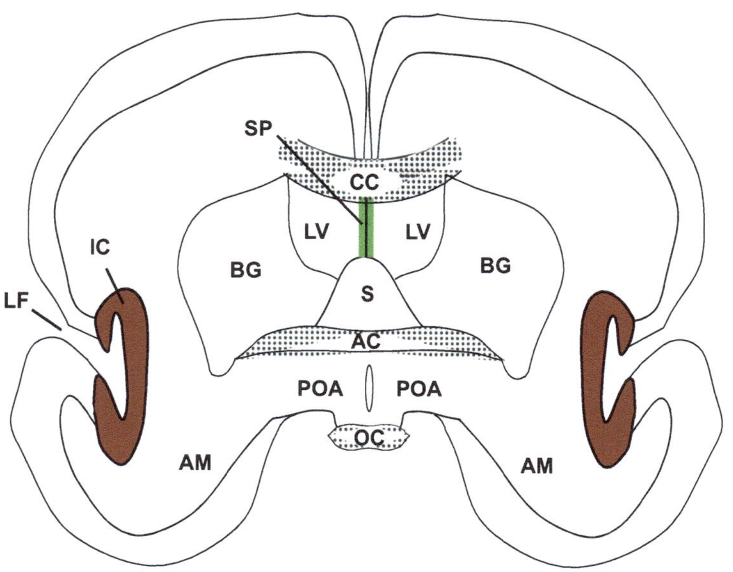

FIGURE 7.1 The location of the insular cortex (IC), shown in red, on a frontal section through the human brain at the level of the preoptic area (POA; includes medial and lateral POA). This section depicts a single level of IC, which extends in an anterior and posterior direction from this site. Since IC is buried within the lateral fissure (LF), one can gain an appreciation of the anterior-to-posterior extent of IC by referring to Figure 1.13 in Chapter 1, which shows the LF on a lateral view of the primate brain. Finally, note the location of the septum pellucidum (SP), shown in green above the septal area (S). SP forms a separation between the two lateral ventricles (LV). Other abbreviations: AC = anterior commissure; AM = amygdala; BG = basal ganglia; CC = corpus callosum; OC = optic chiasm. *Redrawn and modified from Figure II.23 in Martin [608], with permission from the author, Elsevier, and McGraw Hill. The current edition (4th) of this textbook is published by McGraw Hill (2012).*

could influence prosocial behaviors. Since the medial PFC projects to the nucleus accumbens and hypothalamus, this analysis makes sense (see Chapter 1). Finally, the AI also projects strongly to the nucleus accumbens/ventral striatum, and this could be another route through which feeling states affect motivated behaviors.

7.4.2 A Neural Model

In order to effectively organize the functional neuroanatomy described above, a preliminary neural model is presented in Figure 7.2 (also see [698]). This model has two important characteristics. First, it presents the neuroanatomical connectivity that may explain how empathic feeling states, such as those associated with observing someone who needs help, could be translated into aid-giving prosocial behaviors. Second, based on the perspective that the advanced aspects of prosocial behaviors observed in humans, such as cooperation and altruism, have been built on a neural foundation that controls the goal-directed aspects of maternal care, aspects of the maternal circuitry have been incorporated into the neural model (also see [785]).

The model proposes that prosocial behaviors are influenced by MPOA interactions with the mesolimbic DA system, which regulate the way striatal-pallidal circuits respond to amygdala inputs so that goal-directed prosocial behaviors can occur. Affective feeling states, including empathy, may affect prosocial behaviors through AI projections to parts of the medial PFC, nucleus accumbens, and amygdala, and medial PFC inputs to MPOA may be an essential link in the activation of prosocial behaviors. The fact that AI can receive inputs from the lateral PFC indicates that cognitive processes can influence the empathic experiences mediated by the AI. This might be one route through which in-group versus out-group membership could influence the intensity of empathic feeling states.

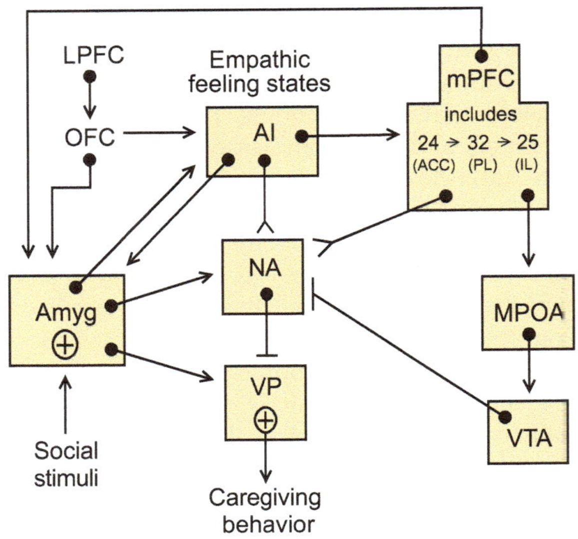

FIGURE 7.2 A model depicting some of the neural circuits that may allow empathic feeling states to be translated into prosocial caregiving behaviors in humans. This model integrates human research, which has emphasized the roles of the anterior insular cortex (AI) and other parts of the prefrontal cortex in the mediation of empathy, with animal research, which has emphasized medial preoptic area (MPOA) interactions with the mesolimbic dopamine system in the control of a variety of prosocial behaviors, particularly those associated with reproduction. The medial prefrontal cortex (mPFC), which includes areas 24 (anterior cingulate cortex (ACC) in rodents), 32 (prelimbic cortex (PL) in rodents), and 25 (infralimbic cortex (IL) in rodents), may serve as a crucial link between empathic feeling states mediated by AI and prosocial behaviors mediated by MPOA. Inputs from the prefrontal cortex to the amygdala (Amyg) may regulate the types of social stimuli that can gain access to positively valent neurons in this region. Such prefrontal cortex regulatory influences may allow in-group members priority access to prosocial circuits. In this figure, the prefrontal cortex is shown as exerting stimulatory effects, which is presumably what occurs during the cognitive appraisal of in-group members. For out-group members, it is likely that prefrontal cortical mechanisms depress the ability of such social stimuli to gain full access to amygdala and other prosocial circuits. See the text for other important details. Axons ending in a bar are inhibitory, and those ending in an arrow are excitatory. The axon terminations (shown as Y) between AI and nucleus accumbens (NA) and between mPFC and NA are left undefined, although in order to promote prosocial behavior, it is assumed that these connections act to potentiate the output of positively valent ventral pallidum neurons (VP+). Other abbreviations: LPFC = lateral prefrontal cortex; OFC = orbital prefrontal cortex; VTA = ventral tegmental area.

Since the medial PFC and the OFC both project to the amygdala, these might represent additional routes through which prosocial responses are modulated based on in-group versus out-group membership. Finally, the model also proposes that positively valent amygdala neurons relay positive social stimuli to nucleus accumbens–ventral pallidum (NA-VP) and to AI. In conclusion, this preliminary model attempts to explain at least certain aspects of how thinking and feeling are translated into prosocial responses in humans. Evidence supporting the model is presented below.

7.4.3 The Anterior Insular Cortex and Empathy

7.4.3.1 Is the AI Essential for Emotional Experiences and Empathy?

Functional magnetic resonance imaging (fMRI) studies clearly show that the AI is active during natural and empathic emotional experiences. However, these are correlational findings, and lesion studies have asked whether the AI is essential for these experiences. Damasio et al. [211] have examined a patient who had severe damage to the AI and other parts of his brain as a result of Herpes simplex encephalitis that occurred when he was 48 years old. The brain damage was documented through MRI analysis. According to Damasio et al., all aspects of emotional feeling states were intact in this subject. This patient reported feeling pleasure, happiness, sadness, apprehension, empathy, compassion, and caring, and he behaved in a manner consistent with such feeling states. The authors argue that these results indicate that the insula and adjoining OFC should not be considered as the exclusive basis for the experience of emotions, and they further propose that subcortical structures, such as the hypothalamus and ventral pallidum, which were intact, may mediate a rudimentary form of emotional experience.

The following should be considered in evaluating these results. First, MRI analysis probably cannot ascertain whether a particular brain region is completely destroyed. Second, since the brain damage occurred at 48 years of age, prior natural and empathic emotional states may have allowed the patient to cognitively appreciate, in an abstract or conceptual sense, an emotional state so that it could be labeled, without actually having a true emotional experience.

In an interesting study that supports the role of the AI in empathic feeling states, Gu et al. [365] examined three patients with discrete lesions of the AI and three patients with lesions outside of AI (control lesions). The lesions were the result of surgical removal of gliomas. The patients were shown pictures of people exposed to painful and nonpainful stimuli, and they were asked to judge whether the person in the photo was suffering from pain or not. As an example, one photo showed a person's toe being crushed under a table, while a similar photo show someone's toe near a table, but not under it. Patients with AI lesions, but not those with control lesions, exhibited significant deficits in determining whether the person in the photo was suffering from pain.

Naqvi and Bechara [676] have provided evidence that the AI is involved in natural emotional experiences. The particular emotional experience that they examined was the craving for cigarettes in subjects who were trying to stop smoking. They found that smokers with brain damage to the insula, in striking contrast to smokers with damage to other brain regions, were able to quit smoking immediately and did not exhibit relapses. Subjects with insula damage reported that they did not experience the urges and cravings that are typically associated with an addicted person's attempt to quit smoking.

In a morphological study, although correlational in nature, evidence has been presented that supports a role for the AI in human socio-emotional processes. Recall from Chapter 1 that patients with the genetic anomaly that results in Williams syndrome (WS) are hyper-social and highly empathic when compared to healthy controls [461]. In an MRI study employing a voxel-based morphometry analysis, Jabbi et al. [447] reported that while patients with WS exhibited an overall decrease in gray matter volume (GMV) within the insula as a whole, a dramatic increase in GMV was detected in the ventral AI in comparison to controls. Further, a significant positive correlation existed between GMV in this part of the AI and the intensity of WS personality traits. Jabbi et al. suggest that the ventral AI may be a critical component of the neural circuitry that underpins human social and emotional processes.

In an fMRI analysis of brain activity during an economic game, Dawes et al. [221] reported a positive correlation between participants' AI BOLD responses and egalitarian behaviors, that is, behaviors that resulted in a more equal distribution of monetary resources among in-group members at a cost to the participant. This example might represent an empathic response to an unfair distribution of resources, whereby AI activity promoted a prosocial response that led to a fairer outcome (a more equal distribution of resources) for group members.

In a nonhuman primate imaging study, Rilling et al. [810] compared the brains of chimpanzees with those of bonobos, which are our two closest living primate relatives. When comparing the social behavior of these two ape species, chimpanzees lean toward greater selfishness, while bonobos are more prosocial [245,810,963]. For example, bonobos display more food sharing, less aggression, and more adult play than do chimpanzees. Although one can never know what a nonhuman animal is experiencing, it has been suggested that bonobos, to a greater extent that chimpanzees, may experience a rudimentary form of empathy that leads to greater prosociality [245]. In this regard, using an MRI procedure, Rilling et al. reported, for the right side of the brain, that the GMV in the dorsal amygdala, AI, medial PFC, and hypothalamus was greater in bonobos when compared to chimpanzees. Such a morphometric analysis fits with the neural circuitry model shown in Figure 7.2 to explain how empathy might lead to prosocial behaviors. Interestingly, even in monkeys, the output of the insular cortex may modulate prosocial behaviors. When the ventral insular cortex was electrically stimulated during the display of a facial threat in rhesus monkeys, the threat grimace was replaced by lip-smacking, which is an affiliative gesture [455].

In conclusion, the fMRI, lesion, stimulation, and anatomical studies described above support the idea that although the AI may not be absolutely necessary for all types of emotional experiences, it is a very important component of the neural circuitry that gives rise to empathic feeling states. Such states, if sufficiently strong, may lead to prosocial behaviors.

7.4.3.2 Modulatory Influences on AI Activity, Empathy, and Prosocial Behaviors

Although a baseline level of empathic feelings for others occurs in most humans, several factors have been shown to influence such feeling states and the prosocial behaviors that they induce, and these include whether the person being observed by the empathizer is a member of one's group, is a fair person, or is a close friend. A particular individual may form stronger social bonds to some individuals than to others, and such strong social bonds may allow high levels of empathy to activate prosocial acts toward the person in need of aid.

In a study by Singer et al. [907], male participants played an economic game with an initially unfamiliar player who turned out to be either fair or unfair. The fair players shared money with the participants during the game, while unfair players did not. Subsequently, while participants were in an fMRI scanner, they observed the fair and unfair players receiving an electric shock to their hand. The BOLD response in AI, and the presumed underlying empathic feeling state, was higher when the participants observed the fair player receiving the electric shock.

In an fMRI study, Meyer et al. [635] measured the neural and emotional responses of participants who observed either their best friend or a stranger either being included or excluded from a computerized ball-tossing game. Social exclusion or rejection is considered to be an example of social suffering. Participants empathized more with the presumed negative emotional experiences of their best friend being excluded from the game when compared to their empathy toward the stranger, and this greater empathy was associated with a greater BOLD response in the AI and ACC. In a related finding, Beeney, Franklin, Levy, and Adams [80] reported that the closeness of one's relationship with a friend also modulates AI and ACC activity. Participants came to the experiment with a friend, and the closeness of each participant's relationship varied so that in some cases the friend was only a work acquaintance while in other cases the friend was a best friend. In the ball-toss social exclusion game, the degree of activation of the AI and ACC when observing one's friend being excluded from the game was positively correlated with the closeness of the observer's relationship to the person being excluded.

In an important study, participants observed members of their in-group or an out-group receiving electric shocks while they were in an fMRI scanner [400]. In-group members were fans of a participant's favorite soccer team, while out-group members were fans of an opposing team. Participants reported significantly more empathy for the painful suffering of in-group than out-group members, and this was correlated with a greater BOLD response in the AI when observing in-group members receiving shock. In a second part of this study, while participants were outside the scanner, they were given the opportunity to engage in prosocial aid behaviors toward either in-group or out-group members—they could receive some shock themselves in order to reduce the shock received by others. Participants were significantly more likely to engage in such altruistic responses toward in-group than out-group members, and importantly, their degree of altruism (the number of trials on which they chose to help by receiving shock themselves) was positively correlated with the intensity of AI activation in the first part of the study when observing others receiving shock. Similar results have been reported by others, with the additional findings that when high levels of empathy were found to be correlated with subsequent prosocial responses, activity not only in AI, but also in ACC and other parts of the medial PFC, was positively correlated with the likelihood that prosocial aid behaviors would be directed toward those who activated high levels of empathy in the participant [610,611]. These results are certainly consistent with the proposal that when empathy-related neural activity in the AI reaches some threshold level, then AI efferents to the ACC/medial PFC participate in the activation of neural circuits that promote prosocial caregiving behaviors (see Figure 7.2).

Mitchell, Macrae, and Banaji [651] asked participants to think about either in-group members (people who shared the political philosophy of the participant) or out-group members (those of an opposing political philosophy) while they were in an fMRI scanner. Thinking about in-group members was associated with activity in the ventral parts of the medial PFC within the anterior cingulate cortex region, while thinking about out-group members was related to activity in the dorsal parts of the medial PFC located outside of and dorsal to the anterior cingulate cortex (outside of areas 24, 25, and 32). Perhaps activity within certain parts of the ventromedial PFC (area 25, which is homologous to rodent IL cortex) promotes prosocial mentalizing and responding in part through projections to the hypothalamus, nucleus accumbens, and amygdala (see Figure 7.2). Perhaps such activity is depressed by activity in the dorsomedial PFC.

Although the output of certain parts of the medial PFC may promote prosocial behaviors, how might in-group versus out-group membership, or variations in either individual fairness or the intensity of a friendship influence one's empathic responsiveness that, in turn, affects the likelihood of prosocial acts? Clearly, cognitive factors contribute to such modulatory influences on empathy and prosociality. One possibility is that regulatory PFC inputs to the amygdala modulate the ability of social stimuli to activate positively valent amygdala neurons that project to AI and other regions ([484,698]; also see Chapter 1). Further, PFC input to AI may also exert modulatory influences (see Figure 7.2). Finally, in an analogy to the formation of pair bonds in voles and strong mother–infant attachments in rats, perhaps the formation of friendships in humans increases the probability of prosocial acts because the neural link between positively valent amygdala neurons and the ventral pallidum has been strengthened.

7.4.3.3 *The Larger Neural Circuitry within Which the AI-ACC-Medial PFC May Operate*

In this selective review, I want to present data that support the larger neural model show in Figure 7.2. Chapters 4, 5, and 6 have provided evidence for the important role of the MPOA in sexual behaviors, maternal behavior, and perhaps in the social bonding process that occurs between monogamous mating partners. Clearly, the MPOA appears to be critically involved in prosocial reproductive motivational processes in mammals, including humans. In Chapter 5, I reviewed the question of MPOA specificity: are different MPOA neurons involved in the appetitive aspects of male and female sexual behavior and maternal behavior, or does a common pool of MPOA neurons generally influence reproductive motivation? Although that issue was not resolved, one can ask whether certain general or specific MPOA neurons also influence prosocial affiliative motivation in humans outside the context of reproduction.

Although human cognitive and affective neuroscience studies typically do not focus on the hypothalamus, there is some evidence to support the idea of MPOA involvement in affiliative prosocial motivational and emotional states in humans. Most of these studies are imaging studies, and the interpretation of MPOA involvement is limited by the spatial resolution of these methods. In a study by Moll et al. [656], described in Chapter 5, human participants, while in an fMRI scanner, read various scenarios that described social interactions between individuals. After reading each scenario, the participant rated the intensity of their feelings of warmth, tenderness, and caring. Activity in the septal-preoptic area was positively correlated with such positive prosocial feeling states. In a related lesion study, Moll et al. [658] reported a positive correlation between the amount of damage to the septal-preoptic region in patients with dementia and the degree to which such patients displayed deficits in prosocial feeling states.

In the studies that I described previously in support of the view that high levels of empathy result in prosocial aid behavior, empathy was measured inside a scanner, while prosocial responding was analyzed when the participants were not being scanned (see [400]). These studies correlated activity in the AI and certain parts of the medial PFC with empathy, which was subsequently associated with prosocial behavior. An important question to ask is whether MPOA interactions with the mesolimbic DA system might have been observed, as a result of intense AI and medial PFC activity, if fMRI data had been collected during prosocial acts. In a relevant study, Decety and Porges [228] examined this issue. While in an fMRI scanner, participants viewed short visual scenes that included scenes where an individual was shown as alleviating another person's pain. The participants were required to mentally simulate being the performer of such prosocial acts. In comparison to the viewing of control scenes that did not involve the performance of prosocial acts, simulating the performance of prosocial acts was associated with an increased BOLD response in several brain regions, which included the septal-preoptic area and nucleus accumbens, and was also associated with an increase in the functional connectivity between the amygdala and the NA-VP region. Decety and Porges propose that their findings highlight those neural responses that are associated with a sense of agency with respect to prosocial acts, by which they mean the subjective awareness that one is initiating and executing a goal-directed (voluntary) prosocial act. Further, they propose that imagining oneself performing an act is likely to trigger neural activity in just those brain regions that would be active during the actual overt act. To the extent that this is the case, this research offers some support for the view that septal-preoptic area interactions with the mesolimbic DA system, which in turn may influence the processing of amygdala inputs to the NA-VP circuit, play a role in human prosocial behaviors.

Moll et al. [657] examined the neural responses that occurred when participants were in a scanner and either received a monetary reward or donated money to a charitable organization of their choice. The nucleus accumbens and ventral tegmental area were activated in association with both pure monetary rewards and with decisions to donate. Importantly, area 25 of the mPFC and the septal-preoptic region were activated, and the NA was more active, when participants chose to donate in comparison to when they received monetary rewards. What these results suggest is that the neural link between the septal-preoptic region and mesolimbic DA system may be critical for prosocial caregiving responses, and that activity across this circuit may be induced by medial PFC (area 25) input to MPOA, as suggested in Figure 7.2. Findings supporting the involvement of the septal-preoptic-hypothalamic region in social cooperation between two individuals (reciprocal altruism) have also been presented [513], and these authors suggested that such involvement might be due to the activation of OT neural systems in the PVN. Indeed, such OT activation may be the result of MPOA input to PVN. Although this may be the case, the role of direct MPOA-mesolimbic connectivity is most certainly also involved. Definitive conclusions with respect to these important anatomical issues await improvements in the spatial resolution characteristics of brain imaging procedures.

Finally, a recent study has specifically emphasized the importance of septal area activity for both empathy and prosocial behavior [664]. It is interesting to speculate that during social interactions, the septal area may primarily be involved in suppressing antisocial responses (see Figure 3.5), while MPOA output promotes prosocial responses.

A study by Cikara, Botvinick, and Fiske [189], which examined the neural events associated with observing the success of in-group or out-group members, provides insights related to the proper interpretation of NA-VP activity (also see [400]). Avid Red Sox or Yankee fans observed baseball plays while in an fMRI scanner. Two weeks later, while outside the scanner, participants took a survey that measured the probability that they would be aggressive toward a rival fan (heckling and threatening behavior). Increases in the ventral striatal-nucleus accumbens BOLD response (presumably representing enhanced DA input) occurred when participants saw successful plays by their favored team or failed plays by their rival team. The magnitude of the ventral striatal BOLD response when observing the rival team fail was positively correlated with the likelihood of exhibiting hostile behaviors toward a rival fan. For those who view the mesolimbic DA system purely as a reward or pleasure circuit, these latter results are usually interpreted as signifying the pleasure one obtains from the failure of an out-group competitor [189,400]. Alternatively, these results can be interpreted as supporting the view that mesolimbic and ventral striatal neural activity can support both prosocial and antisocial responses. As reviewed in Chapter 1, VTA-DA input to the NA is involved in both goal-directed approach/acceptance responses and goal-directed avoidance/rejection responses, with DA input to the anterior NA being more involved in appetitive responses and DA input to the posterior NA being more involved in rejection responses. Once the spatial resolution of imaging procedures improves, perhaps it would be found that observing the success of an in-group member (along with the possible mental simulation of aiding such a success) would be associated with activity in the anterior NA, while failure of the out-group member (along with the possible mental simulation of fostering such a failure) would be associated with activity in the posterior NA. This analysis also emphasizes that there are separate prosocial and antisocial neural circuits over which social stimuli are processed, which influence how we respond to in-group and out-group members, and that these disparate functional circuits may reside within a larger brain region (ROI), such as the NA-VP and the amygdala (see Chapter 1 and cf. [109]).

Finally, to complete the evidence in support of the model shown in Figure 7.2, I want to present research that shows that the amygdala is involved in empathy, possibly through its connections with the AI. Hurlemann et al. [436] studied two patients with Urbach-Weithe (UW) disease, which causes calcification and lesions of the amygdala, as was described in the case of SM in Chapter 1. These patients were shown pictures of people in emotionally salient situations of either a negative or positive nature, such as crying, or hugging and kissing. The emotions were expressed by facial expressions, body language, and context. These authors distinguished cognitive empathy from emotional empathy. In order to assess cognitive empathy, the patients and healthy controls were required to infer the mental state of the individual depicted in each scene by choosing a verbal description from a list of four alternatives. To assess emotional empathy, participants were asked to rate on a 9-point intensity scale how much they were feeling for the individual in each scene. The UW patients did not exhibit deficits in cognitive empathy, but were significantly impaired in emotional empathy. As I have stated previously, because of prior experience while the amygdala was intact, a cognitive or conceptual understanding of the emotional states of others may be intact in individuals without a normally functioning amygdala, but without a normal amygdala, the corresponding ability to actually feel a raw natural or empathic emotion appears to be impaired.

Upon examining Figure 7.2 in the context of feeling empathy for an individual in distress, it may seem surprising that the model proposes that the social stimuli associated with an individual that is in distress, such as after receiving a painful stimulus, are shown as activating *positively* valent amygdala neurons that project to AI, so that the empathizer might vicariously experience emotional distress. However, these amygdala neurons are labeled as positive, even though they might give rise to a negative feeling state, because they activate prosocial approach responses rather than antisocial avoidance/rejection responses.

7.5 OXYTOCIN AND HUMAN PROSOCIALITY

The previous chapters have clearly documented the positive role played by OT neural systems in fostering prosocial behaviors and decreasing antisocial behaviors in both nonhuman animals and humans. OT action on the MPOA, VTA, amygdala, and/or NA promotes maternal motivation and facilitates the formation of long-term social bonds between mothers and infants and between mating partners. In this section, I want to present evidence that indicates that OT is also involved in the broader types of prosociality that are observed in human groups. OT release into the human brain during social interactions between two individuals who are not kin and who are not mating/romantic partners may promote the formation of social bonds that underpin strong friendships, and OT action on the brain may also enhance empathy, cooperation, and altruism. As suggested in Chapter 6, such release of OT into the brain may be directed, in part, by MPOA input to PVN. This literature is quite large and complex [93,230,937], and I will only present a selective review that is most pertinent to the core prosocial neural circuits shown in Figure 7.2.

In a study by Liu, Guastella, and Dadds [558], human participants received either intranasal OT or placebo prior to being introduced to another individual through a short video clip. Twenty-four hours later, they were asked whether they would like to find out more information about that person, or, instead, if they would prefer to learn about a new person who was shown to each participant in a photo. Those subjects who were administered the placebo were more interested in acquiring information about the new person, while those that were administered intranasal OT wanted to find out more about the person they had met 24 h earlier. This effect of OT was not dependent upon the sex of the participant or the sex of the individuals to whom the participants were exposed. One interpretation of these results is that exposure to enhanced levels of brain OT, produced through exogenous administration rather than endogenous release, acted to promote the initial steps of friendship formation that normally would not have occurred after such a brief introduction to a new acquaintance via a video clip.

In an important study, Kosfeld, Heinrichs, Zak, Fischbacher, and Fehr [507] examined the effects of intranasal OT or placebo administration on the behavioral responses of male participants in an economic trust game. Each participant was given a standard amount of money that he could invest with an unfamiliar trustee, who might then share any profits made from the investment with the participant. The problem for the investor, however, is the possibility of selfish behavior on the part of the trustee, who might keep all of the investment and profits without sharing anything with the participant. The dependent variable in this study was the proportion of the original monetary endowment that each participant gave to the trustee during an *initial* investment and without any knowledge of the trustworthiness of the trustee. A second part of this study also examined the initial back transfer of money from trustees to investors. Note that this was not an iterated game—each interacting investor and trustee only made a single decision, how much to invest or how much to give back, respectively.

For the investors, subjects in the placebo group tended to invest only a small amount of money, while OT-treated subjects invested a large percentage of their endowment with the trustee. Kosfeld et al. [507] proposed that OT action in the brain increased the level of trusting behavior shown by the investor. Can these results, in the context of the previous study by Liu et al. [558], be interpreted more generally? Under normal conditions, if one individual, a donor, were to help another individual, the recipient, for the first time, a small-to-moderate amount of aid might be provided. If the recipient reciprocated the favor in the future, then upon subsequent interactions, the original donor might provide greater support to the recipient. This is typically how friendships develop—social ties between individuals are enhanced by reciprocal altruism, which promotes the trust that each individual has in the other. Perhaps as social bonds and trust between individuals become stronger, social stimuli from one individual activate OT neural systems in a friend, with such activation favoring prosocial cooperative behaviors. OT may be producing two effects during such friendship formation: consolidating an enduring social bond between two individuals, perhaps through an action at the level of the NA (see Chapters 5 and 6); and promoting prosocial behaviors, perhaps through actions at the level of the MPOA, VTA, and amygdala (see Chapter 5). In the study by Kosfeld et al. [507], participants that received intranasal OT may have invested a lot of money with a trustee in the absence of a naturally developed friendship with that trustee because the exogenously administered OT replicated an internal state that would normally occur only after a friendship had developed.

Kosfeld et al. [507] present some data that seem to contradict the proposal that I have just presented. They found that when the trustee received intranasal OT, such treatment did not increase the trustee's back transfer of money to the investor above that exhibited by placebo-treated trustees. However, all trustees, irrespective of treatment, did not behave selfishly, and tended to transfer more money back to the investor than was initially invested (because of profits). Therefore, a ceiling effect may have caused this lack of OT effect, in that all trustees may have considered that they were acting in a fair manner.

Additional research has supported the finding that intranasal OT administration enhances trust, altruism, and cooperation as tested in various economic games where participants invest or transfer money to others [65,230,446]. Interestingly, in situations where one's in-group and out-group have been clearly defined, De Dreu et al. [231] present evidence that intranasal OT increases the sharing of money with in-group, but not out-group, members. This latter finding suggests that context is important with respect to OT's effect on prosocial behaviors. In studies like De Dreu's, however, it would be important to not only compare OT effects with a placebo control group, but to also include a group that receives intranasal vasopressin (AVP). Since OT can act at both OT receptors and AVP receptors, and because of the involvement of AVP in aggression (see Chapter 3), perhaps OT action on AVP receptors antagonizes the trust and cooperation-inducing effects of OT action on OT receptors when contextual social stimuli indicate that one is interacting with a competitor.

In an iterated economic game, where participants are able to respond to the back transfer decisions of trustees, Rilling et al. [809] report in an fMRI study that an increased BOLD response in the nucleus accumbens/caudate nucleus was associated with cooperation and reciprocal altruism (the sharing of investments and profits) between participants, and that the nucleus accumbens BOLD response increased in response to reciprocity and also prior to a decision to cooperate. One interpretation of these results is that DA release into NA of an individual that is on the receiving end of a reciprocated act fosters social ties between the donor and the recipient, which then promotes future prosocial responding by the recipient toward the donor. Rilling et al. suggest that such a mechanism reinforces reciprocal altruism and decreases the tendency of an individual to engage in selfish behavior. Drawing an analogy to the mechanisms underlying pair bond formation in prairie voles, perhaps OT co-acts with DA to foster the development of social ties during bouts of reciprocal altruism. Relevantly, in a more recent study, Rilling et al. [808] reported that intranasal OT administration increased the BOLD response that occurs in the caudate nucleus in response to reciprocated cooperation, and they suggest that this effect of OT may increase the reward associated with reciprocated cooperation, which might then facilitate learning that another person can be trusted. To state this more simply, and to relate these findings to the mechanisms described in the nonhuman animal studies reviewed in Chapters 5 and 6, OT and DA action in NA (and caudate) may strengthen the ability of social stimuli derived from the amygdala and other sites to activate attraction/acceptance/prosocial circuits within striatal-pallidal neural networks, which then promotes future cooperative acts.

The various economic games described above are meant to measure trusting behavior, cooperation, and reciprocal altruism, but such games do not capture the variety and depth of natural human social interactions, where most cases of aid behavior, caregiving, and cooperation between individuals do not involve financial decisions. In a perhaps more ecologically valid study that involved a computerized ball-toss game, participants observed two players either including or excluding a third player from the game. When the experimental participants were given an opportunity to toss the ball to one of the three players, they were more likely to toss the ball to the excluded player, and this effect was further enhanced by intranasal OT administration [807]. Riem et al. [807] suggested that participants empathized with the social rejection felt by the excluded player (who they were familiar with), and that such empathy motivated prosocial responses toward that player, which were further enhanced by OT.

The results of the Riem et al. [807] study, along with the other studies reviewed, suggest that there are likely to be several neural sites where OT acts to promote prosocial behaviors in humans. Figure 7.3, which presents a modified version of Figure 7.2, proposes some possibilities, and these proposed sites are based on the literature reviewed in Chapters 5 and 6. OT action on MPOA is proposed to have dual effects. First, it is shown as activating MPOA circuits (perhaps also including septal circuits) that have an inhibitory influence on antisocial behaviors. Second, in conjunction with OT action at the level of the VTA, OT action at MPOA stimulates prosocial motivation toward positively valent social stimuli by enhancing DA release into NA. Next, DA and OT action at the level of the NA may act to cement certain social bonds, so that particular social stimuli have an enhanced and enduring ability to activate prosocial NA-VP neural circuits. Further, OT action at the level of the amygdala may have multiple effects: it may facilitate the ability of social stimuli to activate NA-VP prosocial circuits, it may decrease the ability of social stimuli to activate amygdala-derived antisocial circuits, and it may increase empathy by stimulating the projection of prosocial amygdala neurons to the AI. Empathy-related AI neurons, in turn, further promote prosocial helping behavior through the mechanisms previously described, which may include the activation of medial PFC input to MPOA. Finally, since OTRs are located within the medial PFC of rodents [686,911], that region may be another site of OT action in humans. Perhaps OT action within area 25 increases the receptivity of these medial PFC neurons to empathy-related inputs derived from AI.

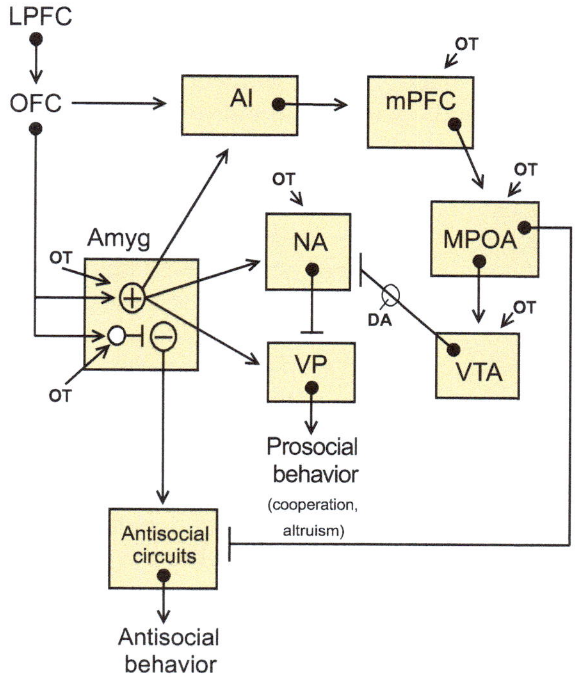

FIGURE 7.3 Based on Figure 7.2, neural sites where oxytocin (OT) might act to promote prosocial feeling states and behaviors in humans while also depressing antisocial responses. Amygdala (Amyg) neurons with a positive valence promote prosociality, and those with a negative valence influence antisocial responses. See text for other details. Axons ending in a bar are inhibitory, and those ending in an arrow are excitatory. Other abbreviations: AI = anterior insular cortex; LPFC = lateral prefrontal cortex; mPFC = medial prefrontal cortex; MPOA = medial preoptic area; NA = nucleus accumbens; OFC = orbital prefrontal cortex; VP = ventral pallidum; VTA = ventral tegmental area.

In the context of the neural model shown in Figure 7.3, cognitive processes in the PFC are likely to bias prosocial tendencies toward in-group, and away from out-group, members through influences on both AI and amygdala. Also, note that the extent to which empathy evokes prosocial behaviors depends on the strength of the other neural connections shown in the model, many of which can be modulated by the presence or absence of OT release. Clearly, one would be more likely to exhibit a costly aid response to a friend with whom one has already formed an enduring bond than to a stranger or competitor. This model awaits research that will provide up-to-date information about the location of OTRs in the human brain and which will be aimed at a more refined and detailed analysis of the role of the MPOA in human prosocial motivation.

With respect to the role of OT action at the level of the amygdala in empathy and prosocial behavior, the following can be offered as initial support. (1) As described previously, patients with damage to the amygdala demonstrate deficits in emotional empathy [436]. Further, this study also showed that intranasal administration of OT increased emotional empathy scores in healthy control subjects. (2) In an fMRI study, Striepens et al. [938] reported that intranasal OT increased the functional connectivity between the basolateral amygdala and the anterior insula. (3) Patients with Williams syndrome, who are hypersocial and highly empathic, show increased amygdala BOLD responses to happy faces and decreased amygdala BOLD responses to fearful or angry faces [372,638]. In healthy subjects, intranasal OT increases amygdala BOLD response to happy faces and decreases such responses to fearful/angry faces [324,492]. To the extent that angry/fearful faces are viewed as threatening and activate antisocial circuits [492], while happy faces activate prosocial circuits, these results support the perspective that OT may enhance amygdala prosocial input to NA-VP to promote social approach and may enhance amygdala input to AI to promote empathy. (4) In Chapter 1, I discussed how neutral stimuli acquire the ability to activate either positively or negatively valent amygdala neurons, based upon whether such stimuli are associated with reward or punishment. In Chapter 6, I discussed the nonhuman animal studies that support a role for OT action on the amygdala in processes related to individual recognition. Perhaps as friendships develop between people, OT action on the amygdala promotes the ability of individual specific social stimuli to activate prosocial amygdala neurons that project to both AI and NA-VP.

Most of the studies reviewed above have shown that the exogenous administration of OT through the intranasal route increases various aspects of prosociality in humans. I will now turn to the evidence that supports a role for naturally occurring endogenous activation of OT neural systems in human prosociality. In a study by Barraza and Zak [66], participants viewed videos of emotional and nonemotional social interactions between people. Self-reported ratings of empathy were higher when subjects viewed the emotional scenes, and this was associated with higher plasma OT levels. Feldman [289] has reported preliminary data on the peripheral levels of OT in saliva while children interact with their friends. Children who were more cooperative with their partners had higher OT levels. To the extent that these peripheral measures reflect endogenous OT release in the brain, which is somewhat controversial [467], these results support a role for endogenous brain OT systems in empathy and prosocial cooperation in humans.

Recall that an SNP of the OXTR gene has been associated with prosocial behavior in humans. Individuals that are homozygous for the G allele at rs53576 are judged as more prosocial in their interactions with others than those who carry the A allele at this site [503,967]. Rodrigues, Saslow, Garcia, John, and Keltner [817] administered a self-reported empathy scale to male and female participants. Individuals with one or two copies of the A allele (AA or AG) at rs53576 exhibited lower empathy scores in comparison to GG participants. Finally, Krueger et al. [514] have reported that individuals who are homozygous for the G allele, in comparison to carriers of the A allele, show higher trust behavior in the economic trust game that was used by Kosfeld et al. [507]. On an initial transaction, GG subjects invested more money with a trustee than did A allele carriers. Importantly, the high trust behavior of the GG subjects matched the level of trust that Kosfeld et al. [507] reported after subjects were administered intranasal OT. (It is assumed that the placebo and OT treated groups in the Kosfeld et al. study did not differ in the distribution of their rs53576 genotypes. An interesting study would be to determine whether intranasal OT might stimulate higher initial investments in GG subjects than in AA subjects.)

7.6 HUMAN PROSOCIALITY: CONCLUSIONS

A religious view of cooperation, altruism, and morality in humans would argue that such human impulses are derived from God. Although there is no doubt that various religious teachings support moral behavior in humans, a scientific analysis indicates that both human prosociality and human selfishness are derived from evolutionary forces that, along with environmental inputs, have shaped the way the human brain operates during social interactions [245,1036]. Although there are important commonalities in the core neural circuits and neurochemicals that regulate prosocial behaviors in nonhuman animals and in humans, the enhanced role of prefrontal and other cortical mechanisms in humans, along with the evolution of empathy, have provided unique mechanisms for the neural control of social behavior in humans, with the result that prosocial behaviors extend beyond mother–infant, mating partner, and kin-related interactions. In the previous paragraphs, I have outlined aspects of what we know about how basic human conscious cognitive and emotional experiences (thinking and feeling) interact with the core neural circuits that regulate prototypical forms of prosociality (e.g., maternal behavior) in mammals to give rise to complex human cooperation and altruism, which is typically directed toward in-group members.

As I sat at my desk while writing these words, a crisis was occurring in Syria because the government of Bashar al-Assad used chemical weapons against his opponents during a civil war in that country. The use of such chemical weapons caused wider concerns, and the United States was considering punitive military action against Syria. Subsequent developments resulted in this dispute being resolved by the United Nations without the need for military intervention. My point, however, is that in-group versus out-group antagonisms, based in part on ethnic, religious, and political differences between groups that compete for resources, have been a root problem throughout human history, perhaps resulting, in part, from the intense group selection that has molded our brains during evolution. In-group social stimuli can gain access to prosocial neural circuits, while out-group stimuli are likely to engage antisocial circuits. Although what I am going to say will probably sound trite and naïve, peace in our world and cooperation among all peoples will be greatly enhanced once we develop an ability to view all of humanity as part of one in-group, and cognitive processes along with friendship formations will play contributory roles to such a developmental process. Since the PFC has connections with the AI, amygdala, nucleus accumbens, and hypothalamus, there are multiple routes through which cognitive processes can modulate the neural circuits regulating social behavior, allowing for appropriate cognition to enlarge the spectrum of individuals toward whom prosocial and cooperative responding is directed [698,1025].

7.7 PSYCHOPATHY: A BREAKDOWN OF PROSOCIAL NEURAL CIRCUITS THAT RESULTS IN ANTISOCIAL BEHAVIOR

7.7.1 Introduction

Wilson [1036] has argued that both group selection and individual selection have jointly influenced the evolution of human personality and behavioral characteristics, giving rise to individuals with a combination of altruistic/cooperative and selfish traits. Individuals tend to form friendships within their particular social group, which promotes prosocial behaviors, while antisocial behaviors, or at least the absence of prosocial behaviors, tend to be directed toward out-group members. However, even within one's social group, some level of competition exists, and in the face of such competition one can discern a continuum where certain individuals are more cooperative and altruistic, while other individuals are more competitive and selfish. One extreme of such a continuum would be a person who is completely selfish and completely guided by his/her own self-interest, without any concern for the welfare of other individuals within or outside their group. For such an individual, all social stimuli would be processed over antisocial circuits, leading to self-serving behaviors and aggressive competition. For such an individual, prosocial circuits would be, in essence, nonfunctional. Such an extreme case seems to define psychopathy.

Psychopathy is a psychiatric personality disorder that is associated with antisocial behaviors and criminality. In its severe form, it affects about 1% of the human population, but as many as 30–50% of violent criminals [30,99,388,389,600,794]. One of the major measurement tools used to assess psychopathy is the revised form of Hare's Psychopathy Checklist (PCL-R; [389]). Psychopathy can be divided into three essential components, which are shown in Table 7.2. The core personality characteristics of psychopathy include a lack of empathy and concern for the welfare of others coupled with an egocentric and manipulative personality style. As described above, it is just such phenotypes that one would expect to occur under intense individual selection for selfish traits without the counterbalancing prosocial effects of group selection. These personality characteristics give rise to antisocial behaviors that include both proactive (planned, premeditated) aggression and reactive (response to a perceived threat) aggression, with the former enabling the psychopath to obtain self-centered goals. An inability to form enduring social bonds, although not usually emphasized in the literature, is another important behavioral characteristic of psychopaths: they have many short-term (nonenduring) social relationships, they tend to be promiscuous rather than monogamous in their sexual relationships, and they are irresponsible parents.

TABLE 7.2 Personality and Behavioral Characteristics Used to Define Psychopathy

PERSONALITY CHARACTERISTICS
Lack of empathy
Callous and unemotional
Egocentric and manipulative
Pathological lying
ANTISOCIAL BEHAVIORS
Proactive aggression
Reactive aggression
Irresponsible behaviors
Criminality
LACK OF PROSOCIAL BONDING
Promiscuous sexual behavior
Parental irresponsibility
Many short-term social relationships

Although most of the personality, behavioral, and neural characteristics of psychopaths have been derived from examining forensic populations within prisons, some studies have also administered the PCL-R to community samples and have defined the participants as scoring either high or low on this psychopathy rating scale [794]. In some cases, the criminal records of individuals in such community samples have also been examined and compared to their psychopathy scores.

It is important to distinguish psychopathy from antisocial personality disorder (ASPD), which is a more general term. The latter is defined by the occurrence of antisocial behaviors or conduct disorders, while psychopathy is defined by both personality and behavioral traits. A good contrast would be between patients with intermittent explosive disorder (IED) and psychopaths. Both disorders could be subsumed under ASPD, because both are associated with antisocial acts. However, patients with IED are hyperemotional, capable of expressing empathy, and engage primarily in reactive aggression (see Chapter 3), while the unique features of psychopathy include a high degree of proactive (planned) aggression coupled with callous unemotional reactions and a corresponding lack of empathy.

In viewing Figure 7.2, one can see how dysfunctions at various nodes within the neural circuits that allow empathic reactions to promote prosocial behaviors might give rise to many of the core characteristics of psychopathy. In subsequent sections of this chapter, I will review the findings that define such dysfunctional circuitry.

In Chapter 3, when I reviewed the developmental aspects of affective (reactive) aggression and IED, the importance of gene–environment interactions were emphasized, and it was shown that early adverse experiences, such as maternal neglect, importantly influenced the development of affective aggression. In contrast, Hare [388] has argued that early mother–infant interactions probably do not play a central role in the development of psychopathy, and he has stressed the importance of genetic factors over environmental factors in the etiology of psychopathy (also see [794]). Indeed, family, twin, and adoption studies have indicated that the heritability for psychopathy is about 0.70, which is quite high [370]. In an interesting study [996], children were divided into groups (based on teacher ratings) that contained individuals that engaged in antisocial behaviors and possessed callous/unemotional traits without empathy (AB/CU) or that contained individuals that were antisocial but did not exhibit CU traits (AB/CU-). The heritability estimate for the AB/CU trait was 0.71, while that for the AB/CU- was 0.36.

Some studies, however, have suggested that early mother–infant bonding may play a role in the development of psychopathy. Kimonis, Cross, Howard, and Donaghue [488] examined a sample of 227 male adolescent criminals in a prison environment. Each participant completed questionnaires that measured their recollection of the maternal care they received, their self-reported aggressive behavior, and their self-reported emotional characteristics. Boys who rated their mothers as being less warm and caring scored higher on callous-unemotional traits, and the frequency of reported aggression was highest in those adolescents that scored high on callous-unemotional traits and were also exposed to low levels of maternal care. In Chapter 5, I reviewed the evidence that less than adequate maternal care may disrupt the full development of oxytocin neural systems and the neural connections between the MPOA and mesolimbic DA system in offspring. Such disruptions could be routes through which poor maternal care might decrease prosocial behaviors in the affected offspring. Although poor parenting may promote the development of a callous and unemotional personality that lacks empathy, it is also possible that genetic factors are the cause of poor parenting (less emotional involvement and concern for one's infant), and that the inheritance of such genes results in the callous and unemotional characteristics of the offspring. Interestingly, even when infants are raised by attentive and caring mothers, the infants' personality characteristics, resulting from the particular genes inherited from *fathers*, may cause the infants to bond less effectively to their parents, and this may be an early manifestation of the poor social bonding characteristics of adult psychopaths [209].

In an interesting study, Ishikawa, Raine, Lencz, Birhle, and LaCasse [445] examined the characteristics of community samples, obtained from temporary employment agencies, of adult successful and unsuccessful psychopaths and healthy controls. All psychopaths had high scores on the PCL-R, but the successful psychopaths did not have previous criminal prosecutions and convictions while the unsuccessful psychopaths had prior criminal convictions. The successful psychopaths scored higher on a test of cognitive executive function than did their unsuccessful counterparts, and they also performed better than the healthy controls. Perhaps such enhanced cognitive function allowed the successful psychopaths to avoid criminal acts that could be easily detected. More importantly, the successful psychopaths experienced greater periods of parental absence during their childhood than did the remaining two groups, while the unsuccessful psychopaths did not differ from the control subjects on this measure. This study suggests that there may be different routes to psychopathy, and that psychopathic traits are composed of multiple aspects that can vary between individuals.

It is probably safe to say that psychopathy is strongly influenced by genetic factors, but that deficiencies in mother–infant bonding may modulate and exacerbate its development [388]. Surprisingly, not much research has been done on the particular genes that might influence psychopathy; I will return to the literature that does exist on this issue at the end of this chapter.

7.7.2 Psychopathy, Amygdala, and Emotion

A dominant view with respect to the neurobiology of psychopathy is that a dysfunctional amygdala is a core deficit [99,600], and an examination of Figure 7.2 helps one understand that proposal. Since the amygdala is involved in emotional responses, emotional learning, and subjective emotional experiences (see Chapter 1), it makes sense for abnormal amygdala function to contribute to the callous and unemotional traits, along with the lack of empathy, that characterize psychopathy. With respect to the specific amygdala-mediated emotional dysfunctions associated with psychopathy, it has been argued that a decrease in fear-related behaviors and an inability to subjectively experience fear are primary [99,600]. Marsh [600] argues that an inability to experience fear in oneself gives rise to a lack of empathy for others who are exhibiting fearful expressions indicative of distress. This, in turn, contributes to a dual effect: a decreased ability to inhibit harming others and a lack of aid-giving behaviors to help those in need (increased antisocial and decreased prosocial responding).

Birmbaumer et al. [97] examined fear-related conditioning in callous/unemotional psychopaths with criminal records and in healthy control subjects. Images of neutral faces were used as conditioned stimuli. Some faces (CS+) were followed by a painful pressure stimulus (US) to the participant's finger, while other faces (CS–) were not followed by pain. For the control subjects, CS+ acquired the ability to elicit increased sweating (SCR; skin conduction response) as a conditioned fear response, and the subjects also rated the faces that were associated with painful pressure as more aversive (negative valence) than those not associated with pain. These aversive learning-induced changes did not occur in the psychopaths—the SCR did not increase to faces associated with pain and such faces were not rated as being more aversive or unpleasant than faces not associated with pain. Interestingly, both the psychopathic group and the healthy controls cognitively understood which faces were associated with the US and which were not. These results suggest a cognitive, but not an emotional, appreciation of the stimulus contingencies. Importantly, a concurrent fMRI analysis during conditioning indicated that while healthy subjects acquired an increased amygdala BOLD response to faces associated with pain, such classical conditioning, measured at the level of the amygdala, did not occur in the psychopathic subjects. Significantly, healthy controls, but not the psychopaths, also showed increases in the BOLD response at the level of the AI and ACC during acquisition training (refer to Figure 7.2). One interpretation of these results is that faces associated with pain came to arouse fearful responses and fearful subjective experiences in the control subjects, but not in the psychopaths. In support of this interpretation with respect to the subjective experience of fear, Marsh et al. [603] report that adolescents with psychopathic traits recollect experiencing fear less frequently and less strongly than control subjects.

Marsh and Blair [601] have presented a meta-analysis of 20 studies that compared the ability of participants with antisocial personality disorder (ASPD; high levels of antisocial behaviors) and control subjects to recognize the facial expressions of anger, disgust, fear, happiness, sadness, and surprise. In most of these studies, images of the various emotional facial expressions were briefly presented, and then, using a multiple-choice format, the subjects had to choose the word that they thought matched the expression. In comparison to controls, participants with ASPD were significantly impaired in recognizing the facial expressions of fear and sadness. Marsh and Blair suggest that distress cues like fear and sadness (somewhat akin to submissive behavioral displays shown by subordinate animals during aggressive interactions with a dominant conspecific) are likely to inhibit aggression and antisocial acts, and this may occur, in part, through the occurrence of empathic responses in the perceiver. Therefore, a deficit in the perception of distress cues should be associated with increases in antisocial behaviors toward others, in this way contributing to psychopathic traits. It should be noted that others have suggested that the deficit in the recognition of emotional facial expressions may be more general than that proposed by Blair and Marsh and not limited to expressions related to distress [220].

In support of the view that amygdala dysfunction is involved in the deficits associated with the recognition of fearful facial expressions, several studies have shown that, in comparison to controls, the amygdala BOLD response to fearful facial expressions is blunted in psychopathic individuals with callous-unemotional traits [99,602,997]. These results, of course, fit nicely with those of Birmbaumer et al. [97] described above. However, the role of the amygdala in psychopathy may be more complex than a simple decrease in responsiveness to facial expressions of fear and distress. A recent study has presented evidence indicating that psychopathic individuals may exhibit a decreased amygdala BOLD response to fearful facial expressions and an increased amygdala BOLD response to angry facial expressions [160]. Such amygdala responses to socially relevant fearful and angry faces may correspond to the personality (lack of empathy) and behavioral (increased aggression) characteristics of psychopathy.

Finally, there is also evidence for structural deficits in the amygdala of psychopaths. In MRI studies, the gray matter volume of the amygdala is decreased significantly in psychopathic individuals, and the locus of the decrease appears to be centered in the basolateral nucleus [102,1052]. Although the GMV of the BLA is significantly decreased, Boccardi et al. [102] have reported that the GMVs of some other amygdala nuclei might actually be increased in psychopathic individuals. Clearly, as reviewed in Chapter 1, the amygdala is heterogeneous in terms of its functions, some of which may be decreased while others are enhanced in psychopathic individuals, and this view corresponds to the findings of Carre et al. [160] described above.

Although a decrease in fearfulness and in the responsiveness of certain amygdala neurons to facial cues signifying fear and distress may contribute to psychopathic traits, they do not seem to be sufficient factors. Recall that SM, the patient with bilateral amygdala damage, exhibits deficits in both fearful behaviors and fearful subjective experiences (see Chapter 1). She also has difficulty recognizing fearful facial expressions [7]. But SM is not a psychopath. Similarly, patients with Williams syndrome are hypersocial and highly empathic, and in comparison to controls, their amygdala BOLD response to happy faces is increased, while their amygdala BOLD responses to both fearful and angry faces are decreased [372,461,638]. The exact anomaly within the amygdala of psychopathic individuals might best be conceived as consisting of a dampening in the function of amygdala neurons that contribute to prosocial neural circuits that are involved in empathy, cooperation, altruism and aid-related behaviors, a dampening of the function of amygdala neurons that contribute to fearfulness, and an enhancement in the function of amygdala neurons that contribute to antisocial neural circuits, particularly those involved in selfish goal-directed antisocial and aggressive acts. Further, dysfunctions in a wider neural circuitry within which the amygdala operates are likely to give rise to the full expression of psychopathic traits.

Studies that have examined the responses of psychopathic individuals to fearful facial expressions have typically utilized frontal facial images without any context. However, a normal individual's response and interpretation of fearful faces is likely to be strongly modulated by the context in which the fearful facial expression is exhibited. Depending on context, fearful facial expressions might lead a normal perceiver to engage in either avoidance behavior or helping behavior. As two simple examples, if a person observed another running away from a scene while expressing fear, the perceiver would probably also run away. However, if a person observed another individual expressing fear and distress as a boulder fell on his leg, the perceiver might attempt to lift the boulder off the leg of the person in distress. When fearful facial expressions are presented in the absence of context, they may activate amygdala neurons that are part of both empathy-prosocial circuits and avoidance circuits. Perhaps the decrease in the amygdala's responsiveness to fearful facial expressions in psychopathic individuals involves a depression of both circuits, leading to a lack of empathy and a lack of fearfulness. In contrast, individuals with Williams syndrome may primarily exhibit a disruption of amygdala avoidance circuits, with prosocial circuits remaining intact and possibly even being enhanced.

7.7.3 Psychopathy and Amygdala-to-Anterior Insular Cortex-to-Medial Prefrontal Cortex Connectivity

In reference to Figure 7.2, several studies have indicated that amygdala connectivity to the AI-medial PFC network is defective in psychopathic persons, and many of these studies have been anatomical in nature and have employed structural MRI procedures. The structural organization of the uncinate fasciculus (UF), which is a myelinated neural tract, is altered in psychopaths when compared to healthy controls [204,669]. Such an alteration may include a decrease in axon density, diameter, and myelination. Importantly, the UF is a neural tract that reciprocally interconnects the anterior temporal lobe, which includes the amygdala, with the OFC, which includes the AI cortex [1002]. These results could certainly be interpreted as supporting the view that the transfer of empathy-inducing social signals from the amygdala to the AI is disrupted in individuals with psychopathy. In line with such a perspective, several MRI studies have shown that the GMV in AI is reduced in individuals with psychopathic traits, and that significant correlations exist between both psychopathy and empathy scores and AI GMV—as GMV decreases, psychopathy scores and aggression increase and empathy scores decrease [200,282,359,863].

The neural model depicted in Figure 7.2 proposes that AI projections to the medial PFC, particularly allocortical areas 24, 25, and 32, contribute to human prosocial neural circuitry. In line with this view, several structural MRI studies have reported decreases in GMV within these medial PFC regions in psychopathic individuals [102,239,582,1051]. Importantly, in an fMRI analysis, Ly et al. [582] have also reported a decrease in the functional connectivity between the insular cortex and these medial PFC regions in psychopathy. De Oliveira-Souza et al. [239] have emphasized that decreases in the GMV of area 25 within the medial PFC were strongly related to decreases in empathy scores, and they suggested that this region of the ventromedial PFC might be particularly important for translating feelings of empathy into prosocial behaviors. In support of this idea, Boes et al. [107] have reported on a case of a 14-year-old boy who had a congenital malformation of the ventromedial PFC that included damage to areas 25 and 32. This boy exhibited severe antisocial behaviors and aggression along with egocentricity, callous and unemotional traits, and a lack of empathy. As I described in Chapter 1, these extreme manifestations of antisocial behavior, which also include a lack of social bonding with others, are much more severe than those that manifest themselves when the ventromedial PFC is damaged in adulthood [29]. How can one explain a lack of empathy in an individual with damage to the more ventral parts of the medial PFC? One explanation is that such damage not only destroys neurons within the medial PFC but also destroys those neurons in the AI with axonal terminations in this part of the medial PFC. Finally, since the AI and medial PFC are reciprocally connected to each other [685], the feedback that the AI receives from the medial PFC may also contribute to the development of empathic feeling states.

In an fMRI study, Decety, Skelly, and Kiehl [229] examined the neural (BOLD) responses of incarcerated prisoners with or without psychopathic traits as they visually observed scenes depicting other individuals being harmed and expressing facial expressions of pain. Participants in the psychopathic group showed a significantly reduced activation of the ventromedial PFC but a significantly *increased* activation of the insular cortex when compared to the nonpsychopathic prisoners. The increased activity in the insular cortex is difficult to interpret. One possibility is that the observed BOLD response in the insular cortex was due to a decrease in modulatory and informative feedback from the PFC. Also recall from Chapter 1 that increases in BOLD responses are difficult to interpret and can occur under conditions of either increased or decreased output from a brain region. Finally, the insular cortex is a heterogenous region that is not only involved in empathy but is also involved in a variety of natural emotional states. Perhaps the psychopathic prisoners felt pleasure when observing pain in others. Once the spatial resolution of fMRI procedures increase, and once we have a better understanding of the particular AI regions that project to the medial PFC in humans, researchers can zero in on particular functional-anatomical units within the AI to examine the BOLD responses in those regions as they relate to psychopathic traits.

7.7.4 The Preoptic Region and Psychopathy

As already mentioned, fMRI and MRI researchers rarely include the hypothalamus as a region of interest, and, therefore, there is no direct evidence linking MPOA dysfunction to psychopathy. However, research has detected some tantalizing neurodevelopmental anomalies associated with psychopathy that may implicate abnormalities in the function of the MPOA (and septal area), along with other hypothalamic nuclei, in the etiology of psychopathy. Two important studies have found that abnormal development of the septum pellucidum (SP), which results in the formation of a cavum septum pellucidum (CSP), is associated with psychopathic personality traits and aggressive behavior [795,1031]. As shown in Figure 7.1, the SP is a band of connective tissue that lies above the septal nuclei and that separates the two lateral ventricles. Early in brain development, during gestation, the SP consists of two separate leaves that ultimately join together to form a single midline structure at the time of birth. If brain development is compromised in some way so that the two leaves do not join together properly, the formation of a CSP occurs, as shown in Figure 7.4. Importantly, the presence of a CSP is a marker of abnormal forebrain development, which in some cases is also associated with abnormalities in the development of the optic nerves, and with hypothalamic dysfunctions, which are usually detected by the presence of neuroendocrinological disturbances [106]. Bodensteiner and Schaefer [106] suggest that the presence of a CSP represents a milder form of what is referred to as septo-optic dysplasia, where individuals never even develop any of the leaves of the SP—the SP is completely absent [616]. The point to emphasize is that both the presence of a CSP or septo-optic dysplasia (the absence of an SP) represents a heterogeneous disorder that can be associated with several different brain anomalies, not all of which may manifest themselves in a particular individual. When a CSP is present, MRI scans can measure the separation of the two leaves, and wider separations (greater than 1 cm) are considered to be representative of greater abnormalities in forebrain development.

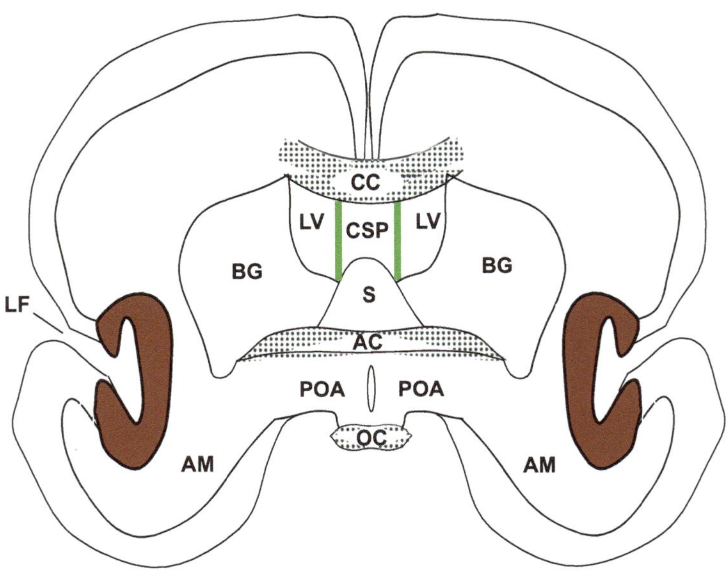

FIGURE 7.4 The location of a cavum septum pellucidum (CSP) within the lateral ventricles (LV) and above the septal area (S) on a frontal section of the human brain. CSP, delimited by green lines, is a marker of abnormal forebrain development. Figure 7.4 is a copy of Figure 7.1, except that in Figure 7.1 a typical septum pellucidum (SP) is shown, while Figure 7.4 represents the atypical formation of a CSP. See text for additional details. Other abbreviations: AC = anterior commissure; AM = amygdala; BG = basal ganglia; CC = corpus callosum; LF = lateral fissure; OC = optic chiasm; POA = preoptic area (both medial and lateral POA).

Raine et al. [795] examined antisocial behaviors and psychopathy scores in a community sample of participants working in temporary employment agencies. MRI analyses separated the participants into those with a CSP and those with a normal SP. Individuals with a CSP had higher PCL-R scores for both the personality traits and antisocial behavioral traits associated with psychopathy. CSP participants also had more criminal charges and convictions for previous crimes. In a more recent study on adolescent boys with conduct disorder (CD) and psychopathic traits, White et al. [1031] reported that a large CSP was present in about 25% of the boys with CD, while a normal SP was present in all of the boys in a healthy control comparison group. This difference between the groups was significant, supporting the work of Raine et al. [795]. However, White et al. also report that within the sample of boys with CD, the level of aggression and psychopathy scores did not differentiate boys with or without a large CSP. The conclusion from these data is that the presence of a CSP is not a necessary marker for the development of psychopathy. This makes sense, as there are likely to be multiple routes through which psychopathy develops. Since the hypothalamus was not directly examined in these studies, future research might focus more selectively on the issue of whether MPOA dysfunctions are associated with psychopathy, since inadequate MPOA function could contribute to a disruption of both prosocial behaviors and the ability to form strong social bonds with others. The presence of a CSP can interfere with the development of many forebrain structures, which can include the septal area, preoptic area, other parts of the hypothalamus, and the amygdala. As we refine our neural models of psychopathy, and as imaging procedures improve, we will begin to focus on more specific regions that may be dysfunctional in psychopaths, whether or not a CSP is present.

7.7.5 The Mesolimbic DA System and Psychopathy

Not much research has been done on the involvement of the mesolimbic DA system in psychopathy (for a review, see [340]). In a structural MRI study, Schiffer et al. [863] detected a significant increase in the GMV of the NA in violent offenders with psychopathic traits, and they found a significant positive correlation between personality scores on the PCL-R and the volume of the NA. Some fMRI studies have reported an increased BOLD response in NA in psychopathic individuals relative to controls during economic games where specific stimuli signaled that a monetary reward would be received [138]; but, confusingly, other studies have reported a decrease in NA BOLD in psychopathic subjects, relative to controls, under analogous testing conditions [160].

To the extent that an increased BOLD response in NA is indicative of increased DA input to NA, one might expect differences in the NA response based on the specific testing conditions that are employed during fMRI scanning. As reviewed in previous chapters, the mesolimbic DA system is involved in goal-directed responses to obtain personal rewards, goal-directed prosocial responses, and goal-directed aggressive responses. The specific stimuli that access the NA-VP circuit, and the particular subcircuits within NA-VP that are engaged, contribute to the type of voluntary response that is observed. In comparison to control subjects, I would predict increased activity within the NA-VP circuit of psychopathic individuals when they engage in behaviors in order to receive personal (self-interested) rewards [138] and when they imagine themselves committing a crime or harming others for their personal gain. In contrast, to the extent that MPOA interactions with the mesolimbic DA system are disrupted in psychopathic individuals, I would predict that activity within the NA-VP circuit would be decreased, relative to controls, when such individuals imagine themselves helping another individual who is in distress. More specifically, fMRI testing conditions similar to those employed by Decety and Porges [228] could be used to test whether activity in the NA-VP circuit is increased in psychopathic subjects relative to controls during the mental simulation of proactive antisocial acts, while such activity is decreased in psychopathic individuals relative to controls during the mental simulation of proactive prosocial acts. Such decreased activity should also be associated with a decrease in the BOLD response within the MPOA region.

7.7.6 The Genetics of Psychopathy

Given the high heritability for psychopathic traits, one would expect that research would be providing significant information on the contributions of specific genes to this serious personality disorder, but this is not the case. Recently, however, it has been proposed that polymorphisms within genes that affect serotonin neural systems may contribute to the etiology of psychopathy ([337,1055] and cf. [927]).

Recall from Chapter 3 that genetic polymorphisms exist within the promoter region of the 5-HT transporter gene, with two main alleles, referred to as either the short (S) or long (L) allele. In Chapter 3, I reviewed the evidence that carriers of the S allele have an increased susceptibility for developing hyperemotionality, increased stress reactivity, and impulsive (reactive) aggression. Since individuals that are homozygous for the L allele exhibit decreases in emotionality and stress reactivity, Glenn [337] has suggested that LL individuals may be at an increased risk for developing psychopathy, particularly with respect to the detached, unemotional, and callous aspects of the disorder. Glenn notes that the distribution of the various 5-HT transporter genotypes in the human population is: LL = 36%, LS = 48%, and SS = 16%. However, the prevalence of psychopathy is only about 1%. Glenn therefore suggests that homozygosity for the L allele may be one factor among several that contributes to the development of psychopathy.

Given the importance of OT for prosocial behaviors and the likelihood that OT acts at several neural sites to increase prosocial behaviors, decrease antisocial behaviors, and promote social bonding (see Figure 7.3), it is very surprising that an association between genetic polymorphisms within the OT system and psychopathic traits has not been examined. Recall that carriers of the A allele at rs53576 of the OXTR gene are less prosocial than their GG counterparts. It is interesting to speculate that specific alleles within the 5-HT transporter gene and the OXTR gene may co-act to increase an individual's vulnerability for developing psychopathy. Perhaps individuals with the following genotypes are at greater risk for developing psychopathic traits: LL for the 5-HT transporter gene and AA at rs53576 for the OXTR gene. Further, another prediction would be that such a genetic risk for the development of psychopathy might be exacerbated by poor parenting.

Considering the potential involvement of deficiencies in OT neural systems with respect to psychopathy, it is also surprising that no one has explored the effects of intranasal OT application on the expression of psychopathic traits. Would such a treatment increase empathy scores in psychopathic individuals as it does in healthy individuals?

It should be noted that there is a sex bias for psychopathy, with the disorder being more prevalent in males than in females. The contributions of perinatal and adult levels of testosterone have been implicated in contributing to this sex bias [1054], although the contribution of genetic sex differences should not be excluded (also see [758]).

To conclude, given the high heritability for psychopathy, much more research needs to be done on the specific genes that contribute to this disorder. Such an analysis, coupled with knowledge of the affected neural circuitry, might lead to therapeutic treatments that could ameliorate some of the antisocial personality traits associated with the disorder.

7.8 EPILOGUE

This chapter culminates a book that has delved into the neurobiology of social behavior. Throughout, I have emphasized the importance of integrating animal and human studies.

To do this, I have concentrated on those neural regions and circuits where important points of commonality exist between animals and humans. For example, when discussing the role of the medial PFC in human social behavior, I have emphasized the allocortical parts (areas 24, 25, and 32), which have presumed homologues in animals, over isocortical parts of the medial PFC. Along these lines, I did not discuss in much detail important cognitive processes that influence social behavior in humans, such as moral decision making, since these processes do not occur in animals. In the context of psychopathy, it is interesting to note that psychopathic individuals seem to cognitively appreciate right from wrong based on societal rules (mediated by isocortical regions), but due to their unemotional and detached personality (mediated by allocortical and subcortical regions), they just don't care, in an emotional sense, if they harm someone to get what they want [338,339]. In other words, although appropriate cognition and mentalizing may be functional (particularly in successful psychopaths), core socioemotional circuits that underpin social behavior are dysfunctional.

Through an integration of animal and human studies, I have described in very concrete terms much of the chemical and functional neuroanatomy that contributes to the core circuitry that regulates social behavior across mammals. I have also shown how dysfunctions in this elemental circuitry, caused by experiential and genetic factors, can have varying degrees of disruptive effects on the social behavior of animals and humans. Prosocial and antisocial circuits have been described throughout this book, and the key players across mammals have included interactions between the PFC, hypothalamus, amygdala, and mesolimbic DA system. Indeed, when comparing the maternal circuits delineated in Figure 5.8 with the human prosocial circuits proposed in Figure 7.3, one can discern a remarkable degree of similarity.

During social interactions, the particular circuits that are active determine whether social stimuli give rise to prosocial or antisocial acts. By understanding the prosocial and antisocial brain, it is hoped that a variety of interventionist strategies, including experiential and pharmacological manipulations, can be created to enhance prosocial behaviors and decrease violence and aggression in humans.

References

[1] Abler B, Seeringer A, Hatrmann A, Gron G, Metzger C, Walter M, et al. Neural correlates of antidepressant-related sexual dysfunction: a placebo-controlled fMRI study on healthy males undr subchronic paroxetine and bupropion. Neuropsychopharmacol 2011;36:1837–47.

[2] Abrams DA, Lynch CJ, Cheng KM, Phillips J, Supekar K, Ryali S, et al. Underconnectivity between voice-selective cortex and reward circuitry in children with autism. Proc Natl Acad Sci USA 2013;110:12060–5.

[3] Acevedo BP, Aron A, Fisher HE, Brown LL. Neural correlates of long-term intense romantic love. Soc Cogn Affect Neurosci 2012;7:145–59.

[4] Adams DB. Brain mechanisms of aggressive behavior: an updated review. Neurosci Biobehav Rev 2006;30:304–18.

[5] Adkins-Regan E. Hormonal mechanisms of mate choice. Am Zool 1998;38:166–78.

[6] Adolphs R, Tranel D, Damasio AR. The human amygdala in social judgment. Nature 1998;393:470–4.

[7] Adolphs R, Tranel D, Damasio H, Damasio A. Impaired recognition of emotion in facial expressions following bilateral damage to the human amygdala. Nature 1994;372:669–72.

[8] Afonso VM, King S, Chatterjee D, Fleming AS. Hormones that increase maternal responsivenesss affect accumbal dopaminergic responses to pup- and food-stimuli in the female rat. Horm Behav 2009;56:11–23.

[9] Afonso VM, King SJ, Novakov M, Burton CL, Fleming AS. Accumbal dopamine function in postpartum rats that were raised without their mothers. Horm Behav 2011;60:632–43.

[10] Afonso VM, Sison M, Lovic V, Fleming AS. Medial prefrontal cortex lesions in the female rat affect sexual and maternal behavior and their sequential orgnanization. Behav Neurosci 2007;121:515–26.

[11] Agmo A, Villalpando A, Picker Z, Fernandez H. Lesions of the medial prefrontal cortex and sexual behavior in the male rat. Brain Res 1995;696:177–86.

[12] Agrati D, Fernandez-Guasti A, Ferreira A. The reproductive stage and experience of sexually receptive mothers alter their preference for pups or males. Behav Neurosci 2008;122:998–1004.

[13] Agrati D, Fernandez-Guasti A, Ferreno M, Ferreira A. Coexpression of sexual behavior and maternal aggression: the ambivalence of sexually active mother rats toward male intruders. Behav Neurosci 2011;125:446–51.

[14] Albers HE. The regulation of social recognition, social communication and aggression: vasopressin in the social behavior neural network. Horm Behav 2012;61:283–92.

[15] Alberts B, Johnson A, Lewis J, Raff M, Roberts K, Walter P. Molecular biology of the cell. New York: Garland Science; 2008.

[16] Alenina N, Kikic D, Todiras M, Mosienko V, Qadri F, Plehm R, et al. Growth retardation and altered autonomic control in mice lacking brain serotonin. Proc Natl Acad Sci USA 2009;106:10332–7.

[17] Alexander BM, Skinner DC, Roselli CE. Wired on steroids: sexual differentiation of the brain and its role in the expression of sexual partner preferences. Front Endocrinol 2011;2:42.

[18] Alia-Klein N, Goldstein RZ, Kriplani A, Logan J, Williams B, Telang F, et al. Brain monoamine oxidase A activity predicts trait aggression. J Neurosci 2008;28:5099–104.

[19] Allaman-Exertier G, Reymond-Marron I, Tribollet E, Raggenbass M. Vasopressin modulates lateral septal network activity via two distinct electrophysiological mechanisms. Eur J Neurosci 2007; 26:2633–42.

[20] Amano T, Unal CT, Pare D. Synaptic correlates of fear extinction in the amygdala. Nat Neurosci 2010;13:489–94.

[21] Amaral DG, Schumann CM, Nordahl CW. Neuroantomy of autism. Trends Neurosci 2008;31:137–45.

[22] Amateau SK, McCarthy MM. A novel mechanism of dendritic spine plasticity involving estradiol induction of prostaglandin-E2. J Neurosci 2002;22:8585–96.

[23] Amateau SK, McCarthy MM. Induction of PGE2 by estradiol mediates developmental masculinization of sex behavior. Nat Neurosci 2004;7:643–50.

[24] Ambroggi F, Ishikawa A, Fields HL, Nicola SM. Basolateral amygdala neurons facilitate reward-seeking behavior by exciting nucleus accumbens neurons. Neuron 2008;59:648–61.

[25] Amemori K, Graybiel AM. Localized microstimulation of the primate pregenual cingulate cortex induces negative decision-making. Nat Neurosci 2012;15:776–85.

[26] Amico JA, Mantella RC, Vollmer RR, Li X. Anxiety and stress responses in female oxytocin deficient mice. J Neuroendocrinol 2004;16:319–24.

[27] Amico JA, Thomas A, Hollingshead DJ. The duration of estradiol and progesterone exposure prior to progesterone withdrawal regulates oxytocin mRNA levels in the paraventricular nucleus of the rat. Endocr Res 1997;23:141–56.

[28] Andari E, Duhamel JR, Zalla T, Herbrecht E, Leboyer M, Sirigu A. Promoting social behavior with oxytocin in high-functioning autism spectrum disorders. Proc Natl Acad Sci USA 2010;107:4389–94.

[29] Anderson SW, Bechara A, Damasio H, Tranel D, Damasion AR. Impairment of social and moral behavior related to early damage in human prefrontal cortex. Nat Neurosci 1999;2:1032–7.

[30] Anderson NE, Kiehl KA. The psychopath magnetized: insights from brain imaging. Trends Cogn Sci 2012;16:52–60.

[31] Andrade R. Serotonergic regulation of neuronal excitability in the prefrontal cortex. Neuropharmacology 2011;61:382–6.

[32] Angenstein F, Kammerer E, Scheich H. The BOLD response in the rat hippocampus depends rather on local processing of signals than on the input or output activity. A combined functional MRI and electrophysiological study. J Neurosci 2009;29:2428–39.

[33] Anstrom KK, Miczek KA, Budygin EA. Increased phasic dopamine signaling in the mesolimbic pathway during social defeat in rats. Neuroscience 2009;161:3–12.

[34] Aou S, Oomura Y, Yoshimatsu H. Neuron activity of the ventromedial hypothalamus and the medial preoptic area of the female monkey during sexual behavior. Brain Res 1988;455:65–71.

[35] Aragona BJ, Liu Y, Curtis JT, Stephan FK, Wang Z. A critical role for nucleus accumbens dopamine in partner-preference formation in male prairie voles. J Neurosci 2003;23:3483–90.

[36] Aragona BJ, Liu Y, Yu YJ, Curtis JT, Detwiler JM, Insel TR, et al. Nucleus accumbens dopamine differentially mediates the formation and maintenance of monogamous pair bonds. Nat Neurosci 2006;9:133–9.

[37] Aragona BJ, Wang Z. Dopamine regulation of social choice in a monogamous rodent species. Front Behav Neurosci 2009;3:15.

[38] Arborelius L, Hawks BW, Ownes MJ, Plotsky PM, Nemeroff CB. Increased responsiveness of presumed 5-HT cells to citalopram in adult rats subjected to prolonged maternal separation relative to brief separation. Psychopharmacology 2004;176:248–53.

[39] Arendash GW, Gorski RA. Effects of discrete lesions of the sexually dimorphic nucleus of the preoptic area or other medial preoptic regions on the sexual behavior of male rats. Brain Res Bull 1983;10:147–54.

[40] Arnold AP. The organizational-activational hypothesis as the foundation for a unified theory of sexual differentiation of all mammalian tissues. Horm Behav 2009;55:570–8.

[41] Arnold AP, Chen X. What does the "four core genotypes" mouse model tell us about sex differences in the brain and other tissues? Front Neuroendocrinol 2009;30:1–9.

[42] Arthurs OJ, Boniface S. How well do we understand the neural origins of the fMRI BOLD signal? Trends Neurosci 2002;25:27–31.

[43] Attwell D, Iadecola C. The neural basis of functional brain imaging signals. Trends Neurosci 2002;25: 621–5.

[44] Atzil S, Hendler T, Feldman R. Specifying the neurobiological basis of human attachment: brain, hormones, and behavior in synchronous and intrusive mothers. Neuropsychopharmacol 2011;36: 2603–15.

[45] Audero E, Milnar B, Baccini G, Skachokova ZK, Corradetti R, Gross C. Suppression of serotonin neuron firing increases aggression in mice. J Neurosci 2013;33:8678–88.

[46] Badrinarayan A, Wescott SE, Vander Weele CM, Saunders BT, Couturier BE, Maren S, et al. Aversive stimuli differentially modulate real-time dopamine transmission dynamics within the nucleus accumbens core and shell. J Neurosci 2012;32:15779–90.

[47] Bagatell CJ, Heiman JR, Rivier JE, Bremmer WJ. Effects of endogenous testosterone and estradiol on sexual behavior in normal young men. J Clin Endocrinol Metab 1994;78:711–6.

[48] Bakermans-Kranenburg MJ, van Ijzendoorn MH. Oxytocin receptor (OXTR) and serotonin transporter (5-HTT) genes associated with observed parenting. Soc Cogn Affect Neurosci 2008;3:128–34.

[49] Bakker J, Brock O. Early oestrogens in shaping reproductive networks: evidence for a potential organisational role of oestradiol in female brain development. J Neuroendocrinol 2010;22:728–35.

[50] Baldo BA, Kelley AE. Discrete neurochemical coding of distinguishable motivational processes: insights from nucleus accumbens control of feeding. Psychopharmacology 2007;191:439–59.

[51] Bale TL, Davis AM, Auger AP, Dorsa DM, McCarthy MM. CNS region-specific oxytocin receptor expression: importance in regulation of anxiety and sex behavior. J Neurosci 2001;21:2546–52.

[52] Bale TL, Dorsa DM. NGF, cyclic AMP, and phorbol esters regulate oxytocin receptor gene transcription in SK-N-SH and MCF7 cells. Brain Res Mol Brain Res 1998;53:130–7.

[53] Bales KL, Kim AJ, Lewis-Reese AD, Carter CS. Both oxytocin and vasopressin may influence alloparental behavior in male prairie voles. Horm Behav 2004;45:354–61.

[54] Bales KL, Mason WA, Catana C, Cherry SR, Mendoza SP. Neural correlates of pair-bonding in a monogamous primate. Brain Res 2007;1184:245–53.

[55] Balfour ME, Brown JL, Yu L, Coolen LM. Potential contributions of efferents from medial prefrontal cortex to neural activation following sexual behavior in the male rat. Neuroscience 2006;137:1259–76.

[56] Balthazart J. Minireview: hormones and human sexual differentiation. Endocrinology 2011;152:2937–47.

[57] Balthazart J, Ball GF. Topography in the preoptic region: differential regulation of appetitive and consummatory male sexual behaviors. Front Neuroendocrinol 2007;28:161–78.

[58] Banerjee SB, Liu RC. Storing maternal memories: hypothesizing an interaction of experience snd estrogen on sensory cortical plasticity to learn infant cues. Front Neuroendocrinol 2013;34:300–14.

[59] Bang SJ, Jensen P, Dymecki SM, Commons KG. Projections and interconnections of genetically defined serotonin neurons in mice. Eur J Neurosci 2012;35:85–96.

[60] Barbas H. Specialized elements of orbitofrontal cortex in primates. Ann NY Acad Sci 2007;1121:10–32.

[61] Barbas H, Zikopoulos B, Timbie C. Sensory pathways and emotional context for action in primate prefrontal cortex. Biol Psychiatry 2011;69:1133–9.

[62] Barofsky AL, Taylor J, Tizabi Y, Kumar R, Jones-Quartey K. Specific neurotoxin lesions of median raphe serotonergic neurons disrupt maternal behavior in the lactating rat. Endocrinology 1983;113:1884–93.

[63] Baron-Cohen S, Lombardo MV, Auyeung B, Ashwin E, Chakrabarti B, Knickmeyer R. Why are autism spectrum conditions more prevalent in males? PLoS Biol 2011;9:e1001081.

[64] Barr CS, Newman TK, Becker ML, Parker CC, Champoux M, Lesch KP, et al. The utility of the non-human primate; a model for studying gene by environment interactions in behavioral research. Genes Brain Behav 2003;2:336–40.

[65] Barraza JA, McCullough ME, Ahmadi S, Zak PJ. Oxytocin infusion increases charitable donations regardless of monetary resources. Horm Behav 2011;60: 148–51.

[66] Barraza JA, Zak PJ. Empathy toward strangers triggers oxytocin release and subsequent generosity. Ann NY Acad Sci 2009;1167:182–9.

[67] Barrett C, Keebaugh AC, Ahern TH, Bass CE, Terwilliger EF, Young LJ. Variation in vasopressin (Avpra1a) expression creates diversity in behaviors related to monogamy in praire voles. Horm Behav 2013;63:518–26.

[68] Barrett J, Wonch KE, Gonzalez A, Ali N, Steiner M, Hall GB, et al. Maternal affect and quality of parenting experiences are related to amygdala response to infant faces. Soc Neurosci 2012;7:252–68.

[69] Barretta S, Pantazopoulos H, Caldera M, Pantazopoulos P, Pare D. Infalimbic cortex activation increases c-Fos expression in intercalated neurons of the amygdala. Neuroscience 2005;132:943–53.

[70] Bartels A, Zeki S. The neural correlates of maternal and romantic love. NeuroImage 2004;21:1155–66.

[71] Basar K, Sesia T, Groenewegen H, Steinbusch HW, Visser-Vandewalle V, Temel Y. Nucleus accumbens and impulsivity. Prog Neurobiol 2010;92:533–57.

[72] Baum MJ. Mammalian animal models of psychosexual differentiation: when is 'translation' to the human situation possible? Horm Behav 2006;50:579–88.

[73] Baum MJ, Everitt BJ. Increased expression of c-fos in the medial preoptic area after mating in male rats: role of afferent inputs from the medial amygdala and midbrain central tegmental field. Neuroscience 1992;50:627–46.

[74] Beach SR, Brody GH, Todorov AA, Gunter TD, Philibert RA. Methylation at SLC6A4 is linked to family history of child abuse: and examination of the Iowa Adoptee sample. Am J Med Genet B Neuropsychiatr Genet 2010;153B:710–3.

[75] Beauregard M, Levesque J, Bourgouin P. Neural correlates of conscious self-regulation of emotion. J Neurosci 2001;21:RC165.

[76] Bechara A, Damasio H, Damasio AR. Emotion, decision making and the orbitofrontal cortex. Cereb Cortex 2000;10:295–307.

[77] Bechara A, Damasio AR, Damasio H, Anderson SW. Insensitivity to future consequences following damage to human prefrontal cortex. Cognition 1994; 50:7–15.

[78] Becker JB, Rudnick CN, Jenkins WJ. The role of dopamine in the nucleus accumbens and striatum during sexual behavior in the female rat. J Neurosci 2001;21:3236–41.

[79] Beckley JT, Evins CE, Fedarovich H, Gilstrap MJ, Woodward JJ. Medial prefrontal cortex inversely regulates toluene-induced changes in markers of synaptic plasticity of mesolimbic dopamine neurons. J Neurosci 2013;33:804–13.

[80] Beeney JE, Franklin RG, Levy KN, Adams RB. I feel your pain: emotional closeness modulated neural responses to empathically experienced rejection. Soc Neurosci 2011;6:369–76.

[81] Beery AK, Lacey EA, Francis DD. Oxytocin and vasopressin receptor distributions in a solitary and a social species of tuco-tuco (*Ctenomys haigi* and *Ctenomys sociabilis*). J Comp Neurol 2008;507:1847–59.

[82] Behrens TE, Hunt LT, Rushworth MF. The computation of social behavior. Science 2009;324:1160–4.

[83] Beiderbeck DI, Reber SO, Havasi A, Bredewold R, Veenema AH, Neumann ID. High and abnormal forms of aggression in rats with extremes in trait anxiety – involvement of the dopamine system in the nucleus accumbens. Psychoneuroendocrino 2012;37:1969–80.

[84] Belova MA, Paton JJ, Salzman CD. Moment-to-moment tracking of state value in the amygdala. J Neurosci 2008;28:10023–30.

[85] Benekareddy M, Goodfellow NM, Lambe EK, Vaidya VA. Enhanced function of prefrontal serotonin 5-HT(2) receptors in a rat model of psychiatric vulnerability. J Neurosci 2010;30:12138–50.
[86] Bennett AJ, Lesch KP, Heils A, Long JC, Lorenz JG, Shoaf SE, et al. Early experience and serotonin transporter gene variation interact to influence primate CNS function. Mol Psychiatry 2002;7:118–22.
[87] Berlin LJ, Appleyard K, Dodge KA. Intergenerational continuity in child maltreatment: mediating mechanisms and implications for prevention. Child Dev 2011;82:162–76.
[88] Berman ME, McCloskey MS, Fanning JR, Schumacher JA, Coccaro EF. Serotonin augmentation reduces response to attack in aggressive individuals. Psychol Sci 2009;20:714–20.
[89] Bernhardt BC, Singer T. The neural basis of empathy. Annu Rev Neurosci 2012;35:1–23.
[90] Berridge KC. Motivation concepts in behavioral neuroscience. Physiol Behav 2004;81:179–209.
[91] Berridge KC. The debate over dopamine's role in reward: the case for incentive salience. Psychopharmacology 2007;191:391–431.
[92] Bester-Meredith JK, Marler CA. Vasopressin and the transmission of paternal care across generations in mated, cross-fostered *Peromyscus* mice. Behav Neurosci 2003;117:455–63.
[93] Bethlehem RA, van Honk J, Auyeung B, Baron-Cohen S. Oxytocin, brain physiology, and functional connectivity: a review of intranasal oxytocin fMRI studies. Psychoneuroendocrino 2013;38:962–74.
[94] Bialy M, Kalata U, Nikolaev-Diak A, Nikolaev E. D1 receptors involved in the acquisition of sexual experience in male rats. Behav Brain Res 2010;206:166–76.
[95] Bian X, Yanagawa Y, Chen WR, Luo M. Cortical-like functional organization of the pheromone-processing circuits in the medial amygdala. J Neurophysiol 2008;99:77–86.
[96] Bielsky IF, Young LJ. Oxytocin, vasopressin, and social recognition in mammals. Peptides 2004;25:1565–74.
[97] Birbaumer N, Veit R, Lotze M, Erb M, Hermann C, Grodd W, et al. Deficient fear conditioning in psychopathy: a functional magnetic resonance imaging study. Arch Gen Psychiatry 2005;62:799–805.
[98] Bisceglia R, Jenkins JM, Wigg KG, O'Connor TG, Moran G, Barr CL. Arginine vasopressin 1a receptor gene and maternal behavior: evidence of association and moderation. Genes Brain Behav 2012;11:262–8.
[99] Blair RJ. The amygdala and the ventromedial prefrontal cortex: functional contributions and dysfunction in psychopathy. Philos Trans R Soc Lond B Biol Sci 2008;363:2557–65.
[100] Blanchard DC, Blanchard RJ. What can animal aggression research tell us about human aggression. Horm Behav 2003;44:171–7.
[101] Blanchard RJ, Wall PM, Blanchard DC. Problems in the study of rodent aggression. Horm Behav 2003;44:161–70.
[102] Boccardi M, Frisoni GB, Hare RD, Cavedo E, Najt P, Pievani M, et al. Cortex and amygdala morphology in psychopathy. Psychiatry Res 2011;193:85–92.
[103] Boccia ML, Petrusz P, Suzuki K, Marson L, Pedersen CA. Immunohistochemical localization of oxytocin receptors in human brain. Neuroscience 2013;253:155–64.
[104] Bocklandt S, Horvath S, Vilain E, Hamer DH. Extreme skewing of X chromosome inactivation in mothers of homosexual men. Hum Genet 2006;118:691–4.
[105] Bocklandt S, Vilain E. Sex differences in brain and behavior: hormones versus genes. Adv Genet 2007;59:245–66.
[106] Bodensteiner JB, Schaefer GB. Wide cavum septum pellucidum: a marker of disturbed brain development. Pediatr Neurol 1990;6:391–4.
[107] Boes AD, Grafft AH, Joshi C, Chuang NA, Nopoulos P, Anderson SW. Behavioral effects of congenital ventromedial prefrontal cortex malformation. BMC Neurol 2011;11:151.
[108] Bonnin A, Levitt P. Fetal, maternal, and placental sources of serotonin and new implications for developmental programming of the brain. Neuroscience 2011;197:1–7.
[109] Borg C, de Jong PJ, Georgiadis JR. Subcortical BOLD responses during visual sexual stimulation vary as a function of implicit porn associations in women. Soc Cogn Affect Neurosci 2014;9:158–66.
[110] Born J, Lange T, Kern W, McGregor GP, Bickel U, Fehm HL. Sniffing neuropeptides: a transnasal approach to the human brain. Nat Neurosci 2002;5:514–6.
[111] Borrow AP, Cameron NM. The role of oxytocin in mating and pregnancy. Horm Behav 2012;61:266–76.
[112] Bortolato M, Chen K, Godar SC, Chen G, Wu W, Rebrin I, et al. Social deficits and perseverative behaviors, but not overt aggression, in MAO-A hypomorphic mice. Neuropsychopharmacol 2011;36:2674–88.
[113] Bosch OJ, Neumann ID. Brain vasopressin is an important regulator of maternal behavior independent of dams' trait anxiety. Proc Natl Acad Sci USA 2008;105:17139–44.
[114] Bosch OJ, Pfortsch J, Beiderbeck DI, Landgraf R, Neurmann ID. Maternal behaviour is associated with vasopressin release into the medial preoptic area and bed nucleus of the stria terminalis. J Neuroendocrinol 2010;22:420–9.

[115] Bouwknecht JA, van der Gugten J, Hijzen TH, Maes RA, Hen R, Olivier B. Absence of 5-HT(1B) receptors is associated with impaired impulse control on male 5-HT(1B) knockout mice. Biol Psychiatry 2001;49:557–68.
[116] Bowles S. Group competition, reproductive leveling, and the evolution of human altruism. Science 2006;314:1569–72.
[117] Brackett NL, Edwards DA. Medial preoptic connections with the midbrain tegmentum are essential for male sexual behavior. Physiol Behav 1984;32:79–84.
[118] Brackett NL, Iuvone PM, Edwards DA. Midbrain lesions, dopamine and male sexual behavior. Behav Brain Res 1986;20:231–40.
[119] Bradley KC, Haas AR, Meisel RL. 6-Hydroxydopamine lesions in female hamsters (*Mesocricetus auratus*) abolish the sensitized effects of sexual experience on copulatory interactions with males. Behav Neurosci 2005;119:224–32.
[120] Bridges RS. Long-term effects of pregnancy and parturition upon maternal responsiveness in the rat. Physiol Behav 1975;14:245–9.
[121] Bridges RS. Parturition: its role in the long term retention of maternal behavior in the rat. Physiol Behav 1977;18:487–90.
[122] Bridges RS. Retention of rapid onset of maternal behavior during pregnancy in primiparous rats. Behav Biol 1978;24:113–7.
[123] Bridges RS. A quantitative analysis of the roles of dosage, sequence, and duration of estradiol and progesterone exposure in the regulation of maternal behavior in the rat. Endocrinology 1984;114:930–40.
[124] Bridges RS, Mann PE, Coppeta JS. Hypothalamic involvement in the regulation of maternal behaviour in the rat: inhibitory roles for the ventromedial hypothalamus and the dorsal/anterior hypothalamic areas. J Neuroendocrinol 1999;11:259–66.
[125] Bridges RS, Numan M, Ronsheim PM, Mann PE, Lupini CE. Central prolactin infusions stimulate maternal behavior in steroid-treated, nulliparous female rats. Proc Natl Acad Sci USA 1990;87:8003–7.
[126] Brischoux F, Chakraborty S, Brierley DI, Ungless MA. Phasic excitation of dopamine neurons in ventral VTA by noxious stimuli. Proc Natl Acad Sci USA 2009;106:4894–9.
[127] Britt JP, Benaliouad F, McDevitt RA, Stuber GD, Wise RA, Bonci A. Synaptic and behavioral profile of multiple glutamatergic inputs to the nucleus accumbens. Neuron 2012;76:790–803.
[128] Broad KD, Hinton MR, Keverne EB, Kendrick KM. Involvement of the medial prefrontal cortex in mediating behavioural responses to odour cues rather than olfactory recognition memory. Neuroscience 2002;114:715–29.
[129] Brock O, Baum MJ, Bakker J. The development of female sexual behavior requires prepubertal estradiol. J Neurosci 2011;31:5574–8.
[130] Bromberg-Martin ES, Matsumoto M, Hikosaka O. Dopamine in motivational control: rewarding, aversive, and alerting. Neuron 2010;68:815–34.
[131] Brown JL. The evolution of behavior. New York: Norton; 1975.
[132] Brown SL, Brown M, Preston SD. The human caregiving system: a neuroscience model of compassionate motivation and behavior. In: Brown SL, Brown M, Penner LA, editors. Moving beyond self-interest: perspectives from evolutionary biology, neuroscience, and the social sciences. New York: Oxford University Press; 2012. pp. 75–88.
[133] Brown JR, Ye H, Bronson RT, Dikkes P, Greenberg ME. A defect in nuturing in mice lacking the immediate early gene fosB. Cell 1996;86:297–309.
[134] Brunetti M, Babiloni C, Ferretti A, Del Gratta C, Merla A, Olivetti Belardinelli M, et al. Hypothalamus, sexual arousal and psychosexual identity in human males: a functional magnetic resonance imaging study. Eur J Neurosci 2008;27:2922–7.
[135] Brunner HG, Nelen M, Breakefield XO, Ropers HH, van Oost BA. Abnormal behavior associated with a point mutation in the structural gene for monoamine oxidase A. Science 1993;262:578–80.
[136] Brunnlieb C, Munte TF, Tempelmann C, Heldman M. Vasopressin modulates neural responses related to emotional stimuli in the right amygdala. Brain Res 2013;1499:29–42.
[137] Buckholtz JW, Meyer-Lindenberg A. MAOA and the neurogenetic architecture of human aggression. Trends Neurosci 2008;31:120–9.
[138] Buckholtz JW, Treadway MT, Cowan RL, Woodward ND, Benning SD, Li R, et al. Mesolimbic dopamine reward system hypersensitivity in individuals with psychopathic traits. Nat Neurosci 2010;13:419–21.
[139] Buhle JT, Kober H, Ochsner KN, Mende-Siedlecki P, Weber J, Hughes BL, et al. Common representation of pain and negative emotion in midbrain periaqueductal gray. Soc Cogn Affect Neurosci 2013;8:609–16.
[140] Burbach JPH, Young LJ, Russell JA. Oxytocin: synthesis, secretion, and reproductive function. In: Neill JD, editor. Physiology of reproduction, vol. 2. San Diego: Academic Press; 2006. pp. 3055–128.
[141] Burkart JM, Fehr E, Efferson C, van Schaik CP. Other-regarding preferences in a non-human primate: common marmosets provision food altruistically. Proc Natl Acad Sci USA 2007;104:19762–6.

[142] Byne W, Tobet S, Mattiace LA, Lasco MS, Kemether E, Edgar MA, et al. The interstitial nuclei of the human anterior hypothalamus: an investigation of variation with sex, sexual orientation, and HIV status. Horm Behav 2001;40:86–92.

[143] Byrd AL, Manuck SB. MAOA, childhood maltreatment, and antisocial behavior: meta-analysis of a gene-environment interaction. Biol Psychiatry 2014;75:9–17.

[144] Byrnes EM, Bridges RS. Endogenous opioid facilitation of maternal memory in rats. Behav Neurosci 2000;114:797–804.

[145] Byrnes EM, Rigero BA, Bridges RS. Dopamine antagonists during parturition disrupt maternal care and the retention of maternal behavior in rats. Pharmacol Biochem Behav 2002;73:869–75.

[146] Caiafa P, Zampieri M. DNA methylation and chromatin structure: the puzzling CpG islands. J Cell Biochem 2005;94:257–65.

[147] Calamandrei G, Keverne EB. Differential expression of Fos protein in the brain of female mice dependent on pup sensory cues and maternal experience. Behav Neurosci 1994;108:113–20.

[148] Calcagnoli F, de Boer SF, Beiderbeck DI, Althaus M, Koolhaas JM, Neumann ID. Local oxytocin expression and oxytocin receptor binding in the male rat brain is associated with aggressiveness. Behav Brain Res 2014;261:315–22.

[149] Caldji C, Diorio J, Meaney MJ. Variations in maternal care alter GABA(A) receptor subunit expression in brain regions associated with fear. Neuropsychopharmacol 2003;28:1950–9.

[150] Caldji C, Tannenbaum B, Sharma S, Francis D, Plotsky PM, Meaney MJ. Maternal care during infancy regulated the development of neural systems mediating fearfulness. Proc Natl Acad Sci USA 1998;95:5335–40.

[151] Caldwell HK, Lee HJ, Macbeth AH, Young WS. Vasopressin: behavioral roles of an "original" neuropeptide. Prog Neurobiol 2008;84:1–24.

[152] Campbell DB, Datta D, Jones JT, Lee EB, Sutcliffe JS, Hammock EAD, et al. Association of oxytocin receptor (OXTR) gene variants with multiple phenotype domains of autusm spectrum disorder. J Neurodev Disord 2011;3:101–12.

[153] Canteras NS. The medial hypothalamic defensive system: hodological organization and functional implications. Pharmacol Biochem Behav 2002;71:481–91.

[154] Canteras NS, Simerly RB, Swanson LW. Organization of projections from the medial nucleus of the amygdala: a PHAL study in the rat. J Comp Neurol 1995;360:213–45.

[155] Cantor JM, Kabani N, Christensen BK, Zipursky RB, Barbaree HE, Dickey R, et al. Cerebral white matter deficiencies in pedophilic men. J Psychiatr Res 2008;42:167–83.

[156] Caramaschi D, de Boer SF, de Vries H, Koolhaas JM. Development of violence in mice through repeated victory along with changes in prefrontal cortex neurochemistry. Behav Brain Res 2008;189:263–72.

[157] Caramaschi D, de Boer SF, Koolhaas JM. Differential role of the 5-HT1A receptor in aggressive and nonaggressive mice: an across-strain comparison. Physiol Behav 2007;90:590–601.

[158] Carlezon WA, Duman RS, Nestler EJ. The many faces of CREB. Trends Neurosci 2005;28:436–45.

[159] Carr DB, Sesack SR. Projections from the rat prefrontal cortex to the ventral tegmental area: target specificity in the synaptic associations with mesoaccumbens and mesocortical neurons. J Neurosci 2000;20:3864–73.

[160] Carre JM, Hyde LW, Neumann CS, Viding E, Hariri AR. The neural signatures of distinct psychopathic traits. Soc Neurosci 2013;8:122–35.

[161] Carre JM, McCormick CM, Hariri AR. The social neuroendocrinology of human aggression. Psychoneuroendocrino 2011;36:935–44.

[162] Carrillo M, Ricci LA, Coppersmith GA, Melloni RH. The effect of increased serotonergic neurotransmission on aggression: a critical meta-analytic review of preclinical studies. Psychopharmacology 2009;205:349–68.

[163] Carter CS. Sex differences in oxytocin and vasopressin: implications for autism spectrum disorders? Behav Brain Res 2007;176:170–86.

[164] Carter CS, DeVries AC, Getz LL. Physiological substrates of mammalian monogamy: the prairie vole model. Neurosci Biobehav Rev 1995;19:303–14.

[165] Cases O, Seif I, Grimsby J, Gaspar P, Chen K, Pournin S, et al. Aggressive behavior and altered amounts of brain serotonin and norepinephrine in mice lacking MAOA. Science 1995;268:1763–6.

[166] Caspi A, McClay J, Moffitt TE, Mill J, Martin J, Craig IW, et al. Role of genotype in the cycle of violence in maltreated children. Science 2002;297:851–4.

[167] Catani M, Howard RJ, Pajevic S, Jones DK. Virtual in vivo interactive dissection of white matter fasciculi in the human brain. NeuroImage 2002;17:77–94.

[168] Centenaro LA, Vierira K, Zimmermann N, Miczek KA, Lucion AB, de Almeida RM. Social instigation and aggressive behavior in mice: role of 5-HT1A and 5-HT1B receptors in the prefrontal cortex. Psychopharmacology 2008;201:237–48.

[169] Chamero P, Leinders-Zufall T, Zufall F. From genes to social communication: molecular sensing by the vomeronasal organ. Trends Neurosci 2012;35:597–606.

[170] Champagne FA. Epigenetic mechanisms and the transgenerational effects of maternal care. Front Neuroendocrinol 2008;29:386–97.

[171] Champagne FA, Chretien P, Stevenson CW, Zhang TY, Gratton A, Meaney MJ. Variations in nucleus accumbens dopamine associated with individual differences in maternal behavior in the rat. J Neurosci 2004;24:4113–23.

[172] Champagne F, Diorio J, Sharma S, Meaney MJ. Naturally occurring variations in maternal behavior in the rat are associated with differences in estrogen-inducible central oxytocin receptors. Proc Natl Acad Sci USA 2001;98:12736–41.

[173] Champagne FA, Francis DD, Mar A, Meaney MJ. Variations in maternal care in the rat as a mediating influence for the effects of environment on development. Physiol Behav 2003;79:359–71.

[174] Champagne FA, Weaver IC, Diorio J, Dymov S, Szyf M, Meaney MJ. Maternal care associated with the methylation of the estrogen receptor-alpha1b promoter and estrogen receptor-alpha expression in the medial preoptic area of female offspring. Endocrinology 2006;147:2909–15.

[175] Champe PC, Harvey RA, Ferrier DR. Lippincott's illustrated reviews: biochemistry. Baltimore: Lippincott Williams & Wilkins; 2005.

[176] Chapple CL, Tyler KA, Bersani BE. Child neglect and adolescent violence: examining the effects of self-control and peer rejection. Violence Vict 2005;20:39–53.

[177] Charara A, Grace AA. Dopamine receptor subtypes selectively modulate excitatory afferents from the hippocampus and amygdala in rat nucleus acccumbens neurons. Neuropsychopharmacol 2003;28:1412–21.

[178] Chen J, Evans AN, Liu Y, Honda M, Saavedra JM, Aguilera G. Maternal deprivation in rats is associated with corticotrophin-releasing hormone (CRH) promoter hypoemethylation and enhances CRH transcriptional responses to stress in adulthood. J Neuroendocrinol 2012;24:1055–64.

[179] Chikama M, McFarland NR, Amaral DG, Haber SN. Insular cortical projections to functional regions of the striatum correlate with cortical cytoarchitectonic organization in the primate. J Neurosci 1997;17:9686–705.

[180] Childress AR, Ehrman RN, Wang Z, Li Y, Sciortino N, Hakun J, et al. Prlude to passion: limbic activation by "unseen" drug and sexual cues. PLoS One 2008;3:e1506.

[181] Cho MM, DeVries AC, Williams JR, Carter CS. The effects of oxytocin and vasopressin on partner preferences in male and female prairie voles (*Microtus ochrogaster*). Behav Neurosci 1999;113:1071–9.

[182] Cho YT, Ernst M, Fudge JL. Cortico-amygdala-striatal circuits are organized as hierarchical subsystems through the primate amygdala. J Neurosci 2013;33:14017–30.

[183] Choi JS, Cain CK, LeDoux JE. The role of amygdala nuclei in the expression of auditory signaled two-way active avoidance in rats. Learn Mem 2010;17:139–47.

[184] Choi GB, Dong HW, Murphy AJ, Valenzuela DM, Yancopoulos GD, Swanson LW, et al. Lhx6 delineates a pathway mediating innate reproductive behaviors from the amygdala to the hypothalamus. Neuron 2005;46:647–60.

[185] Choleris E, Little SR, Mong JA, Puram SV, Langer R, Pfaff DW. Microparticle-based 185. Delivery of oxytocin receptor antisense DNA in the medial amygdala blocks social recognition in female mice. Proc Natl Acad Sci USA 2007;104:4670–5.

[186] Choong CS, Wilson EM. Trinucleotide repeats in the human androgen receptor: a molecular basis for disease. J Mol Endocrinol 1998;21:235–57.

[187] Christensen A, Dewing P, Micevych P. Membrane-initiated estradiol signaling induces spinogenesis required for female sexual receptivity. J Neurosci 2011;31:17583–9.

[188] Christian BT, Wooten DW, Hillmer AT, Tudorascu DL, Converse AK, Moore CF, et al. Serotonin transporter genotype affects serotonin 5-HT1A binding in primates. J Neurosci 2013;33:2512–6.

[189] Cikara M, Botvinick MM, Fiske ST. Us versus them: social identity shapes neural responses to intergroup competition and harm. Psychol Sci 2011;22:306–13.

[190] Ciocchi S, Herry C, Grenier F, Wolff SB, Letzkus JJ, Vlachos I, et al. Encoding of conditioned fear in central amygdala inhibitory circuits. Nature 2010;468:277–82.

[191] Cleare AJ, Bond AJ. Ipsapirone challenge in aggressive men shows an inverse correlation between 5-HT1A receptor function and aggression. Psychopharmacology 2000;148:344–9.

[192] Coccaro EF, Kavoussi RJ, Hauger RL, Cooper TB, Ferris CF. Cerebrospinal fluid vasopressin levels: correlates with aggression and serotonin function in personality disordered subjects. Arch Gen Psychiatry 1998;55:708–14.

[193] Coccaro EF, McCloskey MS, Fitzgerald DA, Phan KL. Amygdala and orbitofrontal reactivity to social threat in individuals with impulsive aggression. Biol Psychiatry 2007;62:168–78.

[194] Coccaro EF, Sripada CS, Yanowitch RN, Phan KL. Corticolimbic function in impulsive aggressive behavior. Biol Psychiatry 2011;69:1153–9.

[195] Coleman K, Robertson ND, Bethea CL. Long-term ovariectomy alters social and anxious behaviors in semi-free ranging Japanese macaques. Behav Brain Res 2011;225:317–27.

[196] Comings DE. Polygenic inheritance and micro/minisatellites. Mol Psychiatry 1998;3:21–31.

[197] Condes-Lara M, Veinante P, Rabai M, Freund-Mercier MJ. Correlation between oxytocin neuronal sensitivity and oxytocin-binding sites in the amygdala of the rat: electrophysiological and histoautoradiographic study. Brain Res 1994;637:277–86.

[198] Conway CC, Keenan-Miller D, Hammen C, Lind PA, Najman JM, Brennan PA. Coaction of stress and serotonin transporter genotype in predicting aggression at the transition to adulthood. J Clin Child Adolesc Psychol 2012;41:53–63.

[199] Coolen LM, Allard J, Truitt WA, McKenna KE. Central regulation of ejaculation. Physiol Behav 2004;83:203–15.

[200] Cope LM, Shane MS, Segall JM, Nyalakanti PK, Stevens MC, Pearlson GD, et al. Examining the effect of psychopathic traits in gray matter volume in a community of substance abusers. Psychiatry Res 2012;204:91–100.

[201] Coplan JD, Smith EL, Altemus M, Scharf BA, Owens MJ, Nemeroff CB, et al. Variable foraging demand rearing: sustained elevations in cisternal cerebrospinal fluid corticotropin-releasing factor concentrations in adult primates. Biol Psychiatry 2001;50:200–4.

[202] Coria-Avila GA, Gavrila AM, Boulard B, Charrin N, Stanley G, Pfaus JG. Neurochemical basis of conditioned partner preference in the female rat: II. Disruption by flupenthixol. Behav Neurosci 2008;122:396–406.

[203] Coria-Avila GA, Pfaus JG. Neuronal activation by stimuli that predict sexual reward in female rats. Neuroscience 2007;148:623–32.

[204] Craig AD. How do you feel-now? The anterior insula and human awareness. Nat Rev Neurosci 2009;10:59–70.

[205] Curley JP, Jensen CL, Mashoodh R, Champagne FA. Social influences in neurobiology and behavior: epigenetic effects during development. Psychoneuroendocrino 2011;36:352–71.

[206] Curley JP, Keverne EB. Genes, brains and mammalian social bonds. Trends Ecol Evol 2005;20:561–7.

[207] Da Costa AP, Guevara-Guzman RG, Ohkura S, Goode JA, Kendrick KM. The role of oxytocin release in the paraventricular nucleus in the control of maternal behaviour in the sheep. J Neuroendocrinol 1996;8:163–77.

[208] Da Cunha-Bang S, Stenbaek DS, Holst K, Licht CL, Jensen PS, Frokjaer VG, et al. Trait aggression and trait impulsivity are not related to frontal cortex 5-HT2A receptor binding in healthy individuals. Psychiatry Res Neuroimaging 2013;212:125–31.

[209] Dadds MR, Allen JL, Oliver BR, Faulkner N, Legge K, Moul C, et al. Love, eye contact and the developmental origins of empathy v. psychopathy. Br J Psychiatry 2012;200:191–6.

[210] Dalley JW, Cardinal RN, Robbins TW. Prefrontal executive and cognitive functions in rodents: neural and neurochemical substrates. Neurosci Biobehav Rev 2004;28:771–84.

[211] Damasio A, Damasio H, Tranel D. Persistence of feelings and sentience after bilateral damage to the insula. Cereb Cortex 2013;23:833–46.

[212] Damasio H, Grabowski T, Frank R, Galaburda AM, Damasio AR. The return of Phineas Gage: clues about the brain from the skull of a famous patient. Science 1994;264:1102–5.

[213] Damsma G, Pfaus JG, Wenkstern D, Phillips AG, Fibiger HC. Sexual behavior increases dopamine transmission in the nucleus accumbens and striatum of male rats: comparison with novelty and locomotion. Behav Neurosci 1992;106:181–91.

[214] Daniels D, Miselis RR, Flanagan-Cato LM. Central neuronal circuit innervating the lordosis-producing muscles defined by transneuronal transport of pseudorabies virus. J Neurosci 1999;19:2823–33.

[215] Darvas M, Fadok JP, Palmiter RD. Requirement of dopamine signaling in the amygdala and striatum for learning and maintenance of a conditioned avoidance response. Learn Mem 2011;18:136–43.

[216] David SP, Murphy NV, Rabiner EA, Munafo MR, Johnstone EC, Jacob R, et al. A functional genetic variation of serotonin (5-HT) transporter affects 5-HT1A receptor binding in humans. J Neurosci 2005;25:2586–90.

[217] Davidson JM. Activation of the male rat's sexual behavior by intracerebral implantation of androgen. Endocrinology 1966;79:783–94.

[218] Davidson RJ, Putnam KM, Larson CL. Dysfunction in the neural circuitry of emotion regulation—a possible prelude to violence. Science 2000;289:591–4.

[219] Davis JF, Loos M, Di Sebastiano AR, Brown JL, Lehman MN, Coolen LM. Lesions of the medial prefrontal cortex cause maladaptive sexual behavior in male rats. Biol Psychiatry 2010;67:1199–204.

[220] Dawel A, O'Kearney R, McKone E, Palermo R. Not just fear and sadness: meta-analytic evidence of pervasive emotion recognition deficits for facial and vocal expressions in psychopathy. Neurosci Biobehav Rev 2012;36:2288–304.

[221] Dawes CT, Loewen PJ, Schreiber D, Simmons AN, Flagan T, McElreath R, et al. Neural basis of egalitarian behavior. Proc Natl Acad Sci USA 2012;109:6479–83.

[222] D'Cunha TM, King SJ, Fleming AS, Levy F. Oxytocin receptors in the nucleus accumbens shell are involved in the consolidation of maternal memory in postpartum rats. Horm Behav 2011;59:14–21.

[223] De Almeida RM, Rosa MM, Santos DM, Saft DM, Bebini Q, Miczek KA. 5-HT(1B) receptors, ventral orbitofrontal cortex, and aggressive behavior in mice. Psychopharmacology 2006;185:441–50.

[224] De Boer SF, Koolhaas JM. 5-HT1A and 5-HT1B receptor agonists and aggression: a pharmacological challenge of the serotonin deficiency hypothesis. Eur J Pharmacol 2005;526:125–39.
[225] De Boer A, van Buel EM, Ter Horst GJ. Love is more than just a kiss: a neurobiological perspective on love and affection. Neuroscience 2012;201:114–24.
[226] De Bruin JP, van Oyen HG, Van de Poll N. Behavioural changes following lesions of the orbital prefrontal cortex in male rats. Behav Brain Res 1983;10:209–32.
[227] Decety J, Norman GJ, Berntson GG, Cacioppo JT. A neurobehavioral evolutionary perspective on the mechanisms underlying empathy. Prog Neurobiol 2012;98:38–48.
[228] Decety J, Porges EC. Imagining being the agent of actions that carry different moral consequences: an fMRI study. Neuropsychologia 2011;49:2994–3001.
[229] Decety J, Skelly LR, Kiehl KA. Brain responses to empathy-eliciting scenarios involving pain in incarcerated individuals with psychopathy. JAMA Psychiatry 2013;70:638–45.
[230] De Dreu CK. Oxytocin modulates cooperation within and competition between groups: an integrative review and research agenda. Horm Behav 2012;61:419–28.
[231] De Dreu CK, Greer LL, Handgraaf MJ, Shalvi S, Van Kleef GA, Baas M, et al. The neuropeptide oxytocin regulates parochial altrusim in intergroup conflict among humans. Science 2010;328:1408–11.
[232] De Jong TR, Chauke M, Harris BN, Saltzman W. From here to paternity: neural correlates of the onset of paternal behavior in California mice (*Peromyscus californicus*). Horm Behav 2009;56:220–31.
[233] De Jonge FH, Louwerse AL, Ooms MP, Evers P, Endert E, van de Poll NE. Lesions of the SDN-POA inhibit sexual behavior of male Wistar rats. Brain Res Bull 1989;23:483–92.
[234] De Kloet ER, Voorhuis DA, Boschma Y, Elands J. Estradiol modulates density of putative 'oxytocin receptors' in discrete rat brain regions. Neuroendocrinology 1986;44:415–21.
[235] Delville Y, De Vries GJ, Ferris CF. Neural connections of the anterior hypothalamus and agonistic behavior in golden hamsters. Brain Behav Evol 2000;55:53–76.
[236] Delville Y, Mansour KM, Ferris CF. Testosterone facilitates aggression by modulating vasopressin receptors in the hypothalamus. Physiol Behav 1996;60:25–9.
[237] De Oliveira-Souza R, Hare HD, Bramati IE, Garrido GJ, Azevedo Ignacio F, Tovar-Moll F, et al. Psychopathy as a disorder of the moral brain: fronto-temporo-limbic grey matter reductions by voxel-based morphometry. NeuroImage 2008;40:1202–13.
[238] Depue RA, Morrone-Strupinsky JV. A neurobehavioral model of affiliative bonding: implications for conceptualizing a human trait of affiliation. Behav Brain Sci 2005;28:313–50.
[239] Devidze N, Mong JA, Jasnow AM, Kow LM, Pfaff DW. Sex and estrogenic effects on coexpression of mRNAs in single ventromedial hypothalamic neurons. Proc Natl Acad Sci USA 2005;102:14446–51.
[240] De Vries GJ, Buijis RM, Van Leeuwen FW, Caffe AR, Swaab DF. The vasopressinergic innervation of the brain in normal and castrated rats. J Comp Neurol 1985;233:236–54.
[241] De Vries GJ, Panzica GC. Sexual differentiation of central vasopressin and vasotocin systems in vertebrates: different mechanisms, similar endpoints. Neuroscience 2006;138:947–55.
[242] De Vries GJ, Simerly RB. Anatomy, development, and function of sexually dimorphic neural circuits in the mammalian brain. In: Pfaff DW, Arnold AP, Etgen AM, Fahrbach SE, Rubin RT, editors. Hormones brain, and behavior, vol. 4. San Diego: Academic Press; 2002. pp. 137–91.
[243] De Vries GJ, Sodersten P. Sex differences in the brain: the relation between structure and function. Horm Behav 2009;55:589–96.
[244] De Waal FB. Putting the altruism back in altruism: the evolution of empathy. Annu Rev Psychol 2008;59:279–300.
[245] De Waal F. The bonobo and the atheist. New York: Norton; 2013.
[246] Dewing P, Chiang CW, Sinchak K, Sim H, Fernagut PO, Kelly S, et al. Direct regulation of adult brain function by male-specifc factor SRY. Curr Biol 2006;16:415–20.
[247] Dhakar MB, Rich ME, Reno EL, Lee HJ, Caldwell HK. Heightened aggressive behavior in mice with lifelong versus postweaning knockout of the oxytocin receptor. Horm Behav 2012;62:86–92.
[248] Dichter GS. Functional magnetic resonance imaging of autism spectrum disorders. Dialogues Clin Neurosci 2012;14:319–51.
[249] Ditzen B, Bradley B, Heim CM. Oxytocin and pair bonding: on possible influences during the life course. Biol Psychiatry 2012;73:e3–4.
[250] Ditzen B, Schaer M, Gabriel B, Bodenmann G, Ehlert U, Heinrichs M. Intranasal oxytocin increases positive communication and reduces cortisol levels during couple conflict. Biol Psychiatry 2009;65:728–31.
[250A] Domes G, Heinrichs M, Kumbier E, Grossmann A, Hauenstein K, Herpertz SC. Effects of intranasal oxytocin on the neural basis of face processing in autism spectrum disorder. Biol Psychiatry 2013;74:164–71.

[251] Domes G, Heinrichs M, Michel A, Berger C, Herpertz SC. Oxytocin improves "mind-reading" in humans. Biol Psychiatry 2007;61:731–3.

[252] Dominguez J, Riolo JV, Xu Z, Hull EM. Regulation by the medial amygdala of copulation and medial preoptic dopamine release. J Neurosci 2001;21: 349–55.

[253] Donaghue C, Westley BR, May FE. Selective promoter usage of the human estrogen receptor-alpha gene and its regulation by estrogen. Mol Endocrinol 1999;13:1943–50.

[254] Donaldson ZR, Spiegel L, Young LJ. Central vasopressin V1a receptor activation is independently necessary for both partner preference formation and expression in socially monogamous male prairie voles. Behav Neurosci 2010;124:159–63.

[255] Dong HW, Petrovich GD, Swanson LW. Topography of projections from the amygdala to bed nuclei of the stria terminalis. Brain Res Rev 2001;38:192–246.

[256] Dong HW, Swanson LW. Projections from the bed nuclei of the stria terminalis, posterior division: implications for the cerebral hemisphere regulation of defensive and reproductive behaviors. J Comp Neurol 2004;471:396–433.

[257] Dulac C, Kimchi T. Neural mechanisms underlying sex-specific behaviors in vertebrates. Curr Opin Neurobiol 2007;17:675–83.

[258] Dulac C, Wagner S. Genetic analysis of brain circuits underlying pheromone signaling. Annu Rev Genet 2006;40:449–67.

[259] Duke AA, Begue L, Bell R, Eisenlohr-Moul T. Revisiting the serotonin-aggression relation in humans: a meta-analysis. Psychol Bull 2013;139:1148–72.

[260] Dumartin B, Doudnikoff E, Gonon F, Bloch B. Differences in ultrastructural localization of D1 receptors between dorsal striatum and nucleus accumbens in the rat. Neurosci Lett 2007;419:273–7.

[261] DuPree MG, Mustanski BS, Bocklandt S, Nievergelt C, Hamer DH. A candidate gene study of CYP19 (aromatase) and male sexual orientation. Behav Genet 2004;34:243–50.

[262] Eblen F, Graybiel AM. Highly restricted origin of prefrontal cortical inputs to striosomes in the macaque monkey. J Neurosci 1995;15:5999–6013.

[263] Ebstein RP, Knafo A, Mankuta D, Chew SH, Lai PS. The contributions of oxytocin and vasopressin pathway genes to human behavior. Horm Behav 2012;61:359–79.

[264] Ecker C, Spooren W, Murphy D. Translational approaches to the biology of autism: false dawn or new era? Mol Psychiatry 2013;18:435–42.

[265] Edwards DA, Einhorn LC. Preoptic and midbrain control of sexual motivation. Physiol Behav 1986;37:329–35.

[266] Edwards DA, Nahai FR, Wright P. Pathways linking the olfactory bulbs with the medial preoptic anterior hypothalamus are important for intermale aggression in mice. Physiol Behav 1993;53:611–5.

[267] Ehrlich I, Humeau Y, Grenier F, Ciocchi S, Herry C, Luthi A. Amygdala inhibitory circuits and the control of fear memory. Neuron 2009;62:757–71.

[268] Eisenberger NI. The pain of social disonnection: examining the shared neural underpinnings of physical and social pain. Nat Rev Neurosci 2012;13:421–34.

[269] Eisenberger NI, Way BM, Taylor SE, Welch WT, Lieberman MD. Understanding genetic risk for aggression: clues from the brain's response to social exclusion. Biol Psychiatry 2007;61:1100–8.

[270] Ellingsen E, Agmo A. Sexual-incentive motivation and paced sexual behavior in female rats after treatment with drugs modifying dopaminergic neurotransmission. Pharmacol Biochem Behav 2004;77:431–45.

[271] Engen HG, Singer T. Empathy circuits. Curr Opin Neurobiol 2013;23:275–83.

[272] Ernst M, Fudge JL. A developmental neurobiological model of motivated behavior: anatomy, connectivity and ontogeny of the triadic nodes. Neurosci Biobehav Rev 2009;33:367–82.

[273] Erskine MS, Lehmann ML, Cameron NM, Polston EK. Co-regulation of female sexual behavior and pregnancy induction: an exploratory synthesis. Behav Brain Res 2004;153:295–315.

[274] Evans S, Shergill SS, Averbeck BB. Oxytocin decreases aversion to angry faces in an associative learning task. Neuropsychopharmacol 2010;35:2502–9.

[275] Everitt BJ. Sexual motivation: a neural and behavioural analysis of the mechanisms underlying appetitive and copulatory responses in male rats. Neurosci Biobehav Rev 1990;14:217–32.

[276] Everitt BJ, Cador M, Robbins TW. Interactions between amygdala and ventral striatum in stimulus-reward associations: studies using a second-order schedule of sexual reinforcement. Neuroscience 1989;30:63–75.

[277] Everitt BJ, Stacey P. Studies of instrumental behavior with sexual reinforcement in male rats (*Rattus norvegicus*): II. Effects of preoptic area lesions, castration, and testosterone. J Comp Psychol 1987;101:407–19.

[278] Faccidomo S, Quadros IM, Takahashi A, Fish EW, Miczek KA. Infralimbic and dorsal raphe microinjection of the 5-HT(1B) receptor agonist CP-93,129: attenuation of aggressive behavior in CFW male mice. Psychopharmacology 2012;222:117–28.

[279] Fahrbach SE, Morrell JI, Pfaff DW. Possible role for endogenous oxytocin in estrogen-facilitated maternal behavior in rats. Neuroendocrinology 1985;40:526–32.

[280] Fahrbach SE, Morrell JI, Pfaff DW. Identification of medial preoptic neurons that concentrate estradiol and project to the midbrain in the rat. J Comp Neurol 1986;247:364–82.
[281] Fahrbach SE, Pfaff DW. Effect of preoptic region implants of dilute estradiol on the maternal behavior of ovariectomized, nulliparous rats. Horm Behav 1986;20:354–63.
[282] Fairchild G, Hagan CC, Walsh ND, Passamonti L, Calder AJ, Goodyer IM. Brain structure abnormalities in adolescent girls with conduct disorder. J Child Psychol Psychiatry 2013;54:86–95.
[283] Fan Y, Duncan NW, de Greck M, Northoff G. Is there a core neural network in empathy? An fMRI based quantitative meta-analysis. Neurosci Biobehav Rev 2011;35:903–11.
[284] Farrar AM, Font L, Pereira M, Mingote S, Bunce JG, Chrobak JJ, et al. Forebrain circuitry involved in effort-related choice: injections of the GABAA agonist muscimol into ventral pallidum alter response allocation in food-seeking behavior. Neuroscience 2008;152:321–30.
[285] Faure A, Reynolds SM, Richard JM, Berridge KC. Mesolimbic dopamine in desire and dread: enabling motivation to be generated by localized glutamate disruptions in nucleus accumbens. J Neurosci 2008;28:7184–92.
[286] Febo M, Felix-Ortiz AC, Johnson TR. Inactivation or inhibition of neuronal activity in the medial prefrontal cortex largely reduces pup retrieval and grouping in maternal rats. Brain Res 2010;1325:77–88.
[287] Feinstein JS, Adolphs R, Damasio A, Tranel D. The human amygdala and experience of fear. Curr Biol 2011;21:34–8.
[288] Feinstein JS, Buzza C, Hurlemann R, Follmer RL, Dahdaleh NS, Coryell WH, et al. Fear and panic in humans with bilateral amygdala damage. Nat Neurosci 2013;16:270–2.
[289] Feldman R. Oxytocin and social affiliation in humans. Horm Behav 2012;61:380–91.
[290] Feldman R, Weller A, Zagoory-Sharon O, Levine A. Evidence for a neuroendocrinological foundation of human affiliation: plasma oxytocin levels across pregnancy and the postpartum period predict mother-infant bonding. Psychol Sci 2007;18:965–70.
[291] Feng X, Wang L, Yang S, Qin D, Wang J, Li C, et al. Maternal separation produces lasting changes in cortisol and behavior in rhesus monkeys. Proc Natl Acad Sci USA 2011;108:14312–7.
[292] Ferguson JN, Aldag JM, Insel TR, Young LJ. Oxytocin in the medial amygdala is essential for social recognition in the mouse. J Neurosci 2001;21:8278–85.
[293] Ferrari PF, van Erp AM, Tornatzky W, Miczek KA. Accumbal dopamine and serotonon in anticipation of the next aggressive episode in rats. Eur J Neurosci 2003;17:371–8.
[294] Ferreira A, Dahlof LG, Hansen S. Olfacotry mechanisms in the control of maternal aggression, appetite, and fearfulness: effects of lesions to olfactory receptors, mediodorsal thalamic nucleus, and insular prefrontal cortex. Behav Neurosci 1987;101:709–17.
[295] Ferretti A, Caulo M, Del Gratta C, Di Matteo R, Merla A, Montorsi F, et al. Dynamics of male sexual arousal: distinct components of brain activation revealed by fMRI. NeuroImage 2005;26:1086–96.
[296] Ferris CF, Delville Y, Miller MA, Dorsa DM, De Vries GJ. Distribution of small vasopressinergic neurons in golden hamsters. J Comp Neurol 1995;360:589–98.
[297] Ferris CF, Melloni RH, Koppel G, Perry KW, Fuller RW, Delville Y. Vasopressin/serotonin interactions in the anterior hypothalamus control aggressive behavior in golden hamsters. J Neurosci 1997;17:4331–40.
[298] Ferris CF, Potegal M. Vasopressin receptor blockade in the anterior hypothalamus suppresses aggression in hamsters. Physiol Behav 1988;44:235–9.
[299] Ferris CF, Snowdon CT, King JA, Duong TQ, Zeigler TE, Ugurbil K, et al. Functional imaging of brain activity in conscious monkeys responding to sexually arousing cues. Neuroreport 2001;12:2231–6.
[300] Ferris CF, Stolberg T, Delville Y. Serotonin regulation of aggressive behavior in male golden hamsters. Behav Neurosci 1999;113:804–15.
[301] Fink S, Excoffier L, Heckel G. Mammalian monogamy is not controlled by a single gene. Proc Natl Acad Sci USA 2006;103:10956–60.
[302] Finn PD, Yahr P. Projection from the ventral bed nucleus of the stria terminalis to the retrorubral field in rats and the effects of cells in these areas on mating in male rats versus gerbils. Horm Behav 2005;47:123–38.
[303] Fiorino DF, Coury A, Phillips AG. Dynamic changes in nucleus accumbens dopamine efflux during the Coolidge effect in male rats. J Neurosci 1997;17:4849–55.
[304] Fleming AS, Cheung U, Myhal N, Kessler Z. Effects of maternal hormones on 'timidity' and attraction to pup-related odors in female rats. Physiol Behav 1989;46:449–53.
[305] Fleming AS, Korsmit M, Deller M. Rat pups are potent reinforcers to the maternal animal: effects of experience, parity, hormones, and dopamine function. Psychobiology 1994;22:44–53.
[306] Fleming AS, Luebke C. Timidity prevents the virgin female rat from being a good mother: emotionality differences between nulliparous and parturient females. Physiol Behav 1981;27:863–8.
[307] Fleming AS, Rosenblatt JS. Olfactory regulation of maternal behavior in rats. II. Effects of peripherally induced anosmia and lesions of the lateral olfactory tract in pup-induced virgins. J Comp Physiol Psychol 1974;86:233–46.

[308] Fleming AS, Vaccarino F, Luebke C. Amygdaloid inhibition of maternal behaviorin the nulliparous female rat. Physiol Behav 1980;25:731–43.

[309] Fleming A, Vaccarino F, Tambosso L, Chee P. Vomeronasal and olfactory system modulation of maternal behavior in the rat. Science 1979;203:372–4.

[310] Flores-Barrera E, Vizcarra-Chacon BJ, Bargas J, Tapia D, Gallarraga E. Dopaminergic modulation of corticostriatal responses in medium spiny projection neurons from direct and indirect pathways. Front Syst Neurosci 2011;5:15.

[311] Foley DL, Eaves LJ, Wormley B, Silberg J, Maes HH, Kuln J, et al. Childhood adversity, monoamine oxidase A genotype, and risk for conduct disorder. Arch Gen Psychiatry 2004;61:738–44.

[312] Fonagy P. Early-life trauma and the psychogenesis and prevention of violence. Ann NY Acad Sci 2004;1036:181–200.

[313] Forger NG. Control of cell number in the sexually dimorphic brain and spinal cord. J Neuroendocrinol 2009;21:393–9.

[314] Forster GL, Pringle RB, Mouw NJ, Vuong SM, Watt MJ, Burke AR, et al. Corticotropin-releasing factor in the dorsal raphe increases medial prefrontal cortical serotonin via type 2 receptors and median raphe nucleus activity. Eur J Neurosci 2008;28: 299–310.

[315] Frankle WG, Lombardo I, New AS, Goodman M, Talbot PS, Huang Y, et al. Brain serotonin transporter distribution in subjects with impulsive aggressivity: a positron emission study with [^{11}C]McN 5652. Am J Psychiatry 2005;162:915–23.

[316] Franzen EA, Myers RE. Neural control of social behavior: prefrontal and anterior temporal cortex. Neuropsychologia 1973;11:141–57.

[317] French JA, Fite JE, Ross CN. Family life in marmosets: causes and consequences of variation in caregiving. In: Bridges RS, editor. Neurobiology of the parental brain. San Diego: Academic Press; 2008. pp. 461–77.

[318] Freund K, Kuban M. Toward a testable developmental model of pedophilia: the development of erotic age preference. Child Abuse Negl 1993;17:315–24.

[319] Fuster JM. The prefrontal cortex. San Diego: Academic Press; 2008.

[320] Gabbott PL, Warner TA, Jays PR, Salway P, Busby SJ. Prefrontal cortex in the rat: projections to subcortical autonomic, motor, and limbic centers. J Comp Neurol 2005;492:145–77.

[321] Gabor CS, Phan A, Clipperton-Allen AE, Kavaliers M, Choleris E. Interplay of oxytocin, vasopressin, and sex hormones in the regulation of social recognition. Behav Neurosci 2012;126:97–109.

[322] Gainer H, Wray S. Cellular and molecular biology of oxytocin and vasopressin. In: Knobil E, Neill JD, editors. The physiology of reproduction, vol. 2. New York: Raven Press; 1994. pp. 1099–129.

[323] Gallagher JP, Orozco-Cabal LF, Liu J, Shinnick-Gallagher P. Synaptic physiology of central CRH system. Eur J Pharmacol 2008;583:215–25.

[324] Gamer M, Zurowski B, Buchel C. Different amygdala subregions mediate valence-related and attentional effects of oxytocin in humans. Proc Natl Acad Sci USA 2010;107:9400–5.

[325] Gao B, Moore RY. The sexually dimorphic nucleus of the hypothalamus contains GABA neurons in rat and man. Brain Res 1996;742:163–71.

[326] Gaspar P, Cases O, Maroteaux L. The developmental role of serotonin: news from mouse molecular genetics. Nat Rev Neurosci 2003;4:1002–12.

[327] Geiman TM, Muegge K. DNA methylation in early development. Mol Reprod Dev 2010;77:105–13.

[328] Geisler S, Derst C, Veh RW, Zahm DS. Glutamatergic afferents of the ventral tegmental area in the rat. J Neurosci 2007;27:5730–43.

[329] Geisler S, Zahm DS. Afferents of the ventral tegmental area in the rat-anatomical substratum for integrative functions. J Comp Neurol 2005;490:270–94.

[330] Georgiadis JR, Kringelbach ML. The human sexual response cycle: brain imaging linking sex to other pleasures. Prog Neurobiol 2012;98:49–81.

[331] Gerfen CR, Surmeier DJ. Modulation of striatal projection systems by dopamine. Annu Rev Neurosci 2011;34:441–66.

[332] Gilbert AN, Pelchat RJ, Adler NT. Postpartum copulatory and maternal behaviour in Norway rats under seminatural conditions. Anim Behav 1980;28:989–95.

[333] Gillath O, Shaver PR, Baek JM, Chun DS. Genetic correlates of adult attachment style. Pers Soc Psychol Bull 2008;34:1396–405.

[334] Gilmore RF, Varnum MM, Forger NG. Effects of blocking developmental cell death on sexually dimorphic clabindin cell groups in the preoptic area and bed nucleus of the stria terminalis. Biol Sex Differ 2012;3:5.

[335] Gingrich B, Liu Y, Cascico C, Wang Z, Insel TR. Dopamine D2 receptors in the nucleus accumbens are important for social attachment in female prairie voles (*Microtus ochrogaster*). Behav Neurosci 2000;114:173–83.

[336] Gizewski ER, Krause E, Karama S, Baars A, Senf W, Forsting M. There are differences in cerebral activation between females in distinct menstrual phases during viewing of erotic stimuli: a fMRI study. Exp Brain Res 2006;174:101–8.

[337] Glenn AL. The other allele: exploring the long allele of the serotonin transporter gene as a potential risk factor for psychopathy: a review of the parallels and findings. Neurosci Biobehav Rev 2011;35:612–20.
[338] Glenn AL, Raine A, Schug RA. The neural correlates of moral decision-making in psychopathy. Mol Psychiatry 2009;14:5–6.
[339] Glenn AL, Raine A, Schug RA, Young L, Hauser M. Increased DLPFC activity during moral decision-making in psychopathy. Mol Psychiatry 2009;14:909–11.
[340] Glenn AL, Yang Y. The potential role of the striatum in antisocial behavior and psychopathy. Biol Psychiatry 2012;72:817–22.
[341] Glickman SE, Short RV, Renfree MB. Sexual differentiation in three unconventional mammals: spotted hyenas, elephants and tammar wallabies. Horm Behav 2005;48:403–17.
[342] Glocker ML, Langleben DD, Ruparel K, Loughead JW, Valdez JN, Griffin MD, et al. Baby schema modulates the brain reward system in nulliparous women. Proc Natl Acad Sci USA 2009;106:9115–9.
[343] Gobrogge KL, Liu Y, Jia X, Wang Z. Anterior hypothalamic neural activation and neurochemical associations with aggression in pair-bonded male prairie voles. J Comp Neurol 2007;502:1109–22.
[344] Gobrogge KL, Liu Y, Young LJ, Wang Z. Anterior hypothalamic vasopressin regulates pair-bonding and drug-induced aggression in a monogamous rodent. Proc Natl Acad Sci USA 2009;106:19144–9.
[345] Gobrogge KL, Wang ZW. Genetics of aggression in voles. Adv Genet 2011;75:121–50.
[346] Gonzalez A, Lovic V, Ward GR, Wainwright PE, Fleming AS. Intergenerational effects of complete maternal deprivation and replacement stimulation on maternal behavior and emotionality in female rats. Dev Psychobiol 2001;38:11–32.
[347] Gonzalez-Liencres C, Shamay-Tsoort SG, Brune M. Towards a neuroscience of empathy: ontogeny, phylogeny, brain mechanisms, context and psychopathology. Neurosci Biobehav Rev 2013;37:1537–48.
[348] Goodson JL. Deconstructing sociality, social evolution and relevant nonapeptide functions. Psychoneuroendocrino 2013;38:465–78.
[349] Goodson JL, Kelly AM, Kingsbury MA. Evolving nonapeptide mechanisms of gregariousness and social diversity in birds. Horm Behav 2012;61:239–50.
[350] Gordon I, Zagoory-Sharon O, Leckman JF, Feldman R. Oxytocin and the development of parenting in humans. Biol Psychiatry 2010;68:377–82.
[351] Gorski RA, Harlan RE, Jacobson CD, Shryne JE, Southham AM. Evidence for the existence of a sexually dimorphic nucleus in the preoptic area of the rat. J Comp Neurol 1980;193:529–39.
[352] Gozzi A, Jain A, Giovannelli A, Bertollini C, Crestan V, Schwarz AJ, et al. A neural switch for active and passive fear. Neuron 2010;67:656–66.
[353] Gray TS. Amygdaloid CRF pathways. Ann NY Acad Sci 1993;697:53–60.
[354] Grebe NM, Gangestad SW, Garver-Apgar CE, Thornhill R. Women's luteal-phase sexual proceptivity and the functions of extended sexuality. Psychol Sci 2013;24:2106–10.
[355] Green L, Fein D, Modahl C, Feinstein C, Waterhouse L, Morris M. Oxytocin and autistic disorder: alterations in peptide forms. Biol Psychiatry 2001;50:609–13.
[356] Greenberg DM, Bradford JM, Curry S, O'Rourke A. A comparison of treatment paraphiias with three serotonin reuptake inhibitors: a retospective study. Bull Am Acad Psychiatry Law 1996;24:525–32.
[357] Gregg TR, Siegel A. Brain structures and neurotransmitters regulating aggression in cats: implications for human aggression. Prog Neuropsychopharmacol Biol Psychiatry 2001;25:91–140.
[358] Gregory SG, Connelly JJ, Towers AJ, Johnson J, Biscocho D, Markunas CA, et al. Genomic and epigenetic evidence for oxytocin receptor deficiency in autism. BMC Med 2009;7:62.
[359] Gregory S, Ffytche D, Simmons A, Kumari V, Howard M, Hodgins S, et al. The antisocial brain: psychopathy matters. Arch Gen Psychiatry 2012;69:962–72.
[360] Griffin GD, Flanagan-Cato LM. Ovarian hormone action in the hypothalamic ventromedial nucleus: remodeling to regulate reproduction. J Neuroendocrinol 2011;23:465–71.
[361] Grillner S, Hellgren J, Menard A, Saitoh K, Wikstrom MA. Mechanisms for selection of basic motor programs—roles for the striatum and pallidum. Trends Neurosci 2005;28:364–70.
[362] Groenewegen HJ, Berendse HW, Haber SN. Organization of the output of the ventral striatopallidal system in the rat: ventral pallidal efferents. Neuroscience 1993;57:113–42.
[363] Gross CT, Canteras NS. The many paths to fear. Nat Rev Neurosci 2012;13:651–8.
[364] Gruber AJ, Powell EM, O'Donnell P. Cortically activated interneurons shape spatial aspects of cortico-accumbens processing. J Neurophysiol 2009;101:1876–82.
[365] Gu X, Hof PR, Friston KJ, Fan J. Anterior insular cortex and emotional awareness. J Comp Neurol 2013;521:3371–88.
[366] Guarraci FA, Clark AS. Ibotenic acid lesions of the medial preoptic area disrupt the expression of partner preference in sexually receptive female rats. Brain Res 2006;1076:163–70.

[367] Guarraci FA, Megroz AB, Clark AS. Effects of ibotenic acid lesions of the nucleus accumbens on paced mating behavior in the female rat. Behav Neurosci 2002;116:568–76.

[368] Guarraci FA, Megroz AB, Clark AS. Paced mating behavior in the female rat following lesions of three regions responsive to vaginocervical stimulation. Brain Res 2004;999:40–52.

[369] Guastella AJ, Einfeld SL, Gray KM, Rinehart NJ, Tonge BJ, Lambert TJ, et al. Intranasal oxytocin improves emotion recognition for youth with autism spectrum disorders. Biol Psychiatry 2010;67:692–4.

[370] Gunter TD, Vaughan MG, Philibert RA. Behavioral genetics in antisocial spectrum disorders and psychopathy: a review of recent literature. Behav Sci Law 2010;28:148–73.

[371] Gutzler SJ, Karom M, Erwin WD, Albers HE. Arginine-vasopressin and the regulation of aggression in female Syrian hamsters (*Mesocricetus auratus*). Eur J Neurosci 2010;31:1655–63.

[372] Haas BW, Mills D, Yam A, Hoeft F, Bellugi U, Reiss A. Genetic influences on sociability: heightened amygdala reactivity and event-related responses to positive social stimuli in Williams syndrome. J Neurosci 2009;29:1132–9.

[373] Halasz J, Toth M, Kallo I, Liposits Z, Haller J. The activation of prefrontal cortical neurons in aggression-a double labeling study. Behav Brain Res 2006;175:166–75.

[374] Halasz J, Toth M, Mikics E, Hrabovszky E, Barsy B, Barsvari B, et al. The effect of neurokinin 1 receptor blockade on territorial aggression and in a model of violent aggression. Biol Psychiatry 2008;63:271–8.

[375] Halasz J, Zelena D, Toth M, Tulogdi A, Mikics E, Haller J. Substance P neurotransmission and violent aggression: the role of tachykinin NK_1 receptors in the hypothalamic attack area. Eur J Pharmacol 2009;611:35–43.

[376] Hale MW, Lowry CA. Functional topography of midbrain and pontine serotonergic systems: implications for synaptic regulation of serotonergic circuits. Psychopharmacology 2011;213:243–64.

[377] Haller J, Kruk MR. Normal and abnormal aggression: human disorders and novel laboratory models. Neurosci Biobehav Rev 2006;30:292–303.

[378] Haller J, Toth M, Halasz J. The activation of raphe serotonergic neurons in normal and hypoarousal-driven aggression: a double labeling study in rats. Behav Brain Res 2005;161:88–94.

[379] Hamani C, Mayberg H, Stone S, Laxton A, Haber S, Lozano AM. The subcallosal cingulate gyrus in the context of major depression. Biol Psychiatry 2011;69:301–8.

[380] Hamann S, Herman RA, Nolan CL, Wallen K. Men and women differ in amygdala response to visual sexual stimuli. Nat Neurosci 2004;7:411–6.

[381] Hamilton WD. The evolution of altruistic behaviour. Am Nat 1963;97:354–6.

[382] Hammock EA, Young LJ. Functional microsatellite polymorphism associated with divergent social structure in vole species. Mol Biol Evol 2004;21:1057–63.

[383] Hammock EA, Young LJ. Microsatellite instability generates diversity in brain and soiobehavioral traits. Science 2005;308:1630–4.

[384] Hannon J, Hoyer D. Molecular biology of 5-HT receptors. Behav Brain Res 2008;195:198–213.

[385] Hansen S. Medial hypothalamic involvement in maternal aggression of rats. Behav Neurosci 1989;103:1035–46.

[386] Hansen S, Harthon C, Wallin E, Lofberg L, Svensson K. Mesotelencephalic dopamine system and reproductive behavior in the female rat: effects of ventral tegmental 6-hydroxydopamine lesions on maternal and sexual responseiveness. Behav Neurosci 1991;105:588–98.

[387] Hansen S, Kohler C, Ross SB. On the role of the dorsal mesencephalic tegmentum in the control of masculine sexual behavior in the rat: effects of electrolytic lesions, ibotenic acid and DSP 4. Brain Res 1982;240:311–20.

[388] Hare RD. Without conscience: the disturbing world of psychopaths amongst us. New York: Guilford Press; 1993.

[389] Hare RD, Neumann CS. Psychopathy as a clinical and empirical construct. Annu Rev Clin Psychol 2008;4:217–46.

[390] Harlow HF. Learning to love. San Francisco: Albion; 1971.

[391] Harlow HF, Dodsworth RO, Harlow MK. Total isolation in monkeys. Proc Natl Acad Sci USA 1965;54:90–7.

[392] Harlow HF, Harlow MK, Hansen EW. The maternal affectional system of rhesus monkeys. In: Rheingold HL, editor. Maternal behavior in mammals. New York: Wiley; 1963. pp. 254–81.

[393] Hart BL, Leddy MG. Neurological bases of male sexual behavior: a comparative analysis. In: Adler N, Pfaff D, Goy RW, editors. Handbook of behavioral neurobiology. Reproduction, vol. 7. New York: Plenum Press; 1985. pp. 373–422.

[394] Hartley CA, Phelps EA. Changing fear: the neurocircuitry of emotion regulation. Neuropsychopharmacol 2010;35:136–46.

[395] Haselton MG, Gangestad SW. Conditional expression of women's desires and men's mate guarding across the ovulatory cycle. Horm Behav 2006;49:509–18.

[396] Hayes UL, De Vries GJ. Role of pregnancy and parturition in induction of maternal behavior in prairie voles (*Microtus ochrogaster*). Horm Behav 2007;51:265–72.

[397] Heatherton TF, Wagner DD. Cognitive neuroscience of self-regulation failure. Trends Cogn Sci 2011;15:132–9.

[398] Heberlein AS, Adolphs R. Impaired spontaneous anthropomorphizing despite intact perception and social knowledge. Proc Natl Acad Sci USA 2004;101:7487–91.

[399] Heim C, Young LJ, Newport DJ, Mletzko T, Miller AH, Nemeroff CB. Lower CSF oxytocin concentrations in women with a history of childhood abuse. Mol Psychiatry 2009;14:954–8.

[400] Hein G, Silani G, Preuschoff K, Batson CD, Singer T. Neural responses to ingroup and outgroup members' suffering predict individual differences in costly helping. Neuron 2010;68:149–60.

[401] Helton TD, Xu W, Lipscombe D. Neuronal L-type calcium channels open quickly and are inhibited slowly. J Neurosci 2005;25:10247–51.

[402] Hendricks TJ, Fyodorov DV, Wegman LJ, Lelutiu NB, Pehek EA, Yamamoto B, et al. Pet-1 ETS gene plays a critical role in 5-HT neuron development and is required for normal anxiety-like and aggressive behavior. Neuron 2003;37:233–47.

[403] Henley CL, Nunez AA, Clemens LG. Estrogen treatment during development alters adult partner preference and reproductive behavior in female laboratory rats. Horm Behav 2009;55:68–75.

[404] Hennessey AC, Camak L, Gordon F, Edwards DA. Connections between the pontine central gray and the ventromedial hypothalamus are essential for lordosis in female rats. Behav Neurosci 1990;104:477–88.

[405] Hennessey AC, Wallen K, Edwards DA. Preoptic lesions increase the display of lordosis by male rats. Brain Res 1986;370:21–8.

[406] Henschen CW, Palmiter RD, Darvas M. Restoration of dopamine signaling to the dorsal striatum is sufficient for aspects of active maternal behavior in mice. Endocrinology 2013;154:4316–27.

[407] Herry C, Ciocchi S, Senn V, Demmou L, Muller C, Luthi A. Switching on and off fear by distinct neuronal circuits. Nature 2008;454:600–6.

[408] Hicks C, Jorgenson W, Brown C, Fardell J, Koehbach J, Gruber CW, et al. The nonpeptide oxytocin receptor agonist WAY 267,464: receptor-binding profile, prosocial effects and distribution of c-Fos expression in adolescent rats. J Neuroendocrinol 2012;24:1012–29.

[409] Higashida H, Jin D, Liu H, Lopatina O, Yokoyama S, Koizumi K, et al. Social memory, maternal care, and oxytocin secretion, but not vasopressin release, require CD38 in mice. In: Bridges RS, editor. Neurobiology of parental behavior. San Diego: Academic Press; 2008. pp. 361–75.

[410] Higley JD, Suomi SJ, Linnoila M. A nonhuman primate model of type II alcoholism? Part 2. Diminished social competence and excessive aggression correlates with low cerebrospinal fluid 5-hydroxyindoleacetic acid concentrations. Alcohol Clin Exp Res 1996;20:643–50.

[411] Hikida T, Kimura K, Wada N, Funabiki K, Nakanishi S. Distinct roles of synaptic transmission in direct and indirect striatal pathways to reward and aversive behavior. Neuron 2010;66: 896–907.

[412] Hines M. Gender development and the human brain. Annu Rev Neurosci 2011;34:68–88.

[413] Ho JM, Murray JH, Demas GE, Goodson JL. Vasopressin cell groups exhibit strongly divergent responses to copulation and male-male interactions in mice. Horm Behav 2010;58:368–77.

[414] Hoffman GE, Lyo D. Anatomical markers of activity in neuroendocrine systems: are we all 'fos-ed out'? J Neuroendocrinol 2002;14:259–68.

[415] Hoistad M, Barbas H. Sequence of information processing for emotions through pathways linking temporal and insular cortices with the amygdala. Neuro Image 2008;40:1016–33.

[416] Holmes A. Genetic variation in cortico-amygdala serotonin function and risk for stress-related disease. Neurosci Biobehav Rev 2008;32:1293–314.

[417] Holmes A, Murphy DL, Crawley JN. Reduced aggression in mice lacking the serotonin transporter. Psychopharmacology 2002;161:160–7.

[418] Holsboer F. The corticosteroid receptor hypothesis of depression. Neuropsychopharmacol 2000;23:477–501.

[419] Holsboer F, Ising M. Stress hormone regulation: biological role and translation into therapy. Annu Rev Psychol 2010;61:81–109.

[420] Holstege G. Descending motor pathways and the spinal motor system: limbic and non-limbic components. Prog Brain Res 1991;87:307–421.

[421] Holtzheimer PE, Mayberg HS. Deep brain stimulation for psychiatric disorders. Annu Rev Neurosci 2011;34:289–307.

[422] Homberg JR, Lesch KP. Looking on the bright side of serotonin transporter gene variation. Biol Psychiatry 2011;69:513–9.

[423] Hoover WB, Vertes RP. Anatomical analysis of afferent projections to the medial prefrontal cortex. Brain Struct Funct 2007;212:149–79.

[424] Horseman ND, Zhao W, Montecino-Rodriguez E, Tanaka M, Nakashima K, Engle SJ, et al. Defective mammopoiesism, but normal hematopoiesis, in mice with a targeted disruption of the prolactin gene. EMBO J 1997;16:6926–35.

[425] Hoshina Y, Takeo T, Nakano K, Sato T, Sakuma Y. Axon-sparing lesion of the preoptic area enhances receptivity and diminishes proceptivity among components of female rat sexual behavior. Behav Brain Res 1994;61:197–204.

[426] Hrabovszky E, Halasz J, Meelis W, Kruk MR, Liposits Z, Haller J. Neurochemical characterization of hypothalamic neurons involved in attack behavior: glutamatergic dominance and co-expression of thyrotropin-releasing hormone in a subset of glutamatergic neurons. Neuroscience 2005;133: 657–66.

[427] Hrdy SB. Mothers and others: the evolutionary origins of mutual understanding. Cambridge: Belknap Press; 2009.

[428] Hsu HK, Shao PL, Tsai KL, Shih HC, lee TY, Hsu C. Gene regulation by NMDA receptor activation in the SDN-POA neurons of male rats during sexual development. J Mol Endocrinol 2005;34:433–45.

[429] Hsu HK, Yang RC, Shih HC, Hsieh YL, Chen UY, Hsu C. Prenatal exposure to testosterone prevents SDN-POA neurons of postnatal male rats from apoptosis through NMDA receptor. J Neurophysiol 2001;86:2374–80.

[430] Hu S, Pattatucci AML, Patterson C, Li L, Fulker DW, Cherny SS, et al. Linkage between sexual orientation and chromosome Xq28 in males but not females. Nat Genet 1995;11:248–56.

[431] Huber D, Veinante P, Stoop R. Vasopressin and oxytocin excite distinct neuronal populations in the central amygdala. Science 2005;308:245–8.

[432] Hull EM, Wood RI, McKenna KE. Neurobiology of male sexual behavior. In: Neill JD, editor. Physiology of reproduction, vol. 2. San Diego: Academic Press; 2006. pp. 1729–824.

[433] Hume JM, Wynne-Edwards KE. Castration reduces male testosterone, estradiol, and territorial aggression, but not paternal behavior in biparental dwarf hamsters (*Phodopus campbelli*). Horm Behav 2005;48:303–10.

[434] Humphries MD, Prescott TJ. The ventral basal ganglia, a slelction mechanism at the crossroads of space, strategy, and reward. Prog Neurobiol 2010;90: 385–417.

[435] Huo L, Straub RE, Roca C, Schmidt PJ, Shi K, Vakkalanka R, et al. Risk for premenstrual dysphoric disorder is associated with genetic variation in ESR1, the estrogen receptor alpha gene. Biol Psychiatry 2007;62:925–33.

[436] Hurlemann R, Patin A, Onur OA, Cohen MX, Baumgartner T, Metzler S, et al. Oxytocin enhances amygdala-dependent, socially reinforced learning and emotional empathy in humans. J Neurosci 2010;30:4999–5007.

[437] Hurtazo HA, Paredes RG, Agmo A. Inactivation of the medial preoptic area/anterior hypothalamus by lidocaine reduces male sexual behavior and sexual incentive motivation in male rats. Neuroscience 2008;152:331–7.

[438] Hyman SE, Nestler EJ. Principles of molecular biology. In: Charney DS, Nestler EJ, editors. Neurobiology of mental illness. New York: Oxford University Press; 2004. pp. 76–86.

[439] Imamura T. Epigenetic setting for long-term expression of estrogen receptor α and androgen receptor in cells. Horm Behav 2011;59:345–52.

[440] Insel TR. The challenge of translation in social neuroscience: a review of oxytocin, vasopressin, and affiliaitive behavior. Neuron 2010;65:768–79.

[441] Insel TR, Harbaugh CR. Lesions of the hypothalamic paraventricular nucleus disrupt the initiation of maternal behavior. Physiol Behav 1989;45:1033–41.

[442] Insel TR, Hulihan TJ. A gender-specific mechanism for pair bonding: oxytocin and partner preference formation in monogamous voles. Behav Neurosci 1995;109:782–9.

[443] Insel TR, Shapiro LE. Oxytocin receptor distribution reflects social organization in monogamous and polygamous voles. Proc Natl Acad Sci USA 1992;89:5981–5.

[444] Insel TR, Young LJ. The neurobiology of attachment. Nat Rev Neurosci 2001;2:129–36.

[445] Ishikawa SS, Raine A, Lencz T, Bihrle S, Lacasse L. Autonomic stress reactivity and executive functions in successful and unsuccessful criminal psychopaths from the community. J Abnorm Psychol 2001;110:423–32.

[446] Israel S, Weisel O, Ebstein RP, Bornstein G. Oxytocin, but not vasopressin, increases both parochial and universal altruism. Psychoneuroendocrino 2012;37:1341–4.

[447] Jabbi M, Kippenhan JS, Kohn P, Marenco S, Mervis CB, Morris CA, et al. The Williams syndrome chromosome 7q11.23 hemideletion confers hypersocial, anxious personality coupled with altered insula structure and function. Proc Natl Acad Sci USA 2012;109:E860–866.

[448] Jahng JW, Ryu V, Yoo SB, Noh Sj, Kim JY, Lee JH. Mesolimbic dopaminergic activity responding to acute stress is blunted in adolescent rats that experienced neonatal maternal separation. Neuroscience 2010;171:144–52.

[449] Jakubowski M, Terkel J. Infanticide and caretaking in nonlactating *Mus musculus*: influence of genotype, family group and sex. Anim Behav 1982;30:1029–35.

[450] Jasinska AJ, Lowry CA, Burmeister M. Serotonin transporter gene, stress and raphe-raphe interactions: a molecular mechanism of depression. Trends Neurosci 2012;35:395–402.

[451] Jenkins WJ, Becker JB. Role of the striatum and nucleus accumbens in paced copulatory behavior in the female rat. Behav Brain Res 2001;121:119–28.

[452] Jenkins WJ, Becker JB. Dynamic increases in dopamine during paced copulation in the female rat. Eur J Neurosci 2003;18:1997–2001.

[453] Jenkins WJ, Becker JB. Female rats develop conditioned place preferences for sex at their preferred interval. Horm Behav 2003;43:503–7.

[454] Jespersen AF, Lalumiere ML, Seto MC. Sexual abuse history aming adult sex offenders and non-sex offenders: a meta-analysis. Child Abuse Negl 2009;33:179–92.

[455] Jezzini A, Caruana F, Stoianov I, Gallese V, Rizzolatti G. Functional organization of the insula and inner perisylvian regions. Proc Natl Acad Sci USA 2012;109:10077–82.

[456] Jhou TC, Fileds HL, Baxter MG, Saper CB, Holland PC. The rostromedial tegmental nucleus (RMTg), a GABAergic afferent to midbrain dopamine neurons, encodes aversive stimuli and inhibits motor responses. Neuron 2009;61:786–800.

[457] Jia R, Tai F, An S, Zhang X. Neonatal paternal deprivation or early deprivation reduces adult parental behavior and central estrogen receptor α expression in mandarin voles (*Microtus mandarinus*). Behav Brain Res 2011;224:279–89.

[458] Johansen JP, Hamanaka H, Monfils MH, Behnia R, Deisseroth K, Blair HT, et al. Optical activation of lateral amygdala pyramidal cells instructs associative fear learning. Proc Natl Acad Sci USA 2010;107:12692–7.

[459] Johansen JP, Wolff SB, Luthi A, LeDoux JE. Controlling the elements: an optogenetic approach to understanding the neural circuits of fear. Biol Psychiatry 2012;71:1053–60.

[460] Johnson AE, Audigier S, Rossi F, Jard S, Tribollet E, Barberis C. Localization and characterization of vasopressin binding sites in the rat brain using an iodinated linear AVP antagonist. Brain Res 1993;622:9–16.

[461] Jones W, Bellugi U, Lai Z, Chiles M, Reilly J, Lincoln A, et al. Hypersociality in Williams syndrome. J Cogn Neurosci 2000;12(Suppl. 1):30–46.

[462] Jordan K, Fromberger P, Stolpmann G, Muller JL. The role of testosterone in sexuality and paraphilia-a neurobiological approach. Part II: testosterone and paraphilia. J Sex Med 2011;8:3008–29.

[463] Jost A, Vigier B, Prepin J, Perchellet JP. Studies on sex differentiation in mammals. Recent Prog Horm Res 1973;29:1–41.

[464] Juarez P, Valdovinos MG, May ME, Lloyd BP, Couppis MH, Kennedy CH. Serotonin2A/C receptors mediate the aggressive phenotype of TLX gene knockout mice. Behav Brain Res 2013;256:354–61.

[465] Kabelik D, Klatt JD, Kingsbury MA, Goodson JL. Endogenous vasotocin exerts context-dependent behavioral effects in a semi-naturalistic colony environment. Horm Behav 2009;56:101–7.

[466] Kaffman A, Meaney MJ. Neurodevelopmental sequelae of postnatal maternal care in rodents: clinical and research implications of molecular insights. J Child Psychol Psychiatry 2007;48:224–44.

[467] Kagerbauer SM, Martin J, Schuster T, Blobner M, Kochs EF, Landgraf R. Plasma oxytocin and vasopressin do not predict neuropeptide concentrations in human cerebrospinal fluid. J Neuroendocrinol 2013;25:668–73.

[468] Kakeyama M, Umino A, Nishikawa T, Yamanouchi K. Decrease of serotonin and metabolite in forebrain and facilitation of lordosis by dorsal raphe nucleus lesions in male rats. Endocr J 2002;49:573–9.

[469] Kakeyama M, Yamanouchi K. Two types of lordosis-inhibiting systems in male rats: dorsal raphe nucleus lesions and septal cuts. Physiol Behav 1994;56: 189–92.

[470] Kalin NH, Shelton SE, Davidson RJ. The role of the central nucleus of the amygdala in mediating fear and anxiety in the primate. J Neurosci 2004;24:5506–15.

[471] Karama S, Lecours AR, Leroux JM, Bourgouin P, Beaudoin G, Joubert S, et al. Areas of brain activation in males and females during viewing of erotic film excerpts. Hum Brain Mapp 2002;16:1–13.

[472] Keebaugh AC, Young LJ. Increasing oxytocin receptor expression in the nucleus accumbens of prepubertal female prairie voles enhances alloparental responsiveness and partner preference formation as adults. Horm Behav 2011;60:498–504.

[473] Keer SE, Stern JM. Dopamine receptor blockade in the nucleus accumbens inhibits maternal retrieval and licking, but enhances nursing behavior in lactating rats. Physiol Behav 1999;67:659–69.

[474] Keller M, Perrin G, Meurisse M, Ferreira G, Levy F. Cortical and medial amygdala are both involved in the formation of olfactory memory in sheep. Eur J Neurosci 2004;20:3433–41.

[475] Kelly AM, Kingsbury MA, Hoffbuhr K, Schrock SE, Waxman B, Kabelik D, et al. Vasotocin neurons and septal V1a-like receptors potently modulate songbird flocking and responses to novelty. Horm Behav 2011;60:12–21.

[476] Kempes MM, Gulickx MMC, van Daalen HJC, Lowerse AL, Sterck EHM. Social competence is reduced in socially deprived rhesus monkeys (*Macaca mulatta*). J Comp Psychol 2008;122:62–7.

[477] Kendrick KM, Dixson AF. Anteromedial hypothalamic lesions block proceptivity but not receptivity in the female common marmoset (*Callithrix jacchus*). Brain Res 1986;375:221–9.

[478] Kendrick KM, Keverne EB, Baldwin BA. Intracerebroventricular oxytocin stimulates maternal behaviour in the sheep. Neuroendocrinology 1987;46: 56–61.

[479] Kendrick KM, Keverne EB, Hinton MR, Goode JA. Oxytocin, amino acid and monoamine release in the region of the medial preoptic area and bed nucleus of the stria terminalis of the sheep during parturition and suckling. Brain Res 1992;569:199–209.

[480] Keyser-Marcus L, Stafisso-Sandoz G, Gerecke K, Jasnow A, Nightingale L, Lambert KG, et al. Alterations of medial preoptic area neurons following pregnancy and pregnancy-like steroidal treatment in the rat. Brain Res Bull 2001;55:737–45.

[481] Kikusui T, Isaka Y, Mori Y. Early weaning deprives mouse pups of maternal care and decreases their maternal behavior in adulthood. Behav Brain Res 2005;162:200–6.

[482] Killcross S, Robbins TW, Everitt BJ. Different types of fear-conditioned behaviour mediated by separate nuclei within amygdala. Nature 1997;388: 377–80.

[483] Kim P, Feldman R, Mayes LC, Eicher V, Thompson N, Leckman JF, et al. Breastfeeding, brain activation to own infant cry, and maternal sensitivity. J Child Psychol Psychiatry 2011;52:907–15.

[484] Kim SH, Hamann S. Neural correlated of positive and negative emotion regulation. J Cogn Neurosci 2007;19:776–98.

[485] Kim P, Leckman JF, Mayes LC, Feldman R, Wang X, Swain JE. The plasticity of human maternal brain: longitudinal changes in brain anatomy during the early postpartum period. Behav Neurosci 2010;124:695–700.

[486] Kim SJ, Young LJ, Gonen D, Veenstra-VanderWeele J, Courchesne R, Courchesne E, et al. Transmission disequilibrium testing of arginine vasopressin receptor 1A (AVPR1A) polymorphisms in autism. Mol Psychiatry 2002;7:503–7.

[487] Kimchi T, Xu J, Dulac C. A functional circuit underlying male sexual behaviour in the female mouse brain. Nature 2007;448:1009–14.

[488] Kimonis ER, Cross B, Howard A, Donoghue K. Maternal care, maltreatment and callous-unemotional traits among urban male juvenile offenders. J Youth Adolesc 2013;42:165–77.

[489] Kinsley CH, Bridges RS. Morphine treatment and reproductive condition alter olfactory preferences for pup and adult male odors in female rats. Dev Psychobiol 1990;23:331–47.

[490] Kirby LG, Freeman-Daniels E, Lemos JC, Nunan JD, Lamy C, Akanwa A, et al. Corticotropin-releasing factor increases GABA synaptic activity and induces inward current in 5-hydroxytryptamine dorsal raphe neurons. J Neurosci 2008;28:12927–37.

[491] Kirkpatrick B, Carter CS, Newman SW, Insel TR. Axon-sparing lesions of the medial nucleus of the amygdala decrease affiliative behaviors in the prairie vole (*Microtus ochrogaster*): behavioral and anatomical specificity. Behav Neurosci 1994;108: 501–13.

[492] Kirsch P, Esslinger C, Chen Q, Mier D, Lis S, Siddhanti S, et al. Oxytocin modulates neural circuitry for social cognition and fear in humans. J Neurosci 2005;25:11489–93.

[493] Kiser D, Steemers B, Branchi I, Homberg JR. The reciprocal interaction between serotonin and social behavioiur. Neurosci Biobehav Rev 2012;36:786–98.

[494] Klasen M, Zvyagintsev M, Schwenzer M, Mathiak KA, Sarkheil P, Weber R, et al. Quetiapine modulates functional connectivity in brain aggression networks. NeuroImage 2013;75:20–6.

[495] Klatt JD, Goodson JL. Oxytocin-like receptors mediate pair bonding in a socially monogamous songbird. Proc Biol Sci 2013;280:20122396.

[496] Klavir O, Genud-Gabai R, Paz R. Low-frequency stimulation depresses the primate anterior-cingulate-cortex and prevents spontaneous recovery of aversive memories. J Neurosci 2012;32:8589–97.

[497] Kleiman DG. Monogamy in mammals. Q Rev Biol 1977;52:39–69.

[498] Kleiman DG, Malcolm J. The evolution of male parental investment in mammals. In: Gubernick DJ, Klopfer PH, editors. Parental care in mammals. New York: Plenum Press; 1981. pp. 347–87.

[499] Klengel T, Mehta D, Anacker C, Rex-Haffner M, Pruessner JC, Pariante CM, et al. Allele-specific FKBP5 DNA demethylation mediates gene-childhood trauma interactions. Nat Neurosci 2013;16: 33–41.

[500] Knafo A, Israel S, Darvasi A, Bachner-Melman R, Uzefovsky F, Cohen L, et al. Individual differences in allocation of funds in the dictator game associated with length of the arginine vasopressin 1a receptor RS3 promoter regions and correlation between RS3 length and hippocampal mRNA. Genes Brain Behav 2008;7:266–75.

[501] Knapska E, Radwanska K, Werka T, Kaczmarek L. Functional internal complexity of amygdala: focus on gene activity mapping after behavioral training and drugs of abuse. Physiol Rev 2007;87:1113–73.

[502] Knobloch HS, Charlet A, Hoffmann LC, Eliava M, Khrulev S, Cetin AH, et al. Evoked axonal oxytocin release in central amygdala attenuates fear response. Neuron 2012;73:553–66.

[503] Kogan A, Saslow LR, Impett EA, Oveis C, Keltner D, Rodrigues Saturn S. Thin-slicing study of oxytocin receptor (OXTR) gene and the evaluation and expression of prosocial disposition. Proc Natl Acad Sci USA 2011;108:19189–92.

[504] Kohlert JG, Rowe RK, Meisel RL. Intromissive stimulation from the male increases extracellular dopamine release from fluoro-gold-identified neurons within the midbrain of female hamsters. Horm Behav 1997;32:143–54.

[505] Kohls G, Schulte-Ruther M, Nehrkorn B, Muller K, Fink GR, Kamp-Becker I, et al. Reward system dysfunction in autism spectrum disorders. Soc Cogn Affect Neurosci 2013;8:565–72.

[506] Kolber BJ, Boyle MP, Wieczorek L, Kelley CL, Orwuzurike CC, Nettles SA, et al. Transient early-life forebrain corticotropin-releasing hormone elevation causes long-lasting anxiogenic and despair-like changes in mice. J Neurosci 2010;30:2571–81.

[507] Kosfeld M, Heinrichs M, Zak PJ, Fischbacher U, Fehr E. Oxytocin increases trust in humans. Nature 2005;435:673–6.

[508] Kow LM, Pfaff DW. The membrane actions of estrogens can potentiate their lordosis behavior-facilitating genomic actions. Proc Natl Acad Sci USA 2004;101:12354–7.

[509] Kravitz AV, Freeze BS, Parker PR, Kay K, Thwin MT, Deisseroth K, et al. Regulation or parkinsonian motor behaviours by optogenetic control of basal ganglia circuitry. Nature 2010;466:622–6.

[510] Kremarik P, Freund-Mercier MJ, Stoeckel ME. Oxytocin and vasopressin binding sites in the hypothalamus of the rat: histoautoradiographic detection. Brain Res Bull 1995;36:195–203.

[511] Kremarik P, Freund-Mercier MJ, Stoeckel ME. Oxytocin and vasopressin binding sites regulated by progesterone in the telencephalon and the hypothalamus of the rat. J Neuroendocrinol 1995;7:281–9.

[512] Kringelbach ML, Lehtonen A, Squire S, Harvey AG, Craske MG, Holliday IE, et al. A specific and rapid neural signature for parental instinct. PLoS One 2008;3:e1664.

[513] Krueger F, McCabe K, Moll J, Kriegekorte N, Zahn R, Strenziok M, et al. Neural correlates of trust. Proc Natl Acad Sci USA 2007;104:20084–9.

[514] Krueger F, Parasuraman R, Iyengar V, Thirnburg M, Weel J, Lin M, et al. Oxytocin receptor genetic variation promotes human trust behavior. Front Hum Neurosci 2012;6:4.

[515] Kruk MR. Ethology and pharmacology of hypothalamic aggression in the rat. Neurosci Biobehav Rev 1991;15:527–38.

[516] Kruk MR, Halasz J, Meelis W, Haller J. Fast positive feedback between adrenocortical stress response and a brain mechanism involved in aggressive behavior. Behav Neurosci 2004;118:1062–70.

[517] Kudwa AE, Bodo C, Gustafsson JA, Rissman EF. A previously uncharacterized role for estrogen receptor beta: defeminization of male brain and behavior. Proc Natl Acad Sci USA 2005;102:4602–12.

[518] Kuepper Y, Grant P, Wielpuetz C, Hennig J. MAOA-uVNTR genotype predicts interindividual differences in experimental aggressiveness as a function of the degree of provocation. Behav Brain Res 2013;247:73–8.

[519] Kuersten S, Goodwin EB. The power of the 3′ UTR: translational control and development. Nat Rev Genet 2003;4:626–37.

[520] Kulikov AV, Osipova DV, Naumenko VS, Terenina E, Mormede P, Popova NK. A pharmacological evidence of positive association between mouse intermale aggression and brain serotonin metabolism. Behav Brain Res 2012;233:113–9.

[521] Kuroda KO, Meaney MJ, Uetani N, Kato T. Neurobehavioral basis of the impaired nurturing in mice lacking the immediate early gene FosB. Brain Res 2008;1211:57–71.

[522] Kuroda KO, Tachikawa K, Yoshida S, Tsuneoka Y, Numan M. Neuromolecular basis of parental behavior in laboratory mice and rats: with special emphasis on technical issues of using mouse genetics. Prog Neuropsychopharmacol Biol Psychiatry 2011;35:1205–31.

[523] Kurth F, Narr KL, Woods RP, O'Neill J, Alger JR, Caplan R, et al. Dimished gray matter within the hypothalamus in autism disorder: a potential link to hormonal effects. Biol Psychiatry 2011;70:278–82.

[524] Kwan M, Greenleaf WJ, Mann J, Crapo L, Davidson JM. The nature of androgen action on male sexuality: a combined laboratory-self-report study on hypogonadal men. J Clin Endocrinol Metab 1983;57:557–62.

[525] Lahey BB, Michalska KJ, Liu C, Chen Q, Hipwell AE, Chronis-Tuscano A, et al. Preliminary genetic imaging study of the association between estrogen receptor-α gene polymorphisms and harsh human maternal parenting. Neurosci Lett 2012;525:17–22.

[526] Lammel S, Lim BK, Ran C, Huang KW, Betley MJ, Tye KM, et al. Input-specific control of reward and aversion in the ventral tegmental area. Nature 2012;491:212–7.

[527] Landgraf R, Neumann ID. Vasopressin and oxytocin release within the brain: a dynamic concept of multiple and variable modes of neuropeptide communication. Front Neuroendocrinol 2004;25:150–76.

[528] Laurent HK, Ablow JC. A cry in the dark: depressed mothers show reduced neural activation to their own infant's cry. Soc Cogn Affect Neurosci 2012;7:125–34.

[529] Laurent HK, Ablow JC. A face a mother could love: depression-related maternal neural responses to infant emotional faces. Soc Neurosci 2013;8:228–39.

[530] LeDoux JE. Emotion circuits in the brain. Annu Rev Neurosci 2000;23:155–84.

[531] Lee AW, Brown RE. Medial preoptic lesions disrupt parental behavior in both male and female California mice (*Peromyscus californicus*). Behav Neurosci 2002;116:968–75.

[532] Lee AW, Brown RE. Comparison of medial preoptic, amygdala, and nucleus accumbens lesions on parental behavior in California mice (*Peromyscus californicus*). Physiol Behav 2007;92:617–28.

[533] Lee TM, Chan SC, Raine A. Strong limbic and weak frontal activation to aggressive stimuli in spouse abusers. Mol Psychiatry 2008;13:655–6.

[534] Lee SS, Chronis-Tuscano A, Keenan K, Pelham WE, Loney J, Van Hulle CA, et al. Association of maternal dopamine transporter genotype with negative parenting: evidence for a gene x environment interaction with child disruptive behavior. Mol Psychiatry 2010;15:548–58.

[535] Lee A, Clancy S, Feming AS. Mother rats bar-press for pups: effects of lesions of the mpoa and limbic sites on maternal behavior and operant responding for pup-reinforcement. Behav Brain Res 2000;108: 215–31.

[536] Lee JH, Durand R, Gradinaru V, Zhang F, Goshen I, Kim DS, et al. Global and local fMRI signals driven by neurons defined optogenetically by type and wiring. Nature 2010;465:788–92.

[537] Lee R, Ferris C, Van de Kar LD, Coccaro EF. Cerebrospinal fluid oxytocin, life history of aggression, and personality disorder. Psychoneuroendocrino 2009;34:1567–73.

[538] Lee RJ, Gollan J, Kasckow J, Geracioti T, Coccaro EF. CSF corticotropin-releasing factor in personality disorder: relationship with self-reported parental care. Neuropsychopharmacol 2006;31:2289–95.

[539] Lee BT, Ham BJ. Monoamine oxidase A-uVNTR genotype affects limbic brain activity in response to affective facial stimuli. Neuroreport 2008;19:515–9.

[540] Lee TL, Raygada MJ, Rennert OM. Integrative gene network analysis provides novel regulatory relationships, genetic contributions and susceptible targets in autism spectrum disorders. Gene 2012;496:88–96.

[541] Lenz KM, Nugent BM, McCarthy MM. Sexual differentiation of the rodent brain: dogma and beyond. Front Neurosci 2012;6:26.

[542] Lerch-Haner JK, Frierson D, Crawford LK, Beck SG, Deneris ES. Serotonergic transcriptional programming determines maternal behavior and offspring survival. Nat Neurosci 2008;11:1001–3.

[543] Lesch KP, Araragi N, Waider J, van den Hove D, Gutknecht L. Targeting brain serotonin synthesis: insights into neurodevelopmental disorders with long-term outcomes related to negative emotionality, aggression and antisocial behaviour. Philos Trans R Soc Lond B Biol Sci 2012;367:2426–63.

[544] Lesch KP, Waider J. Serotonin in the modulation of neural plasticity and networks: implications for neurodevelopmental disorders. Neuron 2012;76:175–91.

[545] Letzkus JJ, Wolff SB, Meyer EM, Tovote P, Courtin J, Herry C, et al. A disinhibitory microcircuit for associative fear learning in the auditory cortex. Nature 2011;480:331–5.

[546] LeVay S. A difference in hypothalamic structure between heterosexual and homosexual men. Science 1991;253:1034–7.

[547] Levita L, Hoskin R, Champi S. Avoidance of harm and anxiety: a role for the nucleus accumbens. NeuroImage 2012;62:189–98.

[548] Levy F. Neural substrates involved in the onset of maternal responsiveness and selectivity in sheep. In: Bridges RS, editor. Neurobiology of the parental brain. San Diego: Academic Press; 2008. pp. 23–37.

[549] Levy F, Keller M. Olfactory mediation of maternal behavior in selected mammalian species. Behav Brain Res 2009;200:336–45.

[550] Lewin B. Genes VI. New York: Oxford University Press; 1997.

[551] Li Q, Luo T, Jiang X, Wang J. Anxiolytic effects of 5-HT1A receptors and anxiogenic effects of 5-HT2C receptors in the amygdala of mice. Neuropharmacology 2012;62:474–84.

[552] Lim MM, Hammock EA, Young LJ. The role of vasopressin in the genetic and neural regulation of monogamy. J Neuroendocrinol 2004;16:325–32.

[553] Lim ET, Raychaudhuri S, Sanders SS, Stevens C, Sabo A, MacArthur DG, et al. Rare complete knockouts in humans: population distribution and significant role in autism spectrum disorders. Neuron 2013;77: 235–42.

[554] Lim MM, Wang Z, Olazabal DE, Ren X, Terwilliger EF, Young LJ. Enhanced partner preference in a promiscuous species by manipulating the expression of a single gene. Nature 2004;429:754–7.

[555] Lim MM, Young LJ. Vasopressin-dependent neural circuits underlying pair bond formation in the monogamous prairie vole. Neuroscience 2004;125: 35–45.

[556] Lin D, Boyle MP, Dollar P, Lee H, Lein ES, Perona P, et al. Functional identification of an aggression locus in the mouse hypothalamus. Nature 2011;470:221–6.

[557] Liu D, Diorio J, Tannenbaum B, Caldji C, Francis D, Freedman A, et al. Maternal care, hippocampal glucocorticoid receptors, and hypothalamic-pituitary-adrenal responses to stress. Science 1997;277:1659–62.

[558] Liu JC, Guastella AJ, Dadds MR. Exploring the role of intra-nasal oxytocin on the partner preference effect in humans. Psychoneuroendocrino 2013;38: 587–91.

[559] Liu C, Maejima T, Wyler SC, Casadesus G, Herlitze S, Deneris ES. Pet-1 is required across different stages of life to regulate serotonergic function. Nat Neurosci 2010;13:1190–8.

[560] Liu Y, Wang ZX. Nucleus accumbens oxytocin and dopamine interact to regulate pair bond formation in female prairie voles. Neuroscience 2003;121:537–44.

[561] Lloyd SA, Dixson AF. Effects of hypothalamic lesions upon the sexual and social behaviour of the male common marmoset (*Callithrix jacchus*). Brain Res 1988;463:317–29.

[562] Lobo MK, Nestler EJ. The striatal balancing act in drug addiction: distinct roles of direct and indirect pathway medium spiny neurons. Front Neuroanat 2011;5:41.

[563] Lodge DJ. The medial prefrontal and orbitofrontal cortices differentially regulate dopamine system function. Neuropsychopharmacol 2011;36:1227–36.

[564] Lonstein JS. Regulation of anxiety during the postpartum period. Front Neuroendocrinol 2007;28:115–41.

[565] Lonstein JS, De Vries GJ. Maternal behaviour in lactating rats stimulates c-fos in glutamate decarboxylase-synthesizing neurons of the medial preoptic area, ventral bed nucleus of the stria terminalis, and ventrocaudal periaqueductal gray. Neuroscience 2000;100:557–68.

[566] Lonstein JS, Stern JM. Somatosensory contributions to c-fos activation within the caudal periaqueductal gray of lactating rats: effects of perioral, rooting, and suckling stimuli from pups. Horm Behav 1997;32:155–66.

[567] Lonstein JS, Stern JM. Role of the midbrain periaqueductal gray in maternal nurturance and aggression: c-fos and electrolytic lesion studies in lactating rats. J Neurosci 1997;17:3364–78.

[568] Lopez HH, Ettenberg A. Haloperidol challenge during copulation prevents subsequent increase in male sexual motivation. Pharmacol Biochem Behav 2000;67:387–93.

[569] Lopez HH, Ettenberg A. Dopamine antagonism attenuates the unconditioned incentive value of estrous female cues. Pharmacol Biochem Behav 2001;68: 411–6.

[570] Lopez HH, Ettenberg A. Exposure to female rats produces differences in c-fos induction between sexually-naïve and experienced male rats. Brain Res 2002;947:57–66.

[571] Loup F, Tribollet E, Dubois-Dauphin M, Dreifuss JJ. Localization of high-affinity binding sites for oxytocin and vasopressin in the human brain. An autoradiographic study. Brain Res 1991;555:220–32.

[572] Loup F, Tribollet E, Dubois-Dauphin M, Pizzolato G, Dreifuss JJ. Localization of oxytocin binding sites in the human brainstem and upper spinal cord: an autoradiographic study. Brain Res 1989;500:223–30.

[573] Lovic V, Gonzalez A, Fleming AS. Maternally separated rats show deficits in maternal care in adulthood. Dev Psychobiol 2001;39:19–33.

[574] Lowry CA, Hale MW. Serotonin and the neurobiology of anxious states. In: Muller C, Jacobs B, editors. Handbook of behavioral neurobiology of serotonin. Amsterdam: Elsevier; 2010. pp. 379–97.

[575] Lu XY, Ghasemzadeh MB, Kalivas PW. Expression of D1 receptor, D2 receptor, substance P and enkaphalin messenger RNAs in the neurons projecting from the nucleus accumbens. Neuroscience 1998;82:767–80.

[576] Lubin M, Leon M, Moltz H, Numan M. Hormones and maternal behavior in the male rat. Horm Behav 1972;3:369–74.

[577] Lucas BK, Ormandy CJ, Binart N, Bridges RS, Kelly PA. Null mutation of the prolactin receptor gene produces a defect in maternal behavior. Endocrinology 1998;139:4102–7.

[578] Lucht MJ, Barnow S, Sonnenfeld C, Rosenberger A, Grabe HJ, Schroeder W, et al. Associations between the oxytocin receptor gene (OXTR) and affect, loneliness and intelligence in normal subjects. Prog Neuropsychopharmacol Biol Psychiatry 2009;33:860–6.

[579] Lukas M, Toth I, Reber SO, Slattery DA, Veenema AH, Neumann ID. The neuropeptide oxytocin facilitates pro-social behavior and prevents social avoidance in rats and mice. Neuropsychopharmacol 2011;36:2159–68.

[580] Lukas M, Toth I, Veenema AH, Neumann ID. Oxytocin mediates rodent social memory within the lateral septum and the medial amygdala depending on the relevance of the social stimulus: male juvenile versus female adult conspecifics. Psychoneuroendocrino 2013;38:916–26.

[581] Lukkes JL, Forster GL, Renner KJ, Summers CH. Corticotropin-releasing factor 1 and 2 receptors in the dorsal raphe differentially affect serotonin release in the nucleus accumbens. Eur J Pharmacol 2008;578:185–93.

[582] Ly M, Motzkin JC, Philippi CL, Kirk GR, Newman JP, Kiehl KA, et al. Cortical thinning in psychopathy. Am J Psychiatry 2012;169:743–9.

[583] Machado CJ, Bachevalier J. The impact of selective amygdala, orbital frontal cortex, or hippocampal formation lesions on established social relationships in rhesus monkeys (*Macaca mulatta*). Behav Neurosci 2006;120:761–86.

[584] Maciag D, Simpson KL, Coppinger D, Lu Y, Wang Y, Lin RC, et al. Neonatal antidepressant exposure has lasting effects on behavior and serotonin circuitry. Neuropsychopharmacol 2006;31:47–57.

[585] Macke JP, Hu N, Hu S, Bailey M, King VL, Brown T, et al. Sequence variation in the androgen receptor gene is not a common determinant of male sexual orientation. Am J Hum Genet 1993;53:844–52.

[586] Macoveanu J, Rowe JB, Hornboll B, Elliott R, Paulson OB, Knudsen G, et al. Serotonin 2A receptors contribute to the regulation of risk-averse decisions. NeuroImage 2013;83:35–44.

[587] Maestripieri D. Early experience affects the intergenerational transmission of infant abuse in rhesus monkeys. Proc Natl Acad Sci USA 2005;102:9726–9.

[588] Maestripieri D, Higley JD, Lindell SG, Newman TK, McCormack KM, Sanchez MM. Early maternal rejection affects the development of monoaminergic systems and adult abusive parenting in rhesus macaques (*Macaca mulatta*). Behav Neurosci 2006;120:1017–24.

[589] Maestripieri D, Hoffman CL, Anderson GM, Carter CS, Higley JD. Mother-infant interactions in free-ranging rhesus macaques: relationships between physiological and behavioral variables. Physiol Behav 2009;96:613–9.

[590] Maestripieri D, Lindell SG, Ayala A, Gold PW, Higley JD. Neurobiological characteristics of rhesus macaque abusive mothers and their relation to social and maternal behavior. Neurosci Biobehav Rev 2005;29:51–7.

[591] Maestripieri D, Lindell SG, Higley JD. Intergenerational transmission of maternal behavior in rhesus macaques and its underlying mechanisms. Dev Psychobiol 2007;49:165–71.

[592] Mah BL, Van Ijzendoorn MH, Smith R, Bakersman-Kranenburg MJ. Oxytocin in postnatally depressed mothers: its influence on mood and expressed emotion. Prog Neuropsychopharmacol Biol Psychiatry 2013;40:267–72.

[593] Malik AI, Zai CC, Abu Z, Nowrouzi B, Beitchman JH. The role of oxytocin and oxytocin receptor gene variants in childhood-onset aggression. Genes Brain Behav 2012;11:545–51.

[594] Mani SK. Signaling mechanisms in progesterone-neurotransmitter interactions. Neuroscience 2006; 138:773–81.

[595] Manning M, Stoev S, Chini B, Durroux T, Mouillac B, Guillon G. Peptide and non-peptide agonists and antagonists for vasopressin and oxytocin V1a, V1b, V2 and OT receptors: research tools and potential therapeutic agents. Prog Brain Res 2008;170:473–512.

[596] Manoli DS, Fan P, Fraser EJ, Shah NM. Neural control of sexually dimorphic behaviors. Curr Opin Neurobiol 2013;23:330–8.

[597] Manuck SB, Kaplan JR, Lotrich FE. Brain serotonin and aggressive disposition in human and nonhuman primates. In: Nelson RJ, editor. Biology of aggression. New York: Oxford University Press; 2006. pp. 65–113.

[598] Maras PM, Petrulis A. Chemosensory and steroid-responsive regions of the medial amygdala regulate distinct aspects of opposite-sex odor preference in male Syrian hamsters. Eur J Neurosci 2006;24:3541–52.

[599] Marby KE, Streatfield CA, Keane B, Solomon NG. avpr1a length polymorphism is not associated with either social or genetic monogamy in free-living prairie voles. Anim Behav 2011;81:11–8.

[600] Marsh AA. What can we learn about emotion by studying psychopathy? Front Hum Neurosci 2013;7:181.

[601] Marsh AA, Blair RJ. Deficits in facial affect recognition among antisocial populations: a meta-analysis. Neurosci Biobehav Rev 2008;32:454–65.

[602] Marsh AA, Finger EC, Mitchell DG, Reid ME, Sims C, Kosson DS, et al. Reduced amygdala response to fearful expressions in children and adolescents with callous-unemotional traits and disruptive behavior disorders. Am J Psychiatry 2008;165:712–20.

[603] Marsh AA, Finger EC, Schechter JC, Jurkowitz IT, Reid ME, Blair RJ. Adolescents with psychopathic traits report reductions in physiological responses to fear. J Child Psychol Psychiatry 2011;52:834–41.

[604] Marsh AA, Yu HH, Pines DS, Gorodetsky EK, Goldman D, Blair RJ. The influence of oxytocin administration on responses to infant faces and potential moderation by OXTR genotype. Psychopharmacology 2012;224:469–76.

[605] Marson L. Lesions of the periaqueductal gray block the medial preoptic area-induced activation of the urethrogenital reflex in male rats. Neurosci Lett 2004;367:278–82.

[606] Marson L, McKenna KE. The identification of a brainstem site controlling spinal sexual reflexes in male rats. Brain Res 1990;515:303–8.

[607] Marson L, McKenna KE. Stimulation of the hypothalamus initiates the urethrogenital reflex in male rats. Brain Res 1994;638:103–8.

[608] Martin JH. Neuroanatomy. New York: Elsevier; 1989.

[609] Martin MM, Liu Y, Wang Z. Developmental exposure to a serotonin agonist produces subsequent behavioral and neurochemical changes in the adult male prairie vole. Physiol Behav 2012;105:529–35.

[610] Masten CL, Morelli SA, Eisenberger NI. An fMRI investigation of empathy for 'social pain' and subsequent prosocial behavior. NeuroImage 2011;55:381–8.

[611] Mathur VA, Harada T, Lipke T, Chiao JY. Neural basis of extraordinary empathy and altruistic motivation. NeuroImage 2010;51:1468–75.

[612] Matsumoto M, Hikosaka O. Two types of dopamine neuron distinctly convey positive and negative motivational signals. Nature 2009;459:837–41.

[613] Matthews HR, Freedland R, Miesfeld RL. Biochemistry: a short course. New York: Wiley-Liss; 1997.

[614] Matthews K, Robbins TW. Early experience as a determinant of adult behavioural responses to reward: the effects of repeated maternal separation in the rat. Neurosci Biobehav Rev 2003;27:45–55.

[615] Mattson BJ, Morrell JI. Preference for cocaine- versus pup-associated cues differentially activates neurons expressing either Fos or cocaine- and amphetamine-regulated transcript in lactating, maternal rodents. Neuroscience 2005;135:315–28.

[616] McCabe MJ, Alatzoglou KS, Dattani MT. Septo-optic dysplasia and other midline defects: the role of transcription factors: HESX1 and beyond. Best Pract Res Clin Endocrinol Metab 2011;25:115–24.

[617] McCarthy MM. Sexual differentiation of brain and behavior. In: Fink G, Pfaff DW, Levine JE, editors. Handbook of neuroendocrinology. Amsterdam: Elsevier; 2012. pp. 393–413.

[618] McCarthy MM, Kleopoulos SP, Mobbs CV, Pfaff DW. Infusion of antisense oligodeoxynucleotides to the oxytocin receptor in the ventromedial hypothalamus reduces estrogen-induced sexual receptivity and oxytocin receptor binding in the female rat. Neuroendocrinology 1994;59:432–40.

[619] McCormack K, Newman TK, Higley JD, Maestripieri D, Sanchez MM. Serotonin transporter gene variation, infant abuse, and responsiveness to stress in rhesus macaque mothers and infants. Horm Behav 2009;55:538–47.

[620] McCullough LD, Sokolowski JD, Salamone JD. A neurochemical and behavioral investigation of the involvement of nucleus accumbens dopamine in instrumental avoidance. Neuroscience 1993;52:919–25.

[621] McDonald AJ. Cortical pathways to the mammalian amygdala. Prog Neurobiol 1998;55:257–332.

[622] McDonald AJ, Mascagni F, Guo L. Projections of the medial and lateral prefrontal cortices to the amygdala: a *Phaseolus vulgaris* leucoagglutinin study in the rat. Neuroscience 1996;71:55–75.

[623] McGinnis MY, Montana RC, Lumia AR. Effects of hydroxyflutaminde in the medial preoptic area or lateral septum on reproductive behaviors in male rats. Brain Res Bull 2002;59:227–34.

[624] McGowan PO, Sasaki A, D'Alessio AC, Dymov S, Labonte B, Szyf M, et al. Epigenetic regulation of the glucocorticoid receptor in human brain associates with childhood abuse. Nat Neurosci 2009;12:342–8.

[625] McGraw LA, Young LJ. The prairie vole: an emerging model organism for understanding the social brain. Trends Neurosci 2010;33:103–9.

[626] McGregor A, Herbert J. Differential effects of excitotoxic basolateral and corticomedial lesions of the amygdala on the behavioural and endocrine responses to either sexual or aggression-promoting stimuli in the male rat. Brain Res 1992;574:9–20.

[627] Meddle SL, Bishop VP, Gkoumassi E, van Leeuwen FW, Douglas AJ. Dynamic changes in oxytocin receptor expression and activation at parturition in the rat brain. Endocrinology 2007;148:5095–104.

[628] Meerts SH, Clark AS. Lesions of the medial preoptic area interfere with the display of a conditioned place preference for vaginocervical stimulation in rats. Behav Neurosci 2009;123:752–7.

[629] Meerts SH, Clark AS. Artificial vaginocervical stimulation induces a conditioned place preference in female rats. Horm Behav 2009;55:128–32.

[630] Meisel RL, Joppa MA, Rowe RK. Dopamine receptor antagonists attenuate conditioned place preference following sexual behavior in female Syrian hamsters. Eur J Pharmacol 1996;309:21–4.

[631] Meisel RL, Mullins AJ. Sexual experience in female rodents: cellular mechanisms and functional consequences. Brain Res 2006;1126:56–65.

[632] Mejia JM, Ervin FR, Baker GB, Palmour RM. Monoamine oxidase inhibition during brain development induces pathological aggressive behavior in mice. Biol Psychiatry 2002;52:811–21.

[633] Melis MR, Melis T, Cocco C, Succu S, Sanna F, Pillolla G, et al. Oxytocin injected into the ventral tegmental area induces penile erection and increases extracellular dopamine in the nucleus accumbens and paraventricular nucleus of the hypothalamus of male rats. Eur J Neurosci 2007;26:1026–35.

[634] Menke A, Klengel T, Rubel J, Bruckl T, Pfister H, Lucae S, et al. Genetic variation in FKBP5 associated with the extent of stress hormone dysregulation in major depression. Genes Brain Behav 2013;12:289–96.

[635] Meyer ML, Masten CL, Ma Y, Wang C, Shi Z, Eisenberger NI, et al. Empathy for social suffering of friends and strangers recruits distinct patterns of brain activation. Soc Cogn Affect Neurosci 2013;8:446–54.

[636] Meyer-Lindenberg A, Buckholz JW, Kolachana B, Hariri A, Pezawas L, Blasi G, et al. Neural mechanisns of genetic risk for impulsivity and violence in humans. Proc Natl Acad Sci USA 2006;103: 6269–74.

[637] Meyer-Lindenberg A, Domes G, Kirsch P, Heinrichs M. Oxytocin and vasopressin in the human brain: social neuropeptides for translational medicine. Nat Rev Neurosci 2011;12:524–38.

[638] Meyer-Lindenberg A, Hariri AR, Munoz KE, Mervis CB, Mattay VS, morris CA, et al. Neural correlates of genetically abnormal social cognition in Williams syndrome. Nat Neurosci 2005;8:991–3.

[639] Meyer-Lindenberg A, Kolachana B, Gold B, Olsh A, Nicodemus KK, Mattay V, et al. Genetic variants in AVPR1A linked to autism predict amygdala activation and personality traits in healthy humans. Mol Psychiatry 2009;14:968–75.

[640] Micevych P, Dominguez R. Membrane estradiol signaling in the brain. Front Neuroendocrinol 2009;30:315–27.

[641] Mickey BJ, Ducci F, Hodgkinson CA, Langenecker SA, Goldman D, Zubieta JK. Monoamine oxidase A genotype predicts human serotonin 1A receptor availability in vivo. J Neurosci 2008;28: 11354–9.
[642] Miczek KA, Fish EW. Monoamines, GABA, glutamate, and aggression. In: Nelson RJ, editor. Biology of aggression. New York: Oxford University Press; 2006. pp. 114–49.
[643] Mikhail N. Does testosterone have a role in erectile function? Am J Med 2006;119:373–82.
[644] Mikics E, Kruk MR, Haller J. Genomic and non-genomic effects of glucocorticoids on aggressive behavior in male rats. Psychoneuroendocrino 2004;29:618–35.
[645] Milad MR, Quirk GJ. Fear extinction as a model for translational neuroscience: ten years of progress. Annu Rev Psychol 2012;63:129–51.
[646] Mileva-Seitz V, Fleming AS, Meaney MJ, Mastroianni A, Sinnwell JP, Steiner M, et al. Dopamine receptors D1 and D2 are related to observed maternal behavior. Genes Brain Behav 2012;11:684–94.
[647] Mill J, Asherson P, Browes C, D'Souza U, Craig I. Expression of the dopamine transporter gene is regulated by the 3′ UTR VNTR: evidence from brain and lymphocytes using quantitative RT-PCR. Am J Med Genet 2002;114:975–9.
[648] Millan MJ, Marin P, Bockaert J, Mannoury LCC. Signaling at G-protein-coupled serotonin receptors: recent advances and future research directions. Trends Pharmacol Sci 2008;29:454–64.
[649] Miller MA, Ferris CF, Kolb PE. Absence of vasopressin expression by galanin neurons in the golden hamster: implications for species differences in extrahypothalamic vasopressin pathways. Mol Brain Res 1999;67:28–35.
[650] Millhouse OE. The organization of the ventromedial hypothalamic nucleus. Brain Res 1973;55:71–87.
[651] Mitchell JP, Macrae CN, Banaji MR. Dissociable medial prefrontal contributions to judgments of similar and dissimilar others. Neuron 2006;50:655–63.
[652] Mizumoto Y, Kimura T, Ivell R. A genomic element within the third intron of the human oxytocin receptor gene may be involved in transcriptional suppression. Mol Cell Endocrinol 1997;135:129–38.
[653] Mizutani H, Hori T, Takahasi T. 5-HT1B receptor-mediated presynaptic inhibition at the calyx of Held of immature rats. Eur J Neurosci 2006;24:1946–54.
[654] Mobbs D, Petrovic P, Marchant JL, Hassabis D, Weiskopf N, Seymour B, et al. When fear is near: threat imminence elicits prefrontal-periaqueductal gray shifts in humans. Science 2007;317:1079–83.
[655] Mogenson GJ. Limbic-motor integration. Prog Psychobiol Physiol Psychol 1987;12:117–70.
[656] Moll J, Bado P, de Oliveira-Souza R, Bramati IE, Lima DO, Paiva FF, et al. A neural signature of affiliative emotion in the human septohypothalamic area. J Neurosci 2012;32:12499–505.
[657] Moll J, Krueger F, Zahn R, Pardini M, de Oliveira-Souza R, Grafman J. Human fronto-mesolimbic networks guide decisions about charitable donation. Proc Natl Acad Sci USA 2006;103:15623–8.
[658] Moll J, Zahn R, de Oliveira-Souza R, Bramati IE, Krueger F, Tura B, et al. Impairment of prosocial sentiments is associated with frontopolar and septal damage in frontotemporal dementia. NeuroImage 2011;54:1735–42.
[659] Moltz H, Lubin M, Leon M, Numan M. Hormonal induction of maternal behavior in the ovariectomized nulliparous rat. Physiol Behav 1970;5:1373–7.
[660] Money J, Ehrhardt AA. Man and women, boy and girl. Baltimore: Johns Hopkins University Press; 1972.
[661] Monti JM. The structure of the dorsal raphe nucleus and its relevance to the regulation of sleep and wakefulness. Sleep Med Rev 2010;14:307–17.
[662] Moore LD, Le T, Fan G. DNA methylation and its basic function. Neuropsychopharmacol 2013;38: 23–38.
[663] Moreines J, Kelton M, Luine VN, Pfaff DW, McEwen BS. Hypothalamic serotonin lesions unmask hormone responsiveness of lordosis behavior in adult male rats. Neuroendocrinology 1988;47:453–8.
[664] Morelli SA, Rameson LT, Lieberman MD. The neural components of empathy: predicting daily prosocial behavior. Soc Cogn Affect Neurosci 2014;9:39–47.
[665] Morgan JI, Curran T. Stimulus-transcription coupling in the nervous system: involvement of the inducible proto-oncogenes fos and jun. Annu Rev Neurosci 1991;14:421–51.
[666] Morris JA, Jordan CL, Breedlove SM. Sexual differentiation of the vertebrate nervous system. Nat Neurosci 2004;7:1034–9.
[667] Moscarello JM, LeDoux JE. Active avoidance learning requires prefrontal suppression of amygdala-mediated defensive reactions. J Neurosci 2013;33: 3815–23.
[668] Moses J, Loucks JA, Watson HL, Matuszewich L, Hull EM. Dopaminergic drugs in the medial preoptic area and nucleus accumbens: effects on motor activity, sexual motivation, and sexual performance. Pharmacol Biochem Behav 1995;51:681–6.
[669] Motzkin JC, Newman JP, Kiehl KA, Koenigs M. Reduced prefrontal connectivity in psychopathy. J Neurosci 2011;31:17348–57.
[670] Moyer KE. The psychobiology of aggression. New York: Harper & Row; 1976.

[671] Munafo MR, Brown SM, Hariri AR. Serotonin transporter (5-HTTLPR) genotype and amygdala acitivation: a meta-analysis. Biol Psychiatry 2008;63:852–7.

[672] Munoz KE, Meyer-Lindenberg A, Hariri AR, Mervis VS, Mattay VS, Morris CA, et al. Abnormalities in neural processing of emotional stimuli in Williams syndrome vary according to social vs. non-social content. NeuroImage 2010;50:340–6.

[673] Murphy AZ, Hoffman GE. Distribution of gonadal steroid receptor-containing neurons in the preoptic-periaqueductal gray-brainstem pathway: a potential circuit for the initiation of male sexual behavior. J Comp Neurol 2001;438:191–212.

[674] Nair HP, Young LJ. Application of adeno-associated viral vectors in behavioral research. Methods 2002;28:195–202.

[675] Nair HP, Young LJ. Vasopressin and pair-bond formation: genes to brain to behavior. Physiology 2006;21:146–52.

[676] Naqvi NH, Bechara A. The hidden island of addiction: the insula. Trends Neurosci 2009;32:56–67.

[677] Nelson RJ. An introduction to behavioral endocrinology. Suderland (MA): Sinauer; 2011.

[678] Nelson RJ, Trainor BC. Neural mechanisms of aggression. Nat Rev Neurosci 2007;8:536–46.

[679] Nephew BC, Bridges RS. Arginine vasopressin V1a receptor antagonist impairs maternal memory in rats. Physiol Behav 2008;95:182–6.

[680] Neuhaus E, Beauchaine TP, Bernier R. Neurobiological correlates of social functioning in autism. Clin Psychol Rev 2010;30:733–48.

[681] New AS, Buchsbaum MS, Hazlett EA, Goodman M, Koenigsberg HW, Lo J, et al. Fluoxetine increases relative metabolic rate in prefrontal cortex in impulsive aggression. Psychopharmacology 2004;176:451–8.

[682] New AS, Hazlett EA, Buchsbaum MS, Goodman M, Mitelman SA, Newmark R, et al. Amygdala-prefrontal disconnection in borderline personality disorder. Neuropsychopharmacol 2007;32:1629–40.

[683] Newman SW. The medial extended amygdala in male reproductive behavior. A node in the mammalian social behavior network. Ann NY Acad Sci 1999;877:242–57.

[684] Newman KS, Halpin ZT. Individual odours and mate recognition in the prairie vole. Anim Behav 1988;36:1779–87.

[685] Nieuwenhuys R. The insular cortex: a review. Prog Brain Res 2012;195:123–63.

[686] Ninan I. Oxytocin suppresses basal glutamatergic transmission but facilitates activity-dependent synaptic potentiation in the medial prefrontal cortex. J Neurochem 2011;119:324–31.

[687] Nishimori K, Takayanagi Y, Yoshida M, Kasahara Y, Yoiung LJ, Kawamata M. New aspects of oxytocin receptor function revealed by knockout mice: socio-sexual behaviour and control of energy balance. Prog Brain Res 2008;170:79–90.

[688] Nishitani S, Doi H, Koyama A, Shinohara K. Differential prefrontal response to infant facial emotions in mothers compared with non-mothers. Neurosci Res 2011;70:183–8.

[689] Nitschke JB, Nelson EE, Rusch BD, Fox AS, oakes TR, Davidson RJ. Orbitofrontal cortex tracks positive mood in mothers viewing pictures of their newborn infants. NeuroImage 2004;21:583–92.

[689A] Noriuchi M, Kikuchi Y, Senoo A. The functional neuroanatomy of maternal love: mother's response to infant's attachment behaviors. Biol Psychiatry 2008;63:415–23.

[690] Normandin JJ, Murphy AZ. Nucleus paragigantocellularis afferents in male and female rats: organization, gonadal steroid receptor expression, and activation during sexual behavior. J Comp Neurol 2008;508:771–94.

[691] Normandin JJ, Murphy AZ. Excitotoxic lesions of the nucleus paragigantocellularis facilitate male sexual behavior but attenuate female sexual behavior in rats. Neuroscience 2011;175:212–23.

[692] Nowak R, Keller M, Levy F. Mother-young relationships in sheep: a model for a multidisciplinary approach of the study of attachment in mammals. J Neuroendocrinol 2011;23:1042–53.

[693] Nugent BM, Schwarz JM, McCarthy MM. Hormonally mediated epigenetic changes to steroid receptors in the developing brain: implications for sexual differentiation. Horm Behav 2011;59:338–44.

[694] Numan M. Brain mechanisms and parental behavior. In: Adler N, Pfaff D, Goy RW, editors. Handbook of behavioral neurobiology. Reproduction, vol. 7. New York: Plenum Press; 1985. pp. 537–605.

[695] Numan M. Maternal behavior. In: Knobil E, Neill JD, editors. The physiology of reproduction, vol. 2. New York: Raven Press; 1994. pp. 221–302.

[696] Numan M. Hypothalamic neural circuits regulating maternal responsiveness toward infants. Behav Cogn Neurosci Rev 2006;5:163–90.

[697] Numan M. Parental behavior. In: Koob G, Le Moal M, Thompson RF, editors. Encyclopedia of behavioral neuroscience, vol. 3. Oxford (UK): Academic Press; 2010. pp. 14–23.

[698] Numan M. Neural circuits regulating maternal behavior: implications for understanding the neural basis of social cooperation and competition. In: Brown SL, Brown M, Penner LA, editors. Moving beyond self-interest: perspectives from evolutionary biology, neuroscience, and the social sciences. New York: Oxford University Press; 2012. pp. 89–108.

[699] Numan M, Bress JA, Ranker LR, Gary AJ, Denicola AL, Bettis JK, et al. The importance of the basolateral/basomedial amygdala for goal-directed maternal responses in postpartum rats. Behav Brain Res 2010;214:368–76.

[700] Numan M, Corodimas KP. The effects of paraventricular hypothalamic lesions on maternal behavior in rats. Physiol Behav 1985;35:417–25.

[701] Numan M, Corodimas KP, Numan MJ, Factor EM, Piers WD. Axon-sparing lesions of the preoptic region and substantia innominata disrupt maternal behavior in rats. Behav Neurosci 1988;102:381–96.

[702] Numan M, Fleming AS, Levy F. Maternal behavior. In: Neill JD, editor. Physiology of reproduction, vol. 2. San Diego: Academic Press; 2006. pp. 1921–93.

[703] Numan M, Insel TR. The neurobiology of parental behavior. New York: Springer; 2003.

[704] Numan M, Numan M. A lesion and neuroanatomical tract-tracing analysis of the role of the bed nucleus of the stria terminalis in retrieval behavior and other aspects of maternal responsiveness in rats. Dev Psychobiol 1996;29:23–51.

[705] Numan M, Numan MJ. Projection sites of medial preoptic area and ventral bed nucleus of the stria terminalis neurons that express Fos during maternal behavior in female rats. J Neuroendocrinol 1997; 9:369–84.

[706] Numan M, Numan MJ, English JB. Excitotoxic amino acid injections into the medial amygdala facilitate maternal behavior in virgin female rats. Horm Behav 1993;27:56–81.

[707] Numan M, Numan MJ, Pliakou N, Stolzenberg DS, Mullins OJ, Murphy JM, et al. The effects of D1 and D2 dopamine receptor antagonism in the medial preoptic area, ventral pallidum, or nucleus accumbens on the maternal retrieval response and other aspects of maternal behavior in rats. Behav Neurosci 2005;119:1588–604.

[708] Numan M, Numan MJ, Schwarz JM, Neuner CM, Flood TF, Smith CD. Medial preoptic area interactions with the nucleus accumbens-ventral pallidum circuit and maternal behavior in rats. Behav Brain Res 2005;158:53–68.

[709] Numan M, Roach JK, del Cerro MC, Guillamon A, Segovia S, Sheehan TP, et al. Expression of intracellular progesterone receptors in rat brain during different reproductive states, and involvement in maternal behavior. Brain Res 1999;830:358–71.

[710] Numan M, Rosenblatt JS, Komisaruk BR. Medial preoptic area and onset of maternal behavior in the rat. J Comp Physiol Psychol 1977;91:146–64.

[711] Numan M, Sheehan TP. Neuroanatomical circuitry for mammalian maternal behavior. Ann NY Acad Sci 1997;807:101–25.

[712] Numan M, Smith HG. Maternal behavior in rats: evidence for the involvement of preoptic projections to the ventral tegmental area. Behav Neurosci 1984;98:712–27.

[713] Numan M, Stolzenberg DS. Hypothalamic interaction with the mesolimbic dopamine system and the regulation of maternal responsiveness. In: Bridges RS, editor. Neurobiology of the parental brain. San Diego: Academic Press; 2008. pp. 3–22.

[714] Numan M, Stolzenberg DS. Medial preoptic area interactions with dopamine neural systems in the control of the onset and maintenance of maternal behavior in rats. Front Neuroendocrinol 2009;30:46–64.

[715] Numan M, Stolzenberg DS, Dellevigne AA, Correnti CM, Numan MJ. Temporary inactivation of ventral tegmental area neurons with either muscimol or baclofen reversibly disrupts maternal behavior in rats through different underlying mechanisms. Behav Neurosci 2009;123:740–51.

[716] Numan M, Woodside B. Maternity: neural mechanisms, motivational processes, and physiological adaptations. Behav Neurosci 2010;124:715–41.

[717] Oei NY, Rombouts SA, Soeter RP, van Gerven JM, Both S. Dopamine modulates reward system activity during subconscious processing of sexual stimuli. Neuropsychopharmacol 2012;37:1729–37.

[718] Ogawa S, Eng V, Taylor J, Lubahn DB, Korach KS, Pfaff DW. Roles of estrogen receptor-alpha gene expression in reproduction-related behaviors in female mice. Endocrinology 1998;139:5070–81.

[719] Olazábal DE, Pereira M, Agrati D, Ferreira A, Fleming AS, González-Mariscal G, et al. New theoretical and experimental approaches on maternal motivation in mammals. Neurosci Biobehav Rev 2013;37:1860–74.

[720] Olazabal DE, Young LJ. Oxytocin receptors in the nucleus accumbens facilitate "spontaneous" maternal behavior in adult female prairie voles. Neuroscience 2006;141:559–68.

[721] Olazabal DE, Young LJ. Oxytocin and individual variation in parental care in prairie voles. In: Bridges RS, editor. Neurobiology of parental behavior. San Diego: Academic Press; 2008. pp. 333–45.

[722] Olivier B, Olivier-Aardema R, Wiepkema PR. Effect of anterior hypothalamic and mammillary area lesions in territorial aggressive behaviour in male rats. Behav Brain Res 1983;9:59–81.

[723] Olivier B, van Oorschot R. 5-HT1B receptors and aggression: a review. Eur J Pharmacol 2005;526:207–17.

[724] Olson VG, Nestler EJ. Topographical organization of GABAergic neurons within the ventral tegmental area of the rat. Synapse 2007;61:87–95.

[725] Olster DH. Ibotenic acid-induced lesions of the medial preoptic/anterior hypothalamus enhance the display of progesterone-facilitated lordosis in male rats. Brain Res 1993;626:99–105.

[726] Ongur D, Price JL. The organization of networks within the orbital and medial prefrontal cortex of rats, monkeys and humans. Cereb Cortex 2000;10:206–19.

[727] Oomura Y, Yoshimatsu H, Aou S. Medial preoptic and hypothalamic neuronal activity during sexual behavior of the male monkey. Brain Res 1983;266:340–3.

[728] Ophir AG. Towards meeting Tinbergen's challenge. Horm Behav 2011;60:22–7.

[729] Ophir AG, Campbell P, Hanna K, Phelps SM. Field tests of cis-regulatory variation at the prairie vole avpr1a locus: association with V1aR abundance but not sexual of social fidelity. Horm Behav 2008;54:694–702.

[730] Ophir AG, Gessel A, Zheng DJ, Phelps SM. Oxytocin receptor density is associated with male mating tactics and social monogamy. Horm Behav 2012;61: 445–53.

[731] Ophir AG, Wolff JO, Phelps SM. Variation in neural V1aR predicts sexual fidelity and space use among male prairie voles in semi-natural settings. Proc Natl Acad Sci USA 2008;105:1249–54.

[732] Orpen BG, Fleming AS. Experience with pups sustains maternal responding in postpartum rats. Physiol Behav 1987;40:47–54.

[733] Ottersen OP. Connections of the amygdala of the rat. IV: coritcoamygdaloid and intraamygdaloid connections as studied with axonal transport of horseradish peroxidase. J Comp Neurol 1982;205:30–48.

[734] Pan H, Epstein J, Silbersweig DA, Stern E. New and emerging techniques for mapping brain circuitry. Brain Res Rev 2011;67:226–51.

[735] Parada M, King S, Li M, Fleming AS. The roles of accumbal dopamine D1 and D2 receptors in maternal memory in rats. Behav Neurosci 2008;122:368–76.

[736] Parker KJ, Lee TM. Central vasopressin administration regulates the onset of facultative paternal behavior in *Microtus pennsylvanicus* (meadow voles). Horm Behav 2001;39:285–94.

[737] Parker KL, Schimmer BP. Embryology and genetics of the mammalian gonads and ducts. In: Neill JD, editor. Physiology of reproduction, vol. 1. San Diego: Academic Press; 2006. pp. 313–36.

[738] Parsey RV, Oquendo MA, Simpson NR, Ogden RT, Van Heertum R, Arango V, et al. Effects of sex, age, and aggressive traits in man on brain serotonin 5-HT1A receptor binding potential measured by PET using C^{11} Way-100635. Brain Res 2002;954:173–82.

[739] Parsons CE, Stark EA, Young KS, Stein A, Kringelbach ML. Understanding the human parental brain: a critical role of the orbitofrontal cortex. Soc Neurosci 2013;8: 525–43.

[740] Paterson LM, Tyacke RJ, Nutt DJ, Knudsen GM. Measuring endogenous 5-HT release by emission tomography: promises and pitfalls. J Cereb Blood Flow Metab 2010;30:1682–706.

[741] Patisaul HB, Bateman HL, Mickens JA. Neonatal agonism of ERalpha masculinizes serotonergic (5-HT) projections to the female rat ventromedial nucleus of the hypothalamus (VMN) but does not impair lordosis. Behav Brain Res 2009;196:317–22.

[742] Paul T, Schiffer B, Zwarg T, Kruger TH, Karama S, Schedlowski M, et al. Brain response to visual sexual stimuli in heterosexual and homosexual males. Hum Brain Mapp 2008;29:726–35.

[743] Pedersen CA. Biological aspects of social bonding and the roots of human violence. Ann NY Acad Sci 2004;1036:106–27.

[744] Pedersen CA, Ascher JA, Monroe YL, Prange AJ. Oxytocin induces maternal behavior in virgin female rats. Science 1982;216:648–50.

[745] Pedersen CA, Boccia ML. Oxytocin maintains as well as initiates female sexual behavior: effects of a highly selective oxytocin antagonist. Horm Behav 2002;41:170–7.

[745A] Pedersen CA, Boccia ML. Oxytocin antagonism alters rat dams' oral grooming and upright posturing over pups. Physiol Behav 2003;80:233–41.

[746] Pedersen CA, Boccia ML. Vasopressin interactions with oxytocin in the control of female sexual behavior. Neuroscience 2006;139:843–51.

[747] Pedersen CA, Caldwell JD, Walker C, Ayers G, Mason GA. Oxytocin activates the postpartum onset of rat maternal behavior in the ventral tegmental and medial preoptic areas. Behav Neurosci 1994;108:1163–71.

[748] Pedersen A, Tomaszycki ML. Oxytocin antagonist treatments alter the formation of pair relationships in zebra finches of both sexes. Horm Behav 2012;62:113–9.

[749] Pedersen CA, Vadlamudi SV, Boccia ML, Amico JA. Maternal behavior deficits in nulliparous knockout mice. Genes Brain Behav 2006;5:274–81.

[750] Pedersen CA, Vadlamudi S, Boccia ML, Moy SS. Variations in maternal behavior in C57BL/6J mice: behavioral comparisons between adult offspring of high or low pup-licking mothers. Front Psychiatry 2011;2:42.

[751] Pehek EA, Thompson JT, Hull EM. The effects of intracranial administration of the dopamine agonist apomorphine on penile reflexes and seminal emission in the rat. Brain Res 1989;500:325–32.

[752] Pennartz CM, Groenewegen HJ, Lopes da Silva FH. The nucleus accumbens as a complex of functionally distinct neuronal ensembles: an integration of behavioral, electrophysiological and anatomical data. Prog Neurobiol 1994;42:719–61.

[753] Pereira M, Morrell JI. The changing role of the medial preoptic area in the regulation of maternal behavior across the postpartum period: facilitation followed by inhibition. Behav Brain Res 2009;205:238–48.

[754] Pereira M, Morrell JI. The medial preoptic area is necessary for motivated choice of pup- over cocaine-associated environments by early postpartum rats. Neuroscience 2010;167:216–31.
[755] Pereira M, Morrell JI. Functional mapping of the neural circuitry of rat maternal motivation: effects of site-specific transient neural inactivation. J Neuroendocrinol 2011;23:1020–35.
[756] Perrin G, Meurisse M, Levy F. Inactivation of the medial preoptic area or the bed nucleus of the stria terminalis differentially disrupts maternal behavior in sheep. Horm Behav 2007;52:461–73.
[757] Perry CJ, McNally GP. A role for the ventral pallidum in context-induced and primed reinstatement of alcohol seeking. Eur J Neurosci 2013;38: 2762–73.
[758] Perry D, Walder K, Hendler T, Shamay-Tsoory SG. The gender you are and the gender you like: sexual preference and empathic neural responses. Brain Res 2013;1534:66–75.
[759] Petrovich GD, Canteras NS, Swanson LW. Combinatorial amygdalar inputs to hippocampal domains and hypothalamic behavior systems. Brain Res Rev 2001;38:247–89.
[760] Petrovich GD, Risold PY, Swanson LW. Organization of projections from the basomedial nucleus of the amygdala: a PHAL study in the rat. J Comp Neurol 1996;374:387–420.
[761] Petrulis A. Neural mechanisms of individual and sexual recognition in Syrian hamsters (*Mesocricetus auratus*). Behav Brain Res 2009;200:260–7.
[762] Peyron C, Petit JM, Rampon C, Jouvet M, Luppi PH. Forebrain afferents to the dorsal raphe nucleus demonstrated by retrograde and anterograde tracing methods. Neuroscience 1998;82:443–68.
[763] Pfaff DW. Estrogens and brain function. New York: Springer-Verlag; 1980.
[764] Pfaff DW. Motivational concepts: definitions and distinctions. In: Pfaff DW, editor. The physiological mechanisms of motivation. New York: Springer-Verlag; 1982. pp. 3–24.
[765] Pfaff DW, Sakuma Y. Deficit in the lordosis reflex of female rats caused by lesions in the ventromedial nucleus of the hypothalamus. J Physiol 1979;288:203–10.
[766] Pfaff DW, Sakuma Y, Kow LM, Lee AWL, Easton A. Hormonal, neural, and genomic mechanisms for female reproductive behaviors, motivation, and arousal. In: Neill JD, editor. Physiology of reproduction, vol. 2. San Diego: Academic Press; 2006. pp. 1825–920.
[767] Pfaff D, Waters E, Khan Q, Zhang X, Numan M. Minireview: estrogen receptor-initiated mechanisms causal to mammalian reproductive behaviors. Endocrinology 2011;152:1209–17.
[768] Pfaus JG, Damsma G, Wenkstern D, Fibiger HC. Sexual activity increases dopamine transmission in the nucleus accumbens and striatum of female rats. Brain Res 1995;693:21–30.
[769] Pfaus JG, Phillips AG. Role of dopamine in anticipatory and consummatory aspects of sexual behavior in the male rat. Behav Neurosci 1991;105:727–43.
[770] Phelps SM, Campbell P, Zheng DJ, Ophir AG. Beating the boojum: comparative approaches to the neurobiology of social behavior. Neuropharmacology 2010;58:17–28.
[771] Phelps EA, LeDoux JE. Contributions of the amygdala to emotion processing: from animal models to human behavior. Neuron 2005;48:175–87.
[772] Phoenix CH, Goy RW, Gerall AA, Young WC. Organizing action of prenatally administered testosterone propionate on the tissues mediating mating behavior in the female guinea pig. Endocrinology 1959;65:369–82.
[773] Phoenix CH, Slob AK, Goy RW. Effects of castration and replacement therapy on sexual behavior of adult male rhesuses. J Comp Physiol Psychol 1973;84:472–81.
[774] Pietrini P, Guazzelli M, Basso G, Jaffe K, Grafman J. Neural correlates of imaginal aggressive behavior assessed by positron emission tomography in healthy subjects. Am J Psychiatry 2000;157:1772–81.
[775] Pobbe RL, Pearson BL, Defensor EB, Bolivar VJ, Young WS, Lee HJ, et al. Oxytocin receptor knockout mice display deficits in the expression of autism-related behaviors. Horm Behav 2012;61:436–44.
[776] Poeppl TB, Nitschke J, Dombert B, Santtila P, Greenlee MW, Osterheider M, et al. Functional cortical and subcortical abnormalities in pedophilia: a combined study using a choice reaction time task and fMRI. J Sex Med 2011;8:1660–74.
[777] Poeppl TB, Nitschke J, Santtila P, Schecklmann M, Langguth B, Greenlee MW, et al. Association between brain structure and phenotypic characteristics in pedophilia. J Psychiatr Res 2013;47:678–85.
[778] Poindron P, Levy F, Keller M. Maternal responsiveness and maternal selectivity in sheep and goats: the two facets of maternal attachment. Dev Psychobiol 2007;49:54–70.
[779] Poling MC, Kaufman AS. Organizational and activational effects of sex steroids on kisspeptin neuron development. Front Neuroendocrinol 2013;34:3–17.
[780] Pomerantz SM, Goy RW, Roy MM. Expression of male-typical behavior in adult female pseudohermaphroditic rhesus: comparisons with normal males and neonatally gonadectomized males and female. Horm Behav 1986;20:483–500.
[781] Pomerantz SM, Roy MM, Thornton JE, Goy RW. Expression of adult female patterns of sexual behavior by male, female, and pseudohermaphroditic female rhesus monkeys. Biol Reprod 1985;33:878–89.

[782] Ponseti J, Bosinski HA, Wolff S, Peller M, Jansen O, Mehdorn HM, et al. A functional endophenotype for sexual orientation in humans. NeuroImage 2006;33:825–33.

[783] Ponseti J, Granert O, Jansen O, Wolff S, Beier K, Neutze J, et al. Assessment of pedophilia using hemodynamic brain response to sexual stimuli. Arch Gen Psychiatry 2012;69:187–94.

[784] Powell RA, Fried JJ. Helping by juvenile pine voles (*Microtus pinetorum*), growth and survival of younger siblings, and the evolution of pine vole sociality. Behav Ecol 1992;3:325–33.

[785] Preston SD. The origins of altruism in offspring care. Psychol Bull 2013;139:1305–41.

[786] Price JL. Free will versus survival: brain systems that underlie intrinsic constraints on behavior. J Comp Neurol 2005;493:132–9.

[787] Price JL. Definition of the orbital cortex in relation to specific connections with limbic and visceral structures and other cortical regions. Ann NY Acad Sci 2007;1121:54–71.

[788] Provencal N, Suderman MJ, Guillemin C, Massart R, Ruggiero A, Wang D, et al. The signature of maternal rearing in the methylome in rhesus macaque prefrontal cortex and T cells. J Neurosci 2012;32:15626–42.

[789] Puig MV, Artigas F, Celada P. Modulation of the activity of pyramidal neurons in rat prefrontal cortex by raphe stimulation in vivo: involvement of serotonin and GABA. Cereb Cortex 2005; 15:1–14.

[790] Puig MV, Watakabe A, Ushimaru M, Yamamori T, Kawaguchi Y. Serotonin modulates fast-spiking interneuron and synchronous activity in the rat prefrontal cortex through 5-HT1A and 5-HT2A receptors. J Neurosci 2010;30:2211–22.

[791] Puts DA, Bailey DH, Cardenas RA, Burriss RF, Welling LLM, Wheatley JR, et al. Women's attractiveness changes with estradiol and progesterone across the ovulatory cycle. Horm Behav 2013;63:13–9.

[792] Raggenbass M. Overview of cellular electrophysiological actions of vasopressin. Eur J Pharmacol 2008;583:243–54.

[793] Ragnauth AK, Devidze N, Moy V, Finley K, Goodwillie A, Kow LM, et al. Female oxytocin geneknockout mice, in a semi-natural environment, display exaggerated aggressive behavior. Genes Brain Behav 2005;4:229–39.

[794] Raine A. The anatomy of violence: the biological roots of crime. New York: Pantheon; 2013.

[795] Raine A, Lee L, Yang Y, Colletti P. Neurodevelopmental marker for limbic maldevelopment in antisocial personality disorder and psychopathy. Br J Psychiatry 2010;197:186–92.

[796] Raine A, Meloy JR, Bihrle S, Stoddard J, LaCasse L, Buchsbaum MS. Reduced prefrontal and increased subcortical brain functioning assessed using positron emission tomography in predatory and affective murderers. Behav Sci Law 1998;16:319–32.

[797] Raineki C, Cortes MR, Belnoue L, Sullivan RM. Effects of early-life abuse differ across development: infant social behavior deficits are followed by adolescent depressive-like behaviors mediated by the amygdala. J Neurosci 2012;32:7758–65.

[798] Rauch A, Rainer G, Logethetis NK. The effect of serotonin-induced dissociation between spiking and perisynaptic activity on BOLD functional MRI. Proc Natl Acad Sci USA 2008;105:6759–64.

[799] Ray RD, Zald DH. Anatomical insights into the interaction of emotion and cognition in the prefrontal cortex. Neurosci Biobehav Rev 2012;36:479–501.

[800] Redoute J, Stoleru S, Pugeat M, Costes N, Lavenne F, Le Bars D, et al. Brain processing of visual sexual stimuli in treated and untreated hypogonadal patients. Psychoneuroendocrino 2005;30:461–82.

[801] Reif A, Rosler M, Freitag CM, Schneider M, Eujen A, Kissling C, et al. Nature and nurture predispose to violent behavior: serotonergic genes and adverse childhood environment. Neuropsychopharmacol 2007;32:2375–83.

[802] Resendez SL, Dome M, Gormley G, Franco D, Nevarez N, Hamid AA, et al. μ-Opioid receptors within subregions of the striatum mediate pair bond formation through parallel yet distinct reward mechanisms. J Neurosci 2013;33:9140–9.

[803] Reynolds SM, Berridge KC. Positive and negative motivation in nucleus accumbens shell: bivalent rostrocaudal gradients for GABA-elicited eating, taste "liking"/"disliking" reactions, place preference/avoidance, and fear. J Neurosci 2002;22:7308–20.

[804] Ribeiro AC, Musatov S, Shteyler A, Simanduyev S, Arrieta-Cruz I, Ogawa S, et al. siRNA silencing of estrogen receptor-α expression specifically in medial preoptic area neurons abolishes maternal care in female mice. Proc Natl Acad Sci USA 2012;109:16324–9.

[805] Richard JM, Berridge KC. Nucleus accumbens dopamine/glutamate interaction switches modes to generate desire versus dread: D_1 alone for appetitive eating but D_1 and D_2 together for fear. J Neurosci 2011;31:12866–79.

[806] Richard JM, Berridge KC. Prefrontal cortex modulates desire and dread generated by nucleus accumbens glutamate disruption. Biol Psychiatry 2013;73:360–70.

[807] Riem MM, Bakermans-Kranenburg MJ, Huffmeijer R, van Ijzendoorn MH. Does intranasal oxytocin promote prosocial behavior to an excluded fellow player? A randomized-controlled trial with Cyberball. Psychoneuroendocrino 2013;38:1418–25.

[808] Rilling JK, DeMarco AC, Hackett PD, Thompson R, Ditzen B, Patel R, et al. Effects of intranasal oxytocin and vasopressin on cooperative behavior and associated brain activity in men. Psychoneuroendocrino 2012;37:447–61.

[809] Rilling J, Gutman D, Zeh T, Pagnoni G, Berns G, Kilts C. A neural basis for social cooperation. Neuron 2002;35:395–405.

[810] Rilling JK, Scholz J, Preuss TM, Glasser MF, Errangi BK, Behrens TE. Differences between chimpanzees and bonobos in neural systems supporting social cognition. Soc Cogn Affect Neurosci 2012;7:369–79.

[811] Ring RH, Malberg JE, Potestio L, Ping J, Boikess S, Luo B, et al. Anxiolytic-like activity of oxytocin in male mice: behavioral and autonomic evidence, therapeutic implications. Psychopharmacology 2006;185:218–25.

[812] Ring RH, Schechter LE, Leonard SK, Dwyer JM, Platt BJ, Graf R, et al. Receptor and behavioral pharmacology of WAY-267464, a non-peptide oxytocin receptor agonist. Neuropharmacology 2010;58:69–77.

[813] Risold PY, Thompson RH, Swanson LW. The structural organization of connections between hypothalamus and cerebral cortex. Brain Res Rev 1997;24:197–254.

[814] Rizvi TA, Ennis M, Shipley MT. Reciprocal connections between the medial preoptic area and the midbrain periaqueductal gray in rat: a WGA-HRP and PHA-L study. J Comp Neurol 1992;315:1–15.

[815] Robertson GS, Jian M. D1 and D2 dopamine receptors differentially increase Fos-like immunoreactivity in accumbal projections to the ventral pallidum and midbrain. Neuroscience 1995;64:1019–34.

[816] Robinson DL, Zitzman DL, Williams SK. Mesolimbic dopamine transients in motivated behaviors: focus on maternal behavior. Front Psychiatry 2011;2:23.

[817] Rodrigues SM, Saslow LR, Garcia N, John OP, Keltner D. Oxytocin receptor genetic variation relates to empathy and stress reactivity in humans. Proc Natl Acad Sci USA 2009;106:21437–41.

[818] Roeling TA, Veening JG, Kruk MR, Peters JP, Vermelis ME, Nieuwenhuys R. Efferent connections of the hypothalamic "aggression area" in the rat. Neuroscience 1994;59:1001–24.

[819] Roggenbach J, Muller-Oerlinghausen, Franke L. Suicidality, impulsivity and aggression – is there a link to 5HIAA concentration in the cerebrospinal fluid? Psychiatry Res 2002;113:193–206.

[820] Romero-Carbente JC, Camacho FJ, Paredes RG. The role of the dorsolateral tegmentum in the control of male sexual behavior: a reevaluation. Behav Brain Res 2006;170:262–70.

[821] Romero-Carbente JC, Hurtazo EA, Paredes RG. Central tegmental field and sexual behavior in the male rat: effects of neurotoxic lesions. Neuroscience 2007;148:867–75.

[822] Romero-Fernandez W, Borroto-Escuela DO, Agnati LF, Fuxe K. Evidence for the existence of dopamine d2-oxytocin receptor heteromers in the ventral and dorsal striatum with facilitatory receptor-receptor interactions. Mol Psychiatry 2013;18:849–50.

[823] Roney JR, Simmons ZI. Hormonal predictors of sexual motivation in natural menstrual cycles. Horm Behav 2013;63:636–45.

[824] Rood BD, De Vries GJ. Vasopressin innervation of the mouse (*Mus musculus*) brain and spinal cord. J Comp Neurol 2011;519:2434–74.

[825] Rood BD, Stott RT, You S, Smith CJ, Woodbury ME, De Vries GJ. Site of origin of and sex differences in the vsasopressin innervation of the mouse (*Mus musculus*) brain. J Comp Neurol 2013;521:2321–58.

[826] Rosa M, Franzini A, Giannicola G, Messina G, Altamura AC, Priori A. Hypothalamic oscillations in human pathological aggressiveness. Biol Psychiatry 2012;72:e33–5.

[827] Rosell DR, Thompson JL, Slifstein M, Xu X, Frankle WG, New AS, et al. Increased serotonin 2A receptor availability in the orbitofrontal cortex of physically aggressive personality disordered patients. Biol Psychiatry 2010;67:1154–62.

[828] Roselli CE, Stormshak F. The neurobiology of sexual partner preference in rams. Horm Behav 2009;55:611–20.

[829] Rosenblatt JS. Nonhormonal basis of maternal behavior in the rat. Science 1967;156:1512–4.

[830] Rosenblatt JS, Ceus K. Estrogen implants in the medial preoptic area stimulate maternal behavior in male rats. Horm Behav 1998;33:23–30.

[831] Rosenblatt JS, Hazelwood S, Poole J. Maternal behavior in male rats: effects of medial preoptic area lesions and presence of maternal aggression. Horm Behav 1996;30:201–15.

[832] Rosenblatt JS, Lehrman DS. Maternal behavior of the laboratory rat. In: Rheingold HL, editor. Maternal behavior in mammals. New York: Wiley; 1963. pp. 8–57.

[833] Rosenblatt JS, Mayer AD. An analysis of approach/withdrawal processes in the initiation of maternal behavior in the laboratory rat. In: Hood KE, Greenberg G, Tobach E, editors. Behavioral development. New York: Garland Press; 1995. pp. 177–230.

[834] Rosenblatt JS, Siegel HI. Hysterectomy-induced maternal behavior during pregnancy in the rat. J Comp Physiol Psychol 1975;89:685–700.

[835] Rosenblum LA, Andrews MW. Influences of environmental demand on maternal behavior and infant development. Acta Paediatr Suppl 1994;397:57–63.

[836] Ross HE, Cole CD, Smith Y, Neumann ID, Landgraf R, Murphy AZ, et al. Characterization of the oxytocin system regulating affiliative behavior in female prairie voles. Neuroscience 2009;162:892–903.

[837] Ross HE, Freeman SM, Spiegel LL, Ren X, Terwilliger EF, Young LJ. Variation in oxytocin receptor density in nucleus accumbens has differential effects on affiliative behaviors in monogamous and polygamous voles. J Neurosci 2009;29:1312–8.

[838] Ross HE, Young LJ. Oxytocin and the neural mechanisms regulating social cognition and affiliative behavior. Front Neuroendocrinol 2009;30:534–47.

[839] Roth TL. Epigenetics of neurobiology and behavior during development and adulthood. Dev Psychobiol 2012;54:590–7.

[840] Roth TL, Lubin FD, Funk AJ, Sweatt JD. Lasting epigenetic influence of early-life adversity on BDNF gene. Biol Psychiatry 2009;65:760–9.

[841] Rubin BS, Barfield RJ. Induction of estrous behavior in ovariectomized rats by sequential replacement of estrogen and progesterone to the ventromedial hypothalamus. Neuroendocrinology 1983;37:218–24.

[842] Rudebeck PH, Behrens TE, Kennerley SW, Baxter MG, Buckley MJ, Walton ME, et al. Frontal cortex subregions play distinct roles in choices between actions and stimuli. J Neurosci 2008;28:13775–85.

[843] Rudebeck PH, Buckley MJ, Walton ME, Rushworth MF. A role for the macaque anterior cingulate gyrus in social valuation. Science 2006;313:1310–2.

[844] Rudebeck PH, Walton ME, Millette BH, Shirley E, Rushworth MF, Bannerman DM. Distinct contributions of frontal areas to emotion and social behaviour in the rat. Eur J Neurosci 2007;26:2315–26.

[845] Ruppenthal GC, Arling GL, Harlow HF, Sackett GP, Suomi SJ. A 10-year perspective of motherless-mother monkey behavior. J Abnorm Psychol 1976;85:341–9.

[846] Rushworth MF, Mars RB, Summerfield C. General mechanisms for making decisions? Curr Opin Neurobiol 2009;19:75–83.

[847] Sa SI, Pereira PA, Malikov V, Madeira MD. Role of estrogen receptor α and β in the induction of progesterone receptors in hypothalamic ventromedial neurons. Neuroscience 2013;238:159–67.

[848] Saddoris MP, Gallagher M, Schoenbaum G. Rapid associative encoding in basolateral amygdala depends on connections with orbitofrontal cortex. Neuron 2005;46:321–31.

[849] Sakuma Y. Estrogen-induced changes in the neural impulse flow from the female rat preoptic regions. Horm Behav 1994;28:438–44.

[850] Sakuma Y, Pfaff DW. Mesencephalic mechanisms for the integration of female reproductive behavior in the rat. Am J Physiol 1979;237:R285–290.

[851] Saltzman W, Maestripieri D. The neuroendocrinology of primate maternal behavior. Prog Neuropsychopharmacol Biol Psychiatry 2011;35:1192–204.

[852] Samuelson CL, Meredith M. Categorization of biologically relevant chemical signals in the medial amygdala. Brain Res 2009;1263:33–42.

[853] Samuelson CL, Meredith M. Oxytocin antagonist disrupts male mouse medial amygdala response to chemical-communication signals. Neuroscience 2011;180:96–104.

[854] Sano K, Tsuda MC, Musatov S, Sakamoto T, Ogawa S. Differential effects of site-specific knockdown of estrogen receptor α in the medial amygdala, medial pre-optic area, and ventromedial nucleus of the hypothalamus on sexual and aggressive behavior of male mice. Eur J Neurosci 2013;37:1308–19.

[855] Sarachana T, Xu M, Wu RC, Hu VW. Sex hormones in autism: androgens and estrogens differentially and reciprocally regulate RORA, a novel candidate gene for autism. PLoS One 2011;6:e17116.

[856] Sari Y. Serotonin 1B receptors: from protein to physiological function and behavior. Neurosci Biobehav Rev 2004;28:565–82.

[857] Sartorius A, Ruf M, Kief C, Demirakca T, Bailer G, Henn FA, et al. Abnormal amygdala activation profile in pedophilia. Eur Arch Psychiatry Clin Neurosci 2008;258:271–7.

[858] Satpute AB, Shu J, Weber J, Roy M, Ochsner KN. The functional architecture of self-reports of affective experience. Biol Psychiatry 2013;73:631–8.

[859] Saudou F, Amara DA, Dierich A, LeMeur M, Ramboz S, Segu L, et al. Enhanced aggressive behavior in mice lacking 5-HT1B receptor. Science 1994;265:1875–8.

[860] Saver JL, Damasio AR. Preserved access and processing of social knowledge in a patient with acquired sociopathy due to ventromedial frontal damage. Neuropsychologia 1991;29:1241–9.

[861] Scalia F, Winans SS. The differential projections of the olfactory bulb and accessory olfactory bulb in mammals. J Comp Neurol 1975;161:31–55.

[862] Scheele D, Striepens N, Güntürkün O, Deutschländer S, Maier W, Kendrick KM, et al. Oxytocin modulates social distance between males and females. J Neurosci 2012;32:16074–9.

[863] Schiffer B, Müller BW, Scherbaum N, Hodgins S, Forsting M, Wiltfang J, et al. Disentangling structural brain alterations associated with violent behavior from those associated with substance use disorders. Arch Gen Psychiatry 2011;68:1039–49.

[864] Schiffer B, Paul T, Gizewski E, Forsting M, Leygraf N, Schedlowski M, et al. Functional brain correlates of heterosexual paedophilia. NeuroImage 2008;41:80–91.

[865] Schiltz K, Witzel J, Northoff G, Zierhut K, Gubka U, Fellmann H, et al. Brain pathology in pedophilic offenders: evidence of volume reduction in the right amygdala and related diencephalic structures. Arch Gen Psychiatry 2007;64:737–46.

[866] Schneider JS, Stone MK, Wynne-Edwards KE, Horton TH, Lydon J, O'Malley B, et al. Progesterone receptors mediate male aggression toward infants. Proc Natl Acad Sci USA 2003;100:2951–6.

[867] Schneiderman I, Zagoory-Sharon O, Leckman JF, Feldman R. Oxytocin during the initial stages of romantic attachment: relations to couples' interactive reciprocity. Psychoneuroendocrino 2012;37:1277–85.

[868] Schneirla TC. An evolutionary and developmental theory of biphasic processes underlying approach and withdrawal. In: Jones MR, editor. Nebraska symposium on motivation, vol. VII. Lincoln: University of Nebraska Press; 1959. pp. 1–42.

[869] Schoenbaum G, Chiba AA, Gallagher M. Neural encoding in orbitofrontal cortex and basolateral amygdala during olfactory discrimination learning. J Neurosci 1999;19:1876–84.

[870] Schoenbaum G, Roesch MR, Stalnaker TA, Takahashi YK. A new perspective on the role of the orbitofrontal cortex in adaptive behaviour. Nat Rev Neurosci 2009;10:885–92.

[871] Schorscher-Petcu A, Dupre A, Tribollet E. Distribution of vasopressin and oxytocin binding sites in the brain and upper spinal cord of the common marmoset. Neurosci Lett 2009;461:217–22.

[872] Schott BH, Minuzzi L, Krebs RM, Elmenhorst D, Lang M, Winz OH, et al. Mesolimbic functional magnetic resonance imaging activations during reward anticipation correlate with reward-related ventral striatal dopamine release. J Neurosci 2008;28:14311–9.

[873] Schwandt ML, Lindell SG, Sjöberg RL, Chisholm KL, Higley JD, Suomi SJ, et al. Gene-environment interactions and response to social intrusion in male and female rhesus macaques. Biol Psychiatry 2010; 67:323–30.

[874] Schwarz JM, McCarthy MM. Steroid-induced sexual differentiation of the developoing brain: multiple pathways, one goal. J Neurochem 2008;105: 1561–72.

[875] Scott-Van Zeeland AA, Dapretto M, Ghahremani DG, Poldrack RA, Bookheimer SY. Reward processing in autism. Autism Res 2010;3:53–67.

[876] Sears RM, Liu RJ, Narayanan NS, Sharf R, Yeckel MF, Laubach M, et al. Regulation of nucleus accumbens activity by the hypothalamic neuropeptide melanin-concentrating hormone. J Neurosci 2010;30:8263–73.

[877] Segovia S, Guillamon A. Sexual dimorphism in the vomeronasal pathway and sex differences in reproductive behaviors. Brain Res Rev 1993;18:51–74.

[878] Seip KM, Morrell JI. Transient inactivation of the ventral tegmental area selectively disrupts the expression of conditioned place preference for pup-but not cocaine-paired contexts. Behav Neurosci 2009;123:1325–38.

[879] Semaan SJ, Dhamija S, Kim J, Ku EC, Kauffman AS. Assessment of epigenetic contributions to sexually-dimorphic Kiss1 expression in the anteroventral periventricular nucleus of mice. Endocrinology 2012;153:1875–86.

[880] Sesack SR, Deutch AY, Roth RH, Bunney BS. Topographical organization of the efferent projections of the medial prefrontal cortex in the rat: an anterograde tract-tracing study with *Phaseolus vulgaris* leucoagglutinin. J Comp Neurol 1989;290:213–42.

[881] Sesack SR, Grace AA. Cortico-basal ganglia reward network: micocircuitry. Neuropsychopharmacol 2010;35:27–47.

[882] Seto MC. Pedophilia. Annu Rev Clin Psychol 2009;5:391–407.

[883] Shabel SJ, Janak PH. Substantial similarity in amygdala neuronal activity during conditioned appetitive and aversive emotional arousal. Proc Natl Acad Sci USA 2009;106:15031–6.

[884] Shahrokh DK, Zhang TY, Diorio J, Gratton A, Meaney MJ. Oxytocin-dopamine interactions mediate variations in maternal behavior in the rat. Endocrinology 2010;151:2276–86.

[885] Shaikh MB, Steinberg A, Siegel A. Evidence that substance P is utilized in medial amygdaloid facilitation of defensive rage behavior in the cat. Brain Res 1993;625:283–94.

[886] Shannon C, Schwandt ML, Champoux M, Shoaf SE, Suomi SJ, Linnoila M, et al. Maternal absence and stability of individual differences in CSF 5-HIAA concentrations in rhesus monkey infants. Am J Psychiatry 2005;162:1658–64.

[887] Sheehan TP, Chambers RA, Russell DS. Regulation of affect by the lateral septum: implications for neuropsychiatry. Brain Res Rev 2004;46:71–117.

[888] Sheehan TP, Cirrito J, Numan MJ, Numan M. Using c-Fos immunocytochemistry to identify forebrain regions that may inhibit maternal behavior in rats. Behav Neurosci 2000;114:337–52.

[889] Sheehan T, Numan M. The septal region and social behavior. In: Numan R, editor. The behavioral neuroscience of the septal region. New York: Springer; 2000. pp. 175–209.

[890] Sheehan T, Numan M. Estrogen, progesterone, and pregnancy termination alter neural activity in brain regions that control maternal behavior. Neuroendocrinology 2002;75:12–23.

[891] Sheehan T, Paul M, Amaral E, Numan MJ, Numan M. Evidence that the medial amygdala projects to the anterior/ventromedial hypothalamic nuclei to inhibit maternal behavior in rats. Neuroscience 2001;106:341–56.

[892] Sheng M, Greenberg ME. The regulation and function of c-fos and other immediate early genes in the nervous system. Neuron 1990;4:477–85.

[893] Shih JC, Ridd MJ, Chen K, Meehan WP, Kung MP, Seif I, et al. Ketanserin and tetrabenazine abolish aggression in mice lacking monoamine oxidase A. Brain Res 1999;835:104–12.

[894] Shughrue PJ, Lane MV, Merchenthaler I. Comparative distribution of estrogen receptor-alpha and -beta mRNA in the rat central nervous system. J Comp Neurol 1997;388:507–25.

[895] Siegel A, Roeling TA, Gregg TR, Kruk MR. Neuropharmacology of brain-stimulation-evoked aggression. Neurosci Biobehav Rev 1999;23:359–89.

[896] Siegel HI, Rosenblatt JS. Hormonal basis of hysterectomy-induced maternal behavior during pregnancy in the rat. Horm Behav 1975;6:211–22.

[897] Siever LJ. Neurobiology of aggression and violence. Am J Psychiatry 2008;165:429–42.

[898] Silk JB, Brosnan SF, Vonk J, Henrich J, Povinelli DJ, Richardson AS, et al. Chimpanzees are indifferent to the welfare of unrelated group members. Nature 2005;437:1357–9.

[899] Silva BA, Mattucci C, Krzywkowski P, Murana E, Illarionova A, Grinevich V, et al. Independent hypothalamic circuits for social and predator fear. Nat Neurosci 2013;16:1731–3.

[900] Simerly RB. Wired for reproduction: organization and development of sexually dimorphic circuits in the mammalian forebrain. Annu Rev Neurosci 2002;25:507–36.

[901] Simerly RB, Chang C, Muramatsu M, Swanson LW. Distribution of androgen and estrogen receptor mRNA-containing cells in the rat brain: an in situ hybridization study. J Comp Neurol 1990;294:76–95.

[902] Simerly RB, Swanson LW. The organization of neural inputs to the medial preoptic nucleus of the rat. J Comp Neurol 1986;246:312–42.

[903] Simerly RB, Swanson LW. Projections of the medial preoptic nucleus: a *Phaseolus vulgaris* leucoagglutinin anterograde tract-tracing study in the rat. J Comp Neurol 1988;270:209–42.

[904] Simmons DA, Neill DB. Functional interaction between the basolateral amygdala and the nucleus accumbens underlies incentive motivation for food reward on a fixed ratio schedule. Neuroscience 2009;159:1264–73.

[905] Simpson KL, Weaver KJ, de Villers-Sidani E, Lu JYF, Cai Z, Pang Y, et al. Perinatal antidepressant exposure alters cortical network function in rodents. Proc Natl Acad Sci USA 2011;108:18465–70.

[906] Sinchak K, Shahedi K, Dewing P, Micevych P. Sexual receptivity is reduced in the female mu-opioid receptor knockout mouse. Neuroreport 2005;16:1697–700.

[907] Singer T, Seymour B, O'Doherty JP, Stephan KE, Dolan RJ, Frith CD. Empathic neural responses are modulated by the perceived fairness of others. Nature 2006;439:466–9.

[908] Skrundz M, Bolten M, Nast I, Hellhammer DH, Meinischmidt G. Plasma oxytocin concentration during pregnancy is associated with development of postpartum depression. Neuropsychopharmacol 2011;36:1886–93.

[909] Skuse DH, Gallagher L. Dopaminergic-neuropeptide interactions in the social brain. Trends Cogn Sci 2009;13:27–35.

[910] Slimp JC, Hart BL, Goy RW. Heterosexual, autosexual and social behavior of adult male rhesus monkeys with medial preoptic-anterior hypothalamic lesions. Brain Res 1978;142:105–22.

[911] Smeltzer MD, Curtis JT, Aragona BJ, Wang Z. Dopamine, oxytocin, and vasopressin receptor binding in the medial prefrontal cortex of monogamous and promiscuous voles. Neurosci Lett 2006;394:146–51.

[912] Smith AS, Agmo A, Birnie AK, French JA. Manipulation of the oxytocin system alters social behavior and attraction in pair-bonding primates, *Callithrix penicillata*. Horm Behav 2010;57:255–62.

[913] Smith CD, Holschbach MA, Olsewicz J, Lonstein JS. Effects of noradrenergic alpha-2 receptor antagonism or noradrenergic lesions in the ventral bed nucleus of the stria terminalis and medial preoptic area on maternal care in female rats. Psychopharmacology 2012;224:263–76.

[914] Smith RJ, Lobo MK, Spencer S, Kalivas PW. Cocaine-induced adaptations in D1 and D2 accumbens projection neurons (a dichotomy not necessarily synonymous with direct and indirect pathways). Curr Opin Neurobiol 2013;23:546–52.

[915] Smith KS, Tindell AJ, Aldridge JW, Berridge KC. Ventral pallidum roles in reward and motivation. Behav Brain Res 2009;196:155–67.

[916] Sodersten P, Boer GJ, De Vries GJ, Buijs RM, Melin P. Effects of vasopressin on female sexual behavior in male rats. Neurosci Lett 1986;69:188–91.

[917] Solomon NG, Richmond AR, Harding PA, Fries A, Jacquemin S, Schaefer RL, et al. Polymorphism at the avpr1a locus in male prairie voles correlated with genetic but not social monogamy in field populations. Mol Ecol 2009;18:4680–95.

[918] Spampanato J, Polepalli J, Sah P. Interneurons in the basolateral amygdala. Neuropharmacology 2011;60:765–73.

[919] Spinelli S, Chefer S, Carson RE, Jagoda E, Lang L, Heilig M, et al. Effects of early-life stress on serotonin$_{1A}$ receptors in juvenile rhesus monkeys measured by positron emission tomography. Biol Psychiatry 2010;67:1146–53.

[920] Spiteri T, Ogawa S, Musatov S, Pfaff DW, Agmo A. The role of the estrogen receptor α in the medial preoptic area in sexual incentive motivation, proceptivity and receptivity, anxiety, and wheel running in female rats. Behav Brain Res 2012;230:11–20.

[921] Stack EC, Balakrishnan R, Numan MJ, Numan M. A functional neuroanatomical investigation of the role of the medial preoptic area in neural circuits regulating maternal behavior. Behav Brain Res 2002;131:17–36.

[922] Stack EC, Numan M. The temporal course of expression of c-Fos and Fos B within the medial preoptic area and other brain regions of postpartum female rats during prolonged mother-infant interactions. Behav Neurosci 2000;114:609–22.

[923] Stanley B, Siever LJ. The interpersonal dimension of borderline personality disorder: toward a neuropeptide model. Am J Psychiatry 2010;167:24–39.

[924] Staples LG, McGregor IS, Apfelbach R, Hunt GE. Cat odor, but not trimethylthiazoline (fox odor), activates accessory olfactory and defense-related brain regions in rats. Neuroscience 2008;151:937–47.

[925] Stern JM. Somatosensation and maternal care in Norway rats. In: Rosenblatt JS, Snowdon CT, editors. Advances in the study of behavior, vol. 25. San Diego: Academic Press; 1996. pp. 234–94.

[926] Stevenson EL, Caldwell HK. The vasopressin 1b receptor and the neural regulation of social behavior. Horm Behav 2012;61:277–82.

[927] Stoltenberg SF, Christ CC, Carlo G. Afraid to help: social anxiety partially mediates the association between 5-HTTLPR tiallelic genotype and prosocial behavior. Soc Neurosci 2013;8:400–6.

[928] Stolzenberg DS, McKenna JB, Keough S, Hancock R, Numan MJ, Numan M. Dopamine D1 receptor stimulation of the nucleus accumbens or the medial preoptic area promotes the onset of maternal behavior in pregnancy-terminated rats. Behav Neurosci 2007;121:907–19.

[929] Stolzenberg DS, Numan M. Hypothalamic interaction with the mesolimbic DA system in the control of maternal and sexual behaviors in rats. Neurosci Biobehav Rev 2011;35:826–47.

[930] Stolzenberg DS, Rissman EF. Oestrogen-independent, experience-induced maternal behaviour in female mice. J Neuroendocrinol 2011;23:345–54.

[931] Stolzenberg DS, Stevens JS, Rissman EF. Experience-facilitated improvements in pup retrieval; evidence for an epigenetic effect. Horm Behav 2012;62:128–35.

[932] Stoop R. Neuromodulation by oxytocin and vasopressin. Neuron 2012;76:142–59.

[933] Strathearn L. Maternal neglect: oxytocin, dopamine and the neurobiology of attachment. J Neuroendocrinol 2011;23:1054–65.

[934] Strathearn L, Fonagy P, Amico J, Montague PR. Adult attachment predicts maternal brain and oxytocin response to infant cues. Neuropsychopharmacol 2009;34:2655–66.

[935] Strathearn L, Iyengar U, Fonagy P, Kim S. Maternal oxytocin response during mother-infant interaction: associations with adult temperament. Horm Behav 2012;61:429–35.

[936] Strathearn L, Mamun AA, Najman JM, O'Callaghan MJ. Does breastfeeding protect against substantiated child abuse and neglect? A 15-year cohort study. Pediatrics 2009;123:483–93.

[937] Striepens N, Kendrick KM, Maier W, Hurlemann R. Prosocial effects of oxytocin and clinical evidence for its therapeutic potential. Front Neuroendocrinol 2011;32:426–50.

[938] Striepens N, Scheele D, Kendrick KM, Becker B, Schäfer L, Schwalba K, et al. Oxytocin facilitates protective responses to aversive social stimuli in males. Proc Natl Acad Sci USA 2012;109:18144–9.

[939] Stuber GD, Sparta DR, Stamatakis AM, van Leeuwen WA, Hardjoprajitno JE, Cho S, et al. Excitatory transmission from amygdala to nucleus accumbens facilitates reward seeking. Nature 2011;475:377–80.

[940] Sturgis JD, Bridges RS. *N*-methyl-DL-aspartic acid lesions of the medial preoptic area disrupt ongoing parental behavior in male rats. Physiol Behav 1997;62:305–10.

[941] Sukikara MH, Mota-Ortiz SR, Baldo MV, Felicio LF, Canteras NS. The periaqueductal gray and its potential role in maternal behavior inhibition in response to predatory threats. Behav Brain Res 2010;209:226–33.

[942] Suomi SJ. Risk, resilience, and gene-environment interplay in primates. J Can Acad Child Adolesc Psychiatry 2011;20:289–97.

[943] Susman EJ. Psychobiology of persistent antisocial behavior: stress, early vulnerabilities and the attenuation hypothesis. Neurosci Biobehav Rev 2006;30:376–89.

[944] Swain A. Sex determination and differentiation. In: Neill JD, editor. Physiology of reproduction, vol. 1. San Diego: Academic Press; 2006. pp. 245–60.

[945] Swain JE. Baby stimuli and the parent brain: functional neuroimaging of the neural substrates of parent-infant attachment. Psychiatry 2008;5:28–36.

[946] Swain JE. The human parental brain: in vivo neuroimaging. Prog Neuropsychopharmacol Biol Psychiatry 2011;35:1242–54.

[947] Swain JE, Lorberbaum JP. Imaging the human parental brain. In: Bridges RS, editor. Neurobiology of the parental brain. San Diego: Academic Press; 2008. pp. 83–100.

[948] Swain JE, Tasgin E, Mayes LC, Feldman R, Constable RT, Leckman JF. Maternal brain response to own baby-cry is affected by cesarean section delivery. J Child Psychol Psychiatry 2008;49:1042–52.

[949] Swanson LW. The projections of the ventral tegmental area and adjacent regions: a combined fluorescent retrograde tracer and immunofluorescence study in the rat. Brain Res Bull 1982;9:321–53.

[950] Swanson LW. The hypothalamus. In: Bjorklund A, Hokfelt T, Swanson LW, editors. Handbook of chemical neuroanatomy, vol. 5. Amsterdam: Elsevier; 1987. pp. 1–124.

[951] Swanson LW. Brain architecture: understanding the basic plan. New York: Oxford University Press; 2003.

[952] Swanson LW, Petrovich GD. What is the amygdala? Trends Neurosci 1998;21:323–31.

[953] Swanson LW, Risold PY. On the basic architecture of the septal region. In: Numan R, editor. The behavioral neuroscience of the septal region. New York: Springer; 2000. pp. 1–14.

[954] Sweatt JD. Experience-dependent epigenetic modifications in the central nervous system. Biol Psychiatry 2009;65:191–7.

[955] Tachikawa KS, Yoshihara Y, Kuroda KO. Behavioral transition from attack to parenting in male mice: a crucial role of the vomeronasal system. J Neurosci 2013;33:5120–6.

[956] Taha SA, Fields HL. Inhibitions of nucleus accumbens neurons encode a gating signal for reward-directed behavior. J Neurosci 2006;26:217–22.

[957] Takayanagi Y, Yoshida M, Bielsky IF, Ross HE, Kawamata M, Onaka T, et al. Pervasive social deficits, but normal parturition, in oxytocin receptor-deficient mice. Proc Natl Acad Sci USA 2005;102: 16096–101.

[958] Terenzi MG, Ingram CD. Oxytocin-induced excitation of neurons in the rat central and medial amygdaloid nuclei. Neuroscience 2005;134:345–54.

[959] Thompson RR, George K, Walton JC, Orr SP, Benson J. Sex-specific influences of vasopressin on human social communication. Proc Natl Acad Sci USA 2006;103:7889–94.

[960] Thornton J, Goy RW. Female-typical sexual behavior of rhesus and defeminization by androgens given prenatally. Horm Behav 1986;20:129–47.

[961] Tobiansky DJ, Roma PG, Hattori T, Will RG, Nutsch VL, Dominguez JM. The medial preoptic area modulates cocaine-induced activity in female rats. Behav Neurosci 2013;127:293–302.

[962] Todd BJ, Schwarz JM, McCarthy MM. Prostaglandin-E2: a point of divergence in estradiol-mediated sexual differentiation. Horm Behav 2005;48:512–21.

[963] Tomasello M, Vaish A. Origins of human cooperation and morality. Annu Rev Psychol 2013;64:231–55.

[964] Torii M, Kubo K, Sasaki T. Influence of opioid peptides on the priming action of estrogen on lordosis in ovariectomized rats. Neurosci Lett 1996;212:68–70.

[965] Torii M, Kubo K, Sasaki T. Differential effects of beta-endorphin and met- and leu-enkephalin on steroid hormone-induced lordosis in ovariectomized female rats. Pharmacol Biochem Behav 1997;58:837–42.

[966] Toscano JE, Bauman MD, Mason WA, Amaral DG. Interest in infants by female rhesus monkeys with neonatal lesions of the amygdala or hippocampus. Neuroscience 2009;162:881–91.

[967] Tost H, Kolachano B, Hakimi S, Lemaitre H, Verchinski BA, Mattay VS, et al. A common allele in the oxytocin receptor gene (OXTR) impacts prosocial temperament and human hypothalamic-limbic structure and function. Proc Natl Acad Sci USA 2010;107:13936–41.

[968] Toth M, Fuzesi T, Halasz J, Tulogdi A, Haller J. Neural inputs of the hypothalamic "aggression area" in the rat. Behav Brain Res 2010;215:7–20.

[969] Toth M, Halasz J, Mikics E, Barsy B, Haller J. Early social deprivation induces disturbed social communication and violent aggression in adulthood. Behav Neurosci 2008;122:849–54.

[970] Toth M, Tulogdi A, Biro L, Soros P, Mikics E, Haller J. The neural background of hyper-emotional aggression induced by post-weaning social isolation. Behav Brain Res 2012;233:120–9.

[971] Trainor BC, Bird IM, Alday NA, Schlinger BA, Marler CA. Variation in aromatase activity in the medial preoptic area and plasma progesterone is associated with the onset of paternal behavior. Neuroendocrinology 2003;78:36–44.

[972] Trainor BC, Marler CA. Testosterone, paternal behavior, and aggression in the monogamous California mouse (*Peromyscus californicus*). Horm Behav 2001;40:32–42.

[973] Trainor BC, Marler CA. Testosterone promotes paternal behaviour in a monogamous mammal via conversion to oestrogen. Proc Biol Sci 2002;269:823–9.

[974] Tripathi A, Prensa L, Cebrian C, Mengual E. Axonal branching patterns of nucleus accumbens neurons in the rat. J Comp Neurol 2010;518:4649–73.

[975] Trivers RL. The evolution of reciprocal altruism. Q Rev Biol 1971;46:35–57.

[976] Tsankova N, Renthal W, Kumar A, Nestler EJ. Epigenetic regulation in psychiatric disorders. Nat Rev Neurosci 2007;8:355–67.

[977] Tsukahara S. Sex differences and the roles of sex steroids in apoptosis of sexually dimorphic nuclei of the preoptic area in postnatal rats. J Neuroendocrinol 2009;21:370–6.

[978] Tsukahara S, Yamanouchi K. Neurohistological and behavioral evidence for lordosis-inhibiting tract from lateral septum to periaqueductal gray in male rats. J Comp Neurol 2001;431:293–310.

[979] Tsuneoka Y, Maruyama T, Yoshida S, Nishimori K, Kato T, Numan M, et al. Functional, anatomical, and neurochemical differentiation of medial preoptic area subregions in relation to maternal behavior in the mouse. J Comp Neurol 2013;521:1633–63.

[980] Turner LM, Young AR, Romper H, Schoneberg T, Phelps SM, Hoekstra HE. Monogamy evolves through multiple mechanisms: evidence from V1aR in deer mice. Mol Biol Evol 2010;27:1269–78.

[981] Tye KM, Deisseroth K. Optogenetic investigation of neural circuits underlying brain disease in animal models. Nat Rev Neurosci 2012;13:251–66.

[982] Tye KM, Prakash R, Kim SY, Fenno LE, Grosenick L, Zarabi H, et al. Amygdala circuitry mediating reversible and bidirectional control of anxiety. Nature 2011;471:358–62.

[983] Ulrich-Lai YM, Herman JP. Neural regulation of endocrine and autonomic stress responses. Nat Rev Neurosci 2009;10:397–409.

[984] Usuda I, Tanaka K, Chiba T. Efferent projections of the nucleus accumbens in the rat with special reference to subdivision of the nucleus: biotinylated dextran amine study. Brain Res 1998;797:73–93.

[985] Van Erp AM, Miczek KA. Aggressive behavior, increased accumbal dopamine, and decreased cortical serotonin in rats. J Neurosci 2000;20:9320–5.

[986] van IJzendoorn MH, Bakermans-Kranenburg MJ, Mesman J. Dopamine system genes associated with parenting in the context of daily hassles. Genes Brain Behav 2008;7:403–10.

[987] van Leengoed E, Kerker E, Swanson HH. Inhibition of post-partum maternal behaviour by injecting an oxytocin antagonist into the cerebral ventricles. J Endocrinol 1987;112:275–82.

[988] Veenema AH. Early life stress, the development of aggression and neuroendocrine and neurobiological correlates: what can we learn from animal models? Front Neuroendocrinol 2009;30:497–518.

[989] Veenema AH, Beiderbeck DI, Lukas M, Neumann ID. Distinct correlations of vasopressin release within the lateral septum and the bed nucleus of the stria terminalis with the display of intermale aggression. Horm Behav 2010;58:273–81.

[990] Veenema AH, Blume A, Niederle D, Buwalda B, Neumann ID. Effects of early life stress on adult male aggression and hypothalamic vasopressin and serotonin. Eur J Neurosci 2006;24:1711–20.

[991] Veenema AH, Neumann ID. Central vasopressin and oxytocin release: regulation of complex social behavior. Prog Brain Res 2008;170:261–76.

[992] Veinante P, Freund-Mercier MJ. Distribution of oxytocin- and vasopressin-binding sites in the rat extended amygdala: a histoautoradiographic study. J Comp Neurol 1997;383:305–25.

[993] Verona E, Joiner TE, Johnson F, Bender TW. Gender specific gene-environment interactions on laboratory-assessed aggression. Biol Psychol 2006;71:33–41.

[994] Vertes RP. Differential projections of the infralimbic and prelimbic cortex in the rat. Synapse 2004;51:32–58.

[995] Vidal-Gonzalez I, Vidal-Gonzalez B, Rauch SL, Quirk GJ. Microstimulation reveals opposing influences of prelimbic and infralimbic cortex on the expression of conditioned fear. Learn Mem 2006;13:728–33.

[996] Viding E, Jones AP, Frick PJ, Plomin R. Heritability of antisocial behavior at 9: do callous-unemotional traits matter? Dev Sci 2008;11:17–22.

[997] Viding E, Sebastian CL, Dadds MR, Lockwood PL, Cecil CA, De Brito SA, et al. Amygdala response to preattentive masked fear in children with conduct problems: the role of callous-unemotional traits. Am J Psychiatry 2012;169:1109–16.

[998] Virkkunen M, Goldman D, Nielsen DA, Linnoila M. Low brain serotonin turnover (low CSF 5-HIAA) and impulsive violence. J Psychiatry Neurosci 1995;20:271–5.

[999] Viviani D, Charlet A, van den Burg E, Robinet C, Hurni N, Abatis M, et al. Oxytocin selectively gates fear responses through distinct outputs from the central amygdala. Science 2011;333:104–7.

[1000] Vochteloo JD, Koolhaas JM. Medial amygdala lesions in male rats reduce aggressive behavior: interference with experience. Physiol Behav 1987;41:99–102.

[1001] Volkmar FR. Understanding the social brain in autism. Dev Psychobiol 2011;53:428–34.

[1002] Von Der Heide RJ, Skipper LM, Klobusicky E, Olson IR. Dissecting the uncinate fasciculus: disorders, controversies and a hypothesis. Brain 2013;136:1692–707.

[1003] Wacker DW, Ludwig M. Vasopressin, oxytocin, and social odor recognition. Horm Behav 2012;61:259–65.

[1004] Wallen K. Desire and ability: hormones and the regulation of female sexual behavior. Neurosci Biobehav Rev 1990;14:233–41.

[1005] Wallen K. Sex and context: hormones and primate sexual motivation. Horm Behav 2001;40:339–57.

[1006] Wallen K. Hormonal influences on sexually differentiated behavior in nonhuman primates. Front Neuroendocrinol 2005;26:7–26.

[1007] Wallen K, Eisler JA, Tannenbaum PL, Nagell KM, Mann DR. Antide (Nal-Lys GnRH antagonist) suppression of pituitary-testicular function and sexual behavior in group-living rhesus monkeys. Physiol Behav 1991;50:429–35.

[1008] Wallen K, Hassett JM. Sexual differentiation of behaviour in monkeys: role of prenatal hormones. J Neuroendocrinol 2009;21:421–46.

[1009] Wallen K, Tannenbaum PL. Hormonal modulation of sexual behavior and affiliation in rhesus monkeys. Ann NY Acad Sci 1997;807:185–202.

[1010] Wallis JD. Cross-species studies of orbitofrontal cortex and value-based decision making. Nat Neurosci 2011;15:13–9.

[1011] Walter M, Bermpohl F, Mouras H, Schiltz K, Tempelmann C, Rotte M, et al. Distinguishing specific sexual and general emotional effects in fMRI-subcortical and cortical arousal during erotic picture viewing. NeuroImage 2008;40:1482–94.

[1012] Walter M, Witzel J, Wiebking C, Gubka U, Rotte M, Schiltz K, et al. Pedophilia is linked to reduced activation in hypothalamus and lateral prefrontal cortex during visual erotic stimulation. Biol Psychiatry 2007;62:698–701.

[1013] Walum H, Lichtenstein P, Neiderhiser JM, Reiss D, Ganiban JM, Spotts EL, et al. Variation in the oxytocin receptor gene is associated with pair-bonding and social behavior. Biol Psychiatry 2012;71: 419–26.

[1014] Walum H, Westberg L, Henningsson S, Neiderhiser JM, Reiss D, Igl W, et al. Genetic variation in the vasopressin receptor 1a gene (AVPR1A) associates with pair-bonding in humans. Proc Natl Acad Sci USA 2008;105:14153–6.

[1015] Wang Z, Bullock NA, De Vries GJ. Sexual differentiation of vasopressin projections of the bed nucleus of the stria terminalis and medial amygdaloid nucleus in rats. Endocrinology 1993;132:2299–306.

[1016] Wang H, Duclot F, Liu Y, Wang Z, Kabbaj M. Histone deacetylase inhibitors facilitate partner preference formation in female prairie voles. Nat Neurosci 2013;16:919–24.

[1017] Wang Z, Ferris CF, De Vries GJ. Role of septal vasopressin innervation in paternal behavior in prairie voles (*Microtus ochrogaster*). Proc Natl Acad Sci USA 1994;91:400–4.

[1018] Wang Z, Smith W, Major DE, De Vries GJ. Sex and species differences in the effects of cohabitation on vasopressin messenger RNA expression in the bed nucleus of the stria terminalis in prairie voles (*Microtus ochrogaster*) and meadow voles (*Microtus pennsylvanicus*). Brain Res 1994;650:212–8.

[1019] Wang Z, Yu G, Cascio C, Liu Y, Gingrich B, Insel TR. Dopamine D2 receptor-mediated regulation of partner preferences in female prairie voles (*Microtus orchrogaster*): a mechanism for pair bonding? Behav Neurosci 1999;113:602–11.

[1020] Wang Z, Zhou L, Hulihan TJ, Insel TR. Immunoreactivity of central vasopressin and oxytocin pathways in microtine rodents: a quantitative comparative study. J Comp Neurol 1996;366:726–37.

[1021] Watts AG, Swanson LW. Anatomy of motivation. In: Pashler H, Gallistel R, editors. Steven's handbook of experimental psychology, vol. 3. New York: Wiley; 2002. pp. 563–631.

[1022] Weaver IC, Cervoni N, Champagne FA, D'Alessio AC, Sharma S, Seckl JR, et al. Epigenetic programming by maternal behavior. Nat Neurosci 2004; 7:847–54.

[1023] Weder N, Yang BZ, Douglas-Palumberi H, Massey J, Krystal JH, Gerlernter J, et al. MAOA genotype, maltreatment, and aggressive behavior: the changing impact of genotype at varying levels of trauma. Biol Psychiatry 2009;65:417–24.

[1024] Weisz A, Rosales R. Identification of an estrogen response element upstream of the human c-fos gene that binds the estrogen receptor and the AP-1 transcription factor. Nucleic Acids Res 1990;18:5097–106.

[1025] Weng HY, Fox AS, Shackman AJ, Stodola DE, Caldwell JZ, Olson MC, et al. Compassion training alters altruism and neural responses to suffering. Psychol Sci 2013;24:1171–80.

[1026] Wenkstern D, Pfaus JG, Fibiger HC. Dopamine transmission increases in the nucleus accumbens of male rats during their first exposure to sexually receptive female rats. Brain Res 1993;618:41–6.

[1027] Whitaker-Azmitia PM. Serotonin and brain development: role in human developmental diseases. Brain Res Bull 2001;56:479–85.

[1028] Whitaker-Azmitia PM, Azmitia EC. Autoregulation of fetal serotonergic development: role of high affinity serotonin receptors. Neurosci Lett 1986;67:307–12.

[1029] Whitaker-Azmitia PM, Zhang X, Clarke C. Effects of gestational exposure to monoamine oxidase inhibitors in rats: preliminary behavioral and neurochemical studies. Neuropsychopharmacol 1994;11:125–32.

[1030] Whitaker-Azmitia P, Zhou F, Hobin J, Borella A. Isolation-rearing of rats produces deficits as adults in the serotonergic innervation of hippocampus. Peptides 2000;21:1755–9.

[1031] White SF, Brislin S, Sinclair S, Fowler KA, Pope K, Blair RJ. The relationship between large cavum septum pellucidum and antisocial behavior, callous-unemotional traits and psychopathy in adolescents. J Child Psychol Psychiatry 2013;54:575–81.

[1032] Whitney JF. Effect of medial preoptic lesions on sexual behavior of female rats is determined by test situation. Behav Neurosci 1986;100:230–5.

[1033] Widom CS. The cycle of violence. Science 1989;244:160–6.

[1034] Wietzikoski EC, Boschen SL, Miyoshi E, Bortolanza M, Dos Santos LM, Frank M, et al. Roles of D1-like dopamine receptors in the nucleus accumbens and dorsolateral striatum in conditioned avoidance responses. Psychopharmacology 2012;219:159–69.

[1035] Williams JR, Slotnick BM, Kirkpatrick BW, Carter CS. Olfactory bulb removal affects partner preference development and estrus induction in female prairie voles. Physiol Behav 1992;52:635–9.

[1036] Wilson EO. The social conquest of earth. New York: Norton; 2012.

[1037] Wilson JD. Androgens, androgen receptors, amd male gender role behavior. Horm Behav 2001;40: 358–66.

[1038] Winslow JT, Hastings N, Carter CS, Harbaugh CR, Insel TR. A role for central vasopressin in pair bonding in monogamous prairie voles. Nature 1993;365:545–8.

[1039] Winslow JT, Noble PL, Lyons CK, Sterk SM, Insel TR. Rearing effects on cerebrospinal fluid oxytocin concentration and social buffering in rhesus monkeys. Neuropsychopharmacol 2003;28:910–8.

[1040] Wirtshafter D, Stratford TR. Evidence for motivational effects elicited by activation of GABA-A or dopamine receptors in nucleus accumbens shell. Pharmacol Biochem Behav 2010;96:342–6.

[1041] Witte AV, Floel A, Stein P, Savli M, Mien LK, Wadsak W, et al. Aggression is related to frontal serotonin-1A receptor distribution as revealed by PET in healthy subjects. Hum Brain Mapp 2009;30:2558–70.

[1042] Witten IB, Lin SC, Brodsky M, Prakash R, Diester I, Anikeeva P, et al. Cholinergic interneurons control local circuit activity and cocaine conditioning. Science 2010;330:1677–81.

[1043] Wittenberger JF. Animal social behavior. Boston: Duxbury Press; 1981.

[1044] Wright CI, Groenewegen HJ. Patterns of convergence and segregation in the medial nucleus accumbens of the rat: relationships of prefrontal cortical, midline thalamic, and basal amygdaloid afferents. J Comp Neurol 1995;361:383–403.

[1045] Wu S, Jia M, Ruan Y, Liu J, Guo Y, Shuang M, et al. Positive association of oxytocin receptor gene (OXTR) with autism in the Chinese Han population. Biol Psychiatry 2005;58:74–7.

[1046] Wynne-Edwards KE, Timonin ME. Paternal care in rodents: weakening support for hormonal regulation of the transition to behavioral fatherhood in rodent animal models of biparental care. Horm Behav 2007;52:114–21.

[1047] Yamaguchi T, Sheen W, Morales M. Glutamatergic neurons are present in the rat ventral tegmental area. Eur J Neurosci 2007;25:106–18.

[1048] Yamasue H. Function and structure in social brain regions can link oxytocin-receptor genes with autistic social behavior. Brain Dev 2013;35:111–8.

[1049] Yamasue H, Yee JR, Hurlemann R, Rilling JK, Chen FS, Meyer-Lindenberg A, et al. Integrative approaches utilizing oxytocin to enhance prosocial behavior: from animal and human social behavior to autistic social dysfunction. J Neurosci 2012;32:14109–17.

[1050] Yang LY, Clemens LG. MPOA lesions affect female pacing of copulation in rats. Behav Neurosci 2000;114:1191–202.

[1051] Yang Y, Raine A. Prefrontal structural and functional brain imaging findings in antisocial, violent, and psychopathic individuals: a meta-analysis. Psychiatry Res 2009;174:81–8.

[1052] Yang Y, Raine A, Narr KL, Colletti P, Toga AW. Localization of deformations within the amygdala in individuals with psychopathy. Arch Gen Psychiatry 2009;66:986–94.

[1053] Yells DP, Hendricks SE, Prendergast MA. Lesions of the nucleus paragigantocellularis: effects on mating behavior in male rats. Brain Res 1992;596:73–9.

[1054] Yildirim BO, Derksen JJ. A review on the relationship between testosterone and the interpersonal/affective facet of psychopathy. Psychiatry Res 2012;197: 181–98.

[1055] Yildirim BO, Derksen JJ. Systematic review, structural analysis, and new theoretical perspectives on the role of serotonin and associated genes in the etiology of psychopathy and sociopathy. Neurosci Biobehav Rev 2013;37:1254–96.

[1056] Yizhar O. Optogenetic insights into social behavior function. Biol Psychiatry 2012;71:1075–80.

[1057] Yizhar O, Fenno LE, Davidson TJ, Mogri M, Deisseroth K. Optogenetics in neural systems. Neuron 2011;71:9–34.

[1058] Yoshida M, Takayanagi Y, Inoue K, Kimura T, Young LJ, Onaka T, et al. Evidence that oxytocin exerts anxiolytic effects via oxytocin receptor expressed in serotonergic neurons in mice. J Neurosci 2009;29:2259–71.

[1059] Young L, Alexander B. The chemistry between us: love, sex, and the science of attraction. New York: Penguin; 2012.

[1060] Young KA, Gobrogge KL, Liu Y, Wang Z. The neurobiology of pair bonding: insights from a socially monogamous rodent. Front Neuroendocrinol 2011;32:53–69.

[1061] Young KA, Gobrogge KL, Wang Z. The role of mesocorticolimbic dopamine in regulating interactions between drugs of abuse and social behavior. Neurosci Biobehav Rev 2011;35:498–515.

[1062] Young LJ, Hammock EA. On switches and knobs, microsatellites and monogamy. Trends Genet 2007;23:209–12.

[1063] Young LJ, Lim MM, Gingrich B, Insel TR. Cellular mechanisms of social attachment. Horm Behav 2001;40:133–8.

[1064] Young LJ, Toloczko D, Insel TR. Localization of vasopressin (V1a) receptor binding and mRNA in the rhesus monkey brain. J Neuroendocrinol 1999;11:291–7.

[1065] Young LJ, Wang Z. The neurobiology of pair bonding. Nat Neurosci 2004;7:1048–54.

[1066] Young LG, Wang Z, Donaldson R, Rissman EF. Estrogen receptor alpha is essential for induction of oxytocin receptor by estrogen. Neuroreport 1998;9:933–6.

[1067] Young LJ, Wang Z, Insel TR. Neuroendocrine bases of monogamy. Trends Neurosci 1998;21:71–5.

[1068] Young LJ, Winslow JT, Nilsen R, Insel TR. Species differences in V1a receptor gene expression in monogamous and nonmonogamous voles: behavioral consequences. Behav Neurosci 1997;111:599–605.

[1069] Zahm DS. The evolving theory of basal forebrain functional-anatomical 'macrosystems'. Neurosci Biobehav Rev 2006;30:148–72.

[1070] Zemlan FP, Kow LM, Pfaff DW. Effect of interruption of bulbospinal pathways on lordosis, posture, and locomotion. Exp Neurol 1983;81:177–94.

[1071] Zhang TY, Hellstrom IC, Bagot RC, Wen X, Diorio J, Meaney MJ. Maternal care and DNA methylation of a glutamic acid decarboxylase 1 promoter in rat hippocampus. J Neurosci 2010;30:13130–7.

[1072] Zhang S, Liu Y, Rao Y. Serotonin signaling in the brain of adult female mice is required for sexual preference. Proc Natl Acad Sci USA 2013;110:9968–73.

[1073] Zhang TY, Meaney MJ. Epigenetics and the environmental regulation of the genome and its function. Annu Rev Psychol 2010;61:439–66.

[1074] Zhao C, Li M. The receptor mechanisms underlying the disruptive effects of haloperidol and clozapine on rat maternal behavior: a double dissociation between dopamine D(2) and 5-HT(2A/2C) receptors. Pharmacol Biochem Behav 2009;93:433–42.

[1075] Zhou L, Furuta T, Kaneko T. Chemical organization of projection neurons in the rat accumbens nucleus and olfactory tubercle. Neuroscience 2003;120: 783–98.

[1076] Zimmerman JM, Rabinak CA, McLachlan IG, Maren S. The central nucleus of the amygdala is essential for acquiring and expressing conditional fear after overtraining. Learn Mem 2007;14:634–44.

[1077] Zink CF, Stein JL, Kempf L, Hakimi S, Meyer-Lindenberg A. Vasopressin modulates medial prefrontal cortex-amygdala circuitry during emotion processing in humans. J Neurosci 2010;30:7017–22.

[1078] Zweifel LS, Parker JG, Lobb CJ, Rainwater A, Wall VZ, Fadok JP, et al. Disruption of NMDAR-dependent burst firing by dopamine neurons provides selective assessment of phasic dopamine-dependent behavior. Proc Natl Acad Sci USA 2009;106:7281–8.

Index

Note: Page numbers followed by f indicate figures; t, tables; b, boxes.

A

Abusive maternal behaviors, 216–217
Adenylate cyclase (AC), 49–50
Adrenocorticotropic hormone (ACTH), 7f, 75
Affective aggression, 72
  in humans, 86–94
    PFC function and, 88–89
    serotonin and, 89–91
    vasopressin/oxytocin and aggression in, 91–94
  neural circuitry of, 72–73
Aggressive behavior
  in cats, 72–73
    affective aggression, neural circuitry of, 72–73
  consummatory *versus* goal- directed mechanisms, 74–75
  early adverse life experiences/ genes and development, 94–105
    CRH, potential role of, 99–100
    gene-environment interactions, 100–105
    serotonin neural systems dysregulation, 96–99
  gene-environment interactions, 100–105
  5-HTT polymorphisms, 104–105
  in humans, 86–94
    PFC function and, 88–89
    serotonin and, 89–91
    vasopressin/oxytocin and aggression in, 91–94
  neural circuitry of, 72–73
  in nonhuman mammals, 65–86
  in old world monkeys, 65
  prefrontal cortex (PFC) and, 78–80
  rats, SP and offensive aggression in, 73–74
  in rodents, 64–65
  serotonin and, 80–86
  social experiences and development, 94–95
  vasopressin, anterior hypothalamus, 75–78
Alloparental behavior
  female laboratory mouse and, 209–210
  paternal and, 209–214
Altruistic behavior, 273
Androgen insensitivity syndrome (AIS), 158–159
Anogenital sniffing, 245
Anterior cingulate cortex, 138–139
Anterior cingulate cortex (ACC), 122, 278f
Anterior hypothalamic nucleus (AHN), 65–66, 248, 249f
Anterior medial hypothalamus (AMH), 72–73, 73f
Antisocial personality disorder (ASPD), 291, 293
Appetitive-consummatory distinction, 111
Appetitive responses, 4
Appetitive reward-seeking behaviors, 5
Arcuate nucleus (ARC), 125f
ASD. *See* Autism spectrum disorder (ASD)
Autism spectrum disorder (ASD), 263–269
  occurrence, sex differences in, 268–269
  OT
    and brain mechanisms of, 268
    plasma and, 266
    related genes and, 265–266
    treatment as therapy for, 266–268
AVPR1A
  gene, 253
  polymorphism, 256

B

Basic epigenetics
  heterochromatin and euchromatin, 53–54
  physiological and behavioral development, epigenetic effects on, 54–61
  physiological stress response, maternal care on, 56–60
  stress reactivity and psychiatric disorders, genetic polymorphisms interactions, 60–61
Basic genetics
  exons and introns, 46–47
  general transcription and translation processes, 45–46
  genetic polymorphisms, 51
  gene transcription, regulation of, 47–51
  knockout mutations, 52
  structure of nucleotides, 43–45
Basolateral amygdala (BLA), 25–26, 26f–27f, 29–30, 29f, 190–191, 191f, 248, 249f, 276–278
Basomedial amygdala (BMA), 25–26, 26f–27f, 29–30, 29f, 97, 98f, 190–191, 191f, 248, 249f, 276–278
Behavioral estrus, 123–124
Blood-oxygen-level dependent (BOLD) responses, 20b–21b, 29–30, 135–137, 259, 276
Brain mechanisms, 169–170, 169f
  additional defeminization processes, 154–156
  demasculinization, feminization associated with, 156
  MPOA, 149–154
Brainstem motor area (BSMA), 2–3

C

Cavum septum pellucidum (CSP), 295–296, 296f
CeAm. *See* Central nucleus of the amygdala (CeAm)
Central nucleus of the amygdala (CeAm), 12f–13f

Cerebrospinal fluid (CSF), 59–60, 247–248
CFR. *See* Conditioned fear response (CFR)
Chromatin, 53–54, 53f
Cocaine-related cues, 138
Coding region, 47, 48f
Codons, 45
Cognitive motor system, 28
Cognitive processes, 288
Conditioned aversive stimulus (CAS), 182, 182f–183f
Conditioned fear response (CFR), 12f–13f, 36f, 182, 182f–183f
Conditioned fear responses, 34–38, 35f
Conduct disorder (CD), 297
Consummatory copulatory behavior, 117
Consummatory phase, 4–5
Contact-return latency, 124
Cooperative breeding, 272
Cooperative genes, 272–273
Core neural circuits, 275
Corticotropin-releasing hormone (CRH), 6–8, 7f
CRH. *See* Corticotropin-releasing hormone (CRH)
Cyclic adenosine monophosphate response element (CRE), 47–49, 48f–49f
Cyclic AMP response element binding protein (CREB), 49f, 50

D

15-day HO females (15HO), 174
Deoxyribonucleic acid (DNA), 43–45, 44f–45f
Dihydrotestosterone (DHT), 144–145, 145f, 157
Dopamine transporter gene (DAT1 gene), 229
Dorsolateral tegmentum (DLT), 119–120

E

Electrode location, reconstruction of, 17
Emotion, 5
ERE. *See* Estrogen response element (ERE)
Estradiol acts, 128–129
Estrogen receptor (ER), 174–175, 175f
Estrogen receptors (ER), 126, 162
Estrogen response element (ERE), 47–49, 174–175
Estrous female, 111–112
Euchromatin, 53–54, 53f
Excitatory postsynaptic potential (EPSP), 49–50
Exons, 46–47, 46f–47f

F

Female rat, 124
Follicle-stimulating hormone (FSH), 6–8, 7f
Fos-expressing cells, 178
Functional magnetic resonance imaging (fMRI), 1, 88–89
Functional neuroanatomy, 3–41

G

Gene-environment interactions, 60
Genetic polymorphisms, 274
Globus pallidus (GPi), 21–22
Glucocorticoid receptors (GR), 55, 56f
Glucocorticoids (GC), 55
Glucocorticoids, circulating blood levels of, 55
Glutamate receptor (GluR) antagonist, 30
Glutamic acid decarboxylase (GAD), 58–59
Goal-directed approach behaviors, 172–173
Goal-directed behavior, 4
Goal-directed motivational processes, 9
Gonadal steroids, 146–147, 146f
Gonadotropin-releasing hormone (GnRH), 6–8
G-protein-coupled receptor, 49–50
Gray matter volume (GMV), 268, 279–280
Group selection, 273–274

H

HAA. *See* Hypothalamic aggression area (HAA)
Heterochromatin, 53–54, 53f
High licking and grooming (HLG) females, 217
Histone acetyltransferases (HATs) transfers, 54
Histone deacetylases (HDAC), 54
Histone proteins, 54
Hormone-treated nulliparous females, 178
5-HTT polymorphisms, 104–105
Human sociality
- epilogue, 298–299
- evolutionary perspective, 271–274
- human prosociality, 289
- insular cortex and empathy, 276–284
  - neural model, 278–279
- OT and MPOA, 275–276
- oxytocin and prosociality, 284–289
- psychopathy, 290–298

6-Hydroxydopamine (6-HD), 29–30
5-Hydroxyindole acetic acid (5-HIAA), 81–82
Hydroxyl (OH) group, 44–45
Hyperemotionality, 298
Hypothalamic aggression area (HAA), 65–72, 66f, 68f
- male mice, VMNvl and offensive aggression in, 66–70
- rats, efferent and afferent neural connections of, 71–72, 71f

Hypothalamic-amygdala-striatal-pallidal behavioral circuits, 275
Hypothalamic-pituitary-adrenal (HPA) axis, 55
Hypothetical neural model, 68

I

IED. *See* Intermittent explosive disorder (IED)
Insular cortex, 137
Intercalated nuclei (ITC), 12f–13f
Intermittent explosive disorder (IED), 64, 291
Intracerebroventricular (ICV), 239
Introns, 46–47, 46f–47f
Isocortex, 2–3

K

Kin selection theory, 273

L

Lateral orbital (LO), 79
Lateral prefrontal cortex (LPFC), 142–143, 142f
Lithium chloride (LiCl), 122
Lordosis quotient (LQ), 123–124
Low foraging demand (LFD), 59–60
Low licking and grooming (LLG), 217
Luteinizing hormone (LH), 6–8

M

Magnetic resonance imaging (MRI) analysis, 17
Main olfactory bulb (MOB), 177, 177f
Male sexual behavior
  appetitive phase of, 111–112
  consummatory aspects of, 112
  MPOA and, 112–114
Mammalian brain, schematic overview of, 2–3
Marital discord, 260–261
Maternal behavior
  consummatory aspects of, 195–196
  hormonal stimulation of, 166–167, 171–172
  in humans, 224–232
  intergenerational continuity of normal variations in, 218–221
  maintenance of, 198–202
  of mammals, 214–224
  mesolimbic DA system, 183–192
  motivation, 167–168
  MPOA, 172–176
    specificity of, 196–198
  neurobiology studies of, 168
  in nonhuman primates, 207–209
  sensitization of, 166
  serotonin and, 202–203
  sheep, neural systems and, 203–207
  in species, 166
Maternal separation (MS), 216
Mating partners
  OT/vasopressin neural systems and birds/primates, social monogamy in, 257–262
  vole model system, 236–257
    mesolimbic DA system and partner preference formation, 241–243
    pair bond formation, neural model of, 248–253
    pair bond formation, oxytocin and vasopressin for, 239–241
    pair bond formation regulation, OT and AVP in, 244–248
    pair bonding and partner preference formation, 238–239
    social organization and mating systems of, 243–244
    vasopressin V1a receptor, genetics of species differences in, 253–257
MDT. *See* Mediodorsal thalamic nucleus (MDT)
Medial amygdala (MeA), 240
Medial prefrontal cortex (mPFC), 278f
Medial preoptic area (MPOA), 112–114, 114f, 120f, 237–238
  appetitive female sexual behavior, 128–129
  blood-oxygen-level dependent (BOLD) response, 135
  electrical lesions of, 134
  mesolimbic DA system, interactions with, 132, 133f
  neurons, 134–135
  VMN of, 135
Mediodorsal thalamus (MDT), 31, 71–72
Medium spiny neurons (MSNs), 21–22, 49–50, 246–247
Mesolimbic dopamine (DA) system, 28, 28f
Mesotocin receptors, 257–258
Mgnetic resonance imaging (MRI) analysis, 17
Microanatomy, 250
Microtus genus, 236
Midbrain, 19–21
Monoamine oxidase B (MAOB), 101
Monoamine oxidase enzymes, 100–102
  genetic polymorphisms in, 102–104
Monogamous mating system, 236–237
Monogamy, 257–258
  in birds, 257–258
Mother-infant bond, 238
MPOA. *See* Medial preoptic area (MPOA)

N

Natural emotional experience, 276
Negative odor, 18
Neural activity, 18
Neural circuitry, 125, 125f
  analysis, 14b–15b
  regulating lordosis, 125–128
    neural model, evidence for, 126–127
    sex/aggression and VMN, 127–128
Neural systems
  amygdala, 9–19
  dorsal and ventral basal ganglia, 19–30
  functional magnetic resonance imaging, 1
  functional neuroanatomy, 3–41
  mammalian brain, 2–3
  nigrostriatal and mesolimbic dopamine systems, 19–30
  prefrontal cortex, 30–41
Neuron-specific excitotoxic amino acid lesions, 121
Neurophysiological research, 29
Nonprimate animals, sexual behaviors in, 111–123
  medial preoptic area, 112–114
  mesolimbic DA system, MPOA interactions with, 116–120
  neurobehavioral model, 114–116
Nonprimate mammals, brain and behavior in, 146–157
  additional defeminization processes, 154–156
  behavioral analysis, 147–149
  MPOA, brain mechanisms, 149–154
Nucleus accumbens (NA), 112, 241–242
Nucleus accumbens-ventral pallidum (NA-VP), 115, 192, 193f, 240–241, 278–279
Nucleus paragigantocellularis (nPGi), 112
Nulliparous control females, 178

O

Optogenetic methods, 12–13, 14b–15b
Orbital frontal cortex (OFC), 276–278
OTA. *See* Oxytocin receptor antagonist (OTA)
OXTR gene, 260
Oxytocin (OT), 8, 75, 171
Oxytocin receptor (OTR), 11–12, 47–49, 49f, 125, 125f, 171, 174–175, 175f, 251f, 252–253
Oxytocin receptor antagonist (OTA), 239

P

Paced mating behavior, 124
PAG. *See* Periaqueductal gray (PAG)
Pair bonding, 272
  autism spectrum disorder (ASD), OT neural systems to, 263–269
  in birds, 257–258
  humans, 259–262
  nonhuman primates, 258–259
  OT and vasopressin, 262–263
  in primates, 258–262

Paraventricular nucleus (PVN), 6–8, 7f, 75, 180, 181f, 243, 248–250, 249f
Parental behavior
female laboratory mouse and allomaternal behavior, 209–210
humans maternal behavior, 224–232
postpartum women, brain systems associated with, 226–232
mammals maternal behavior, developmental influences on, 214–224
behavioral studies on, 214–218
developmental neurobiology of, 218–224
maternal behavior
MPOA neurons for, 172–176, 196–198
oxytocin role, hormonal stimulation of, 171–172
in virgin females, sensitization of, 170–171
maternal circuits, maternal behavior/maternal memory and plasticity within, 198–202
mesolimbic DA system
MPOA Interaction with, 183–192
virgin females, active avoidance of pups by, 192–194
nonhuman primates, maternal behavior in, 207–209
parturient females, defensive inhibition/rejection circuit in, 180–183
paternal behavior, 211–213
postpartum period, changes in MPOA functions changes, 194–195
postpartum rats, consummatory aspects of maternal behavior, 195–196
rats, neural systems regulating maternal motivation in, 169–203
serotonin and maternal behavior, 202–203
sheep, neural systems and maternal behavior in, 203–207
brain mechanisms regulating maternal responsiveness, selective maternal attachment formation, 205–207
virgin females, defensive-rejection circuit in, 176–180
Partner preference tests, 254–255
Paternal behavior, 211–213
Pathological aggression, 87–88
Pedophilia, 139–143
BOLD responses, 141
definition, 140
etiology of, 140
visual association cortex, 141–142
Periaqueductal gray (PAG), 2–3, 112, 125, 125f, 133f, 179–180, 181f, 196f, 248, 249f
Periventricular zone (PVZ), 3–4, 3f
Peromyscus genus, 257
Perplexing issue, 242
Positive odor, 18
Positron emission tomography (PET), 87, 103
Postejaculatory interval (PEI), 111–112
Prefrontal cortex (PFC), 122–123
anatomy of, 31
anterior cingulate cortex within the, 137
conditioned fear responses and goal-directed responses, regulation of, 34–38
emotional and motivational downregulation neurons, 138–139
orbital prefrontal cortex (OFC), 142–143
regulatory mechanisms, 143
in sexual motivation, 139
social behavior, mechanisms to, 39–41
value-based decision making, 38–39
Progesterone receptors (PRs), 126
Proposed synaptic plasticity, 200f
Protein kinase A (PKA), 49–50, 49f
Psychopathy, 290
Pup-induced maternal behavior, 170
PVZ. *See* Periventricular zone (PVZ)

R

Reading the Mind in the Eyes Task (REMT), 267–268
Reciprocal altruism, 273–274
Recombinant adeno-associated viral (rAAV) vector, 247
Regular neurons (RN), 7f, 8–9
Regulatory region, 47, 48f
Repetitive sequence polymorphisms, 51
Resident-intruder test, 66
Reward-seeking responses, 4
Ribonucleic acid (RNA), 43–45, 44f–45f

S

Saporin, 74
Selective photostimulation, 12–13
Selective serotonin reuptake inhibitor (SSRI), 98–99
Selfish genes, 272–273
Self-reported ratings, 288
Sensitized maternal behavior, 170
Septum pellucidum (SP), 277f
Serotonin (5-HT)
disparate effects of, 82, 82f
potential inhibitory and excitatory effects of, 85–86, 86f
Serotonin transporter-linked polymorphic region (5-HTTLPR), 104
Serotonin transporter pump (SERT), 81–82
SERT. *See* Serotonin transporter pump (SERT)
Sexual behaviors
heterosexual females, 139
heterosexual males, 135–139
homosexual males and females, 139
hormonal basis of, 110–111
mesolimbic DA system and, 129–132
MPOA interactions with, 132
MPOA, 128–129
and consummatory aspects of, 120–121
neural circuitry regulating lordosis, 125–128
neural model, evidence for, 126–127
sex/aggression and VMN, 127–128
neurobiology of, 123–134
in nonprimate animals, 111–123
medial preoptic area, 112–114
mesolimbic DA system, MPOA interactions with, 116–120
neurobehavioral model, 114–116
prefrontal cortex and, 122–123
in primates, 134–143
humans, 135–139
monkeys, 134–135
pedophilia, 139–143

sexual differentiation, 143–163
working neural model, 132–134
Sexual differentiation
of behavior in primates, 157–163
nonhuman primates, 157–158
perinatal steroids and human psychosexual differentiation, 158–163
nonprimate mammals, brain and behavior in, 146–157
additional defeminization processes, 154–156
behavioral analysis, 147–149
MPOA, brain mechanisms, 149–154
sexual behaviors, 143–163
Single nucleotide polymorphisms (SNPs), 51
SNPs. *See* Single nucleotide polymorphisms (SNPs)
Social behaviors, 3–4
SSRI. *See* Selective serotonin reuptake inhibitor (SSRI)
Steroid-primed control females, 128
STN. *See* Subthalamic nucleus (STN)
Stria terminalis (BST), 77–78, 78f
Suboptimal hormone regimens, 173–174
Substantia nigra pars compacta (SNc), 19–21
Subthalamic nucleus (STN), 22–23
Supraoptic nucleus (SON), 8, 75
Synaptic plasticity, 250–252

T
Telencephalic mechanisms, 5–6
Transcription factors (TF), 162, 162f
Transgenic mice, 52

U
Uncinate fasciculus (UF), 294–295
Urbach-Weithe (UW) diseases, 284

V
V1a receptor antagonist (VPA), 239
V1a receptor expression, 255
Variable foraging demand (VFD), 59–60
Variable-number tandem repeat (VNTR), 102
Ventral pallidum (VP), 112, 185, 243
Ventral tegmental area (VTA), 29, 29f, 112, 180, 181f, 186–189, 188f–189f, 246–247
Ventromedial nucleus (VMN), 65–66, 66f, 112, 125, 125f, 180, 181f
sex/aggression and, 127–128
Ventromedial nucleus (VMNvl)
ER and PR, 126
estradiol acts on, 128–129
female consummatory responses, main stimulatory population for, 133, 133f
Voxel-based morphometry (VBM) analysis, 228

W
Wanderers, 255

X
X chromosome, 161–162

FSC
www.fsc.org
MIX
Papier | Fördert
gute Waldnutzung
FSC® C083411

Kolibri 360 GmbH
Ferdinand-Jühlke-Straße 7
99095 Erfurt, Deutschland
produktsicherheit@kolibri360.de